Primate Cognition

Primate Cognition

Michael Tomasello

Josep Call

Department of Psychology and Yerkes Primate Center
Emory University
Atlanta, Georgia

New York Oxford
OXFORD UNIVERSITY PRESS
1997

Oxford University Press

Oxford New York
Athens Auckland Bangkok Bogota Bombay Buenos Aires
Calcutta Cape Town Dar es Salaam Delhi Florence Hong Kong
Istanbul Karachi Kuala Lumpur Madras Madrid Melbourne
Mexico City Nairobi Paris Singapore Taipei Tokyo Toronto Warsaw

and associated companies in
Berlin Ibadan

Published by Oxford University Press, Inc.
198 Madison Avenue, New York, New York 10016

Library of Congress Cataloging-in-Publication Data

Tomasello, Michael.
Primate cognition / Michael Tomasello and Josep Call.
p. cm.
Includes bibliographical references and index.
ISBN 978-0-19-510624-4

1. Primates—Behavior. 2. Primates—Psychology.
3. Cognition in animals.
I. Call, Josep. II. Title.
QL737.P9T65 1997
599.8'0451—dc20 96-41424

Printed in the United States of America

on acid-free paper

Preface

In this book we attempt to review all that is scientifically known about nonhuman primate cognition. Our primary audience is thus students of animal behavior and cognition, and at least some students of human evolution, whom we hope will view this topic as interesting and important in its own right. We also hope, however, that the topic will be of some interest to students of human cognition. Although mainstream cognitive science has discovered much about the cognitive processes of adult human beings and the computer programs they create, it has not been much concerned with the phylogenetic origins of these processes. As the limitations of artificial intelligence become increasingly apparent, cognitive science would do well to pay more attention to forms of natural primate cognition that serve to situate the human version in its appropriate evolutionary context.

A book of this sort can only be produced by scientists who receive much support from both institutions and colleagues. In our case, the very generous institutional supports were as follows. Michael Tomasello's research with nonhuman primates has been supported in various ways over the years by the Yerkes Regional Primate Research Center, especially by the Director of Psychobiology, Tom Gordon, and the two directors of the center during that time, Fred King and Tom Insel. His most recent research has been supported by a grant from the National Science Foundation (9507418). During the writing of the book Tomasello was supported by an academic leave from Emory University (David Bright, Dean of Emory College), a grant (06016) from the National Institutes of Health (Duane Rumbaugh, principle investigator), and by employment as instructor of a course entitled "Primate Cognition" at the First International Cognitive Science Institute at the University of Buffalo (Robert van Valin and Leonard Talmy, directors). Josep Call's research has been supported over the years by the Yerkes Regional Primate Research Center, especially by Tom Gordon; the Fundacio "La Caixa" de Barcelona, Catalunya, Spain; and by the Graduate School of Arts and Sciences of

Emory Unviersity. We are deeply grateful to all of these institutions and to the people who represent them for their generous support.

Support from our colleagues in the study of primate cognition has been extremely generous as well. Most important, a number of colleagues have read protions of the manuscript and provided useful suggestions and criticisms. We thank those who read large portions of the manuscript and in some cases, the whole manuscript: Dorothy Cheney, Andy Whiten, Daniel Povinelli, Elisabetta Visalberghi, Dario Maestripieri, Filippo Aureli, and an anonymous reviewer for Oxford University Press. We also thank those who read selected chapters: Frans de Waal, Harold Gouzoules, Dick Byrne, Roger Thomas, Patrizia Potì, Giovanna Spinozzi, Sue Boinski, and Philippe Rochat. We extend out sincerest gratitude to all these colleagues for their time and expertise, which improved the final product immensely. We of course hold none of them responsible for any errors or omissions that may remain.

We would also like to thank our many colleagues who shared with us their pictures, diagrams, figures, and tables as illustrations for the book. There persons are acknowledged officially in the appropriate places, but we would like to single out for their generosity Mike Seres, Frank Kiernan, Frans de Waal, Tetsuro Matsuzawa, Roger Thomas, Sally Boysen, Andy Whiten, Dick Byrne, Elisabetta Visalberghi, Daniel Povinelli, Sue Savage-Rumbaugh, Craig Stanford, Harold Gouzoules, and Duane Rumbaugh. Thanks also to Jaime Call Bonet for his line drawings of primates and to Victor Balaban for his photographs of children.

Finally, we are also grateful to very professional editors and production staff at the Oxford University Press, including especially MaryBeth Branigan who produced the book and Joan Bossert who oversaw the project from start to finish with great skill and care.

We hope that everyone who reads and profits from this book recognizes the contributions made by these silent partners.

July 1996 M. T.
Atlanta, Georgia J. C.

Contents

Chapter One Introduction, 3

1.1 Historical Background, 4
1.2 The Ecological Approach to Cognition, 7
1.3 Primates and Their Lives, 12
1.4 Plan of the Book, 21

PART I KNOWLEDGE OF THE PHYSICAL WORLD

Chapter Two Space and Objects, 27

2.1 Cognitive Mapping, 28
2.2 Searching for Hidden Objects, 39
2.3 Tracking Invisible Displacements, 42
2.4 Other Forms of Spatial Understanding, 46
2.5 What Primates Know about Space and Objects, 53

Chapter Three Tools and Causality, 57

3.1 Object Manipulation, 59
3.2 Tool Use, 71
3.3 Causal Understanding, 88
3.4 What Primates Know about Tools and Causality, 96

Chapter Four Features and Categories, 100

4.1 Discrimination Learning, 102
4.2 Natural Categories, 113
4.3 Relational Categories, 118
4.4 Classification, 125
4.5 What Primates Know about Features and Categories, 133

Chapter Five Quantities, 136

5.1 Estimating Numerousness, 138
5.2 Ordinality and Transitivity, 142
5.3 Counting, Summation, and Proportions, 148
5.4 Conservation of Quantities, 155
5.5 What Primates Know about Quantities, 160

Chapter Six Theories of Primate Physical Cognition, 162

6.1 Summary of Primate Physical Cognition, 162
6.2 Theories of Proximate Mechanism, 169
6.3 Theories of Ultimate Causation, 174
6.4 Directions for Future Research, 179
6.5 Conclusion, 185

PART II KNOWLEDGE OF THE SOCIAL WORLD

Chapter Seven Social Knowledge and Interaction, 191

7.1 The Social Field, 193
7.2 Coalitions and Alliances, 206
7.3 Reciprocity and Interchange, 210
7.4 Cooperative Problem-Solving, 219
7.5 What Primates Know about Others in Social Interaction, 228

Chapter Eight Social Strategies and Communication, 231

8.1 Social Strategies: Deception, 233
8.2 Intentional Communication: Gestures, 243
8.3 Intentional Communication: Vocalizations, 249
8.4 Communication with Humans, 260
8.5 What Primates Know about Others in Communication, 269

Chapter Nine Social Learning and Culture, 273

9.1 Behavioral Traditions in the Wild, 275
9.2 Social Learning of Instrumental Activities, 284
9.3 Social Learning of Communicative Signals and Gestures, 296

9.4 Teaching, 304
9.5 What Primates Know about Others in Social Learning, 308

Chapter Ten Theory of Mind, 311

10.1 Understanding Behavior and Perception, 312
10.2 Understanding Intentions and Attention, 318
10.3 Understanding Knowledge and Beliefs, 324
10.4 Understanding Self, 330
10.5 What Primates Know about Others' Mental States, 338

Chapter Eleven Theories of Primate Social Cognition, 342

11.1 Summary of Primate Social Cognition, 343
11.2 Theories of Proximate Mechanism, 350
11.3 Theories of Ultimate Causation, 354
11.4 Directions for Future Research, 357
11.5 Conclusion, 363

PART III A THEORY OF PRIMATE COGNITION

Chapter Twelve Nonhuman Primate Cognition, 367

12.1 Uniquely Primate Cognition, 368
12.2 Issues of Proximate Mechanism, 379
12.3 Issues of Ultimate Causation, 395
12.4 The Structure of Primate Cognition, 399

Chapter Thirteen Human Cognition, 401

13.1 Human Cognitive Development, 403
13.2 Ontogenetic Processes, 416
13.3 Phylogenetic Processes, 423
13.4 The Structure of Human Cognition, 427

Chapter Fourteen Conclusion, 430

14.1 Theory, 430
14.2 Research, 432
14.3 The Preseration of Primates, 434

Appendix, 436

References, 443

Author Index, 496

Species Index, 504

Subject Index, 508

Primate Cognition

1

Introduction

The Western intellectual tradition was created by people living on a continent with no other indigenous primates. It is therefore not surprising that for more than 2,000 years Western philosophers characterized human beings as utterly different from all other animals, especially with regard to their mental capacities. If Europe had been generously populated by nonhuman primates during this time—if Aristotle and Descartes had encountered chimpanzees and capuchin monkeys routinely on their daily rounds—the belief that humans are the only rational animals might not be so deeply entrenched in our philosophical heritage. Indeed, it was not until the nineteenth century that this biological elitism began to lose its grip on the European world view, as the public became familiar both with nonhuman primates in zoological gardens and with Darwin's (1859, 1871, 1872) theories of the evolutionary continuity between humans and other animal species.

The first scholar to seriously contemplate the scientific study of primate cognition was, not surprisingly, Charles Darwin. In 1838, a full 21 years before publication of *The Origin of Species*, he jotted in the margin of his notebook: "Origin of man now proved. Metaphysic must flourish. He who understands baboon would do more towards metaphysics than Locke" (Notebook M 84e). In concert with his later published writings on human evolution (1871, 1872), this note clearly demonstrates Darwin's understanding that minds (what modern scholars prefer to characterize in terms of cognitive processes) are biological adaptations. As such, minds have evolutionary histories that may be studied in part by making systematic comparisons among related species, using the basic methods of comparative biology in which the aim is to reveal both similarities and differences. Darwin's theories thus set the stage for the comparative study of all types of cognitive processes in all types of animal species. His provocative note of 1838 sets a special challenge for scientists interested in the cognition of nonhuman primates.

In the century and a half since Darwin began formulating his theories, much research has been conducted on primate cognition. The aim of this book is to review that research and to make an assessment of the current state of our knowledge. In the remainder of this introductory chapter, we set the stage for our review with four preliminaries. First, we provide a brief historical background for the modern study of primate cognition. Second, we explicate the ecological approach to cognition that constitutes the theoretical framework of our investigation. Third, we provide an overview of the order Primates and describe some features of their daily lives that are relevant to an understanding of their cognition. Finally, we lay out the plan of the rest of the book, setting the stage for what we intend to be an enlightening exploration of the cognitive capacities of our nearest primate relatives.

1.1 Historical Background

The first Western scientist to seriously embrace Darwin's challenge was George Romanes, who collected many observations of animal behavior and published them in his famous books *Animal Intelligence* (1882) and *Mental Evolution in Animals* (1883). The first Western scientists to embrace Darwin's challenge with respect to primates in particular were Wolfgang Köhler and Robert M. Yerkes.

Wolfgang Köhler was a German psychologist who studied the intelligence of apes at a laboratory on the island of Tenerife, in the Canary Islands, from 1913 to 1917. As a Gestalt psychologist, Köhler was interested in intelligence as something that took organisms beyond punctate sensations and blind trial-and-error learning. He thus posed a variety of problems for his nine chimpanzees to solve, most of which involved the presentation of obstacles to food rewards along with various possibilities for surmounting those obstacles. For example, he suspended bananas from the ceiling of the apes' cages and gave them stacking boxes or long sticks; he placed apples on the ground outside their cages and gave them sticks that could be connected so as to create a rake of the necessary length; and he constructed physical barriers to block the most direct spatial route to observed fruits and provided them with various possibilities for detours. Köhler's aim was to investigate whether and in what ways the chimpanzees would perceptually restructure their world, that is, show "insight" into the problem so as to overcome the obstacle and obtain the reward (Köhler, 1925). A decade later, the American psychologist Heinrich Klüver (1933) conducted similar experiments with several other species of nonhuman primates, mostly macaques and capuchin monkeys.

Robert Yerkes was an American psychologist with fewer theoretical pretensions than Köhler. Yerkes was interested in studying all aspects of the behavior of nonhuman primates, especially apes, and for that purpose he set up a laboratory that still exists and bears his name. His first cognitive studies were carried out during the same time as Köhler's and paralleled them in many ways (Yerkes, 1916). Yerkes, however, continued his studies for many more years and attracted a variety of students and associates who investigated in detail particular aspects of primate cognition; for example, Tinklepaugh (1928, 1932) studied the spatial memory of several primate species in locating hidden food, and Crawford (1937) studied chimpanzees' skills at cooperation in several problem-solving tasks. Associates of Yerkes attempted for the first time to raise apes in ways similar to human children to see how such an upbringing would affect their skills of cognition and communication (Hayes & Hayes, 1951; Kellogg & Kellogg, 1933).

Unfortunately, the pioneering work of Köhler and Yerkes was not vigorously pursued in the middle part of the century. This neglect was due mainly to the dominance

Figure 1.1.
Wolfgang Köhler.

in the study of animal behavior of the theoretical paradigms of behaviorism and ethology during this period. As is well known, behaviorists generally eschew the study of cognition, preferring instead to focus on "observable behavior" and the learning processes that might underlie it (Watson, 1925). Because these learning processes were thought to be the same for all animal species, the majority of behaviorist research in the middle part of the century was done, for the sake of convenience, with rats and pigeons. When behaviorist research was conducted with primate species during this period, it mainly used rhesus monkeys as subjects (again for convenience), and when comparisons among species were made, virtually the only competencies studied were the speed with which different species could be taught via reward-based training in the laboratory to make discriminations, form learning sets, and the like (Harlow, 1959; Rumbaugh, 1970).

Ethologists are the intellectual descendants of natural historians and thus observe a wide variety of animal species, including many nonmammalian species, in their natural habitats. Ethologists are concerned mainly with the "biology of behavior," that is, the ways in which processes of evolution shape behavior, just as they shape the morphological characteristics of organisms (Lorenz, 1965; Tinbergen, 1951). For the most part, the early ethologists were not concerned with cognition, preferring instead to

Figure 1.2. Robert M. Yerkes. (Photo courtesy of F. Kiernan.)

focus on various "innate" and species-specific behavior patterns and how they contribute to the survival and reproduction of particular species (including such things as "innate releasing mechanisms" and "fixed action patterns"). They also explicitly ignored primates. However, the general spirit of ethology did pave the way for much of the early fieldwork on primate behavior, most notably that of Jane Goodall concerning tool use and tool making of a community of chimpanzees in east–central Africa (e.g., van Lawick-Goodall, 1968a).

In the 1960s, there was a cognitive revolution in psychology and the behavioral sciences in which the theoretical shackles of behaviorism were removed from the study of human cognition. The paradigm-defining assumptions of the new field of cognitive psychology, or cognitive science, all revolved around the idea that in human functioning there is something called cognitive representation that serves to organize both the perceptual input and the behavioral output of organisms in functionally meaningful ways (Neisser, 1967). Perhaps due to the dominance of behaviorism and ethology, or perhaps due to the hesitation of many people to attribute mental processes to animals, students of animal behavior were somewhat slow to embrace the cognitive revolution. However, within a decade, even behaviorists and ethologists had gotten into the act. Thus, in the late 1970s, a number of behavioristically oriented psychologists began holding symposia to study animal cognition (Hulse, Fowler & Honig, 1978; Roitblat, Bever, & Terrace, 1984), and at least some ethologists had begun to recognize that the study of cognition should be an important part of their enterprise (Griffin, 1978).

As the cognitive approach became more prevalent in the study of animal behavior, the study of nonhuman primates played a prominent role. Of special note, in the early 1970s Emil Menzel studied a variety of interesting chimpanzee behaviors at the Delta Primate Center, including their communicative skills and knowledge of space, for which he offered some unabashedly cognitive descriptions and explanations (Menzel, 1971a).

Focusing on human-raised apes, David Premack studied a variety of phenomena of chimpanzee functioning that he also explained cognitively, involving everything from mathematical concepts to "theory of mind" (Premack, 1976; Premack & Woodruff, 1978). The Gardners also invoked cognitive explanations in their pioneering study of a chimpanzee's acquisition of sign language (Gardner & Gardner, 1969), as did the Rumbaughs and their colleagues in explaining the behavior of their two chimpanzees raised together by humans (e.g., Savage-Rumbaugh, Rumbaugh, & Boysen, 1978a). Also noteworthy during this period was the work of Marler and colleagues on some cognitive aspects of the communication of several primate species (e.g., Gouzoules, Gouzoules & Marler, 1984; Seyfarth, Cheney, & Marler, 1980a, b); the work of Kawai and his colleagues on the social learning ("culture") of Japanese monkeys in a seminatural setting (Kawai, 1965); the work of Kummer on primate societies (Kummer, 1971); and the work of de Waal on chimpanzee "politics" (de Waal, 1982). Of particular note theoretically were Parker and Gibson's (1979) hypotheses concerning the evolution of primate cognition based on Piaget's theory of human cognitive development and Humphrey's (1976) competing theory based on processes of social cognition as manifest in various naive or "natural psychologies."

In the two decades since these early forays into the study of the minds of animals, primate cognition has become one of the fastest growing and most exciting areas of research in the behavioral and cognitive sciences. The theoretical roots of the scientists who study primate cognition may be traced to ethology, behavioral ecology, behaviorism, and cognitive psychology, as well as to anthropology and developmental psychology. This theoretical mix has produced a field of study in which there is a constant questioning of some of the most basic issues of cognition—how it works and what it is good for—as well as a constant supply of new and interesting empirical findings. The study of primate cognition may be thought of as a subdiscipline within the newly emerging paradigm of cognitive ethology (Cheney & Seyfarth, 1990b; Ristau, 1991).

1.2 The Ecological Approach to Cognition

Organisms adapt to changes in their environments in many different ways. These may be classified quite broadly into two classes: morphological adaptations, involving physical changes of the body, and behavioral adaptations, involving changes in the way organisms interact with their environments behaviorally. For example, if for some reason a species can no longer exploit its normal sources of food (e.g., due to drought or new interspecies competition), we may envision one class of adaptations that involve such things as changes in the way the body breaks down foods chemically so that new foods become nutritionally sustaining, and another class of adaptations that involve such things as finding new sources of the familiar foods by developing more efficient skills of foraging and learning (Milton, 1993). Of course, many (perhaps most) adaptations involve both behavior and morphology in some way, since evolving significant new ways of interacting with the environment behaviorally often requires morphological adaptations (e.g., wings for flying), and morphological adaptations quite often have behavioral consequences (e.g., long hair may lead to different migration patterns). In this context the ecological approach to animal behavior, sometimes called behavioral ecology, focuses on the adaptive problems of a species and how it goes about solving them, with special reference to the behavioral aspects of those solutions.

Figure 1.3. Aye-aye engaging in extractive foraging by means of its specialized middle finger. (Drawing by Margaret C. Nelson.)

Within the class of behavioral adaptations are cognitive adaptations (see Menzel & Wyers, 1981, for a particularly interesting discussion). Cognitive adaptations may be either narrowly focused on a specific adaptive problem or more widely applicable to a number of different adaptive problems. Although there is not a clearcut line of demarcation between cognitive and other behavioral processes, we may simplify the differences by focusing on prototypical cases that highlight key features. The prototype of a cognitive adaptation is a behavioral adaptation in which perceptual and behavioral processes (1) are organized flexibly, with the individual organism making decisions among possible courses of action based on an assessment of the current situation in relation to its current goal; and (2) involve some kind of mental representation that goes "beyond the information given" to direct perception. Complexity in the decisions to be made is also characteristic of the prototype of a cognitive adaptation. Each of these features requires brief discussion.

The sine qua non of cognitive adaptations is *flexibility*, and this must be flexibility that the individual organism to some degree controls. Thus, many behavioral systems are quite complex but not cognitive because they consist of a set of behavioral possibilities that are rigidly connected to perceived situations, and this connection is not susceptible to individual modification to meet novel exigencies. For example, there are some species of wasp that bring dead prey (mostly crickets) to the entrance to a tunnel, placing them carefully on their backs with their antennae just touching the entrance. They then enter and inspect the tunnel, return to the entrance, and bring the prey inside.

Figure 1.4. Chimpanzee engaging in extractive foraging by means of a tool. (Drawing by Margaret C. Nelson based on a photograph by R. W. Byrne.)

The appearance is of intelligent and planned behavior: the wasp seems to be assessing the tunnel for its suitability before depositing its prey. But if a human moves the cricket ever so slightly while the wasp is inside inspecting the tunnel, upon its return the wasp will reposition the prey in the preordained manner and enter and reinspect the just-inspected tunnel. It will continue doing this practically ad infinitum, never understanding that it could skip the pointless extra positionings and inspections with no negative consequences (Gould & Gould, 1986). When behavioral adaptations show such inflexibility, we assume that they have been programmed into the organism's nervous system in a relatively invariant manner in anticipation of certain recurrent situations, and we therefore call them species-specific behavioral adaptations. In such cases, the cognition or intelligence involved is due to the workings of evolution, not to the individual organism. These kinds of species-specific adaptations are an important aspect of the behavioral repertoires of all species, and they are the subject of much research in the field of behavioral ecology.

Cognitive adaptations, on the other hand, are adaptations in which evolution has relinquished its micromanagment of the behavioral interactions an organism has with its environment in deference to the individual and its judgment; these adaptations are the province of cognitive ethology. Given a particular functional context (i.e., a context in which an individual has a relatively well-defined goal), a cognitive adaptation involves the individual's active control of both the behavioral and the perceptual aspects of an interaction with the environment. Thus, in a control systems view (e.g., Powers, 1973), we may say that a cognitive adaptation is one in which the organism first has a state of affairs it wishes to bring about (the goal), and it must then decide which of several courses of action to pursue behaviorally based on its perception/understanding of the current situation in relation to the goal. A cognitive adaptation thus presupposes both a flexible way of perceiving/understanding a situation and a flexible choice of behavioral means based on that perception/understanding in relation to the goal. The result is what are often called "strategies," and these may vary from simple to complex.

It should be noted that such flexibility often derives from individually acquired knowledge or strategies because most things that are individually learned may be unlearned or modified. For example, during their early ontogenies, the youngsters of some animal species have to learn about the particular foods available in their particular environments and how to find, acquire, and process them, or they have to learn about the different individuals in their groups and how to cooperate or compete with them effectively. Because they are learned, these strategies may be unlearned or modified to meet changing circumstances, such as finding food after a flood or finding a mate after a radical shift in the dominance hierarchy of the group. In some cases such learning even leads to individually formulated goals, for instance, goals that must be achieved before another course of action may be undertaken (subgoals). Such flexibility is a relative matter, and there is no precise dividing line that tells us when something is flexible enough to be considered cognitive. The point is simply that without some flexibility, there is no justification for talking about cognitive processes at all (Bruner, 1972).

The other defining characteristic of cognitive processes is some form of mental *representation*: In its decision making process, the individual relies on information from a source other than direct perception, for example, from memory, inference, categorization, or insight. Thus, when an animal is faced with a problem and does not engage in overt behavioral trial-and-error but solves the problem "intelligently" nonetheless, in many cases the basis of that intelligence is some form of mental assessment of the situation. The organism compares the current situation to a previously experienced situation or infers what will happen if certain aspects of the situation are changed. In these cases the individual seems to make an assessment by manipulating the elements of some internal mental model of the problem (i.e., its knowledge) and observing the imagined consequences in that model, as, for instance, when a rat decides on the direction in which it should forage based on an assessment of the likelihood of finding food in the different directions and the possible obstacles to its path that each direction presents. When representation is combined with behavioral flexibility in a dynamic form of mental assessment, we get the prototype of cognitive processes called "thinking." There is no reason to doubt that many animals engage in thinking when it is defined in this way (Tolman, 1948).

This notion of mental representation as an internal model of a perceptual situation does not imply representation in the narrow sense of symbolic representation such as "pictures in the head" or the use of conventional symbols. What is implied is only that the organism can go beyond direct perception and show insight, or even foresight, in anticipating possible difficulties in solving a problem: what will happen if this is done, or what will change in one object if another moves to a new location. This process is what Piaget (1952) called the "prevision of errors" or "mental trial-and-error" because the individual mentally tests alternative scenarios, sometimes avoiding overt errors as a result. It is what Dennett (1995) called "Popperian learning," because failure means not that individual organisms die but that their hypotheses are incorrect. One problem with invoking mental representation of this type, however, is that once an organism has "thought through" a problem, its future encounters with the same or similar problems may show insight and foresight on the immediate perceptual level (Gibson, 1979). Experts who have learned much about certain types of problems by thinking them through soon come to perceive similar problems from the outset in ways that differ from the perception of novices, that is, they show insight into the problems from the outset. Cognitive skills that come to operate on the perceptual level by "automatizing" the representational processes involved so that they operate perceptually still involve mental representation.

The cognitive perspective of behavioral adaptations is in opposition not only to radical behavioristic accounts, which attempt to deny cognitive processes, but also to cognitive approaches that explicitly contrast processes of cognition to processes of instrumental conditioning as if they were different things. According to some theorists, if a behavior can be explained by instrumental conditioning—an organism is doing something because it has been rewarded for doing it in the past—we need not invoke the specter of cognition. But this dualism is not the only theoretical option. The alternative is to take the cognitive revolution seriously and to replace all instrumental conditioning explanations with cognitive explanations. In this view, behavioral adaptations that are typically explained via instrumental conditioning are simply adaptations in which fairly simple and straightforward cognitive processes are at work. A rat in a Skinner box is attempting to determine what it should do in which kinds of situations to obtain food given its experience in similar situations in the past; it is just that experimenters have restricted the possible effective behaviors to one (pressing a bar) and the environmental conditions to tones or lights that they control. Consequently, only relatively simple knowledge and strategies may come into play. If to obtain food the rat had to remember the location of various food items, assess and avoid potential dangers, make novel detours through the foraging space, and infer the presence of food from various cues or categories of cues, as takes place in nature, the flexibility and complexity of the cognitive processes involved would be more readily apparent. Said another way, the fact that an organism obtains food by means of a given behavior does not imply ipso facto that the behavioral process is not cognitive. Rats in Skinner boxes employ all kinds of strategies to obtain food, including climbing the walls, based on their assessment of how they might best go about obtaining food in the current situation.

Also important to cognitive processes is some notion of *complexity*. Complexity is not a sufficient condition for cognitive adaptations, as innate and inflexible behavioral systems may be extremely complex. It is also not necessary that we add complexity to the core definition of cognition, since some degree of complexity is implicit in the notion of flexibility in the choice of alternative courses of action. Nevertheless, given a flexible behavioral system that allows the organism to choose, greater complexity of these choices implies a more complex cognitive adaptation. Another dimension of complexity might be the range of different domains to which a cognitive adaptation applies; the more different situations to which a cognitive adaptation is applied, the more powerful it must be. We may thus think of a continuum of cognitive complexity constituted by some quantitative evaluation of the perceptual and behavioral choices an individual needs to make to behave effectively with respect to a particular adaptive problem, as well as the complexities presented by extensions to other adaptive problems.

Overall, therefore, the behavioral adaptations that are of most interest in cognitive ethology are not those that are innate and inflexible, but those by which an individual organism copes flexibly with a situation by choosing courses of action based on its understanding of the current situation in relation to its current goal, often based on individually acquired knowledge and strategies. These behavioral adaptations may vary in complexity, and explicit mental representation may be involved to a greater or lesser degree. Although the issue is far from settled, it is presumably the case that cognitive adaptations have evolved most frequently in situations in which environmental conditions change with some rapidity during the lifetime of the individual (Boyd & Richerson, 1985). When such rapid ecological changes within individual lifetimes are the rule, it is difficult for nature to predict recurrent situations in which rigidly programmed behavioral systems would accomplish their adaptive ends reliably, and so the best solution is to give to individuals some powers of learning, assessment, and control.

To conclude this theoretical discussion it should be noted that in comparing the cognitive processes of different species, the ecological approach is fundamentally opposed to all theoretical perspectives in which cognition is viewed as a single, unidimensional characteristic, called "intelligence" or "learning," of which particular species may have greater or lesser quantities. Instead, the ecological approach attempts to view the individual species in its ecological context and to investigate what behavioral problems it has to solve and how it solves them (see Neisser, 1984, for a similar approach to human cognition). The organization of cognitive adaptations may differ qualitatively from species to species, and not always in a linear scale on which all species can be ordered (Maier & Schneirla, 1935). Organisms may possess special-purpose cognitive skills, or they may have more broadly based knowledge and strategies. In either case, in the ecological approach to cognition, we do not focus our attention on the classical, general-purpose definitions of cognitive functions in terms of such things as learning or memory or intelligence, but rather we focus on particular ecological problems and how they are solved, with special attention to solutions consisting of flexible and complex behavioral adaptations in which individual organisms make informed choices based on a mental representation of key aspects of the situation.

1.3 Primates and Their Lives

Primates are an order of mammal whose closest living relatives are the Insectivores (e.g., moles, shrews, and hedgehogs). In general, primates are a highly diverse order in virtually all of their behavioral characteristics. This may be attributed, at least in part, to a general mammalian life-history pattern, which primates display in exaggerated form. Thus, primates have a relatively long period of immaturity in which they must learn much about their individual physical and social environments in order to survive and procreate. They also reach sexual maturity at a relatively late age, have few offspring at a time, and invest much parentally in each offspring. In all of these particulars, primates are extreme even for mammals (Richard, 1985). One implication of this life-history pattern is that even when the adaptive problems faced by primates are the same type as those faced by other species in some behavioral domain, their longer period of immaturity with much parental investment may lead to solutions incorporating more learning about the particular environmental exigencies involved and thus to more flexible and complex cognitive skills in that domain.

In addition to this overall life-history pattern, primates would seem to need cognitive adaptations to face specific problems of special complexity as they engage in the two essential activities of all mammals: foraging for food and interacting with conspecifics socially (including mating). In the domain of foraging, theorists such as Milton (1981, 1993) have pointed out the special difficulties that many primates face in maintaining an adequate diet given their high dependence on certain types of plant foods that are patchily distributed in space and time in their particular environments (e.g., fruit trees are located sporadically throughout the forest, and their fruits are only ripe during certain seasons). In the social domain, theorists such as Cheney, Seyfarth, Smuts, and Wrangham (1987) have pointed out that most primates live in highly complex social groups involving various types of feeding competition, display complex forms of emotional expression and communication, and seem to form a variety of long-term social relationships with groupmates based on individual experience. These two sets of adaptive problems, foraging and social interaction, may thus present special difficulties for primates. The result is an order of mammal with a large brain relative to body size, with an expanded neocortex (Dunbar, 1992; Jerison, 1973), that learns much of what it needs to know during ontogeny and deploys that knowledge flexibly in all kinds of complex foraging and social situations.

Depending on how they are counted, there are more than 180 species of primates, almost all of which (with the exception of humans) live exclusively in tropical climates (see figure 1.5). The four main groups of primates are prosimians (36 species), New World monkeys (64 species), Old World monkeys (73 species), and apes (12 species). These four groups are not formally recognized entities—they are not even on the same level of the formal taxonomy of primates—but they have come to be used as a convenient way to group species based on important similarities and differences. Appendix A contains a full listing of all the primate species currently known, with their Latin names, common names, and proper classifications. For current purposes, we simply note that primates emerged in evolution some 60–80 million years ago, with the 4 groups we have identified becoming distinct entities in the approximate chronology depicted in figure 1.6. Some general characteristics of the four groups are described below (this discussion follows very closely Martin, 1990).

1. *Prosimians* are a diverse group of small-bodied primates. The term "prosimian" reflects the fact that many researchers believe these species are more "primitive," that is, more like the original primates of tens of millions of years ago, than are the simian species (New World monkeys, Old World monkeys, and apes). Compared to the simian species, prosimians are generally much smaller in size, less socially oriented, less herbivorous (more insectivorous), and more nocturnal. Within prosimians, as body size increases, there is a trend toward the simian pattern of more sociality, herbivory, and diurnality. There are approximately 18 genera and 36 species of prosimians that may be categorized into three broad groups: lorises, tarsiers, and lemurs.

(i) The 11 species of lorises are all nocturnal, mostly arboreal, mostly small, and mostly omnivorous. All but two species live in Africa, and many species live

Prosimians

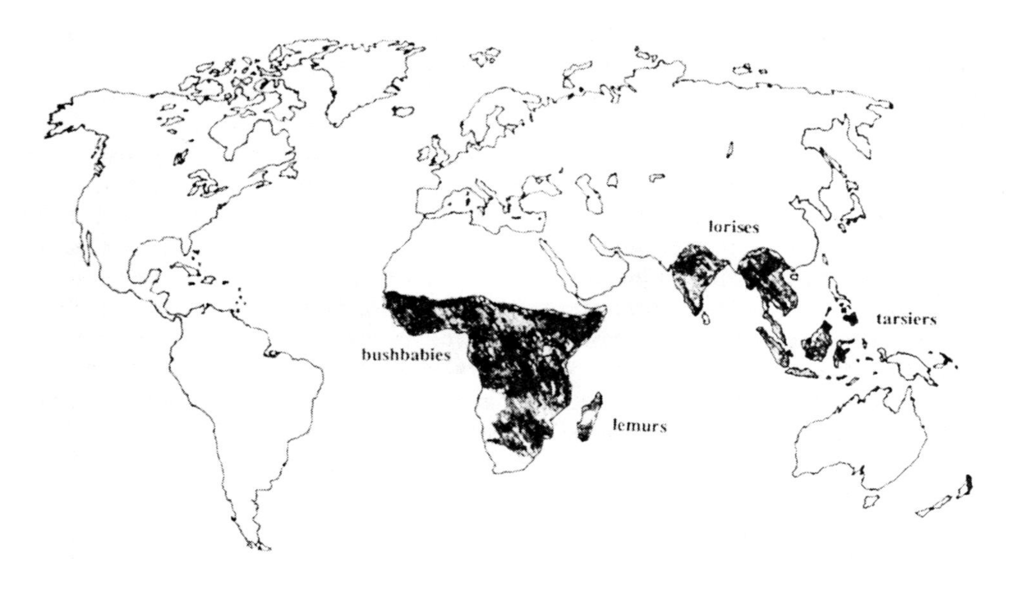

Monkeys

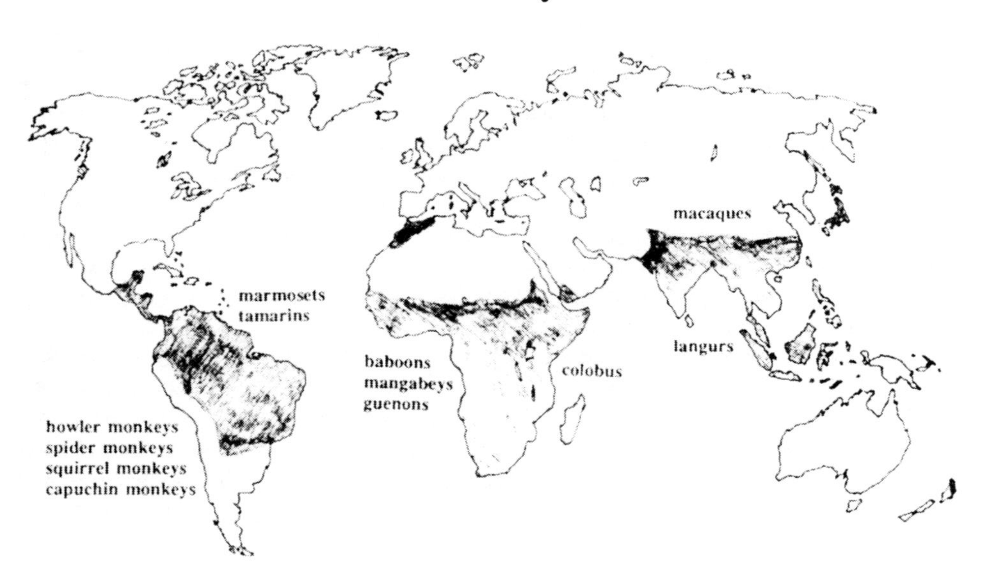

Figure 1.5. The distribution of primate groups across the world.

Apes

Figure 1.5. *(continued)*

fairly solitary lives as compared with the majority of primate species. The best-known species are the bushbabies.

(ii) The four species of tarsiers all live arboreally and nocturnally in the island areas of Southeast Asia. All species are relatively small and survive exclusively on insects and other small animals. They spend some time in monogamous pairs, but they mostly forage alone.

(iii) The approximately 21 species of lemurs live exclusively on the island of Madagascar and live in diverse ways. This diversity is perhaps due to the fact

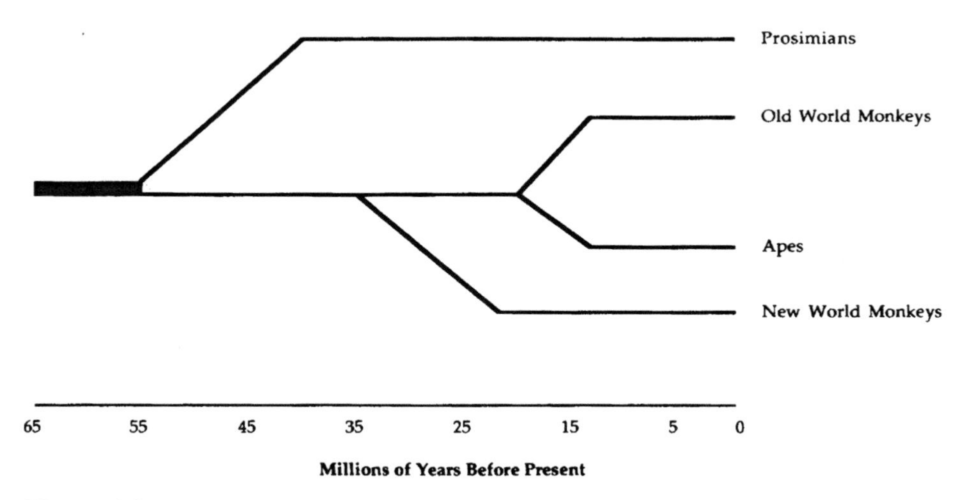

Figure 1.6. Approximate chronology of primate evolution (based on Fleagle, 1988).

that there is less competition on this island from other mammalian species than in most other primate habitats. There are nocturnal and diurnal species, living mostly arboreally (but with some species showing some terrestriality), with a relatively wide range of body sizes, diets of all types, and a wide range of social structures. The best-known species is the ringtailed lemur.

2. *New World monkeys* are so called because they live exclusively in Central America and the tropical parts of South America. They have been reproductively isolated from other primates for some 30–40 million years, from the time of the geological separation of South America and Africa. All species of New World monkeys are mainly arboreal and, with one exception, diurnal. Their flat noses with well-separated nostrils have led to the appellation "platyrrhines." There are approximately 16 genera and 64 species of New World monkeys that may be placed into two broad groups: the marmosets/tamarins and the cebid monkeys.

(i) The 20 or so species of marmosets and tamarins are relatively small simian primates who are basically omnivorous. Marmosets (and to some degree tamarins) feed extensively on tree sap and gums. Most species live as monogamous pairs.

(ii) The 40 or so species of cebid monkeys are markedly bigger than marmosets and tamarins, and the largest of them have developed prehensile tails for use in arboreal locomotion. Cebid monkeys eat mostly fruits and insects, with some leaf eating by some species as well. All cebid monkeys live in social groups, with no species living a predominantly solitary life. The best-known exemplars of the cebid group are various species of howler monkeys, spider monkeys, squirrel monkeys, and capuchin monkeys.

3. *Old World monkeys* are so called because they live in Africa and southern Asia. All species are diurnal, and as a group they are generally larger than the New World monkeys. Their less flat noses with closely-spaced nostrils have led to the appellation "catarrhines." There is considerable variety in the constitution of catarrhine social groups, but all species live in groups of one kind or another. There are approximately 16 genera and 75 species of Old World monkeys that may be placed into two broad groups: the colobines and the cercopithecines.

(i) The 25 or so colobine species live mostly in southern Asia. All species are arboreal and eat leaves as a fair proportion of their diets (some species are referred to as "leaf eaters"). As they forage, they often locomote through trees by brachiation (hanging from branches by the hand/arm and swinging to others). The best-known species of this group are langurs and colobus monkeys.

(ii) The 50 or so cercopithecine species live mostly in tropical Africa (a few in Asia). They are relatively large-bodied relative to the colobines and are distinguished from other monkey species mainly in the terrestrial adaptations of a small number of species who have moved into savannahs, desserts, and other open country grasslands (with most returning to trees to sleep at night). The smaller species generally survive on insects and fruits, whereas the larger species supplement their diet with a sizeable proportion of leaves. The best-known exemplars of the cercoptihecines are the macaques, baboons, mangabeys, and guenons.

4. *Apes* (also catarrhines) are close relatives of the cercopithecine monkeys and inhabit the same parts of the world (Africa and southern Asia), and also exhibit many terrestrial adaptations. All are relatively large bodied and exclusively diurnal. There are 5 genera and 12 species of apes in three broad groups: lesser apes, great apes, and humans.

(i) The eight species of lesser apes live in southeast Asia and are the smallest of the apes. They are mainly arboreal and eat mainly fruit and leaves. Distinctive features of the lesser apes are their arboreal locomotion by means of "true brachiation" (hanging from branches by the arm and swinging to others with a phase of free-flight through the air) and their loud and melodious vocalizations. Lesser apes live mostly in monogamous family groups. The best-known exemplars of this group are the gibbons.

(ii) There are four species of great apes, distinguished most clearly from other nonhuman primates by their large size. Chimpanzees, gorillas, and bonobos (pygmy chimpanzees) show some terrestriality and live in tropical Africa, whereas orangutans are mainly arboreal and live in southeast Asia. All four species are primarily herbivorous, exploiting especially fruits and leaves, but some species eat a variety of other foods including some insects and meat. Three of the great ape species live in social groups of one sort or another, with adults of the other species (orangutans) living mostly solitary lives.

(iii) Human beings are close relatives of the great apes. In the current context, human beings are distinguished from other apes mainly in their wide geographical range (including regions outside the tropics), their exclusively terrestrial life involving bipedality, their extensive use of tools in foraging and hunting, their relatively high consumption of meat, their relatively large brains (especially the neocortex), and their complex social groups involving complex forms of communication and division of labor.

It should be acknowledged at the outset that little is known about the cognitive capacities of the majority of primate species. For a variety of reasons, both practical and theoretical, there has been great selectivity in the species scientists have chosen to study. The majority of what we know about primate cognition comes from the intensive investigation of two to three dozen of the 180+ species, and there are significant taxonomic gaps. This is a major problem if we wish to make comparisons among large groups of species such as Old World monkeys versus New World monkeys, or monkeys versus apes, or prosimians versus other primates, as is quite common in the current research literature. The fact that there is often great diversity of behavior and cognition within these large taxonomic groups only magnifies the problem. Large-scale taxonomic comparisons, therefore, should be made only with the greatest caution, and the data on which they are based, including the particular species involved, should always be clearly specified.

A related problem is that scientists study at least three different kinds of primates in terms of the environments in which they have spent their ontogenies. First are primates in their natural environments studied by fieldworkers. The observations made of these primates are obviously primary to investigators taking an ecological approach to cognition, as they show cognitive adaptations operating in the environments in which they evolved. But naturalistic observations are often indeterminate when the research

Figure 1.9. Cotton-top tamarin. (Photo courtesy of M. Seres.)

Figure 1.10. Spider monkeys. (Photo by J. Call.)

Figure 1.11. Colobus monkeys. (Photo by J. Call.)

Figure 1.12. Baboon. (Photo courtesy of J. Call.)

Figure 1.13. White-handed gibbons. (Photo courtesy of M. Seres.)

question concerns the precise nature of the cognitive mechanisms at work. Second are primates in captivity studied by experimentalists who wish to determine more specifically the precise cognitive mechanisms involved in some behavioral adaptation. Powerful experimental methods that are not easily transferred to field situations are often used in these studies. The problem is that many of these animals have spent their ontogenies in species-atypical circumstances, sometimes in environments that are clearly impoverished in terms of the cognitive problems an individual may encounter. Finally are the few primate individuals, mostly apes, who have been raised in humanlike cultural environments with extensive instruction in various cognitive skills, including in some cases humanlike systems of communication. These individuals too have spent their ontogenies in species-atypical environments, although in this case the environments seem particularly rich for the development of cognitive skills, especially those that are most highly prized by humans (e.g., language and object manipulation). In general, given the assumption that the cognitive skills of primates are dependent in important ways on the nature of their ontogenies, comparisons among species should be contextualized based on a consideration of the nature of the ontogenies of the particular individuals involved.

1.4 Plan of the Book

The topic of this book is not primate behavior in general, but rather primate cognition as a subset of primate behavior. We thus focus our attention on those behavioral adaptations of primates that show a flexible organization involving individual decision making based on some form of mental representation. The comparisons we make across large

Figure 1.14. Lowland gorilla. (Photo by J. Call.)

taxonomic groups take account of both the particular species and individuals involved and attempt to be sensitive to their differing ecologies.

The remainder of the book is divided into three parts: Knowledge of the Physical World, Knowledge of the Social World, and A Theory of Primate Cognition. The first part of the book concerns primate physical knowledge. It includes chapters on space and objects, tools and causality, features and categories, and quantities. The theoretical perspective that has informed the majority of this work is the Piagetian paradigm, which focuses mostly on the individual's knowledge of objects. Some neobehavioristic theories of discrimination and categorization are also prominent in much of this work. We are well aware of the limitations of both these theoretical frameworks, so we do not adhere slavishly to either of them; our focus is not on evaluating theories, but on using them to understand the research findings. Where appropriate, we use research with humans and other animals to clarify key issues. We end with some discussion of the evolution of primate physical cognition, based mainly on various theories of primate foraging strategies.

The second part of the book concerns primate social knowledge. It contains chapters on social knowledge and interaction, social strategies and communication, social learning and culture, and theory of mind. A mixture of theoretical perspectives has informed this work—all involving ideas taken from various theories of "natural psychology"—and, indeed, we have forged something of a synthesis of our own in trying to bring order to this rapidly growing area of research. Again, however, we try not to let the theory become the primary focus of attention and, where appropriate, use research with humans and other animals to clarify key issues. We end with some discussion of the evolution of primate social cognition, based mainly on theories of the respective roles cooperation and competition in primate social life.

The third and final part of the book represents our attempt to construct a general theory of primate cognition. Although premature in some respects, our review of the totality of extant research has led us to propose some hypotheses about primate cognition that we hope will be useful in formulating future research. In particular, we propose a hypothesis about what distinguishes the cognition of primates from that of other mammals that may be readily tested and falsified. We also propose in the final substantive chapter of the book some hypotheses about how human cognition differs from that of other primates—again in a way that may be readily tested and falsified. We believe that this account of human cognition is useful in the current context because so much current research on nonhuman primate cognition assumes as a background, either explicitly or implicitly, something of the nature of human cognitive processes. We conclude our investigation with some suggestions for future directions in the study of primate cognition.

PART I

KNOWLEDGE OF THE PHYSICAL WORLD

The classic view of cognition revolves around an organism's understanding of objects and their various spatial, causal, and featural interrelations. Although such understanding is an integral part of virtually all of an organism's important activities, for many animal species the most pressing problems arise in locating and obtaining food. It is thus widely accepted among behavioral scientists that many important cognitive skills for many animal species evolved in the context of foraging. This is almost certainly true of primates, many of whom spend the greater part of their daily lives searching for and ingesting food. Although primates must face a number of other problems in the physical domain as well (e.g., avoiding predators), we follow tradition and assume that primate physical cognition evolved mainly in the context of foraging. We organize this part of our investigation accordingly.

Two important problems presented by foraging are locating food and manipulating it to make it edible. Accordingly, chapter 2 concerns primates' understanding of large-scale foraging space (cognitive mapping) and the movements of objects from one location to another in small-scale manipulative space (object permanence). Chapter 3 then focuses on primate skills of object manipulation and tool use that are sometimes required to process food or to extract it from a substrate (e.g., digging up roots, probing for insects with sticks, and breaking open nuts). The focus on manipulative skills also introduces the question of how primates understand the causal relations among objects in their environments. It should be noted that in these first two chapters most of the studies reported have been conducted within the framework of Piaget's theory of sensory–motor cognition (e.g., Piaget, 1952, 1954). In this theory the basic categories of knowledge that are cognitively constructed during sensory–motor development are space, time, causality, and objects. Although almost nothing is known about primates' understanding of time, in these first two chapters we attempt to report all that is known about their understanding of objects and their spatial and causal interrelations.

The other two major problems presented by foraging are identifying food items and quantifying the amount of food involved. Consequently, chapter 4 focuses on primates' skills of discrimination learning and categorization, and chapter 5 focuses on their abilities of quantification. It should be noted in this case that most of the research has been conducted in the behaviorist or learning theory paradigm. Because behaviorism and learning theory are general process accounts—assuming that cognitive processes, insofar as they exist, are of the same type for all animals in all domains of learning—most of these studies have been conducted in the laboratory and have concerned one of a small number of primate species making one of a small number of types of discrimination (most of which have no interest or relevance for the animal other than the fact that a human rewards them). In any case, there are facts to be learned about primate cognition from these studies, and in each case there is also some relevant research from the Piagetian paradigm as well, namely, studies of the classification of objects (sorting) and the conservation of quantities. In the Piagetian framework, these skills go beyond sensory–motor cognition in requiring an understanding of more abstract (less perceptually based) dimensions of experience.

Because Piaget's theory is so important to this part of our investigation, we should make clear at the outset that we are aware of its many limitations. Most importantly, the theory places excessive importance on the manipulation of objects, to the neglect of cognition that may take place in the absence of manipulation, and this is decidedly inappropriate for many nonhuman primate species (as well as for human infants in some situations; Baillargeon, 1995). Furthermore, in Piaget's theory of sensory–motor cognition, the organism's knowledge of space, time, objects, and causality all develop within a single framework; his six sensory–motor substages apply uniformly across all four cognitive categories because the categories make up the framework. In addition to the fact that recent empirical work has not corroborated this hypothesis for human children (Gelman & Baillargeon, 1983), in an ecological approach to cognition it is important to look at these categories separately without prejudging whether and in what ways they are related. There is no reason why a particular species might not have an understanding of space characterized by one of Piaget's stages and an understanding of causality characterized by another. In any case, in chapter 6 we discuss both Piagetian and learning theory approaches to primate physical cognition, each from the point of view of both proximate mechanisms and ultimate (evolutionary) causation. We then propose some directions for future research, based on gaps in the existing empirical research and on what we perceive as limitations in current theoretical perspectives.

2

Space and Objects

Nothing is as critical to the survival and procreation of primates, or indeed to animals in general, as the ability to deal effectively with space and the objects that occupy it. Food and mates must be located, and predators must be avoided. Consequently, many animal species, including numerous species of birds, fish, and insects, have evolved highly specialized devices for finding particular locations on the planet, sometimes at great distances, so that they can mate, lay eggs, avoid cold weather, seek safety, and perform other such vital functions (see Alcock, 1984, for a review). These specialized devices seem to involve little learning of any sort, however, and do not seem to be organized flexibly so that the individual may modify them to meet novel exigencies. Seemingly more flexible spatial skills are displayed by many animal species as they navigate in complex local environments; for example, lions pursue a fleeing prey by adjusting their direction of travel to match that of the prey, and mountain goats navigate effectively on extremely uneven terrain in totally novel contexts. However, spatial skills such as these are primarily perceptual, since they rely almost totally on an organism's "direct perception" of a spatial layout or event without any need to go beyond the information given (Gibson, 1979).

If our concern is with the spatial cognition of primates—their behavioral adaptations for dealing with space that show flexible organization and rely to some degree on mental representation—there are two phenomena that merit special attention: cognitive mapping (concerning locomotor space) and object permanence (concerning manipulative space). First, individuals of many primate species must learn during the course of their ontogenies something of the spatial layout of the local environments into which they are born for purposes of locating food, conspecifics, and predators. Beginning with Tolman (1932), behavioral scientists have been especially interested in the flexible way in which some animals are able to learn and understand spatial layouts and to mentally represent them in the form of "cognitive maps." Cognitive maps contain information about what is located at particular places in the environment and are available to organisms in the absence of any direct perception of those objects or locations. Thus, individuals with cognitive maps of their local environments can remember whether a location normally has food and whether they have visited and depleted a food site in the immediate past (sometimes called spatial memory). Furthermore, individuals with cognitive maps do not have to directly perceive and learn each individual spatial path connecting every location in their environment (e.g., by actually traveling from every location to every other), but may fill in pieces of missing spatial information by means of inference. This is evidenced most clearly when an individual flexibly devises a novel shortcut or other creative route from one location to another.

In addition to acquiring knowledge about their relatively large-scale locomotor environments, many primates must also acquire some knowledge of objects and their spatial possibilities in more small-scale, manipulative space. The questions here are whether primates are capable of locating objects hidden from view if they have witnessed the hiding, and, given the appropriate information, if they can infer the possible locations of an object that has changed locations outside their view. These skills are most often observed by primatologists in the context of Piaget's (1954) object permanence tasks, in which individuals must search for objects after experimenters have hidden them in various ways. From their behavior in these tasks, it is possible to determine whether primates know that an object exists when it is not in their immediate perceptual field, that an object may sometimes be found in new places, or that an object can only be in one of the locations through which it, or the container in which it is occluded, has passed. Some of these tasks thus assess not only the mental representation of objects when they are not being perceived, but also the mental representation of object movements in small-scale space.

In this chapter we investigate the spatial cognition of nonhuman primates. We survey first what is known about primates' ability to construct and employ cognitive maps in large-scale locomotor space and then what is known about primates' skills in object permanence tasks in more small-scale manipulative space. We conclude by surveying other studies of primate spatial cognition, especially those concerned with their understanding of transformations in the spatial orientation of objects and the movement of objects through complex spaces involving detours and mazes.

2.1 Cognitive Mapping

Like many mammals, most primates live in social groups that tend to remain within a particular home range whose boundaries they recognize (see papers in Clutton-Brock, 1977a; Holloway, 1974). This space is sometimes defended from other groups of conspecifics (at which point it is called a territory), but often it is not. Primate home ranges vary in size from less than 1 km^2 (some species of prosimians) to more than 50 km^2 (patas monkeys, *Erythrocebus patas*). The general pattern is that larger and more frugivorous species have the largest home ranges (fruits being more widely spaced in the environment than other foods) (Clutton-Brock, 1977b). Within the home range, species also differ in the distances they need to cover each day to forage effectively, referred to as the "day range." Day ranges vary from less than 100 m (some prosimians) to more than 10 km per day (hamadryas baboons, *Papio hamadryas*). Again, diet and body size are important factors in determining day-range size, as is the typical size of the foraging group (larger groups need a larger day range because there are more animals to feed).

Figure 2.1. Rainforest in Santa Rosa National Park, Costa Rica. (Photo by J. Call.)

In this context, primates clearly have a practical knowledge of how to deal effectively with space in obtaining their daily sustenance.

If our concern is with the cognitive mechanisms involved in primate understanding of locomotor space, the primary focus should be on skills of cognitive mapping. There are two main issues: (1) the extent to which primates are able to remember what is located where in an environment, as evidenced by their tendency to travel selectively to sites based on their previous experience of which of these do and do not contain food (or other desired items); and (2) the extent to which primates "know where they are" in an environment, as evidenced by their ability to travel from one site to another via novel routes, especially using shortcuts that minimize travel distance. The prototype of a cognitive map is thus an understanding of space in which there is mental representation of what is where in an environment and flexible use of that information in making decisions about how to travel efficiently from one location to another.

2.1.1 Monkeys

About a dozen species of monkeys have been observed, either in the wild or in laboratory experiments, for signs of cognitive mapping skills. Although the majority have shown positive signs of such skills, the evidence for some New World monkey species is mixed.

Looking first at Old World monkeys, the classic study is that of Tinklepaugh (1932), who studied one rhesus macaque (*Macaca mulatta*) and one longtail macaque (*Macaca fascicularis*) in a simulated foraging task in the laboratory. Each subject saw a pair of containers in a room and then observed as food was hidden in one of them. Several rooms of a barn were visited in this manner before subjects were returned to a neutral site. The macaques were then taken back to the rooms, either in the same order as on their previous visits or in reverse order, to test whether they could identify the containers with food. The order with which they were exposed to the containers had no effect, and subjects showed an overall accuracy of 79% when there were five pairs of containers in five rooms. In a follow-up study, when pairs of containers were placed in close proximity to one another within a single room and then baited while subjects watched, the monkeys experienced more difficulty. They were 75% accurate with three pairs of containers but only 61% accurate with eight pairs. In a related study, Tinklepaugh (1928) found that with a single pair of containers, the same two subjects could accurately remember which one had food after 15- to 20-hour delays, although with multiple pairs of containers, a 24-hour delay disrupted their performance. In this same study Tinklepaugh also found that when the monkeys saw a preferred food hidden under one of two containers, but then found a less preferred food under that container (due to a surreptitious substitution), they showed obvious signs of "disappointment" by continuing to look for the food that they had seen hidden (spending much more time searching in the substitution trials). They often did not eat the substituted food, even though on other occasions when they had seen that same food hidden they retrieved and eaten it without delay. They apparently remembered what specific food was at the key location.

MacDonald and Wilkie (1990) used a similar type of simulated foraging task to study the spatial memory of yellow-nosed monkeys (*Cercopithecus ascanius whitesidei*). They constructed eight food sites in a large enclosure, four of which were baited with food for each search trial. There were two experimental conditions. In the win–stay condition, the food was always placed in the same four locations for a given individual across trials. In the win–shift condition, the food was first found in four locations, but then on subsequent trials was shifted to the four locations that had not previously contained food. In both conditions the monkeys quickly learned where the food was located. This finding is important because it rules out a simple conditioning explanation in which monkeys simply return to the place where they have found food in the past; their learning is much more flexible. An additional important finding was that animals in both conditions generally visited the four sites with food first and used a minimum distance strategy to do so, thus demonstrating their ability to use spatial memory to make flexible decisions about travel routes (see Coussi-Korbel, 1994, for a similar finding with white-collared mangabeys, *Cercocebus torquatus*).

Perhaps the most impressive demonstration of the spatial inferential skills of primates was reported by Menzel (1991), who used experimental techniques to study Japanese macaques (*Macaca fuscata*) in a free-ranging field situation. Immediately adjacent to a well-known monkey trail, Menzel placed either a ripe akebi fruit, native to the area (but out of season for part of the period of observation), or a piece of chocolate; there was also a control condition in which nothing was placed beside the trail. He then waited until an individual discovered the provisioned food (or approached the area in the control condition) and then followed that individual as it ranged for the next 20 minutes. What he found was that when the monkeys found the ripe akebi fruit, they proceeded to search for more akebi fruit. They immediately stared upward from the provisioning site into the trees where akebi vines might be growing (more than in the other two experimental conditions), and they went to locations where akebi vines grew, and actually inspected or touched akebi vines (more often than did subjects in the other two experimental conditions). When subjects found chocolate, on the other hand, they manually foraged on the ground near the provisioning site and returned to that site during the 20-minute trials more often than did monkeys in the other two conditions. Menzel concluded that the monkeys who found the akebi fruit took this as an indication that they should search for other akebi fruit whose location they knew from previous experience. Those who found chocolate, on the other hand, took this as an indication that this might be the specific location of a novel and desirable food that might be exploited in the future. "Taken together, the observations suggest detailed knowledge of the feeding environment, attentiveness to novel ecological details, and considerable flexibility in organizing an effective search routine" (Menzel, 1991: p. 401).

Hemmi and Menzel (1995) demonstrated a related spatial inferential ability in long-tail macaques. Each of five subjects was individually presented with sets of three trials in which food was hidden along an invisible linear trajectory across the three trials in an experimentally constructed sandbox. On the fourth trial, food was hidden either along this same trajectory or at a randomly chosen site. Results indicated that individuals found the food hidden on the trajectory more quickly than they found the food hidden at a random location. This would seem to indicate that the macaques were able to predict or infer the location of the food based on the trajectory established in the previous three trials.

It is important that the findings of these experiments are corroborated by field studies of the natural foraging behavior of Old World monkeys. In his classic field studies of hamadryas baboons (*Papio hamadryas*), Kummer (1968, 1971) noted a number of ways in which these baboons remembered what is where in their environments, and noted that they seem to travel to specific locations by the most efficient route. For example, because of the arid climate of northeastern Africa in which they live, these baboons must search for water on a regular basis (Sigg, 1986). In the process, these animals not only use least distance strategies to go from one place to another (i.e., take shortcuts), but they also have been observed to speed up their travel as they approach a source of water, well before there could be any perceptual evidence of its location, showing that they know the water is nearby (Sigg & Stolba, 1981; see also Altmann & Altmann, 1970). The baboons also seem to use a mental map strategy in locating their nightly sleeping sights on cliffs, although the fact that cliffs can be seen from relatively long distances means that vision may play an important role in locating these sites as well.

It is also noteworthy that Cheney and Seyfarth (1982b), employing field experiments, found that vervet monkeys (*Cercopithecus aethiops*) seem to know the appropriate locations of individuals in different groups of conspecifics inhabiting the Amboseli National Park in southern Kenya. When individuals from one group heard the call of a female played back to them from the territory of a group with whom she was not normally associated, they showed more attention and vigilance than when they heard that individual's call coming from her home range. In a similar vein, Gouzoules, Gouzoules, and Marler (1984) found that when free-ranging infant rhesus monkeys need help from their mothers, they look more in the direction in which they know her to be than in other directions, even when she is out of sight. These observations suggesting that Old World monkeys know the whereabouts of conspecifics are important because they show skills of mental mapping that extend beyond the foraging situation. It should be noted in this context, however, that all the evidence for cognitive mapping in Old World monkeys, involving either food or conspecifics, comes from cercopithecine species, and little is known about the spatial skills of the colobines (other than Marsh's, 1981, general observations of the flexible foraging strategies of red colobus monkeys, *Colobus badius,* in central Africa).

The cognitive mapping skills of New World monkeys present a more mixed picture than that of Old World monkeys. The major field evidence for cebid monkeys is Milton's (1980) study of the foraging of mantled howler monkeys (*Alouatta palliata*) in the rainforests of South America. Although there are suggestions in this work that some form of cognitive mapping might be occurring, Milton observed that for the most part, howler monkeys make heavy use of certain "pivotal trees" in their home ranges, returning to them almost daily, and they seem habitually to use the same travel routes to go from one food source to another (many emanating from a pivotal tree). Another cebid species, titi monkeys (*Callicebus moloch*), also tends to use habitual routes in most of its travel (Mason, 1968).

The most systematic field work is that of Garber (1989, 1993) and Garber and Hannon (1993), who studied the foraging behavior of two species of tamarins (*Saguinus mystax* and *Saguinus fuscicollis*) in the Amazon rainforests of northeastern Peru. Tamarins feed from hundreds of different trees, approximately 12 trees per day, and since the trees are separated from one another by an average of 92 m in a dense rainforest, travel routes cannot be based directly on sight. In their daily observations over a 1-year period, covering most of the foraging time of each day, the investigators noted the location of the group every 2 minutes. Using this information, foraging routes of the troop were computed relative to a number of previously identified reference points in the troop's home range. The investigators compared three computer models to the data obtained: a random foraging model; two different models based on olfactory cues (one including wind direction and the ability of animals to follow a strength of odor gradient); and a model based on cognitive mapping that predicted visitations based on minimum distance to the nearest tree with available fruit (i.e., that they had not recently depleted). Results indicated that the cognitive mapping model best matched the tamarins' behavior (although olfaction may be used in some close-range situations). The investigators estimated that tamarins know the locations of hundreds of feeding trees in their home range, and they remember for days, or even weeks, which trees they have recently depleted.

Laboratory evidence for the mental mapping of New World monkeys in foraging situations is provided by Menzel and Juno (1985). Their subjects were five saddleback tamarins (*Saguinus fuscicollis*) who visited a large room adjacent to their home cage for 15 minutes per day over a period of several months. Two novel objects were added to the room every other day, one of which contained food (and the other smelled like the same food); on the next day neither contained food. There were 30 objects used over 15 days of hiding, and all objects remained in the room after they were placed there, although they sometimes were moved to new locations on subsequent days. Which locations and objects individuals visited on each day was observed and recorded. Results showed that the tamarins were excellent at identifying which objects were novel each day and at remembering the locations of objects at which they had found food on previous days. Menzel and Menzel (1979) also found excellent spatial memory in a study of the same species in which novel objects that never contained food were introduced to the group; animals visited the locations with novel objects and ignored those with familiar objects. See Joubert and Vauclair (1986) for a similar study with Guinea baboons *(Papio papio)*, a species of Old World monkey.

MacDonald, Pang, and Gibeault (1994) also found strong evidence of cognitive mapping in a simulated foraging study with common marmosets (*Callithrix jacchus*). Using the same basic procedure that McDonald and Wilkie (1990) used with yellow-nosed monkeys (see above), these investigators found that the marmosets remembered which locations had contained food on previous trials and were capable of employing either a win-stay or win-shift strategy across trials as needed (although they seemed to prefer a win-stay strategy). There was also some evidence that the marmosets used a least distance strategy in their foraging, again indicating flexibility in the use of information from spatial memory.

In a variation on this theme, De Lillo, Visalberghi, and Aversano (in press) presented four tufted capuchin monkeys (*Cebus apella*) with a simulated foraging environment in which there were three clusters of food locations, each containing three specific sites for food. For each subject a peanut was hidden in each of these nine locations to determine the order in which they would search in these locations and what kinds of mistakes they would make in terms of searching already depleted locations. Subjects were also presented with a control condition in which the nine locations were evenly distributed, unclustered, within the subjects' enclosure. Three of the four subjects searched more efficiently in the clustered condition than in the unclustered condition, using something approaching a minimal distance strategy in which they depleted the food at all or most of the three sites within a cluster before moving on to the next cluster. Relatively few errors were made within clusters and most errors (about three-quarters of the total) occurred as subjects erroneously returned to a cluster whose sites they had already depleted to some degree.

Evidence that not all New World monkeys have cognitive mapping skills of this type comes from two studies. First, Andrews (1988) tested titi monkeys and squirrel monkeys (*Saimiri sciureus*) in a radial-arm maze containing eight arms. When left on their own to forage, the monkeys visited the different arms in a less than systematic way, often visiting sites they had previously depleted. Second, Roberts, Mitchell, and Phelps (1993) constructed a more naturalistic simulated foraging environment, similar to that of De Lillo et al. (in press), in which to test two squirrel monkeys. Their environment was a fairly large room with four simulated trees connected by runways. Each tree had 12 holes in it, 4 on each of 3 sides. When all the holes in all the trees were baited, both monkeys foraged enthusiastically and depleted the food. When only some holes were baited (the same holes across trials), neither subject learned to preferentially visit these specific holes, however, or even to favor the tree that habitually had the most food overall (except when one tree had no holes baited). Attempts to make the food locations more salient and more naturalistic did not facilitate the performance of the monkeys. When some holes were baited with bitter food, the monkeys still visited holes indiscriminately over trials.

This conflicting pattern of findings for New World monkeys may be due to some differences of aims and methodology. For example, the Roberts et al. (1993) findings with squirrel monkeys were mainly concerned with learning to remember the locations of baited and unbaited sites across trials, whereas the De Lillo et al. (in press) findings with capuchins concerned search efficiency when all locations were baited. Nevertheless, both groups of researchers raise the possibility that perhaps some primate species do not need cognitive mapping skills in their natural foraging, since many species (including howler monkeys and some other cebid species) forage in groups that move

Figure 2.2. Experimental apparatus used by Roberts et al. (1993) to test the spatial skills of squirrel monkeys. (Reprinted by permission of Lawrence Earlbaum Associates.)

through trees using habitual routes and eat all the appetizing fruit or leaves en route. These monkeys may thus have evolved spatial memories adequate for locating recurrent sources of food at widely separated sites, perhaps aided by habitual routes, but not for locating and remembering particular sites in more detail (as required in the Roberts study). Because marmosets and most tamarins (as studied experimentally by Menzel & Juno and McDonald et al.) feed to some degree on tree sap and gums that may be easily obtainable at particular locations on particular trees, and capuchins (as studied by De Lillo et al.) sometimes feed on insects residing inside the bark of trees, these species may need spatial memories more attuned to smaller scale details.

2.1.2 Apes

All great ape species observed have shown some skills of cognitive mapping (both spatial memory and the choice of novel and efficient travel routes), although chimpanzees have been studied much more extensively than other species. No species of lesser ape has been observed or tested for skills of cognitive mapping.

First, using the same basic situation in which he tested two monkeys (see section

2.1.1), Tinklepaugh (1932) tested the spatial memory of two adolescent chimpanzees (*Pan troglodytes*). When foods were hidden in one member of a pair of containers in different rooms of a barn, chimpanzees found them easily and, like the monkeys, were not affected by the order in which they returned to the pairs of containers. In this test, the chimpanzees were accurate in choosing the container with food on 90% of the trials with 10 pairs of containers in different rooms. In the test in which the pairs were all in the same room, the subjects were accurate 88% of the time with 8 pairs of containers and 79% of the time with 16 pairs, better than the monkeys and better than the human children who were used as a comparison. Although the subjects were accurate in choosing the correct container of a pair after some delay (up to 1 week), a 24-hour delay using the 16 pairs in a single room caused a total disruption in their performance. It is also interesting that Tinklepaugh (1928, 1932; see also Yerkes & Yerkes, 1928) reports that in several follow-up studies on both rhesus macaques and chimpanzees, by far the most important cue was the position of the pair of containers in the room and the position of hiding in the left or right container. When experimenters used containers that could be identified by their shape or color, changes in their positions (either changes in the location of the pair or changes in which of a pair contained the food) confused the subjects. Individuals of both species mostly treated a novel pair of containers in the location of an old pair as if it were the old pair, and when the members of a pair changed places, subjects continued to search in the location they had seen the food hidden, not the specific, well-marked container. As did the monkeys in the study (Tinklepaugh, 1928), the chimpanzees showed great "disappointment" and "surprise" when they went to find the food they had seen hidden under a container only to find a different food at that location, indicating that they knew what was at the location.

In a more naturalistic setting similar to that of Menzel and Menzel (1979), Menzel (1971b) placed 20 objects in particular locations in the 1-acre enclosure of a captive colony of 6 wild-born chimpanzees. After 10 half-hour sessions of familiarization with these objects, the chimpanzees immediately detected any new objects placed in the enclosure, usually identifying them within 15 seconds of being released into the area. Menzel (1973a) used this same colony to test spatial memory more directly. Keeping 5 chimpanzees in an indoor enclosure, he showed a single chimpanzee where food was hidden in 18 different locations in the 1-acre enclosure and then returned that individual to the enclosure to wait at least 2 minutes. On different trials, different chimpanzees were observers, different hiding places were used, and different routes were used in the hiding process. Overall, across 16 trials, the chimpanzees who had observed the hiding found an average of 12.5 pieces of food, whereas the 5 chimpanzees who had not observed the hiding found among them only 1 piece of food. Observer animals tended to move from one location to another quite deliberately, whereas nonobserver animals seemed to wander about searching randomly. Moreover, observer animals followed a least distance principle in which they visited locations based on their closeness to them, with their travel route bearing no relation to the route used by the human experimenter in the hiding process (see figure 2.3).

In a series of follow-up studies, Menzel (1973a) observed several interesting phenomena. First, when the 18 locations were baited with 9 preferred foods (fruits) and 9 less preferred foods (vegetables), observer animals tended to use a least distance strategy for the preferred foods before moving on to the less preferred foods; in the first 9 trials, observers visited fruits an average of 6.75 times and vegetables an average of 2.25

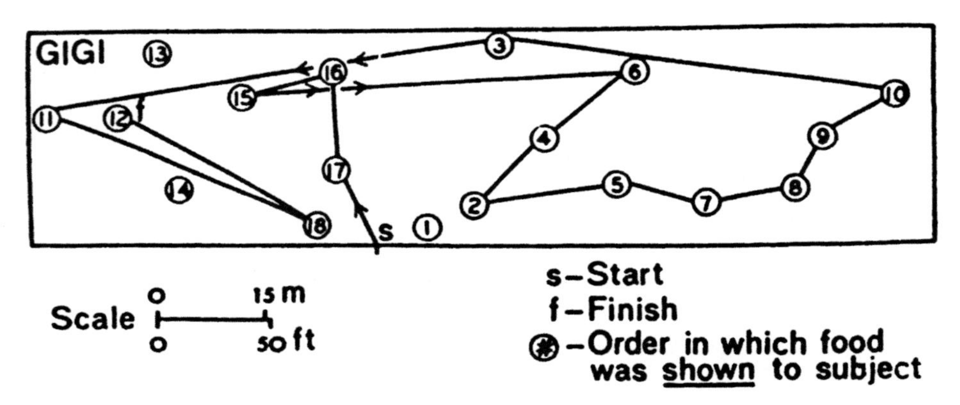

Figure 2.3. Trajectory followed by one of Menzel's chimpanzees to collect food inside the enclosure. Arrows indicate the subject's travel direction; circled numbers depict the order in which the different locations were baited while the subject watched. Note the minimum-effort strategy employed by the subject. (Reprinted from *Science* with permission from the American Association for the Advancement of Science.)

times (see also Gust, 1989, for a report of chimpanzees' flexible foraging strategies as a function of the probability of food at a particular site). Second, when three pieces of food were hidden on one side of the compound and only two pieces on the other, observers almost invariably began with the side containing more food. Finally, Menzel found that when two pieces of food were hidden on each side of the compound, observers almost invariably used a spatially efficient search strategy of going to the two pieces on one side before proceeding across to the other side of the compound (regardless of the order in which they had seen them hidden). Overall, it is clear that subjects simply knew from their observations the locations of the food items and what kinds of food items were at particular locations, and so they then proceeded to travel to those locations in a relatively efficient manner, given their food preferences. It is also interesting that Menzel, Premack, and Woodruff (1978) found that chimpanzees were very skillful at locating people who hid from them in a large enclosure simply by watching the hiding on a black-and-white television monitor (see Vauclair, 1990, for a similar finding with two Guinea baboons).

It is likely that chimpanzees use these same skills in foraging for food in their natural environments (Wrangham, 1977). Although detailed studies of chimpanzees' foraging routes have not been done, Boesch and Boesch (1984) performed a relevant study that is of special interest because the objects found were not food but tools used to obtain food. In the Tai forest in western Africa, chimpanzees routinely use stones and clubs to crack open nuts (see chapter 3 for details of this behavior). Some types of nuts can be cracked open with stones of a certain size and weight only, and such stones are rare in the forest. When individuals finish a nut cracking session, they typically leave the stone in place, although they sometimes carry it with them in search of more nuts. This means that when, on a subsequent occasion, an individual needs a stone, it must choose among all stones whose location is known. The question is whether in doing so they take into account the distance they need to travel to retrieve the stone and its weight, assuming that they do not want to carry a heavy stone too far. Analysis of 76 transport episodes revealed that in more than one-half of the instances, individuals chose to go for the stone that could be obtained in the most energy-efficient manner; that is, the stone that on the whole was both closest and lightest. On another quarter of the occasions, the individual chose the stone that was either closest or lightest, with distance on the whole being more important than weight. Again, the inference is that chimpanzees know what is where, and they adjust their travel behavior to this knowledge in adaptive ways.

With respect to other ape species, MacDonald (1994) investigated the spatial memory of two gorillas (*Gorilla gorilla*), one adult and one 1-year-old infant, using techniques similar to those used with two species of monkeys (see section 2.1.1). Food was hidden in four of eight sites (three of six for the infant), and subjects were allowed to find and consume the food on one trial only. After a delay (24–48 hours for the adult, 1–10 minutes for the infant), subjects were given a second trial with the food in the same four locations. Both gorillas found all the pieces of food by visiting just more than the minimum four sites required. However, in terms of efficient traveling, subjects showed little evidence of using a minimal distance search strategy. This may have been because, as MacDonald points out, the way the containers were arranged in the enclosure may have encouraged simple sequential searches. An interesting additional finding is that the adult gorilla almost always stopped searching after finding the last food item, whereas the infant searched all of his six containers on every trial. This strategy may reflect something of the adult gorilla's understanding both of the permanence of objects (see sections 2.2 and 2.3) and his ability to estimate quantities (see chapter 5). Fischer and Kitchener (1965) report that on tests of delayed response for spatial locations (similar to those of Tinklepaugh, 1928), gorillas performed in a highly competent manner.

With respect to orangutans (*Pongo pygmaeus*), Galdikas and Vasey (1992) report informal observations suggesting that in their natural habitats orangutans use cognitive mapping strategies. They seem to use least distance strategies in travel to different food and water sites, and like the baboons observed by Sigg and Stolba (1981), they speed up in anticipation as they approach a food or water site well before it is perceptually evident, suggesting some form of mental representation of the location of the food or water. Support for these informal observations comes from Fischer and Kitchener (1965), who report that on tests of delayed response for spatial locations, orangutans performed at a very high level. The only information on bonobos (*Pan paniscus*) is the report of Savage-Rumbaugh, et al. (1986) that their human-raised bonobo easily learned to navigate to and communicate about several different locations in a large outdoor environment.

2.1.3 Prosimians

Little information is available on the cognitive mapping skills of prosimians. In their tap-scanning for cavities in trees (indicating the presence of certain kinds of insects), aye-ayes (*Daubentonia madagascariensis*) seem to be able to project the trajectory of a labyrinthine cavity by first locating the central parts of the cavity and then extrapolating in a direction consistent with those parts of the cavity whose location they already know (Erickson, 1994). Picq (1993) reported that gray mouse lemurs (*Microcebus murinus*) were able to master some aspects of efficient performance in a radial-arm maze.

2.1.4 Comparison of Species

Cognitive mapping is not a uniquely primate behavior; many rodent species, some bird species, and even some bees have excellent spatial mapping skills in foraging situations (e.g., Brown & Demas, 1994; Olton & Samuelson, 1976; Shettleworth & Krebs, 1982). Based on available research, it might seem that primate skills of spatial memory and cognitive mapping are especially broad-based because primates seem to know and remember the location of particular types of food in foraging situations, the location of tools needed for extractive foraging, and the locations of things other than food such as other groups of conspecifics or other individuals. The spatial memory skills of other species have not been studied in all of these different circumstances; however, there are some studies suggesting broad-based spatial memory in nonprimates as well. Of most direct relevance, individuals of one species of bird differentiate between sites at which they have previously hidden more and less preferred foods (Sherry, 1984), individuals of one species of rat have been shown to use olfactory cues on group mates to locate preferred food sites (Galef & Wigmore, 1983), and individuals of one species of bee do not forage at locations indicated by experimental manipulations if these sites are "impossible" locations for food (e.g., in the middle of a lake; Gould, 1987).

Within the primates, the cognitive mapping skills of monkeys and apes seem to be of the same general quality (there is not enough information on prosimians to make comparisons). In the only direct comparison of monkeys and apes in identical experimental situations, Tinklepaugh (1932) found that both macaques and chimpanzees remembered locations where they had seen food hidden, and when a different food was substituted they knew it; they both remembered what was where. The chimpanzees seemed to remember the location of more hidden items than the macaques, however. On the other hand, no such differences between monkeys and apes were found in the studies of MacDonald (MacDonald, 1994; MacDonald & Wilkie, 1990; MacDonald, Pang, & Gibeault, 1994) using different monkey and ape species: yellow-nosed monkeys, common marmosets, and gorillas (although it may be that not enough pieces of food were hidden to detect the kinds of differences Tinklepaugh reported). Also arguing for no systematic monkey–ape differences is the study of Menzel and Juno (1985), which found that saddleback tamarins detected novel objects in their environments as well as the chimpanzees in Menzel's (1971b) similar studies. Menzel and Juno (1985) expressed skepticism about large differences within the order Primates in skills of spatial cognition. Interestingly, the strongest suggestion of differences within primates is within New World monkeys. There seem to be differences among some species in these skills in a way that may correlate with the foraging demands experienced by particular species. Although hypotheses about the cognitive mapping skills of primates as a whole will remain premature until more research is done (in particular with prosimians, colobines [of the Old World monkeys], cebidae [of the New World monkeys], and hylobates [of the apes]) our overall conclusion is that there are no major qualitative differences within primates in skills of cognitive mapping, although particular species may have modified the general primate model in the light of species-specific adaptive pressures.

Although the cognitive mapping skills of human children have not been tested in situations precisely comparable to those in which nonhuman animals have been tested, human children do show an ability to navigate in large-scale space in ways consistent with cognitive mapping and to remember where items have been hidden in large-scale space during their second and third years of life (Cornell & Heth, 1983; Foreman, Arber, & Savage, 1984).

2.2 Searching For Hidden Objects

Primates have also demonstrated complex skills of spatial cognition in tasks in which they must search for hidden objects or food in small-scale, manipulative space. These tasks have mainly been Piaget's (1954) series of object permanence tasks in which the hiding process becomes increasingly more complex over trials. First is the stage 4 task in which an experimenter gains the subject's attention to an object and then covers it. The question is whether the subject will remove the cover to retrieve the object, with successful retrieval indicating some knowledge that objects continue to exist when they are not perceived. Stage 4 is typically achieved by human infants at around 8 months of age. It is worth noting that this is essentially a "delayed response" task of the type behaviorists have been studying for years (reviewed in section 4.1.3), and that the kind of spatial memory reviewed in the immediately preceding section would seem to require stage 4 object permanence skills as well.

The stage 5 task involves having the subject find the object first under cover A (perhaps on several trials), and then on a subsequent trial watch the experimenter move it from under A to under another cover, B. Tracking this visible displacement from A to B and finding the object under B indicates a knowledge that objects can move to new locations independently of one's own actions, and is achieved by human infants at around 12 months of age. The A-not-B error in the stage 5 task occurs when subjects continue to search where the object was last found (under A), even though they have just seen it disappear under B. The interpretation of this error is controversial: Piaget (1954) believes that it shows the subject's lack of differentiation of the object from its own actions, Bremner (1980) believes that it indicates the subject's reliance on an egocentric (rather than an allocentric) spatial code, and Diamond (1991) believes that it is simply a matter of the subject learning to inhibit its tendency to conduct the search at the place where it was successful previously.

2.2.1 Prosimians

In a study of different object-related tasks with a number of different prosimian species, Jolly (1964a) reported that one slender loris (*Loris tardigradus*) did not search for food that visibly disappeared into or under an object until it was intensively trained to do so. There is some question, however, whether the reaching and manipulative skills of this species is adapted well to the search task as presented—the subject mostly struck at the occluding object (which it did not succeed in displacing), just as it normally strikes at its food, mainly insects, in its natural environment.

2.2.2 Monkeys

Stages 4 and 5 of object permanence—searching for hidden objects and tracking their visible displacements to new locations—have been reliably reported for a wide range of monkey species. Among these are squirrel monkeys (*Saimiri sciureus*: Vaughter,

Smotherman & Ordy, 1972); woolly monkeys (*Lagothrix flavicauda*: Mathieu et al., 1976); capuchin monkeys (*Cebus apella*: Dumas & Brunet, 1994; Schino, Spinozzi & Berlinguer, 1990; Spinozzi, 1989; *Cebus capucinus:* Mathieu et al., 1976); stumptail macaques (*Macaca arctoides*: Parker, 1977); longtail macaques (Natale, 1989c; Potì, 1989); Japanese macaques (Antinucci, Spinozzi, Visalberghi, & Volterra, 1982; Potì, 1989); and rhesus macaques (de Blois & Novak, 1994; Wise, Wise & Zimmerman, 1974). Because the species represented cover a wide range of both New World and Old World monkeys, it is reasonable to assume that stages 4 and 5 object permanence skills are a part of the cognitive repertoires of all monkey species. It is interesting to note that developmentally, for the species that have been tested, there is typically a period in which the A-not-B error is made before success in the stage 5 task (as there is in human infants).

2.2.3 Apes

Stages 4 and 5 of object permanence have also been reliably demonstrated in two species of great ape: chimpanzees (e.g., Hallock & Worobey, 1984; Mathieu & Bergeron, 1981; Mathieu et al., 1976; Potì & Spinozzi, 1994; Spinozzi & Potì, 1993; Wood et al., 1980) and gorillas (Natale, 1989c; Natale, Antinucci, Spinozzi, & Potì, 1986; Redshaw, 1978). As in monkeys and human infants, typically there is a period in which apes make the A-not-B error prior to success in the stage 5 task. In a more specific test of this developmental progression, Visalberghi (1986) found that a 17-month-old gorilla followed visible displacements of objects, but when the gorilla was physically moved 180° to the other side of the A and B hiding places, it perseverated with responses relative to its own body (e.g., continuing to reach to the right). Human infants often do the same thing in the early phases of stage 5, and this behavior has been interpreted to mean that their spatial understanding relies egocentrically on their own bodies and not allocentrically on landmarks in the environment (e.g., the color of the cloth) (Bremner, 1980). It has also been interpreted to mean that subjects cannot use knowledge to inhibit performing a behavior that was successful in the immediate past. However, it is possible that this behavior on the part of this gorilla was due to its young age; Tinklepaugh (1928) reported that his longtail macaque subject was able to keep track of the location of food after a similar bodily displacement.

There is a single study of object search with gibbons (*Hylobates lar*; Snyder, Birchette & Achenbach, 1978), but the results are difficult to interpret because extensive training regimes were employed. No other ape species to our knowledge has been tested systematically in object permanence tasks. However, the behavior of orangutans in a task in which they had to point to one of three opaque containers in which food might be found indicated at least stage 4, and likely stage 5, object tracking skills (since the food was sometimes transferred to different containers; Call & Tomasello, 1994a); a similar inference may be drawn from the delayed-response study of Fischer and Kitchener (1965). Also, captive bonobos have been observed to engage in object search skills daily, indicating stages 4 and 5 of object permanence knowledge (e.g., hiding-object games in which objects are hidden in different places; Savage-Rumbaugh, personal communication). As with monkeys, therefore, it is reasonable to assume that searching for hidden objects and following their visible displacements (stages 4 and 5 object permanence skills) are a part of the cognitive repertoires of all ape species.

Figure 2.4. Stumptail macaque in an object permanence task. (Photo by J. Call.)

2.2.4 Comparison of Species

The object permanence skills of a number of nonprimate species have also been investigated (see Doré and Dumas, 1987, for a review). Although many species of animals have food gathering or capturing techniques that increase the probability that they will reinstate contact with the food or prey after it has disappeared (e.g., the prey-capturing behaviors of some insects), these are often quite stereotypic movements and are only used in situations of immediate survival value (Etienne, 1984). Other animals learn through experience to use certain cues indicating that they should continue the behavior even after food has disappeared, but they do not engage in active searching (e.g., some chickens). Most mammals do engage in active search processes, however, indicating more flexible cognitive skills in small-scale manipulative space. The best studied species are domestic cats and dogs, which both perform stages 4 and 5 tasks quite readily (Doré, 1986, 1990; Gagnon and Doré, 1992, 1993).

Because of the presence of stage 4 and 5 object permanence skills in monkeys, apes, nonprimate mammals, and human infants at 1 year of age, and because there is a ready motory explanation for the failure of Jolly's (1964a) single loris in a stage 4 search task, it is likely that all primates have stage 4 and 5 skills of object permanence; that is, all primates understand that objects exist even when they are not perceptible and that objects can move to new locations independently of their own movements. Indeed, the behavior of the loris may be similar to that of 5-month-old human infants who know that objects still exist when they are not perceiving them, even though they cannot actively remove a cover to find them (Baillargeon, Spelke & Wasserman, 1985). This interaction of knowledge of the world and the motor skills to express such knowledge (which scientists mostly use to assess knowledge) must be kept constantly in mind. Indeed, Vauclair (1984) makes the point that in many cases object permanence skills are attained earlier in development in nonhuman primates than in human infants and that this may be due to their more precocious motor skills. Again, however, it is important to note that any general hypotheses about the object permanence skills of primates will remain premature until more research is done with prosimians, colobines (of the Old World monkeys), marmosets and tamarins (of the New World monkeys), and hylobates (of the apes).

2.3 Tracking Invisible Displacements

In the stage 6 object permanence task, the experimenter first covers an object by placing it under a small opaque box (or some other small occluder). This box is then transported under several covers, and the object is deposited under one of them. Although the subject has no way of knowing where the object was deposited, successful stage 6 behavior consists of a "systematic search" under all and only those places where the object might be, given the trajectory of the small box's path (perhaps in the order, or reverse order, in which the container traversed them). This is usually achieved by human infants between 18 and 24 months of age. In Piaget's (1954) theory, systematic search in this task indicates a knowledge of space as a "closed system" of locations in which the object *must* be located somewhere along its perceptible or imaginary trajectory and *cannot* be anywhere else (e.g., it cannot be located beneath a cover under which the small box did not pass).

Piaget (1954) claims that stage 6 of object permanence is the only stage of object permanence that requires mental representation. Although stages 4 and 5 obviously require that the subject retain some knowledge that an object still exists when not being perceived, a fairly simple static image that simply continues the perception of the object as it disappears behind the occluder will suffice. In stage 6 tasks, on the other hand, to imagine the object's possible movements when out of sight requires a more dynamic and constructive mental representation of an independent object moving in an independent space. Another way to conceptualize the relation between the different tasks is to recognize that stage 6 is essentially about the understanding of space, in that it is the only stage of object permanence development that requires a mental representation of object movements. Thus, if stage 4 object permanence may be taken to indicate some form of mental representation of nonperceptible objects (with stage 5 only a variant of this since subjects still search where they saw the object disappear last), stage 6 may be taken to indicate some form of mental representation of the nonperceptible spatial displacements of objects.

2.3.1 Monkeys

Claims for stage 6 object permanence have been made for a number of monkey species, including squirrel monkeys (Vaughter et al., 1972), capuchin monkeys (Mathieu et al., 1976), rhesus macaques (Wise et al., 1974), and stumptail macaques (Parker, 1977). But the methodologies of these studies have been criticized by both Gagnon and Doré (1992) and Natale et al. (1986). The problem is that these studies take as their criterion of stage 6 that the animal successfully find the hidden object. But there are a number of simple strategies that subjects may adopt that lead to success in this task, and these studies did not contain the kinds of controls necessary to eliminate them. For example, an animal may employ a simple strategy to search under the last of the covers under which the small box passed or the last of the covers the experimenter touched. Or it may always search under the cover to its right or even under all of the covers available on each trial. For accurate assessment of stage 6 search skills, controls must be constructed to attempt to identify the different strategies subjects are using.

Three sets of studies have tested for the stage 6 object permanence skills of monkeys and have employed all the necessary controls. First, using a single Japanese macaque, Natale et al. (1986) administered a series of stage 6 tasks in which they sought to eliminate the possibility of simple search strategies. The subject saw two large boxes and one small box. A food item was placed under the small box, the small box was moved under one of the large boxes (and the food surreptitiously left there), and then the small box was taken out and left beside the others. Which of the large boxes the food was left under was varied systematically. The idea is that the food could either be under the small box or the large box under which it passed, but it could not be in the other large box (under which it did not pass), and so the subject should not search there. The subjects responses were scored as follows: (1) responses in which the subject searched under the incorrect box were scored as errors (no matter what else it did); (2) responses in which the subject first searched under the small box where it had seen the food disappear and only then under the large box containing the food were scored as "sequential responses" and were considered stage 6 responses because all and only the places the food could have been were systematically searched (called representational search strategies); (3) responses in which the subject searched directly under the large box containing the hidden food were scored as "direct responses" and were considered equivocal with respect to the subject's stage 6 status.

What was found was that of the 27 trials administered, the monkey made 8 errors, 16 direct responses, and used a sequential search strategy on only 3 trials—a result considered inconclusive by the investigators. Natale et al. (1986) therefore administered two sets of control trials. In the first set the subject was given some "false trials" in which the food was placed under the small box as before, but then the small box was left out to the side of the two large boxes without ever being placed under either of them. The experimenter then lifted one of the large boxes as before but without passing anything under it. The idea was to test whether the subject had developed the simple strategy of choosing the box the experimenter lifted regardless of whether the small box had passed under it. Indeed, on half of these false trials, the macaque searched first in the large box that the experimenter lifted, even though the food or its container never passed under it. In a second set of control trials, the normal invisible displacement task was administered, but instead of leaving the small box directly beside the one in which the food had been left (as was done in the original trials), the small box was moved to the opposite side so that the empty box (never touched) was now between the small box and the large box containing the food. This was to make systematic search more difficult by requiring the subject to "skip over" the middle box to systematically search in the small box and the large one it had passed under (now on the two extremes of the row of three boxes). On these so-called nonlinear trials, the monkey searched in the empty box (even though it had never been touched or raised and the food had never gone under it) on 70% of the trials. In similar studies with three longtail macaques and two capuchin monkeys (*C. apella*), Natale and Antinucci (1989) reported similarly negative results: all five subjects were fooled regularly by the false and nonlinear trials into choosing the box under which neither the food nor the small box containing the food had ever passed (see figure 2.5).

The second set of appropriately controlled studies were conducted by two groups of investigators using similar sets of control trials. De Blois and Novak (1994) studied six adult rhesus macaques and found passage of stage 5 but not stage 6 tasks when all the appropriate controls were used. Using a group of capuchin monkeys (*C. apella*) different from those of Natale and colleagues, Dumas and Brunet (1994) also found negative results for a stage 6 task with stringent controls. The conclusion of all these investigators who studied monkeys under tightly controlled conditions is that in stage 6 tasks monkeys acquire various practical strategies for being successful, but they do not use systematic, representational search strategies requiring the mental representation of nonperceptible object displacements.

The third study using highly controlled procedures contradicts these negative findings. Schino, Spinozzi, and Berlinguer (1990) administered stage 6 object permanence tasks to four longtail macaques and two capuchin monkeys (*C. apella*). Five of the six subjects did not succeed in displaying systematic search strategies in the control tasks (including false trials and nonlinear trials). One capuchin subject (male, 9 years of age) did succeed, however. He typically used a direct strategy in regular trials, but on false trials he did not search under the box that was "sham manipulated"; he searched under the small box, the only place the food could be given a stage 6 understanding of space, on 29 of 32 trials. This subject's performance was significantly better than chance on the nonlinear trials as well, although he made more mistakes in this condition (43% of the trials). The investigators concluded that this one adult capuchin monkey was operating with a representational understanding of space. (It should be noted that the one capuchin monkey who failed in this task and the two who failed in the Natale & Antinucci study, 1989, were all juveniles; in fact, one subject was common to both studies, as were three of the longtail macaques.)

2.3.2 Apes

Claims of stage 6 object permanence have been made for two ape species: chimpanzees (e.g., Mathieu, et al. 1976; Mathieu & Bergeron, 1981; Wood et al., 1980) and gorillas (e.g., Redshaw, 1978). However, Natale et al. (1986) and Gagnon and Doré (1992) have criticized these studies in the same way they criticized the monkey studies claiming stage 6 object permanence, that is, there were no controls to ensure that subjects were not finding the hidden food on the basis of simple, practical search strategies.

Natale et al. (1986) tested one juvenile female gorilla in the same tasks as those they used with monkeys and found that the gorilla seemed to be using a representational search strategy. Whereas she began by using a direct search strategy, the gorilla soon switched to a sequential search strategy after a slight change in the procedure (going to the small box and then to the large one under which it had passed), whereas the monkeys with the same procedural change did not change their strategies. In addition, the gorilla was given both the false and nonlinear control trials. In the false trials, the subject never once in 12 trials searched in the box that the experimenter sham lifted. On the nonlinear trials in which a search strategy required the subject to pass over the

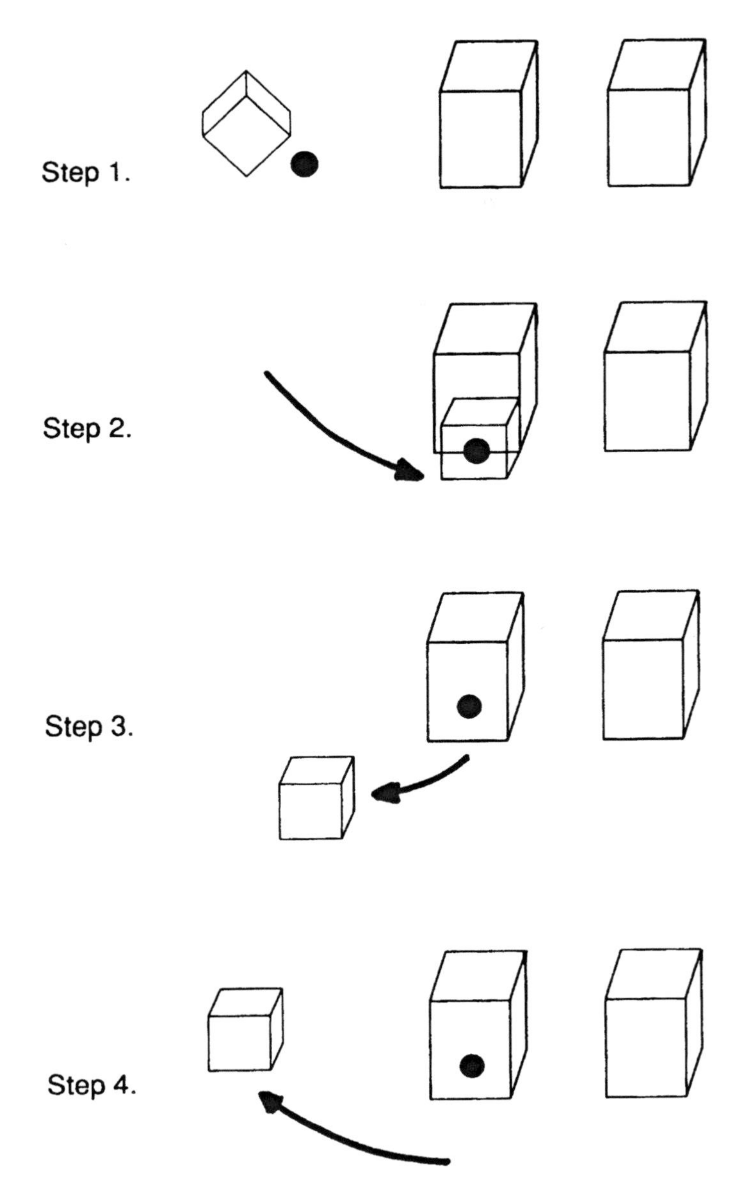

Figure 2.5. Natale et al. (1986) procedure to test object permanence in primates. Note that the container on the right side is not touched by the experimenter and consequently cannot contain the reward.

empty box in its systematic search, the gorilla passed over it and searched correctly at well above chance levels, but with an error rate (38%) similar to that of the successful monkey of Schino et al. (1990). The investigators concluded that this gorilla employed a representational search strategy in which it mentally imagined the object as it passed from location to location in its opaque container.

2.3.3 Comparison of Species

It is difficult to know what to make of these stage 6 findings. When appropriate control procedures have been used, more than a dozen monkeys (representing three Old World and one New World species) have failed to provide evidence of the systematic representational search strategies characteristic of stage 6. However, one capuchin monkey displayed such a strategy, although he made some errors on one type of control trial. One ape (a gorilla) also showed a stage 6 strategy when tested rigorously, but she is the only ape who has been tested appropriately, and her performance also showed some errors on one type of control trial. In all, the extent of stage 6 search skills in nonhuman primates is a question in dire need of future research attention.

These two primate individuals are not alone among nonhuman animals in possessing stage 6 object permanence skills. Although cats do not solve stage 6 problems despite repeated opportunities (Doré, 1986, 1990), dogs (*Canis familiaris;* mostly terriers) show systematic, representationally based search strategies in stage 6 tasks equipped with all the necessary controls (Gagnon & Doré, 1992, 1993). The studies in which stage 6 performance has been claimed for other nonprimate species failed to employ appropriate controls, and so their status is indeterminate (Doré & Dumas, 1987; Gagnon & Doré, 1992).

2.4 Other Forms of Spatial Understanding

In both of the domains of spatial knowledge reviewed so far, the question has been locating objects in space. Three other spatial skills have also been studied in primates: (1) understanding the changing orientations of objects as they rotate in space, (2) understanding complex spatial relations in the "patterned string problem," and (3) understanding the possible trajectories of self and moving objects in complex situations involving obstacles and mazes. Because they seem to operate mostly on the perceptual level, these skills are less prototypically cognitive than cognitive mapping and object permanence.

2.4.1 Mental Rotation of Object Orientation

In a follow-up study to one of their spatial memory tasks, Menzel and Menzel (1979) looked at the reaction of saddleback tamarins to various novelties in the spatial orientations in which they placed hidden objects. If the same object was always found in the same location and its orientation in the hiding place was always constant, the interest the subjects had in that object decreased over trials; since it matched expectations, it was uninteresting. When the hidden objects changed locations, interest increased again, and if new objects were hidden in old locations, interest increased as well. However, when the same object was hidden in the same location but it was simply rotated so that it was oriented differently in that location, presenting a different visual stimulus to the subject, interest continued to decrease nonetheless. The

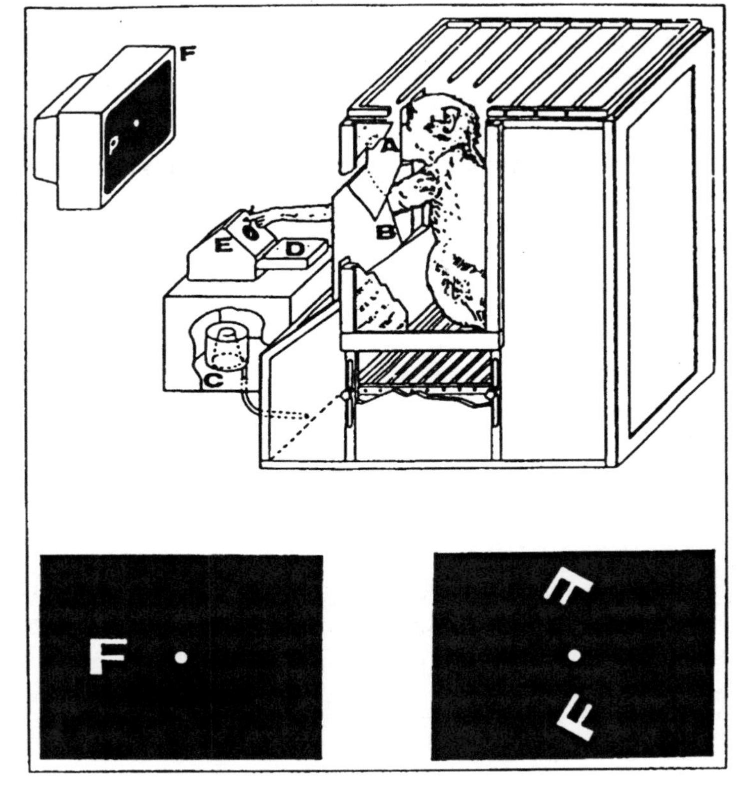

Figure 2.6. Apparatus and stimuli used by Vauclair et al. (1993) in a mental rotation task. (Reprinted from *Psychological Science* with the permission of Cambridge University Press.)

investigators inferred from these findings that the subjects considered it to be the same familiar object no matter what its visual orientation, and thus they inferred form constancy in tamarin visual perception and memory. (A similar conclusion may be drawn from studies of primate recognition of faces irrespective of their orientation; e.g., Dittrich, 1990 for longtail macaques; Phelps & Roberts, 1994 for squirrel monkeys).

Vauclair, Fagot, and Hopkins (1993) also investigated the ability of a monkey species to identify objects irrespective of their orientations, but they did so in a paradigm that focused on mental representation more directly (see also Hopkins, Fagot & Vauclair, 1993). Six wild-born juvenile Guinea baboons were administered a number of experimental trials in which they first were shown a sample stimulus on a TV monitor (e.g., the letter F or its mirror image) and then immediately were asked to choose one of a pair of target stimuli that matched it. One of the target pair matched the sample stimulus exactly, except in orientation (which could vary by 60°, 120°, 180°, 240°, or 300°), whereas the other was the mirror image of the sample stimulus rotated the same amount as the matching target (i.e., it was flipped over) (see figure 2.6). The question was whether the subjects could find the target stimulus that matched the sample stimulus despite its different orientation. Four of the six subjects did so reliably. Moreover, in an analog to the Shepard and Metzler (1971) research with human subjects, the question was whether the time it took these four subjects to make the match would be a function of the difference in the degrees of orientation of the sample and target stimuli. If there was such a match, it could be inferred that they were "mentally rotating" the stimuli before comparing them, since stimuli requiring more mental rotation should take longer, the same way that actually scanning them perceptually would. Results showed that differences in the degree of orientation of the sample and target stimuli (when presented to the right visual half-field) led to decision time differences exactly paralleling those of human subjects, who were also tested in this same experimental paradigm. For both species, longer mental rotations took more time. These findings contrast with findings in pigeons, who discriminate various stimuli despite their orientation but do not show a relation between decision time and degree of rotation (Hollard & Delius, 1982). The authors thus concluded that in performing the matching operation required in this task, the baboons were rotating mental images of the sample and target stimuli.

2.4.2 Patterned-String Problems

Beginning with the study of Harlow and Settlage (1934), a number of different primate species have been tested for their abilities to deal with "patterned string problems." The patterned string problems form a series of 12 (or more) well-standardized tasks in which an individual is shown food tied to a string. The difficulty arises because the string is presented to the animal along with several other strings leading to no reward, and all the strings are crossed and tangled in various systematic ways. For example, in pattern number 6 the food is tied to one string and nothing tied to another, and they are presented to the subject in the pattern of an X. Other items are even more complex, using up to four strings. To solve these types of problems, subjects presumably must understand various complex spatial relationships. The degree to which this skill is perceptual, rather than a conceptual, is an open question. Indeed, in a careful analysis of various types of transfer trials administered to squirrel monkeys, Cha and King (1969) concluded that their subjects were not learning the spatial properties of the strings involved, but rather were simply learning certain perceptual configurations of strings as discriminative stimuli indicating what should be done for a reward.

The main advantage of the standardization of these tasks is that many different species may be reliably compared. First, Harlow and Settlage (1934) studied a number of different monkey species and found them to be skillful in various ways. Finch (1941) studied chimpanzees, and when he compared his results to those of Harlow and Settlage concluded that chimpanzees are more skillful than the monkeys these investigators studied. He also noted that the chimpanzees seemed to improve over trials, whereas the monkeys did not. Fischer and Kitchener (1965) studied gorillas and orangutans. When they compared their results to those of previous studies, they concluded that apes as a whole were more skillful than monkeys as a whole, especially on the most difficult items. The problem is that in coming to this conclusion Fischer and Kitchener did not consider all the data from all the relevant studies. When the performance of all the species that have been studied in this paradigm are directly compared, as in Figure 2.7, it may be seen that several species of monkeys are as skillful as the apes even on the most difficult problems (although the orangutan is the most skillful individual species). Especially skillful are the one pigtail macaque (*Macaca nemestrina)* and the six rhesus macaques studied by Settlage (1939), along with the one spider monkey (*Ateles geoffroyi*) studied by Harlow and Settlage (1934). It should also be noted that the mandrills (*Maudrillus sphinx*) studied by Balasch, Sabater Pi, and Padrosa (1974) are skillful overall and also showed apelike improvement over trials.

Also relevant to the question of possible monkey–ape differences on the patterned string problem is a study by Beck (1967), who found that gibbons did as well or better than Köhler's (1925) chimpanzees on a series of similar spatial tasks involving pulling appropriate strings to obtain food. This finding is especially interesting because gibbons, like orangutans, inhabit Southeast Asia and exhibit highly arboreal lives for which such spatial skills might be especially well suited.

2.4.3 Detours and Mazes

In a number of classic studies, Köhler (1925) gave chimpanzees problems requiring spatial understanding. For example, he showed them food behind a fence that could be circumvented only by moving sideways for some distance (away from the food) and then going around the fence (see also Kellogg & Kellogg, 1933). The chimpanzees were very good at this detour, as were humans and domestic dogs (domestic chickens were not so skillful, however). He also presented his chimpanzees with a basket of food suspended over their heads by a rope hanging from the limb of a tree and extending to its trunk. They were also skillful at obtaining the food in this situation, by climbing the tree and doing one thing or another with the rope. Yerkes (1916, 1927) found that one gorilla and one orangutan were also very good at obtaining food in similar situations.

Primates have also been presented with tasks in which they maneuver not their own bodies but other objects through difficult spatial trajectories. That these are not the same skill is demonstrated by Köhler's (1925) finding that chimpanzees have much more difficulty in pushing objects around barriers toward goals than they do in negotiating similar barriers in their own locomotor activities. The first systematic study of this type was reported by Bingham (1929a), who created a mechanical maze in which chimpanzees had to maneuver food in a variety of complex ways in order to obtain it. He found that chimpanzees were good at devising creative routes and anticipating the outcomes of particular maneuvers with the food (see also Rensch & Döhl, 1968). Similarly, Guillaume and Meyerson (1930) required different primate species to reach through a fence for food and maneuver it around a barrier so that it could be brought inside the cage and found them all to be quite capable in this task. Davis, McDowell, and Nissen (1957) compared rhesus macaques with chimpanzees on the "bent-wire detour problem," requiring subjects to extract objects impaled on various types of bent wires, and found that although both species learned to be successful, the chimpanzees were a bit more skillful at maneuvering objects around obstacles than were the monkeys (see also Davis, 1958). Davis and Leary (1968) found that several species of Old World monkeys (mostly macaques) were more skillful at bent-wire detour problems than several species of New World monkeys (all cebids), who were a bit better than one prosimian species, *Lemur catta*.

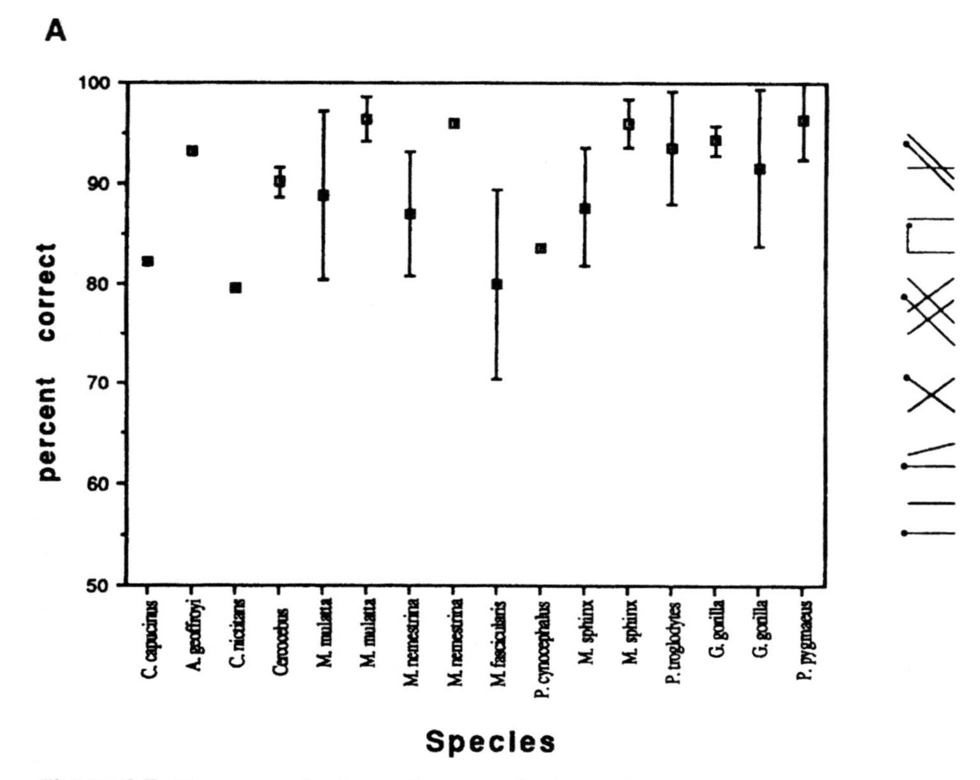

Figure 2.7. The patterned string performance of primates for (a) all patterns and (b) the four most difficult patterns. Only studies that used the same methodology have been included. SOURCES: *Ateles geoffroyi* (Harlow & Settlage, 1934), *Cebus capucinus* (Harlow & Settlage, 1934), *Cercocebus* sp. (Harlow & Settlage, 1934), *Cercopithecus nictitans* (Balasch et al., 1974), *Gorilla gorilla* (Fischer & Kitchener, 1965; Riesen et al., 1953), *Macaca fascicularis* (Harlow & Settlage, 1934), *Macaca mulatta* (Harlow & Settlage, 1934; Riopelle et al., 1964), *Macaca nemestrina* (Harlow & Settlage, 1934), *Mandrillus sphinx* (Balasch et al., 1974; Harlow & Settlage, 1934), *Pan troglodytes* (Finch, 1941), *Papio cynocephalus* (Harlow & Settlage, 1934), *Pongo pygmaeus* (Fischer & Kitchener, 1965). See also Klüver (1933), Warden et al. (1940c), Cha & King (1969) for other possible tests.

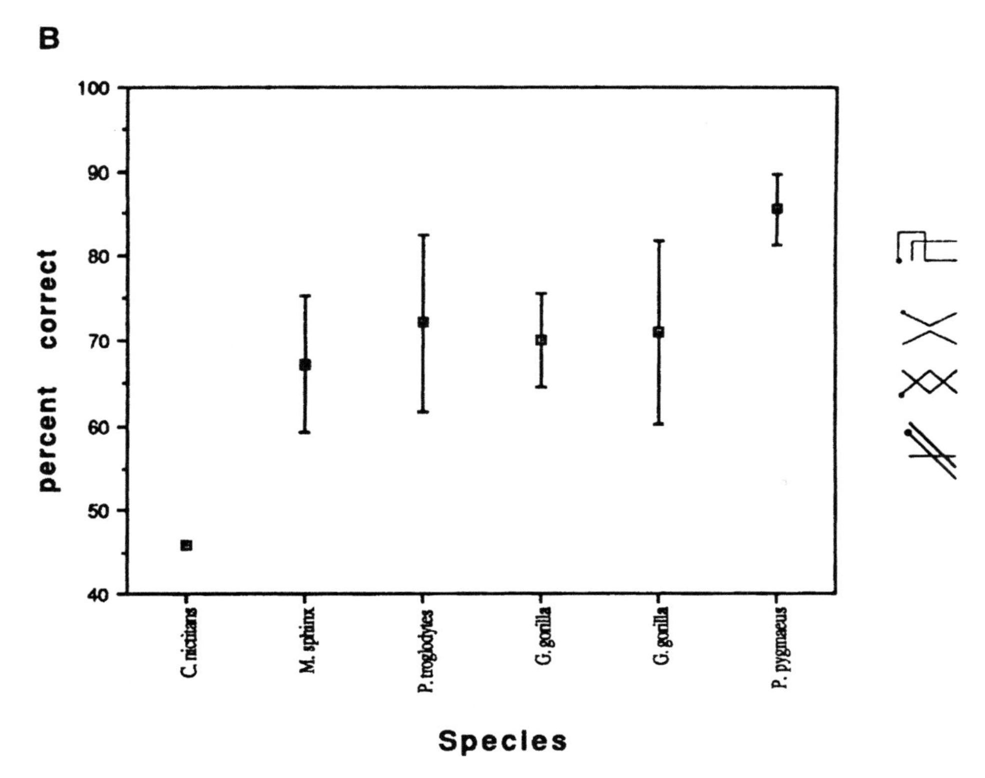

Figure 2.7. *(continued)*

The modern version of maze tasks involves computer technology. In a video format in which subjects learn to manipulate joysticks that control cursors on a TV screen, Washburn, Rumbaugh, and colleagues have investigated a number of different phenomena in rhesus monkey spatial learning (Washburn & Rumbaugh, 1992; Washburn, Hopkins & Rumbaugh, 1991). Once subjects have mastered the joystick technique, they are able to play video games for food rewards that depend on their ability to track moving video objects, anticipate their regular movements, and even anticipate their trajectory as they disappear behind occluders on the screen (i.e., they move the cursor so as to anticipate the location from which a moving object will emerge from behind the occluder, given a constant trajectory). Other species have yet to be tested.

Using the same general video-joystick methodology, Washburn (1992) observed rhesus macaques solving mazes by moving a cursor through various complex pathways. In analyzing the trajectories that monkeys used to move the cursor toward the goal, Washburn found that they were not using a simple trial-and-error method; that is, they were not simply moving the cursor directly toward the goal and then maneuvering it around obstacles by trial-and-error movements as it bumped into them. They were

Figure 2.8. Chimpanzee using a televised image to reach for an out-of-sight target in the Menzel et al. (1985) study. (Reprinted from *Journal of Comparative Psychology* with the permission of the American Psychological Association.)

anticipating the trajectory needed to avoid obstacles by surveying the paths available, at least to some extent, before they started the cursor moving through the maze. Again, it is important that in these tasks the animals are not moving themselves through space, for which they are specifically adapted, but rather they are moving an external object (the cursor) through space in a way for which evolution could not have prepared them. It should be noted that apes, especially one bonobo (*Pan paniscus*), have shown some very skillful navigation of video mazes using the joystick technique as well (personal observations of authors at the Rumbaugh laboratory).

The flexibility of primate spatial skills in mazelike situations is demonstrated most clearly in a study by Menzel, Savage-Rumbaugh, and Lawson (1985). They found that two laboratory-trained chimpanzees were very good at locating objects they could not see directly but could see only using a closed-circuit television picture of their hand and the object (on the other side of a wall through which the chimpanzees put their arms and groped for the objects; see figure 2.8). They quickly learned to use the video images to locate the objects on the backside of the wall even when these images were taken from a number of different angles, and even with images that were reversed and/or inverted.

2.4.4 Comparison of Species

Within primates, demonstrations of mental rotation skills have only been attempted for one species of baboon and humans. However, there is some evidence that this skill is

Figure 2.9. Gorillas foraging. (Photo by J. Call.)

also possessed by another mammal, the dolphin (Herman, Pack, & Morrel-Samuels, 1993). Navigating around detours is accomplished by chimpanzees and other mammals such as dogs. Monkeys have not been systematically tested for this skill, although there is no reason to doubt their adeptness in detour tasks. Perceptually or mentally disentangling strings differentially leading to rewards is accomplished by several species of monkeys and apes, with no striking differences in taxa above the level of individual species. The ability to maneuver objects (including everything from one's own body to video cursors) through complex spaces has been demonstrated for both monkeys and apes, again with no striking differences apparent above the species level (except that the one prosimian species tested on the bent-wire detour problem in one study did not perform as well as monkeys and apes).

2.5 What Primates Know about Space and Objects

Primates have many skills of spatial perception in their locomotion, their foraging, and, in some cases, their tracking and capturing of prey. Indeed, the "where" system of direct spatial perception involving creative, dynamic anticipations seems to be widespread in the animal kingdom. Related skills that have a cognitive component concern the mental rotation of objects and the navigation of self and objects through complex spaces. There is no evidence that primates have special skills in these domains relative to other mammals (although there are few comparative data in precisely comparable situations) nor is there evidence of significant differences within primate species. Perhaps the most interesting studies are those in which individuals use joysticks to navigate computer cursors through mazes because in these tasks individuals must clearly use their skills in new and flexible ways for which they are not specifically adapted. The sense in which the video screen is a "representation" of real objects in the environment, however, is still an open question.

A more prototypically cognitive skill concerning space is cognitive mapping: the ability to learn the particulars of a local environment, what is located where in that environment, and then to use that knowledge flexibly in making travel decisions. Cognitive mapping seems to be fairly widespread in the animal kingdom (although see Bennett, 1996, for a dissenting view). Rats, squirrels, and some bird species have shown phenomenal skills at remembering where many different food items are located in a single foraging space (e.g., Shettleworth & Krebs, 1982). Monkeys and apes have excellent spatial memories for the location of multiple food items as well, and in addition they have clearly shown the ability to remember what type of food, or what type of object, is located at particular sites. They then use this knowledge flexibly in deciding on travel routes: they search first where they know preferred foods to be (sometimes on the basis of inference from finding ripe fruits of a particular species in another location), and they use a minimum distance strategy between different food locations. In addition, at least two monkey species have demonstrated that they know where other individuals should be located, and one ape species has shown a memory for where tools needed for extractive foraging are located. In what sense this same flexibility is true of nonprimate species is not known at this time, but at least some foraging species seem to have very flexible cognitive mapping skills as well (see section 2.1.4).

Differences among primates species in cognitive mapping skills are not striking. What differences there are seem to be small and quantitative; for example, chimpanzees seem to be able to remember more different food locations than macaques (Tinkelpaugh, 1928). There is certainly no indication that monkeys and apes, or any other taxonomic oppositions within primates, understand or mentally represent their foraging spaces in qualitatively different ways. The largest differences observed are among some species of New World monkeys. Marmosets, tamarins, and capuchin monkeys display some close-range cognitive mapping skills within trees that not all New World monkeys display. One possibility is that the cognitive skills for dealing with space have been tailored to the specific foraging needs of each species, the major factor in this case being the need for marmosets, tamarins, and capuchins to locate sources of sap, gum, and insects at particular sites on particular trees. In any case, it would be useful to have a rigorous comparative study in which investigators make a priori predictions about the cognitive mapping skills of different primate species based on their foraging patterns in the wild and then test those predictions against their performance on some standardized set of foraging tasks.

The situation in regard to object permanence is in a general sense similar to that of cognitive mapping skills. Many nonprimate mammals have stage 4 and 5 skills of object permanence (Gagnon & Doré, 1992), and indeed these skills are part and parcel of the kinds of spatial memory and cognitive mapping skills that primates and other mammals use in their foraging activities. The one negative finding is for a prosimian species (the slender loris), but many of the species-specific foraging behaviors of this species seem to hinder performance on this task (i.e., it is not well adapted to grasping and removing a cover from food). It is thus possible, indeed likely, that prosimians understand that objects still exist after they have disappeared, but they simply do not have the motor skills to display that knowledge (as is true of human infants at early stages of development).

In terms of stage 6 object permanence, primates as a group do not seem to have unique skills, as domestic dogs have been shown to have stage 6 skills (skills that, so far, only two primate individuals have been shown unequivocally to have, and several other primate species have been shown not to have). It is unlikely that any motoric factor leads to this mixed picture for primates (monkeys should be better able to manipulate covers than dogs), but it is still possible that there is some other performance factor at work. For example, the fact that human infants continue to make the A-not-B error well after they seemingly "know better" (i.e., searching where the object was found previously, rather than where it just disappeared), may be due at least in part to a number of performance issues having to do with the difficulty of inhibiting previously successful reaching behaviors in the situation (Diamond, 1991). It is thus possible that if a stage 6 task were devised in which subjects did not have to inhibit responses to particular locations in control trials (either to locations the experimenter touched on false trials or locations adjacent to the initial search location on sequential trials), more nonhuman primates would display this skill.

On the other hand, it is also possible that stage 6 skills are not widespread among primates. This might be because in most foraging situations such skills are simply not needed. Stage 6 skills might have evolved in some species not for foraging, as stage 4 and 5 skills presumably did, but rather for tracking other objects in the environment that engage in significant nonperceptible movements on their own; that is, other animate beings such as conspecifics, predators, or highly mobile prey. If this hypothesis were true, it would suggest that stage 6 skills would not be especially prevalent in primates as compared with other mammalian species, but rather they would be prevalent among species who needed to find objects that move in important ways when out of sight. To date, not enough species have been tested in appropriate ways to evaluate this hypothesis. In any case, it is likely that differences among primates in object permanence skills are of the same general nature as differences in their cognitive mapping skills in the sense that there is general continuity in the order, with only minor variations across species. All primates understand that objects continue to exist when they are not being perceived and that objects can be found in different locations (stages 4 and 5); this is required for efficient foraging. Those who need to understand the nonperceptible movements of objects (representational stage 6 search strategies) to deal with some adaptive problem, whether in foraging or for some other purpose, will be more likely to have developed them.

Overall, primates have the general mammalian spatial skills of cognitive mapping and object permanence, with some species having more specialized skills, as do some nonprimate species. It is thus not likely that spatial cognition is a major domain of difference in the cognitive skills of primates and other mammals, which makes evolutionary sense given that all mammals must locate important items in their environment in order to survive and reproduce. Nor are there strong and reliable qualitative differences within the primate species, although there are only a few situations in which a wide variety of species have been explicitly compared. In any case, in this domain of primate cognition, there is no evidence of a linear "scale of intelligence" for mammals, which primates sit atop, with apes atop the primates (nor will we find this in other domains). In spatial cognition, all mammals are likely created equal, with different species evolving slightly different cognitive skills as needed to deal with specific adaptive problems.

It is also unlikely that humans have any special skills in these domains of spatial cognition. They too possess the general mammalian skills of cognitive mapping and object permanence (with clear stage 6 skills early in ontogeny). However, humans are able to amplify their skills of spatial cognition with all kinds of cultural artifacts from maps to compasses, and human languages may be a rich source of spatial knowledge as well, as children are exposed via language to the habitual ways that their communities structure space (for example, learning which landmarks in the environment are reliable, the names for certain places and directions, the best paths to take to valued locations, and so forth).

3

Tools and Causality

Before primates can ingest the food they have located, most of them need to do something further to obtain or process the food. For example, they must pick fruit or leaves off of tree limbs, dig roots out of the ground, uncover insects in a pile of dead leaves or under a dead tree, take the hard shell off nuts, or probe into a hole for honey. Primates have thus evolved a major morphological adaptation for use in their foraging activities: their hands. Some primate species have evolved specialized hands designed for specific foraging tasks (e.g., the tap-scanning and probing that aye-ayes accomplish with their elongated middle digit). Other primate species have evolved hands that support more general and flexible skills of object manipulation, including the use of tools to extract food from substrates (e.g., chimpanzees). On the whole, primates, especially simian primates, have evolved hands that can accomplish object manipulations impossible for any other order of animal (Passingham, 1981). Engaging in these object manipulations sometimes involves complex cognitive processes.

Much of the study of primate object manipulation has been conducted within the framework of Piaget's (1952) theory of sensory–motor development. In Piaget's theory the attainment of object permanence skills (see chapter 2) depends on the manipulation of objects, and thus his series of six stages of sensory–motor development—the six stages of object permanence being a special case of these—are defined in terms of object manipulation skills. Empirical research with human infants has not found the intimate relation Piaget hypothesized; many studies have shown a developmental asynchrony between object manipulation and object permanence (Uzgiris & Hunt, 1974), and some studies have shown knowledge of some aspects of object permanence in infants before they had any substantial experience manipulating objects (Baillargeon et al., 1985). The fact that mammals such as dogs, who manipulate objects very little, have good object permanence skills would also argue against object manipulation as a necessary prerequisite. In this context, Vauclair (1982, 1984) has argued that object permanence skills evolved primarily in the context of locomotion and search, engaged in at some level by all mammals for purposes of locating food, whereas object manipulation skills evolved for purposes of extracting and processing food and other objects, which is more particular to primates as "hand-feeders," especially some species.

In discussions of the evolution of cognition, tool use has often been treated as a special case of object manipulation, for at least two reasons. First, tool use manifests a certain degree of cognitive complexity in that it is an indirect means of goal attainment. Tools are intermediary means in a complex means–ends sequence and thus involve relating objects not just to one's own actions (i.e., tool and self), but also relating objects to one another (e.g., tool and food). The fact that humans and their closest relative, the chimpanzee, are the most proficient tool users in the animal kingdom only reinforces the view that tool use is cognitively complex. Second, from an evolutionary point of view, the use of tools has been viewed as opening up unique adaptive possibilities because it often results in organisms making substantial modifications to the niche to which they are adapting. This process seems to have been especially important in the case of human evolution. Combining these two features, cognitive complexity and adaptive significance, tool use has sometimes been considered a general marker of "level of intelligence" for different animal species. This is clearly a mistake, however, because most instances of tool use by most animal species are fairly narrowly specialized feeding adaptations involving only limited flexibility of use; for example, some bird species use sticks to probe in holes for insects and some fish spit water on insects to immobilize them (Beck, 1980).

Parker and Gibson (1977) attempted to differentiate primate tool use from such narrow species-specific adaptations. They noted that the tool use of primates (mainly of chimpanzees and capuchin monkeys) consists of multiple tool use skills which are more broadly and flexibly applied across a variety of contexts ("intelligent tool use"). These more flexible uses are singled out because they seem to rely more heavily on cognitive processes, especially on an understanding of causality defined as the dynamic interrelations of objects with one another (as opposed to the relation of self to objects). In this view, animals are able to use multiple tools in a flexible manner to the extent that they perform a causal analysis of how the tool creates its effects on the environment. Such analyses pave the way for appropriate and effective modifications in the way tools are used, leading in some cases to the manufacture of tools when novel circumstances arise. For example, if an individual habitually uses a stick to probe into a hole for insects but this strategy is not working on a particular occasion, the animal can only know what to do if it understands *why* the normally successful behavior is not working, which then suggests the change of behavior to be made. Thus, if the tool does not reach deep enough into the hole, perhaps a longer one is needed; if it does not fit down the hole, perhaps a modification or a thinner tool is needed; or if there is an impediment, perhaps the same tool can be used but with more force. When the animal performs such causal analyses, without engaging in a lengthy process of overt trial-and-error learning, some form of mental representation is at work ("thinking" in the form of insight, or even foresight).

In this chapter we investigate primate skills of object manipulation, including tool use, and their understanding of causality. We first review what is known about the object manipulation skills of primates in noninstrumental contexts, often in play and often with human objects in captivity (and even including some observations that have been interpreted as "symbolic play"). We then review what is known about primate tool

Table 3.1 Piagetian stages of sensory–motor development

Stage 1: *Reflexes*—actions under little voluntary control
Stage 2: *Primary schemes*—actions directed to one's own body
Stage 3: *Secondary schemes*—actions aimed at reproducing interesting effects on external entities
Stage 4: *Coordination of secondary schemes*—hierarchical embedding of secondary schemes (differentiation of means and ends; intentionality)
Stage 5: *Tertiary schemes*—actions aimed at relating external entities to one another

Note: Stage 6 of sensory–motor development involves explicit mental representations and thus is not, strictly speaking, sensory–motor (Piaget, 1952).

use, both in the wild, where it is rare outside of chimpanzees, and in captivity where it is more common and widespread. We conclude with an examination of primates' understanding of causality, mainly in object manipulation and tool use situations.

3.1 Object Manipulation

From soon after birth, human infants begin exploring aspects of their own body using what Piaget calls *primary* sensory–motor schemes. Around the third or fourth month of life, they begin reaching for objects and grasping them using what Piaget calls *secondary* schemes (since they are actions directed at objects outside the body). These soon develop into a number of procedures designed to reinstate interesting effects that infants accidentally produce in objects, for example, when they accidentally produce a noise in a rattle by moving their arm, they often repeat this action with seeming attention to the noise and how it is produced (stage 3 of sensory–motor development; Piaget, 1952). From about 8 months of age, infants are capable of embedding one secondary scheme within another in a hierarchical relation, for example, by grasping and displacing one object that is blocking access to their grasping of another object (stage 4). Piaget sees this as the first clear differentiation and coordination of means and ends in the infant's behavior, since in another context the grasping and displacing of an object might be an end in itself. Around 1 year of age, infants begin for the first time to relate objects to one another in what Piaget calls *tertiary* schemes (because they involve not relations between one's own action and an object but rather the relations among objects); for example, by stacking them, by putting one inside another, or by using one as a tool to reach another (stage 5). At this age infants also begin actively exploring the relations among objects; for example, by simply dropping them to see the effect, pouring them out of a container to see the effect, and in many ways playing with object relations. At 18 months of age (stage 6), infants can imagine the relations and transformations of objects mentally, and thus they use mental manipulations, rather than overt trial-and-error, to solve many sensory–motor problems. This means, technically, that they are no longer operating on a purely sensory–motor basis (Piaget thus designates stage 6 as a transitional phase out of the sensory–motor period). Table 3.1 provides a brief summary of the Piagetian stages of sensory–motor developement.

Nonhuman primates, as well as many other animals, commonly investigate objects by visually orienting to them, smelling them, and so forth (Glickman & Sroges, 1966). Exploring objects by manipulating them with the hands is less common, especially in the wild where other needs are often more pressing. Although few systematic observations have been made of primates' object manipulations in their natural habitats, in captivity where they are freer from the demands of everyday life and are exposed to a wide variety of human objects and artifacts, many nonhuman primates engage in object manipulations that provide evidence for their understanding of objects and their properties and relations.

3.1.1 Prosimians

Jolly (1964a,b) presented a number of different complex human objects (sometimes containing food, sometimes not) to several dozen, mostly wild-caught, individuals representing three genera and six species of prosimians. Based on their interactions with the objects in this laboratory setting, Jolly divided her animals into three groups. Lorises (*Loris tardigradus*) and lesser bushbabies (*Galago senegalensis*) stared at the food-baited object and then struck it with their hands, which is typical of their behavior when facing their normal food prey, insects. They basically ignored the unbaited objects (although in the studies of Ehrlich, 1970, and Renner, Bennett, Ford, & Pierre, 1992, greater bushbabies, *Galago garnettii*, were more exploratory and engaged in some manipulations in the form of pushing, poking, and grasping human objects). Thick-tailed bushbabies (*Galago crassicaudatus*) and pottos (*Perodicticus potto*), typically omnivorous, manipulated the objects a bit more, especially with their mouths, but only when the object was food-baited. (Ehrlich 1970, found that the slow loris, *Nycticebus coucang*, a close relative of pottos, was also indifferent to novel human objects not associated with food.) Finally, two species of lemurs (*Lemur fulvus* and *L. catta*) interacted readily with unbaited objects, using both the mouth and the hand, and investigated food-baited objects immediately. Lemurs are leaf- and fruit-eating.

Jolly (1964b) interpreted these results in terms of the ecological niches occupied by the different species. She points out, for example, that lorises depend entirely on their eyes and hands to capture their insect prey, but this hand–eye system is much too narrowly specialized to be of use in manipulating other objects. Lemurs, on the other hand, must manipulate leaves and fruit to some extent to gather and eat them. Overall, Jolly characterizes her object manipulation results as follows:

> These results show how the approach to objects is part of a species' biology, which will determine success on "psychological" tests. Animals whose food is likely to fly away must watch, bide their time, and pounce with precision. Animals whose food waits dangling from a branch will haul in the branch, rather than touch the food and perhaps cause it to drop. Differences in attention span, orientation, and manipulation thus relate directly to food habits. (1964b, p. 568)

3.1.2 Monkeys

There have been few systematic studies of the object manipulation of monkeys in their natural habitats. The major exception is capuchin monkeys, who have been observed in the wild with a focus on their possible use of tools (see section 3.2.3). In one study of chacma baboons (*Papio ursinus*), Hamilton, Buskirk, and Buskirk (1978) observed three troops living in two different environments for their object manipulations other than direct food processing. They observed such behaviors as digging in the ground, stripping bark off trees, turning over logs, lifting debris, and so forth, all in the context of searching for food. In other contexts the baboons carried objects, shook objects while they were in their mouths, and tweaked at twigs. Interestingly, of all monkey species, capuchin monkeys and baboons have also shown the most complex object manipulations in captivity.

In a laboratory study, Parker (1977) studied the object manipulations of a captive stumptail macaque from the point of view of Piaget's stages of sensory–motor cognition. She found that the macaque manipulated objects using a variety of behaviors. He rubbed, pulled apart, rotated, and otherwise manipulated a variety of human artifacts and objects, sometimes embedding one scheme within another by removing obstacles in a way typical of stage 4. But he did not seem interested in the effects he produced on objects. For example, he engaged in no procedures to repeat or reinstate effects on objects such as the noise a rattle makes or the swinging of an object on a string, which in human children are typical of many of the secondary schemes of stage 3. Moreover, the macaque in his manipulations did not relate objects to one another in any systematic way (e.g., there was no putting objects in other objects or using tools), nor did he engage in the disinterested explorations of objects; that is, he did not show any signs of tertiary schemes (stage 5). Antinucci et al. (1982) observed almost identical behaviors in a single Japanese macaque, who also did not seem to manipulate objects with a consideration of the effects produced or their relations with one another. Parker and Gibson (1979) speculate that this type of manipulative pattern is to be expected of species that are specifically adapted for hand-foraging and feeding.

Potì (1989) investigated two other macaque species in this same framework. In observing three Japanese macaques and a longtail macaque longitudinally, she essentially replicated Parker's results. Her subjects all engaged in the complex means–ends coordinations typical of human infants in stage 4 (removing obstacles), but they did not engage in simple secondary procedures for repeating interesting effects on objects (stage 3). Nor did these macaques engage in tertiary explorations of object relations or tool use (stage 5). Potì points out that her subjects in almost all cases assimilated objects to their own body by sucking or grasping them, with little interest in object properties themselves (see also Torigoe, 1986, 1987, for further studies of Japanese macaques). Chevalier-Skolnikoff (1977) reported several examples of secondary schemes directed toward reproducing effects on objects in stumptail macaques, but Potì (1989) reexamined those observations and found no evidence that the behaviors were directed toward object effects; for example, the branch shaking of the stumptail macaques apparently was done without them paying specific attention to the effects produced. (See Huffman, 1984, for observations of Japanese macaques manipulating stones.)

Chevalier-Skolnikoff (1989) reported observations of the object manipulations of three species of spider monkey (*Ateles belzebuth, A. fusciceps, A. geoffroyi*) and two species of capuchin monkey (*Cebus apella, C. albifrons*), all New World monkeys in zoo settings. Also using a Piagetian framework, she found that spider monkeys, like the macaques of previous studies, did not engage in any tertiary manipulations involving object–object relations or tool use. She did report observation of some secondary schemes; however, they all consisted of repetitions of behaviors on objects, and it is unclear from the reports the extent to which these monkeys were interested in the effects produced. The capuchin monkeys, on the other hand, engaged in a variety of both secondary schemes for producing object effects and tertiary exploratory behaviors (e.g., spinning, rolling, and bouncing objects against a substrate and watching the result). They also engaged in a variety of tertiary actions relating objects to one another in specific ways (e.g., placing objects in or on one another) and a variety of tertiary tool-use behaviors (e.g., digging with a stick and using a stick to try to crack open a nut).

Parker and Gibson (1977) reported similar observations of secondary and tertiary exploratory behaviors in capuchin monkeys in three different zoo settings, as did Spinozzi (1989) for three captive capuchins. In a systematic comparison of capuchins and macaques using a common set of relatively complex objects, Natale (1989b) found that 41% of the sensory–motor schemes produced by three capuchin subjects consisted of secondary procedures to maintain or reinstate interesting object effects. The corresponding number for two macaques was 11%. In looking at the subset of those schemes that he judged have the greatest implications for learning cause-and-effect relations (that is, those schemes in which object effects vary as a function of small changes in the subject's behavior), Natale found that macaques almost never produce these, whereas these behaviors were relatively frequent in capuchins. Fragaszy and Adams-Curtis (1991) found that capuchins explicitly related objects to one another with some frequency as well.

Perhaps the most complex object manipulations of capuchin monkeys were reported by Westergaard and Suomi (1994b), who presented 10 subjects with a series of 4 nesting cups (which fit into one another, in order of size) and observed their manipulations. Following Greenfield's (1991) analysis of the manipulation of nesting cups by human infants and children, these investigators found that 2 of the 10 subjects placed cups within one another in what Greenfield called the pairing method. Two other subjects used this method and, in addition, sometimes repeated the procedure to place more than one cup into a larger one (the pot method). Three subjects used these two methods and, in addition, used a third method called the subassembly method, in which one cup is first placed into another and then this pair (or perhaps more) is placed into a larger cup. The subassembly method is thought to be especially complex cognitively because it involves the hierarchical organization of actions on objects. It may also require some mental representation in the form of planning (imagining the outcome of particular sequences of actions with the cups).

With regard to baboons, Westergaard (1992, 1993) observed in olive baboons (*Papio cynocephalus anubis*) various types of evidence for tertiary manipulations involving relations between objects. In a series of sessions, he provided five infant baboons (all with relatively extensive human contact) with various human objects. Four of the five infants combined objects in tertiary schemes such as placing an object in a bowl and touching or rubbing one object with another—for no obvious instrumental goal other than object exploration. All five of the infants used paper towels to sponge up liquids as well (a tertiary scheme in the form of tool use; see section 3.2.3 for more details).

The overall picture is thus that the majority of monkey species observed in captivity, both New World and Old World, manipulate various kinds of objects and artifacts, but they do not seem particularly interested in the effects their interactions produce. They also do not seem to relate objects to one another with regularity. The two major exceptions are capuchins, of New World monkeys, and baboons, of Old World monkeys. They seem to be much more interested in the effects their behaviors produce on objects and on the relations between objects when they manipulate them together in some way. As we shall see in section 3.2.3, these two species are also the most proficient tool users of all monkey species.

3.1.3 Apes

Apes have the same need as monkeys to process food, and so many of their object manipulations are of the same type. It is perhaps important, however, that, in addition, all four species of great apes build nests in their natural habitats, in some cases on a nightly basis (Bernstein, 1969; Groves & Sabater Pi, 1985). This activity involves the manipulation of objects outside the context of food and does not involve tools in any direct way.

Like capuchin monkeys and baboons, chimpanzees engage in a variety of secondary procedures for repeating or maintaining interesting effects on objects. Mathieu and Bergeron (1981) observed four nursery-raised chimpanzees at 3–4 years of age interacting with human objects and artifacts (Mathieu, 1982, reports similar results when they were 5–6 years). These researchers report some secondary schemes in which the chimpanzees discovered an effect of an object by accident and then deliberately repeated it, observing the object for the effect; for example, by rubbing, banging, swinging, or rolling the objects while observing the result. They also observed some tertiary relating of objects to substrates and to one another, mostly as subjects dropped them and observed the result. Much more common, however, were secondary procedures in which individuals coordinated means and ends in a hierarchical fashion for a concrete goal (stage 4; usually to obtain food or to get out of the room they were in). Overall, the researchers concluded that: "Chimpanzees do not seem to be very much interested in the functioning of things: it seems that they very seldom ask the question 'How does it work?' They are more interested in 'What can I do with this object?' or 'How can I get access to that goal?'" (Mathieu, 1982, pp. 3–4). The comparison they had in mind when drawing this conclusion was, of course, human children who engage in disinterested exploration much more often.

Potì and Spinozzi (1994) followed four chimpanzees longitudinally for varying periods during their first year of life as they interacted with human objects and artifacts. They also reported a number of secondary procedures for repeating interesting object effects of the same type as Mathieu. They also saw a number of embeddings of secondary schemes in pursuit of a goal (stage 4). In comparing these observations to published observations of human infants, these researchers (unlike Mathieu) saw few differences from human children. Where these researchers saw the most differences from human infants was in tertiary schemes; that is, they saw very few instances of the use of novel schemes to explore object relations. The same may be said of Hallock and Worobey's (1984) similar observations of two infant chimpanzees in their first year of life using the Uzgiris-Hunt standardization of Piaget's tasks. Although all of these subjects were less than 1 year old (and it is likely that they will show at least some of these behaviors later in development), Potì's (1996) study of older chimpanzees also found very few instances of the active manipulation of object relations other than one subject stacking objects on top of one another (e.g., there were no manipulations involving containment or other explicit spatial alignments as in human infants in their second year; see also Köhler, 1925, and Matsusawa, 1990, for other observations of chimpanzees stacking objects).

Vauclair and Bard (1983) observed two chimpanzees, a bonobo, and a human infant at similar ages in their first year of life in a standard object manipulation situation. They observed a number of secondary procedures to repeat interesting object effects and

Figure 3.1. Young orangutan manipulating a set of seriated containers. (Photo courtesy of F. Kiernan.)

stage 4 means–ends coordinations in all three subjects. The major differences they observed came in tertiary schemes involving exploring the characteristics of the objects in relation to substrates and to one another. The human infant spent 6% of its time exploring the objects' properties compared to only 1% each for the chimpanzee and bonobo. Similarly, the human infant spent 12% of its time overtly relating objects to one another compared to 5% and 0% for the chimpanzees and bonobo, respectively. These researchers also noted that whereas both of the apes typically acted on the objects while leaving them on the ground, the human infant much more frequently extracted the object from the background and manipulated it off the ground, and did this more often than the apes using bimanual coordination. This is consonant with McGrew's (1977) observation that in the wild the vast majority of the objects chimpanzees manipulate are leaves, twigs, tree bark, and other objects that are still attached to their respective substrates. Vauclair (1984) also reports on chimpanzees, bonobos, and a human infant in interaction with a single complex object that had several effects to be produced (a light, a tone) by various manipulations (a switch, a lever). Human infants explored the object and produced the effects more than both infant and adult bonobos and chimpanzees by about four to five times. Even Kellogg and Kellogg (1933) noticed less exploration and manipulation of objects by a chimpanzee as compared with a human child.

More recently, Brakke and Savage-Rumbaugh (1994) observed a chimpanzee, a bonobo, and a human infant in their first year of life in a similar situation. Overall, they found results similar to those of Vauclair and Bard, but with a bit more manipulative activity from the apes. They observed interesting object manipulations in all three subjects, but they found more tertiary relating and combining of objects (as well as more bimanual coordination) in the human infant. They also observed a greater facility in the human infant to switch attention from object to object, or to quit acting on an object and then come back to it with the same action some time later. It should be emphasized that in both of these studies comparing chimpanzees and bonobos to human infants, however, the apes were less than 1 year of age when they were observed, and it may be that they are slower to develop manipulative behaviors than human infants, even though they are more precocious in locomotion. In a similar study with these same two ape subjects at approximately 6 years of age (plus two more apes: one chimpanzee and one bonobo), Brakke (1996) found a strong effect for rearing environment. Apes who had been raised with extensive human contact manipulated objects in more complex ways than apes who had had less contact with humans and their artifacts.

In a replication of Greenfield's (1991) studies with human infants (and parallel to Westergaard & Suomi's study with capuchin monkeys reported in the previous subsection), Matsuzawa (1991) gave nine chimpanzees a series of five nesting cups (see figure 3.2). Seven of the subjects (all aged 4 or less) used only the pairing and pot methods of assembly, placing one or more cups into a larger one. However, two adult chimpanzees, Sarah and Ai, both of whom had received much human interaction and training (including some language training), learned to use the subassembly method. Both did so relatively quickly and without any specific training in this task. Again, the subassembly method is thought to be cognitively complex because it requires a hierarchical organization of actions on objects, and it may involve mental representation in the form of planning by imagining potential outcomes of specific actions with the cups.

Gorillas are the only apes whose object manipulations have been observed in a detailed manner in their natural environment. Byrne and Byrne (1991, 1993; see also Byrne, 1994) observed the complex manipulations that gorillas perform in gathering plant foods and processing them for eating. Different plants require different techniques due to a number of factors, among them the size, location, and texture of the plants, and how the edible parts are attached to the inedible parts (including sometimes the presence of painful thorns). These researchers report a number of complex sequences of behaviors with different plants. For example, in eating nettles, gorillas use one hand to pull a bunch by the base into range of the other hand, which then slides up the stem to grip the base of the bunch of leaves at the top; the first hand then moves up to the leaves and grips and twists them off the stem; the second hand picks out unwanted debris from the bunch of leaves and then folds them over (so as to keep the nettles inside); the first hand then puts them into the mouth. Different techniques of equal complexity are used with other plant species. In all cases, Byrne and Byrne observed various types of hierarchical organization (often with clear subroutines) in the execution of these manipulations, again with the implication that such organization involves mental representation and planning.

Obviously, feeding is not the ideal context in which to explore the disinterested exploration of objects by gorillas. Knobloch and Pasamanick (1959) observed a gorilla for the first 2 years of life in interaction with human objects and found that this individual never related objects to one another and could not be induced to put one thing inside another at all. Redshaw (1978) tested four nursery-reared gorillas in the Uzgiris-Hunt standardization of Piaget's sensory–motor tasks in their first 18 months of life and

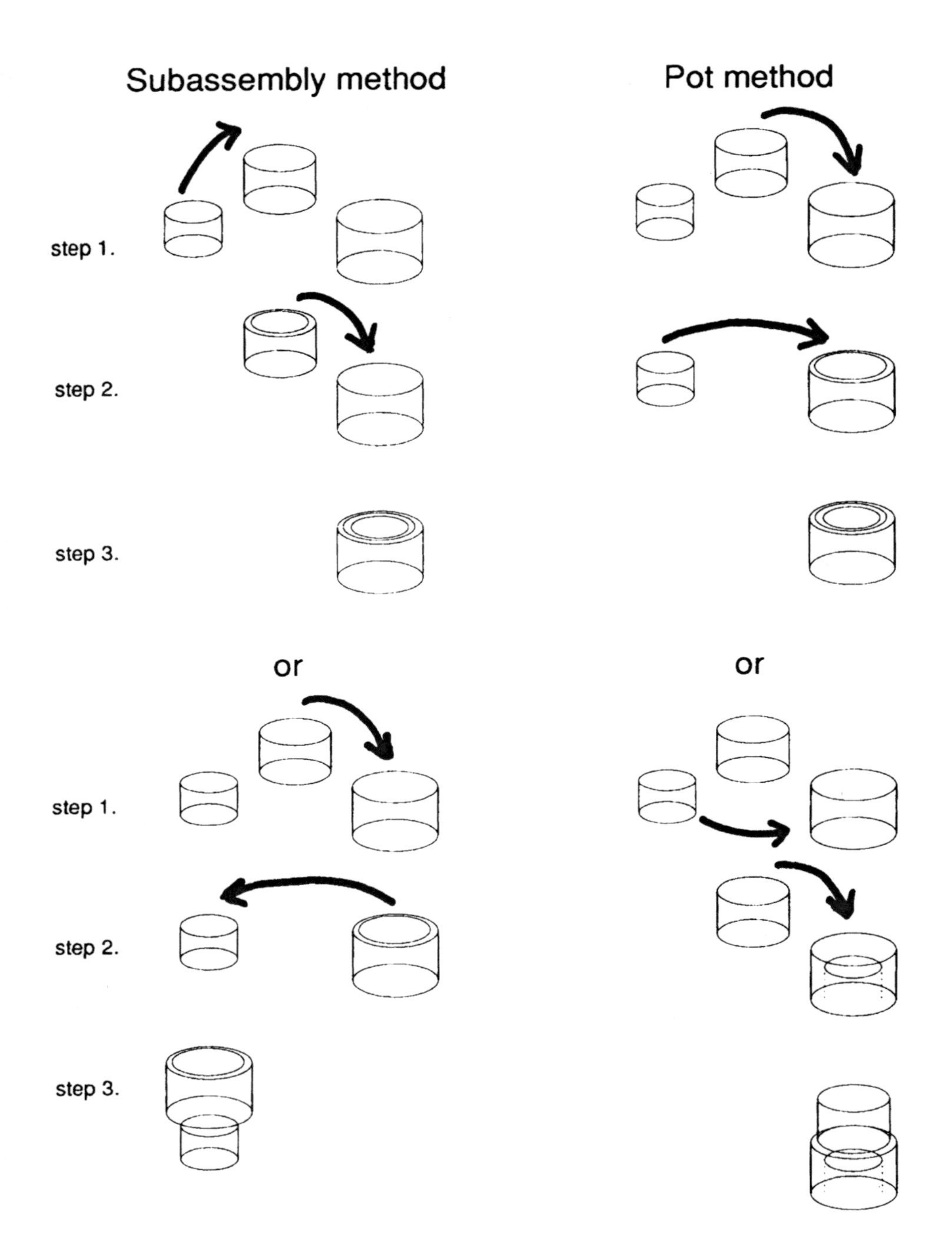

Figure 3.2. Greenfield et al. (1971) methods for combining seriated cups. The pot method is thought to be less cognitively demanding than the subassembly method.

observed that the gorillas were similar to human infants in many ways, but they did not combine objects with one another in a single case (including no use of objects as tools). Redshaw did not report specifically on the interest of her gorillas in the effects their behavior produced in objects, but Spinozzi and Natale (1989b) were interested in just this focus on objects. They observed a single gorilla in the first 15 months of life. Whereas they saw many complex behaviors indicating a coordination of differentiated means and ends such as removing an obstacle (stage 4), they did not see either the secondary or tertiary version of the exploration of objects; that is, they did not see the gorilla repeat a behavior that had accidentally produced an interesting effect on an object, nor did they see any active relating of objects to one another. This was in spite of the fact that on a number of occasions the gorillas did accidentally produce "interesting" sounds and other effects in the objects present. It must be stressed, however, that all of these observations of gorillas in captivity were of infants less than 2 years of age.

With respect to orangutans, systematic research is sparse, but almost all observers of primates in captivity report that orangutans are skillful manipulators of objects. In the only systematic study, Chevalier-Skolnikoff (1983) observed four orangutans at various ages up to adulthood. She observed two subjects engaging in secondary procedures to reinstate interesting object effects and a few cases of subjects engaging in tertiary procedures of relating objects to one another by putting one object in another or using a tool. She also reported some exploration of objects, but the observations are not given in enough detail to know if subjects were indeed exploring their relations. It should also be mentioned that Parker (1969) compared the object manipulation of orangutans with that of gorillas and chimpanzees and found that orangutans showed a wider range of specific behaviors used to manipulate objects than the other two ape species, as well as more proficiency in two tool-use tasks. In terms of orangutans in their natural habitats, Galdikas (1982) and Bard (1995) have observed orangutans manipulating objects in various ways, although it is difficult to determine from their reports the extent to which these manipulations involved exploring the effects of actions on objects or relating objects to one another systematically. Russon and Galdikas (1993) reported different types of complex object manipulation by rehabilitant (living in both a forest and a human camp) orangutans using human artifacts.

If there is not an inordinate focus on studies in which infant apes are observed, or in which apes are compared with human children, it is clear that great apes are more interested in and skillful at manipulating objects than most other primate species. Adult great apes are somewhat interested in the effects of their actions on objects and in the relations of objects to one another, although less so than human children. How apes compare with capuchins and baboons, the most interested and skillful of monkey species, is not known at this time (but see section 3.1.4 for some relevant observations).

3.1.4 Comparative Studies

Two sets of studies are of special note because they tested a wide range of primate species in an identical set of object manipulation situations. First, Parker (1973) observed four individuals from each of five primate genera: four prosimians (two species of lemur), four New World monkeys (one species of spider monkey), eight Old World Monkeys (four silver leaf monkeys and four pigtail macaques), and four lesser apes (one species of gibbon). Each individual was exposed to two objects on separate occasions: a nylon rope with a loose knot tied in it and an aluminum bar with an open slot and a steel cable coming out. He found that the gibbons and macaques contacted the object more often, in a greater variety of ways, for longer periods of time, and for longer strings of responses than the silver leaf monkeys and spider monkeys; lemurs were intermediate on all these measures. Parker's explanation for these results is that the gibbons and macaques have hands that are better adapted for object manipulation than the two arboreal monkey species. Following up on these results, Parker (1974b) found that when chimpanzees, gorillas, and orangutans were given these same objects, they interacted with them in more diverse ways than any of the monkey or prosimian species observed in the 1973 study (and a few others tested in this second study as well, including capuchin monkeys, *Cebus capucinus*). Looking specifically at subjects' relating of the experimental objects to surfaces and to other objects in their cages in these two studies, Parker (1974a) found that chimpanzees and orangutans engaged in this tertiary behavior more frequently than gorillas and all monkey and prosimian species tested.

The second similar study was reported by Torigoe (1985). He gave subjects a nylon rope and a wooden cube at separate times. Subjects were from 74 different primate species, representing 24 genera and 6 families: three species of prosimian (all lemurs), 18 species of New World monkey, 44 species of Old World monkey (including several species of baboon), and 8 species of ape (5 lesser apes and 3 great apes). Results on a number of measures revealed three groups of subjects, ranked from least manipulative to most manipulative: (1) lemurs, New World monkeys (except capuchins), and leaf-eating Old World monkeys; (2) Old World monkeys (except leaf-eaters); and (3) capuchin monkeys (all four species) and apes. Individuals from the first group picked up, mouthed, and transported the object only; they did not relate the object to any other substrate or object in the cage area. Individuals from the second group had more varied modes of manipulation, including rolling, rubbing, and sliding the object against a substrate (floor or cage bars). The third group displayed a more varied repertoire still, including drape, drop, swing, strike, and throw, and, with the exception of the lesser apes, they engaged in relating the object to other objects more often (e.g., to the cage bars, the water in the cage, food in the cage, or feces).

It should be noted that a serious limitation of both of these comparative studies is a lack of other objects to which subjects could relate the experimental objects. It may also be important that the situation did not include any obvious goals with respect to the objects. The results of these studies should thus be taken as indications of primates' tendencies to manipulate fairly simple objects in fairly simple situations in which there is no immediate goal.

3.1.5 Symbolic Play

As can be seen in the previous four subsections, primates, especially some species, physically manipulate and explore objects with some flexibility and complexity. But physical manipulations of this type do not fit the prototype of cognitive processes because they do not appear to involve explicit mental representations. However, some object manipulations do seem to involve explicit mental representations. In Piaget's theory, stage 6 of sensory–motor development involves "mental combinations"; that is, the mental manipulation of objects and their relations ("insight" or "thinking"). These typ-

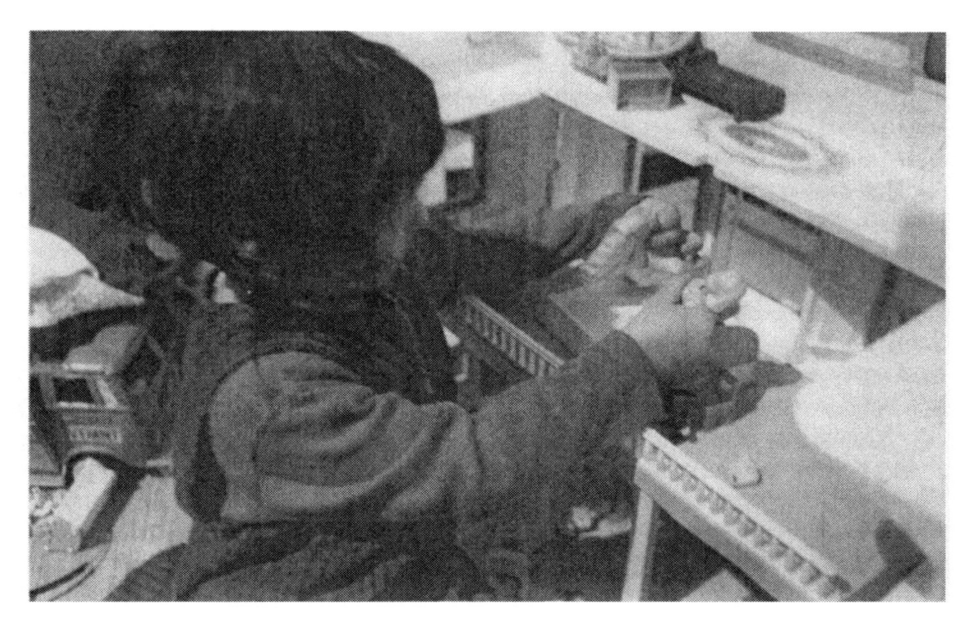

Figure 3.3. Human child engaged in symbolic play. (Photo courtesy of Victor Balaban.)

ically occur when the individual is working toward some goal (as, perhaps, in the two studies with nesting cups cited above). Consequently, most of the research focused on this issue has been conducted in the context of tool use, and it will be reviewed in the following two sections.

Another way that primates may employ mental representations in their manipulations of objects is in using them "symbolically," or in pretense, to represent other objects. The problem is that symbolic play is difficult to identify in nonverbal organisms (including human infants). For example, Goodall (1986) reported that a young chimpanzee "fished" for ants with a twig at a place where no ants were present, raising the possibility that he was imagining the presence of ants. Other interpretations of this behavior in terms of simple manipulative play with sticks are also possible, however.

Some more compelling possible instances of symbolic play have been reported in apes that have been raised by humans in the context of human artifacts, although, again, other interpretations are possible. For example, when raised by humans, all four great ape species have been observed to play with dolls, engaging in a number of different activities including bathing, feeding, and tickling (see Call & Tomasello, 1996, for a review). Also, Patterson and Linden (1981) reported that the home-raised gorilla Koko pretended to be an elephant by using a fat rubber tube as a substitute for the trunk; Hayes (1951) reported that the home-raised chimpanzee Viki played with a nonexistent pull toy; and the home-raised bonobo Kanzi was observed to engage others in an eating game involving imaginary food (Savage-Rumbaugh & Mcdonald, 1988). In addition, Jensvold and Fouts (1993) reported six instances of imaginary play in three chimpanzees who used some American Sign Language signs (based on 15 hours of observation of five chimpanzees). These mostly involved the use of signs typically used with animate objects for inanimate objects (e.g., signing "tickle" in association with a stuffed bear). In contrast, Mignault (1985) reported that symbolic play was not observed in four nursery-reared juvenile chimpanzees who had received regular exposure to humans and a variety of toys, and Miles (1990) reported the same negative result in her home-raised orangutan Chantek.

The main problem in these observations is that it is unclear what these behaviors mean for these individuals. Many investigators are even skeptical that the so-called symbolic play of human infants is something other than manipulative play with objects such as dolls that adults see as symbolic (e.g., Lillard, 1993). In some cases this play may be supplemented with the mimicking of adult behaviors with these objects; for example, dipping dolls into water (as the adult says "You're giving him a bath"). Experimental research showing young children's difficulties with using a doll's room as symbolic of a real room also casts doubt on the symbolic nature of young children's object play (DeLoache, 1995), and indeed when chimpanzees are given this same opportunity to use a doll room as a symbolic representation of a real room, they seem to fail totally (Ettlinger, 1983; Premack & Premack, 1983). When Kanzi or a 1-year-old human infant offers an empty spoon to the mouth of a doll, therefore, it is unclear that they have a mental representation of food on the spoon or that they imagine that the doll is representing a real person eating, they may just be playing with the spoon in ways they have seen others play with it. The overall picture is thus that while some human-raised great apes engage in some of the same kinds of behaviors with toys and other artifacts as do human infants, the symbolic or representational status of these behaviors is unclear (see Harris, 1993, for an enlightening discussion). Apparently other species of primates have not been given the opportunity to display these same "symbolic" behaviors.

3.1.6 Comparison of Species

Complex object manipulations are not frequent in the behavior of most animal species. Raccoons wash and process their food in various ways, and a number of species, from insects to sea otters, use tools in restricted contexts. But the primate hand is clearly more adapted for hand-feeding and other manipulative activities than the appendages of virtually all nonprimate species, and some primates' behavior with objects takes full advantage of this morphological adaptation in manipulating objects. Within primates, however, investigators have found relatively little interest in object manipulation in prosimians and some leaf-eating monkeys (both New World and Old World). Other species of monkeys, especially capuchin monkeys and baboons, show more complex manipulative tendencies and skills, even relating objects to one another systematically in some contexts. The great apes are also interested in and skillful at manipulating objects, although how they compare with capuchins and baboons in all manipulative contexts is not totally clear at this point. There are some indications that apes are more interested in and skillful with objects in relatively simple, non–goal-directed situations. It is important in this context that all four great ape species build nests in the wild, whereas no monkey species do so with regularity. The role of mental representation in primate object manipulations, in the form of symbolic play, is uncertain.

Nonhuman primates do not engage in the kinds of disinterested object manipulations, involving the complex relating of objects to one another, that human infants and children do with regularity from early in development (Piaget, 1952). Vauclair (1984) emphasizes the fact that human infants develop complex object manipulation skills at least partly because their locomotion is retarded relative to that of other primates, and thus to explore objects they need to manipulate them rather than go to them. Vauclair also emphasizes that it is in the social context, in interaction with encouraging adults, that much of human infants' complex object manipulations take place. It is interesting in this regard that most of primates' humanlike manipulative skills are observed in human captivity with human objects and artifacts.

3.2 Tool Use

Investigation of the object manipulation tendencies of the various primate species revealed that only great apes and a few monkey species seem interested in exploring the properties of objects for their own sake and in relating objects to one another (and all these species are considerably less interested than human children). This fact is important in investigating the tool use of nonhuman primates because in tool use an individual uses an object in relation to another object (tertiary relation), but does so in a situation in which there is ample motivation beyond simply exploring the properties of objects. Perhaps with ample motivation, some additional species will relate objects to one another in the form of tool use, indicating that their failure to display such skills in manipulative play does not mean that they are unable to understand object–object relations altogether.

The competence-versus-motivation issue is complicated by the fact that in its natural environment a species may have many perfectly adequate ways of obtaining food without the use of tools. Experimental situations that make tools necessary in the acquisition of food are therefore an important source of information about a species' abilities as well. Thus, there is only one nonhuman primate species, the chimpanzee, whose members virtually all use tools regularly as a natural part of their everyday foraging activities in the wild; but in captivity, when tools are more often needed, many species have been observed to use tools, some quite proficiently. The use of tools is a tertiary stage 5 ability in the sense that it requires the relating of objects to one another, and so the flexible use of tools is the main subject matter of this section. But stage 6 mental manipulations may be inferred if there is evidence that the individual understands certain tool–object relations without actually manipulating the tool or object physically; for example, by choosing a tool, or even modifying or making a tool, to meet the demands of a novel situation without overt trial and error. This process is touched upon in this section, but it is a main focus of section 3.3 on causality, which reports experiments designed to address this issue directly.

3.2.1 Chimpanzees

In their natural habitats in tropical Africa, chimpanzees use a wide variety of tools for a wide variety of purposes, mostly for the acquisition of food but for some other purposes as well (e.g., for grooming and agonism). In the latest count, individuals from 32 populations or groups have been observed to use a tool of some type, and virtually all groups who have been observed over a long period of time have shown some tool use. If attention is restricted to habitual patterns of tool use (i.e., tool use engaged in by several individuals on repeated occasions), the number drops to 12 populations, with the num-

Figure 3.4. Chimpanzee extracting syrup from an artificial termite mound. (Photo by J. Call.)

ber of habitual uses ranging from 1 to 11 per group (although this may be due to the different lengths of time that different populations have been observed). These facts and many of those that follow about chimpanzee tool use are taken from McGrew's (1992) excellent and thorough review.

The types of tool use that are common and widespread in chimpanzees mostly fall into four broad functional categories (see Table 3.2 for more detail). The most common of these is the use of implements to extend the reach, or to probe into places where the hand cannot go, usually for food. The best known example is termite fishing in which the individual pokes a twig or stem down a hole in a termite mound, sometimes wiggling it, until some termites bite and grab hold of the stick; the individual then extracts the stick and eats the termites. Similar but different techniques are used to get ants out of holes in trees, or holes in the ground, and to pry bone marrow or the meat of nuts out of difficult places (Goodall, 1986). The second most common type of tool use is the use of sticks and stones as weapons. Individuals often brandish or throw sticks and stones at opponents in fights (Goodall, 1986). The third type of tool use is the use of objects to amplify force, again usually for food. The best known example of this is nut cracking, in which an individual places a nut on a wood or stone anvil and cracks it open with a wood or stone "hammer" (Boesch & Boesch, 1990). The fourth major function of chimpanzee tool use is to sponge up liquids, for example, crumpling up leaves and inserting them into crevices containing water and then sucking the leaves and dipping sticks into bee hives for honey (Goodall, 1986). Chimpanzees on occasion use tools for other purposes as well, for example, for personal hygiene (e.g., wiping their bodies clean with leaves) and in communicative situations as attention-getters (e.g., stripping a brittle leaf to make a loud noise) (McGrew, 1992).

In focusing on the cognitive processes involved in chimpanzee tool use, four phenomena seem especially important. The first two have to do with flexibility and complexity, and the second two have to do with mental representation. First, chimpanzees have been observed on occasion to use two or more tools in sequence for a single problem. For example, Boesch and Boesch (1990) observed that chimpanzees who open a particular kind of nut by cracking it with a rock will often have to use a stick to then pry the embedded meat out of the nut from the crevices inside the shell. Brewer and McGrew (1990) observed a semi–free-ranging chimpanzee use a series of four tools, each with a different function, to break open a bees' nest in the stump of a tree and extract honey from it. Sugiyama (1995) made similar observations with regard to chimpanzees obtaining liquids from difficult-to-reach places. These observations demonstrate both flexibility and complexity in chimpanzee tool use.

Second, Kitahara-Frisch, Norikoshi, and Hara (1987) observed two captive chimpanzees use a tool to make a tool ("metatools"). These chimpanzees first learned to use a rock to break open a bone for food inside, and they then used a chip of the bone to pierce a skin covering some desirable drink. It must be noted, however, that this behavior, as well as similar behavior by an orangutan (Wright, 1972) and a bonobo (Toth, Schick, Savage-Rumbaugh, Sevcik & Rumbaugh, 1993), was aided in various ways by humans who structured the problem, reinforced the subjects, and demonstrated aspects of the solution. However, Matsuzawa (1991) reported three observations of three different wild chimpanzees at Bossou, Guinea, using a metatool in the cracking of nuts (while interacting with experimentally provided stones and nuts). In all three cases the individuals needed to make the anvil level on which cracking was to take place, and they

Table 3.2 Tool-using in apes.

Context	Material	Action	Goal	Wild	Laboratory
Grooming	Leaves	Rubbing	Clean	*Pan troglodytes:* McGrew, 1992; Nishida, 1980; van Lawick-Goodall, 1970 *Pan paniscus:* Ingmanson, 1996 *Pongo pygmaeus:* Mackinnon, 1974	*Pongo pygmaeus:* Galdikas, 1982 *Pan paniscus:* Jordan, 1982
	Sticks	Dipping	Clean	*Pan troglodytes:* Nishida & Nakamura, 1993 *Pan paniscus:* Ingmanson, 1996	*Pan troglodytes:* Fouts & Fouts, 1989; McGrew & Tutin, 1973 *Gorilla gorilla:* Fontaine et al., 1995
	Leaves	Covering	Protect	*Pan paniscus:* Kano, 1982; Ingmanson, 1996 *Pongo pygmaeus:* Rijksen, 1978	
Aggression	Branches	Shaking	Intimidate	*various:* van Lawick-Goodall, 1970	*Various:* van Lawick-Goodall, 1970
	Sticks, stones	Throwing	Intimidate, injure	*various:* Galdikas, 1982; Goodall, 1986; McGrew, 1992; van Lawick-Goodall, 1970	*Pan troglodytes:* Köhler, 1925; Takeshita & van Hooff, 1996; van Lawick-Goodall, 1970 *Pan paniscus:* Jordan, 1982 *Pango pygmaeus:* Galdikas, 1982 *Gorilla gorilla:* Fontaine et al., 1995
	Sticks	Hitting	Injure	*Pan troglodytes:* Goodall, 1986; Korlandt, 1967	*Pan troglodytes:* Köhler, 1925; Takeshita & van Hooff, 1996 *Pongo pygmaeus:* Galdikas, 1982
Feeding	Strings, supports	Pulling	Bring within reach		*Pan troglodytes:* Crawford, 1937; Finch, 1941; Hayes, 1951; Köhler, 1925; Rumbaugh, 1970 *Pongo pygmaeus:* Chevalier-Skolnikoff, 1983; Fischer & Kitchener, 1965; Lethmate, 1982

				Gorilla gorilla: Fischer & Kitchener, 1965; Knoblock & Passamanick, 1959; Redshaw, 1978; Riesen et al., 1953; Spinozzi & Natale, 1989b; Yerkes, 1927
Stones, logs, pestle	Hitting	Open	*Pan troglodytes:* Boesch & Boesch, 1983, 1984, 1990; Sakura & Matsuzawa, 1991; Sugiyama & Koman, 1979; Yamakoshi & Sugiyama, 1995	*Pan troglodytes:* Hannah & McGrew, 1987; Sumita et al., 1985 *Pongo pygmaeus:* Lethmate, 1982; Rijksen, 1978; Wright, 1972 *Pan paniscus:* Toth et al., 1993
Stone	Cutting	Open		*Pan paniscus:* Toth et al., 1993 *Pongo pygmaeus:* Wright, 1972
Sticks, leaves	Inserting	Extract	*Pan troglodytes:* Boesch & Boesch, 1990; Brewer & McGrew, 1990; McGrew, 1974, 1977; McGrew et al., 1979; Nishida, 1973; Nishida & Hiraiwa, 1982; van Lawick-Goodall, 1968a *Pongo pygmaeus:* van Schaik et al., 1996	*Pan troglodytes:* Bard et al., 1995; Finch, 1941; Hayes, 1951; Köhler, 1925; Nash, 1982; Paquette, 1992a; Parish, 1994; Savage-Rumbaugh et al., 1978a; Visalberghi et al., 1995; Yerkes, 1943 *Pongo pygmaeus:* Chevalier-Skolnikoff, 1983; Galdikas, 1982; Haggerty, 1913; King, 1986a; Lethmate, 1982; McEwen 1987; Savage & Snowdon, 1982; Visalberghi et al., 1995; Yerkes, 1916 *Pan paniscus:* Jordan, 1982; Parish, 1994 *Gorilla gorilla:* Yerkes, 1927

(*continued*)

Table 3.2 *(continued)*

Context	Material	Action	Goal	Wild	Laboratory
	Sticks	Inserting	Perforate	*Pan troglodytes:* Brewer & McGrew, 1990; Sabater Pi, 1974; Tutin et al., 1995	*Pan troglodytes:* Kitahara-Frisch et al., 1987
	Leaves	Inserting	Absorb	*Pan troglodytes:* Goodall, 1986; Sugiyama, 1995; Tutin et al., 1995 van Lawick-Goodall, 1968a	*Pan paniscus:* Jordan, 1982 *Pongo pygmaeus:* Lethmate, 1982; Mackinnon, 1974 *Hylobates:* Tingpalong et al., 1981 *Pan troglodytes:* Kitahara-Frisch & Norikoshi, 1982; Takeshita & van Hooff, 1996 *Gorilla gorilla:* Fontaine et al., 1995
	Sticks	Sweeping	Bring within reach		*Pan troglodytes:* Brent et al., 1995; Hallock & Worobey, 1984; Hayes, 1951; Kellogg & Kellogg, 1933; Köhler, 1925; Nagell et al., 1993; Parker, 1969; Takeshita & van Hooff, 1996; Tomasello et al., 1987 *Pongo pygmaeus:* Call & Tomasello, 1994b; Chevalier-Skolnikoff, 1983; Lethmate, 1982; Parker, 1969; Savage & Snowdon, 1982; Yerkes, 1916 *Gorilla gorilla:* Fontaine et al, 1995; Knoblock & Passamanick, 1959; Parker, 1969; Redshaw, 1978; Spinozzi & Natale, 1989b; Yerkes, 1927

Stone, log	Throwing	Knock down		*Pan troglodytes:* Takeshita & van Hooff, 1996
Sticks	Hitting	Knock down	*Pan troglodytes:* Sugiyama & Koman, 1979 *Pongo pygmaeus:* Borner, 1979	*Pan troglodytes:* Köhler, 1925; Kohts, 1935 *Pan paniscus:* Jordan, 1982 *Pongo pygmaeus:* Chevalier-Skolnikoff, 1983; Lethmate, 1982 *Gorilla, gorilla:* Yerkes, 1927
Poles	Climbing	Reach		*Pan troglodytes:* de Waal, 1982; Hayes & Nissen, 1971; Köhler, 1925; Menzel, 1972; Takeshita & van Hooff, 1996 *Pongo pygmaeus:* Chevalier-Skolnikoff, 1983; Lethmate, 1982; Yerkes, 1916 *Gorilla gorilla:* Fontaine et al., 1995; Gómez, 1990; Yerkes, 1927
Sticks, poles	Levering	Pry open	*Pan troglodytes:* Goodall, 1986	*Pan troglodytes:* Takeshita & van Hooff, 1996
Boxes	Climbing	Reach		*Pan troglodytes:* Bingham, 1929a; Hayes & Nissen, 1971; Köhler, 1925 *Pongo pygmaeus:* Lethmate, 1982; Yerkes, 1916 *Gorilla gorilla:* Yerkes, 1927
Leaves, various objects		Containing	*Pongo pygmaeus:* Rogers & Kaplan, 1994	*Pan troglodytes:* Takeshita & van Hooff, 1996 *Pan paniscus:* van Elsacker & Walraven, 1994

did so by propping one side up with a smaller stone. Again, flexibility and complexity are readily apparent.

The third observation is that in their everyday use of tools, chimpanzees often show insight or foresight by choosing a tool for the particular situation without overt behavioral trial and error and even, on occasion, by making their tools (see Table 3.3 for types of tool making). For example, in termite fishing, chimpanzees only choose sticks that are long enough, and they strip the leaves and twigs away as needed to make them fit down the holes (Goodall, 1986). Similarly, when they are cracking nuts on the ground, they often break branches to make them the needed size for use as hammers without overtly trying out a larger branch (Boesch & Boesch, 1983); when they are going to crack nuts in a tree, they bring along the appropriate tool on their way to cracking; and when they wish to crack an especially hard species of nut, they always get a stone (not a branch) because only a stone will work (Boesch-Achermann & Boesch, 1993). Assuming that they have not had extensive trial and error in each of these cases, these behaviors indicate that the chimpanzees are mentally representing the needs of the situation before they go about choosing or shaping the tool (and their skill at doing this is indicated by the fact that after they have chosen or made a tool, modifications are seldom made after first use; Boesch & Boesch, 1990).

A fourth and related observation is that when a new circumstance arises requiring chimpanzees to adapt their normal pattern of tool use, they are able to do so readily in at least some cases. For example, in a nut cracking situation at Bossou, Sakura and Matsuzawa (1991) observed that when the usual stone substrates were not available for use as anvils, chimpanzees who had in the past only used stones as anvils used a fallen tree instead, and they did so skillfully, without overt trial and error. Goodall (1986) reported, similarly, that the chimpanzees at Gombe used sticks to open human-made boxes, a task they had clearly not faced previously. Also, in captive situations, groups of chimpanzees have been observed to lean logs up against fences and trees to serve as "ladders" (de Waal, 1982; Menzel, 1972) and to make "stairs" with boxes to reach food suspended from the ceiling (Bingham, 1929a; Köhler, 1925). In many of these observations, the skills were acquired so quickly that there was little time for behavioral trial and error, implying the possibility of some kind of insight or foresight into the relation between the needs of the situation and the tools available.

It should also be noted that, in general, chimpanzees raised in humanlike cultural environments learn an incredibly wide array of tool use skills and become quite proficient with a wide range of human artifacts, including all kinds of implements and even machines (Hayes & Hayes, 1951; Savage-Rumbaugh, 1986). Some of the utilized artifacts widely reported by researchers who have raised chimpanzees in their homes are spoons, cups, knives, matches, hammers, keys, combs, toothbrushes, video games, and various toys—basically all the objects and artifacts that engage human infants and toddlers. In one of the more impressive instances of the use of human tools, Döhl's (1968) chimpanzee Julia solved a task in which she had to look at a transparent, locked box to determine what kind of key was needed, and then find that key in another transparent locked box. This second box also required a key, and Julia had to find it in still another transparent locked box, and so on up to five boxes, all presented simultaneously. Although Julia was prepared for this task with various pretraining regimes, in general it appears that her planning was all done mentally, without overt trial and error.

All of these variations of tool-use skills, among both wild and captive chim-

Table 3.3. Tool-making in apes.

Context	Material	Action	Goal	Wild	Laboratory
Grooming	Leaves	Pressing	Modify Shape	*Pan troglodytes:* Nishida, 1980; van Lawick-Goodall, 1970	
	Sticks	Breaking, biting, cutting	Reduce length or diameter	*Pan troglodytes:* Nishida & Nakamura, 1993	*Pan troglodytes:* Fouts & Fouts, 1989; McGrew & Tutin, 1973 *Gorilla, gorilla:* Fontaine et al., 1995
Feeding	Branches, leaves, sticks	Breaking, biting, cutting	Reduce length or diameter	*Pan troglodytes:* Boesch & Boesch, 1990; Goodall, 1986; McGrew, 1974; Nishida, 1973, 1980; Nishida & Hiraiwa, 1982; Sabater Pi, 1984; van Lawick-Goodall, 1968a, 1970 *Pongo pygmaeus:* van Schaik et al., 1996	*Pan troglodytes:* Nash, 1982 *Pan paniscus:* Bard et al., 1995; Visalberghi et al., 1995 *Pongo pygmaeus:* Galdikas, 1982; Lethmate, 1982; Visalberghi et aL, 1995 *Gorilla gorilla:* Fontaine et al., 1995
	Leaves	Chewing	Increase absorption	*Pan troglodytes:* Nishida, 1980; Sugiyama, 1995; van Lawick-Goodall, 1968a; 1970	*Gorilla gorilla:* Fontaine et al., 1995
	Sticks	Inserting two sticks	Increase length		*Pan troglodytes:* Birch, 1945: Köhler, 1925; Schiller, 1957 *Pongo pygmaeus:* Lethmate, 1982
	Stones	Hitting stone	Make a sharp object		*Pan paniscus:* Toth et al., 1993 *Pongo pygmaeus:* Wright, 1972
	Stones	Hitting bone	Make a pointed object		*Pan troglodytes:* Kitahara-Frisch et al., 1987

panzees, argue for the flexibility and complexity of chimpanzee tool use. In some cases they also suggest stage 6 mental manipulations. However, in none of these cases is it known precisely what experience individuals have had with the tools involved (although see Inoue-Nakamura and Matsuzawa, in press, for a study of the category of stone tool use in wild chimpanzees). This makes the attribution of insight and foresight somewhat problematic, as their past opportunities for overt trial-and-error learning are not usually known. In section 3.3 on causality, experiments are reported in which the individual's exposure to a problem and tool are precisely controlled, so that their understanding and mental representation of the causal relations between tool and goal may be determined more precisely.

3.2.2 Other Apes

The tool use of the other apes is very surprising. None of them has been observed to use tools extensively in the wild on a species-wide basis, despite long-term observations at close range for most species. Gibbons sometimes drop branches on intruders, and there is a single observation of leaf-sponging behavior by gibbons similar to that of chimpanzees (Tingpalong et al., 1981). Gorillas seemingly do not use tools at all in the wild. Bonobos sometimes drag branches behind them in a social display and have been observed on some occasions to use leafy twigs as partial shelter from the rain (Kano, 1982). Bornean orangutans sometimes drop branches on human or conspecific intruders (Galdikas, 1982). Recently, van Schaik, Fox, and Sitompul (1996) reported the first observations of habitual, group-wide tool making and use by a species other than chimpanzees: All of the closely observed adults of a group of Sumatran orangutans repeatedly used sticks to probe for insects or to pry seeds out of the husk of some fruits, sometimes modifying the stick to fit the specific circumstance. Van Schaik et al. (1996) speculated that the swampy environment in which this group lives makes the use of these kinds of tools necessary in a way that is not true of Bornean orangutans.

The story of ape tool use in captivity is a completely different story. Despite their limited use of tools in the wild, in captivity orangutans are considered by many researchers to be the most adept tool users of all the apes, including chimpanzees. Yerkes (1916) and Haggerty (1913) both found orangutans to be skillful tool users in Köhler-type problem-solving tasks. Lethmate (1982) reported that the captive orangutans he observed were skillful tool users in a variety of circumstances, and they even engaged in a number of tool-making episodes, some of which were not equaled by the captive chimpanzees he observed. For example, orangutans stripped leaves off twigs so they would fit holes, broke sticks to the right length for a task, made swings, and attached sticks to one another to make them long enough to reach food. In a comparison of chimpanzees, gorillas, and orangutans in two tool-use problems (sponging for liquids and using a stick to rake in out-of-reach food), Parker (1969) found the orangutans superior to the other two species on both problems. In only a few hours of demonstration and shaping, Wright (1972) taught an orangutan to make flake tools for use in cutting a rope to give him access to food (i.e., the use of a metatool). Galdikas (1982) reported that her rehabilitant orangutans (raised by humans and then reintroduced to the wild via a human campsite) engaged in a number of tool-use activities involving both natural objects and human artifacts (see also Russon & Galdikas, 1993).

Captive bonobos also use tools in surprising ways, given their almost total lack of

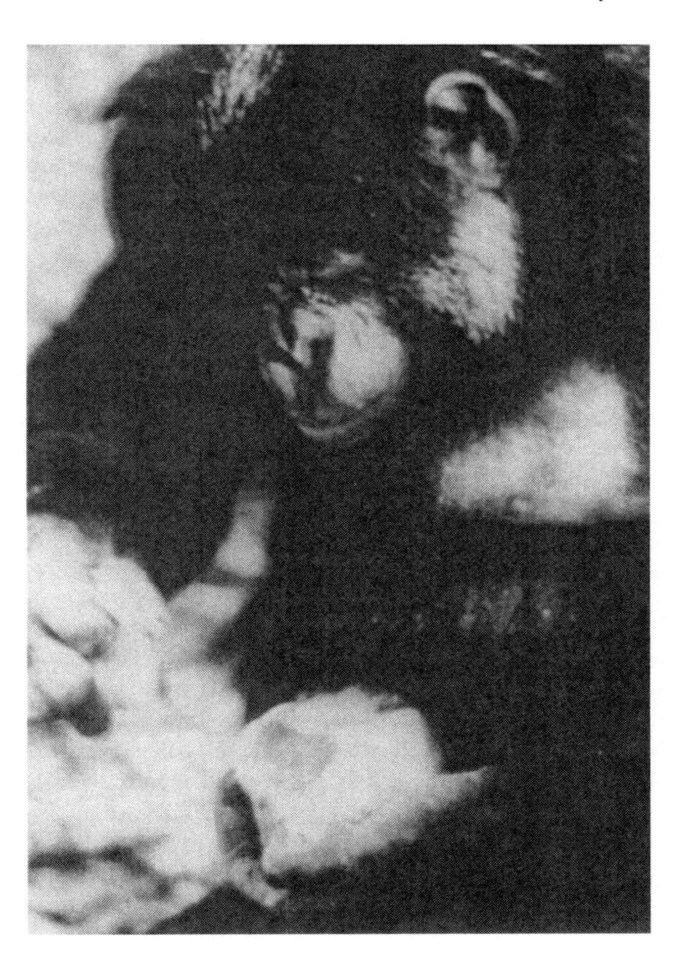

Figure 3.5. The bonobo Kanzi creating a sharp flake stone by striking two stones together. (Photo courtesy of E. S. Savage-Rumbaugh.)

tool use in the wild. In the only study of tool use in captive bonobos, Jordan (1982) reported that these apes engage in a variety of tool-use practices, including throwing sticks and other objects in agonistic displays; the use of sticks to probe crevices for food, to rake in out-of-reach objects, as a lever to try to break the cage mesh; the use of leaves to sponge water and to clean the body; and the use of long twigs as swings. The bonobo Kanzi, raised in a humanlike cultural environment with some training in a humanlike system of communication, shows a wide variety of tool use skills with a wide variety of human tools and artifacts, including tool manufacture (Savage-Rumbaugh, 1991). Recently, Kanzi learned from a human to make a stone flake by hitting one stone on another (or throwing the stone against a concrete floor), which he then used to cut a rope that prevented his access to food—an instance of a metatool use (Toth et al., 1993; see figure 3.5).

There is one study of gibbon tool use in captivity. Rumbaugh (1970) observed one captive gibbon use a cloth as a sponge and a rope to make a swing (McGrew, 1992, reports an anonymous observation that a siamang also made a swing from a rope).

There has been little research on the use of tools by captive gorillas, but Yerkes (1927) observed very little spontaneous tool use in his highland gorilla. However, Wood (1984) reported that lowland gorillas in a zoo setting learned to use sticks to rake in food from outside their cage; Natale (1989c) reported that a gorilla used a towel in an attempt to rake in food; and Gómez (1990) reported some use of human artifacts by the lowland gorilla Muni. The most proficient gorilla individual seems to be the lowland gorilla Koko, raised in a humanlike cultural setting (with sign language training), who has been observed to use a variety of human tools and artifacts (Patterson, 1978).

3.2.3 Capuchin Monkeys

Capuchin monkeys mostly do not use tools in the wild, but they are quite proficient using tools in captivity. In the wild, the only behaviors reliably observed are a few instances of dropping objects on intruders (Chevalier-Skolnikoff, 1990), the use of sticks in aggressive encounters with conspecifics (Chevalier-Skolnikoff, 1990), the use of objects to crack open oysters (Fernandes, 1991; Hernandez-Camacho & Cooper, 1976), and the use of a club on a snake (Boinski, 1988). In captivity, on the other hand, capuchin monkeys show a wide variety of tool use skills. For example, Klüver (1933, 1937) exposed a single capuchin monkey to a wide array of Köhler-type tool-use tasks, mostly involving out-of-reach objects (beyond the cage bars, hanging from the ceiling, etc.) and various objects that might be used to reach them (stick, rope, boxes, etc.). His subject learned to be skillful at almost all of these tasks. More recent research has shown captive capuchin monkeys to be good in tool-use tasks such as cracking nuts (e.g., Visalberghi, 1987), sponging liquids with paper towels (e.g., Westergaard & Fragaszy, 1987a), using a stick as a rake (e.g., Natale, 1989a), probing crevices with sticks (Costello, 1987; Westergaard & Suomi, 1994f), using sticks and stones to cut through encasements of various sorts (Westergaard & Suomi, 1994c,e, 1995; Jalles-Fihlo, 1995), and using sticks to push food out of a clear tube (Visalberghi & Trinca, 1989).

It should be noted that in several studies the capuchins actually modified, or "made" tools for use in a food acquisition task, and in one study they used a sequence of tools. For example, Westergaard and Fragaszy (1987a) observed capuchins crumpling the paper towels for use as sponges, Westergaard and Suomi (1994a) found that capuchins hit rocks on one another or on a surface apparently in order to produce stone flakes that would help them cut through animal skin, and Westergaard and Suomi (1995) observed capuchins modifying bamboo to make it more effective in poking holes in plastic. Although these skills are impressive and the animals were very successful, whether they were using trial-and-error learning or some more insightful strategies is currently under debate (Visalberghi, 1993, and see section 3.3 on causality).

3.2.4 Other Monkeys

A number of other monkey species have been observed to use tools in captivity, especially baboons and macaques (see Table 3.4). Baboons, who have been observed on only a few sporadic occasions to use tools in the wild (Beck, 1974; Oyen, 1979; see Hamilton, Buskirk & Buskirk, 1975, on the use of stones for defensive purposes), use tools with reasonable proficiency in captivity. Beck (1973b) observed several captive Guinea baboons learn to use a stick to rake in out-of-reach food. Beck (1973a) reported the

Table 3.4. Tool-using in monkeys.

Context	Material	Action	Goal	Wild	Laboratory
Grooming	Leaves	Rubbing	Clean	*Papio anubis:* van Lawick-Goodall et al., 1973 *Macaca radiata:* Sinha, in press	
	Sticks	Dipping	Clean		*Cebus apella:* Ritchie & Fragaszy, 1988; Westergaard & Fragaszy, 1987b *Macaca fuscata:* Weinberg & Candland, 1981 *Macaca tonkeana:* Bayart, 1982 *Mandrillus sphinx:* Vincent, 1973
Aggression	Branches	Shaking	Intimidate	*Various:* van Lawick-Goodall, 1970	*Various:* van Lawick-Goodall, 1970
	Sticks, stones	Throwing	Intimidate, injure	*Various:* van Lawick-Goodall, 1970 *Papio anubis:* Hamilton et. al, 1975; Pettet, 1975; Pickford, 1975 *Cebus capucinus:* Boinski, 1988	*Various:* van Lawick-Goodall, 1970 *Cebus apella:* Westergaard & Suomi, 1994a
	Sticks	Hitting	Injure	*Cebus capucinus:* Boinski, 1988	*Cebus:* Cooper & Harlow, 1961
Feeding	Strings, supports	Pulling	Bring within reach		*Cebus:* Cooper & Harlow, 1961; Gibson, 1990; Harlow, 1951; Klüver, 1933; Spinozzi & Potì, 1989; Warden et al., 1940a *Macaca fascicularis:* Spinozzi & Potì, 1989 *Macaca fuscata:* Natale et al., 1988; Spinozzi & Potì, 1989 *Papio anubis:* Bolwig, 1964

(continued)

Table 3.4. *(continued)*

Context	Material	Action	Goal	Wild	Laboratory
Feeding (*continued*)	Stones, logs, pestle	Hitting	Open	*Cebus:* Carpenter, 1887; Fernandes, 1991; Izawa & Mizuno, 1977; Hernandez-Camacho & Cooper, 1976; *Papio:* Marais, 1969; *Various:* Kortlandt & Kooij, 1963; *Various:* Terborgh, 1983	*Cebus:* Anderson, 1990; Antinucci & Visalberghi, 1986; Fragaszy & Visalberghi, 1989; Gibson, 1990; Jalles-Filho, 1995; Vevers & Weiner, 1963; Visalberghi, 1987, 1990; Westergaard, 1994; Westergaard et al., 1995; *Papio papio:* Petit & Thierry, 1993
	Stone, sticks	Digging	Extract	*Various:* Kortlandt & Kooij, 1963	
	Stones, bones	Cutting	Extract		*Cebus apella:* Westergaard & Suomi, 1994f, 1995a
	Stones	Throwing	Extract		*Cebus apella:* Westergaard & Suomi, 1995b
	Sticks	Inserting	Extract	*Various:* Kortlandt & Kooij, 1963; *Cebus capucinus:* Chevalier-Skolnikoff, 1990; *Papio anubis:* Oyen, 1979	*Cebus:* Fragaszy & Visalberghi, 1989; Harlow, 1951; Klüver, 1933; Visalberghi & Limongelli, 1994; Visalberghi & Trinca, 1989; Westergaard, 1993, 1994; Westergaard & Fragaszy, 1985, 1987; *Macaca fuscata:* Tokida et al., 1994; *Macaca silenus:* Westergsaard, 1988; *Macaca tonkeana:* Anderson, 1985; *Papio anubis:* Westergaard, 1989, 1992
	Leaves	Inserting	Absorb	*Macaca silenus:* Fitch-Snyder & Carter, 1993	

Feeding (*continued*)	Sticks	Sweeping	Bring within reach		*Cebus:* Klüver, 1933; Natale, 1989c; Parker & Potì, 1990; Spinozzi & Potì, 1989; Warden et al., 1940b *Macaca fascicularis:* Natale, 1989c; Zuberbühler et al., 1996 *Macaca fuscata:* Natale, 1989c *Macaca mulatta:* Shurcliff et al., 1971 *Macaca nemestrina:* Beck, 1976 *Papio hamadryas:* Beck, 1972, 1973b *Papio papio:* Beck, 1973a *Papio anubis:* Benhar & Samuel, 1978; Bolwig, 1964 *Papio ursinus:* Hall, 1962 *Erythrocebus patas:* Gatinot, 1974
	Sticks	Hitting	Knock down		*Cebus:* Harlow, 1951; Klüver, 1933
	Poles	Climbing	Reach		*Cebus:* Harlow, 1951; Klüver, 1933 *Macaca fuscata:* Machida, 1990
	Boxes	Climbing	Reach		*Cebus:* Harlow, 1951; Klüver, 1933
	Leaves	Rubbing	Clean food	*Macaca fascicularis:* Chiang, 1967 *Macaca silenus:* Hohmann, 1988	

same behavior in a captive hamadryas baboon, who had to procure the tool from his neighbor in an adjoining cage before he could use it. Westergaard (1992, 1993) observed several olive baboons using towels and other materials to soak up liquids. He found that on occasion they modified the towels by crumpling them up as needed, and they used other means of obtaining the liquid when the towels were not available. Petit and Thierry (1993) observed several captive Guinea baboons using stones to break through a soft cement slab so that they could dig in the ground beneath.

Japanese macaques have been observed to use tools in two ways. First, in a manner similar to Menzel's (1972) chimpanzees, 3 individuals in a colony of 39 learned to make "ladders" by leaning poles against the side of their caged enclosure (Machida, 1990). They then climbed to the top and explored the upper reaches of the wall. Second, Tokida, Tanaka, Takefushi and Hagiwara (1994) observed four individuals extract food from a transparent tube by throwing rocks into it and knocking the food out the other side. These researchers claimed that the tool use was insightful because the choice of stones, like chimpanzees' choice of termite-fishing twigs and stone hammers, was appropriate to the task without overt trial and error. However, the systematic recording of which stones were used was not done until after the animals had become skillful, so it is not clear if there was a period of overt trial and error before the "insightful" uses (and, indeed, these individuals had learned to use sticks to poke the food through the tube previously). Beck (1976) observed the spontaneous use of a stick to rake in out-of-reach food in several captive pigtail macaques. The most interesting observations of rhesus macaques were reported by Shurcliff, Brown, and Stollnitz (1971), who presented individuals with a stick too short to reach food, but then helped them learn to rake in a second stick that could then be used to obtain the food.

Westergaard and Lindquist (1987) observed liontail macaques in a captive group make "ladders" in much the same way as chimpanzees and Japanese macaques and make perches by inserting sticks into the cage mesh and sitting on the protruding end (see Table 3.5 for an account of monkey tool making). Westergaard (1988) observed individuals of this same group learn to both make and use a tool. Four of nine group members probed into holes for syrup, breaking off twigs and stripping them of leaves before doing so. (When given a similar opportunity, three mandrills did not manufacture or use probing tools, even when sticks were introduced part way into the holes by experimenters.) Westergaard argued that the macaques in this study were using stage 6 insightful solutions involving mental representation and combination. The two pieces of evidence he adduces are, first, that the macaques chose and/or modified tools appropriate to the task, in some cases without simultaneous visual access to the apparatus, and, second, that three of the four successful animals acquired the behavior "rather suddenly," with basically no failed attempts before they exhibited proficiency. It is noteworthy that while liontail macaques in the wild do not make or use tools, they do extract insects often by doing such things as stripping bark from a tree and carefully turning up grass blades (Johnson, 1985; see also Sinha, in press, for observations of a wild bonnet macaque, *Macaca radiata*, making and using a tool for personal hygiene).

3.2.5 Comparison of Species

A number of nonprimate animals use tools (Beck, 1980), and some do so in fairly flexible ways. Especially interesting in this regard are Hunt's (1996) recent observations of

Table 3.5. Tool-making in monkeys.

Context	Material	Action	Goal	Wild	Laboratory
Grooming	Sticks	Breaking, biting, cutting	Reduce length or diameter	*Macaca radiata:* Sinha, in press	*Cebus apella:* Ritchie & Fragaszy, 1988; Westergaard & Fragaszy, 1987b *Macaca tonkeana:* Bayart, 1982
Feeding	Leaves, sticks, pestles	Breaking, biting, cutting	Reduce length or diameter		*Cebus apella:* Klüver, 1933; Visalberghi & Limongelli, 1994; Visalberghi & Trinca, 1989; Westergaard & Fragaszy, 1987a; Westergaard & Suomi, 1995; Westergaard et al., 1995 *Macaca tonkeana:* Anderson, 1985 *Macaca silenus:* Westergaard, 1988 *Macaca fuscata*: Tokida et al., 1994 *Papio anubis:* Westergaard, 1992
	Stones	Hitting stone	Make a sharp object		*Cebus apella:* Westergaard & Suomi, 1994d,e,f
	Stones	Breaking, biting	Make a sharp object		*Cebus apella:* Westergaard & Suomi, 1995

a species of wild crow that uses, and perhaps even fashions, hook-shaped twigs and leaves to help capture insect prey in various locations inside trees. Therefore, the question of whether primates are the only "intelligent" users of tools, as hypothesized by Parker and Gibson (1977), is still open.

All of the most systematic observations of primate tool use that have been published, for both wild and captive animals, are presented in tables 3.1–3.5. It is clear from these tables that few primate species use tools habitually in their natural habitats (only chimpanzees habitually use tools among all populations), and so there are no large taxonomic differences, such as between monkeys and apes. In captivity, however, most simian primates show some form of tool use, and in this case there may be some differences between great apes and monkeys, although the cause of this difference may be differential exposure to human interaction and training. Thus, similar to great apes, several monkey species have been observed to both use and make tools in captivity in flexible ways, without specific instruction, sometimes in ways that may imply mental representation in the form of insight and foresight (Westergaard, 1988). Great apes raised and trained by humans have learned to use a wide variety of human tools and artifacts, a much wider variety than has been reported for any monkey species, but in no case is it clear if the most skillful monkey species (capuchins, baboons, and macaques) have been given the same opportunities and training.

Why primates that do not use tools in their natural habitats should be so proficient with them when given the opportunity, or perhaps when trained to some degree by humans, is still a puzzle at this point. In a paper appropriately entitled "Why is ape tool use so confusing?" McGrew (1989) could find no obvious differences in the ecology of the four great ape species to account for their widely varying tool use skills in wild and captive settings. Bonobos are close relatives of humans and chimpanzees, but they show basically no sustained tool use in the wild. Perhaps wild bonobos do not need to use tools to obtain food, but many chimpanzee uses of tools are not for food acquisition. On the other hand, orangutans are the most distant great ape relative of humans and chimpanzees, but they show some tool use in the wild and unparalleled tool use skills in captivity. As folivores, gorillas do not need tools in the wild, but even in captivity their skills are not impressive unless they receive extensive instruction from humans. In all, there must be an interesting evolutionary story to tell about the tool use of nonhuman primates, but at present we do not know what that story is. One possibility is that all simian primates have some general sensory–motor skills for means–ends coordination and that these skills can, in the appropriate ecological situations (either in the wild or in captivity), manifest themselves in the use of tools.

3.3 Causal Understanding

Some researchers are not only interested in whether a particular primate species uses a tool, but they are also interested in how that species uses the tool and how it understands the functioning of the tool in terms of the causal relations (i.e., dynamic tertiary relations) between tool and goal. Several investigators have therefore taken as their starting point the suggestive observations of field workers and experimentalists who have claimed that several primate species use tools insightfully and have then probed more deeply into the types of causal understanding that may underlie these uses. The basic goal in all cases is to determine precisely what kinds of experience individuals have had with particular tools and problems so that trial-and-error learning can be distinguished from causal insight or foresight based on mental manipulations, planning, and other forms of cognitive representation.

The classic observation of this type is that of Köhler (1925). Köhler presented the chimpanzee Sultan with an out-of-reach banana and two sticks, neither of which was long enough to reach the banana. However, the sticks could be fitted together into one stick that was sufficiently long. Sultan tried for more than an hour to reach the desired food in many different ways, all in vain. Finally, after Köhler had given up and left the room, a caretaker observed Sultan idly playing with the sticks and then discovering that they fit together. As soon as this happened, Sultan immediately went to rake in the banana. It should be noted that Sultan did not mentally manipulate the two sticks ahead of time to see that they could be attached and so solve the problem; he discovered that they attached in playing with them. What he did was recognize the potential usefulness of the newly made implement once he saw it. Following Köhler, we might call this behavior "insight" (Sultan "saw" the relation between tool and task), reserving the term "foresight" for cases in which an individual sees what is needed ahead of time and chooses or fashions a tool accordingly. (See Birch, 1945; Chance, 1960; Schiller, 1952, 1957, for interesting discussions on the role of species-specific behavior patterns and previous learning in solving tool-use problems.)

3.3.1 Monkeys

In a series of studies, Visalberghi and colleagues have probed the ability of monkeys and apes to adjust to novel tool-use situations by applying their understanding of the causal relations involved. Their first study investigated the causal understanding of capuchin monkeys, who are known to achieve success quite frequently in tool-use problems (see figure 3.6). Visalberghi and Trinca (1989) observed that when food is present, capuchins behave in rapid-fire fashion, trying all kinds of acquisition strategies, some clearly inappropriate. They presented four captive capuchin monkeys with a transparent tube with food inside. To acquire the food, subjects had to use a stick and poke the food out the opposite end of the tube. Three subjects were successful relatively quickly (from 38 to 101 minutes). The investigators then presented each successful subject with three variations of the problem involving different types of tools that required a modified strategy: (1) a bundle of sticks taped together that as a whole was too wide to fit in the tube (the subject thus had to break them apart); (2) three short sticks that together were long enough to remove the food (the subject thus had to put all the sticks in the same end serially to displace the food out the other side); and (3) a stick with a transverse piece on either end that blocked its entrance into the tube (the subject thus had to remove the offending piece). All three subjects solved these variations on the task, but they made a number of telling errors in the process. They did such things as inserting one short stick in one end of the tube and another in the other end; breaking off one transverse piece, but trying to insert the other end with a piece still blocking entrance; breaking open the bundle of sticks and attempting to use a splinter that was too short and flexible to work (often discarding sticks that would have been suitable) or attempting to insert the whole bundle. Moreover, these errors did not decrease significantly over trials, indicating little learning of the causal structure of the task, and, indeed, these errors were still present when the same subjects were tested again years later (Visalberghi, personal communication).

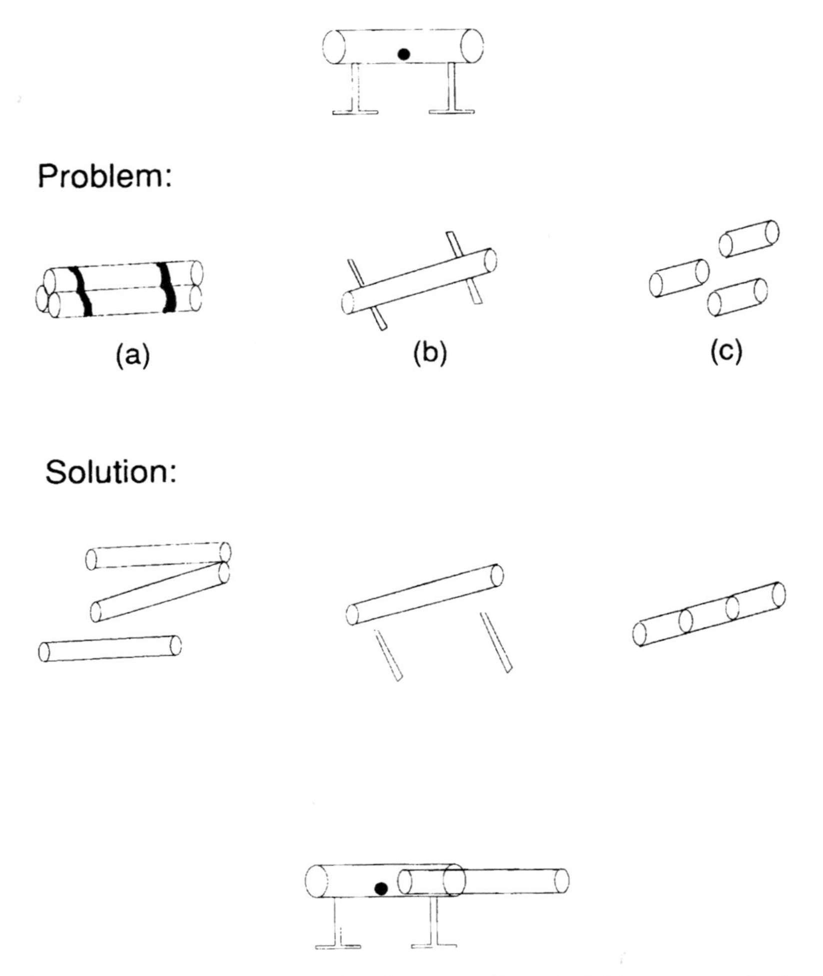

Figure 3.6. Visalberghi and colleagues' tool-use causality task. Subjects are required to extract the reward located inside the tube by pushing it out the opposite end in three different problem situations: (a) bundle, (b) H-tool, and (c) short sticks. Solutions to each of these problems are also depicted.

Visalberghi believes that these errors argue against the notion that the monkeys are operating with insight or foresight; that is, with a mental representation of the causal relations involved. Further evidence for this view comes from two other experiments. First, Visalberghi (1993) placed a number of appropriate and inappropriate tools in a room adjacent to the tube with food in it. If the monkeys were operating with a mental representation of the causal relations of the task, they should have been able to go to the adjoining room and choose an appropriate tool. They were unable to do this. Second, Visalberghi and Limongelli (1994) made a new tube that had a trap on the bottom of the tube, on one side of the midline (see fugure 3.7). If subjects pushed the food in the direction of the trap, it would fall in and they would lose it; to get the food, they

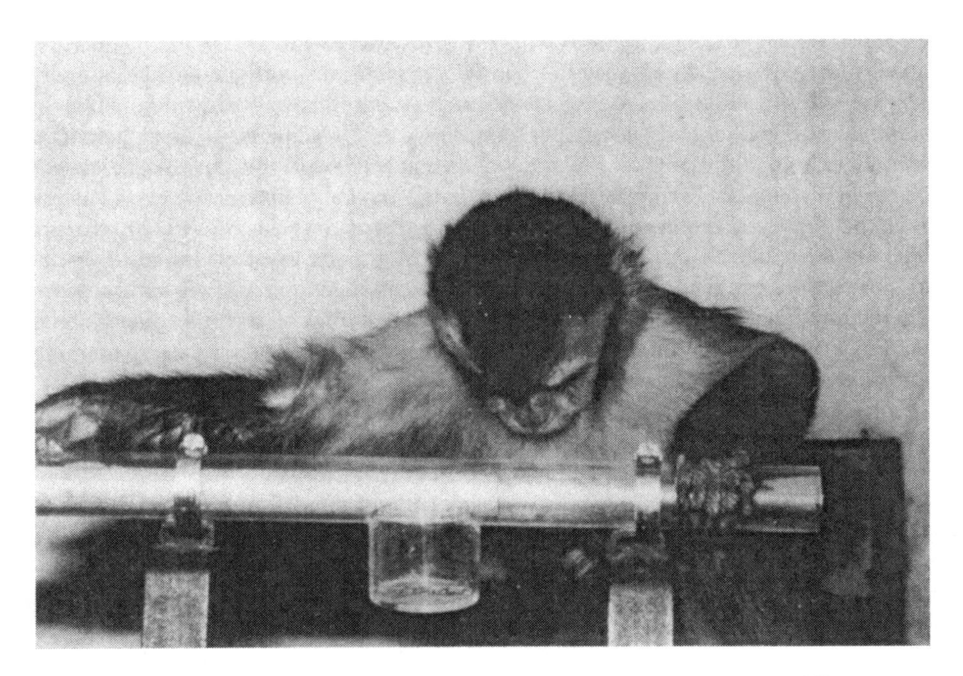

Figure 3.7. Capuchin monkey engaged in the trap tube task. (Photo courtesy of E. Visalberghi.)

had to push the food away from the trap and out the other end. This is an important extension of the Visalberghi and Trinca (1989) study because in that study all subjects eventually succeeded, and so it is possible they simply did not bother to do the task in the "smartest" or most efficient way; in the current study ineffective behaviors would lead to failure, and so there was more pressure on subjects to find the most intelligent and effective solution. However, although all four subjects inserted sticks and attempted to get the food on most trials, only one of four capuchin monkeys solved the task successfully. However, this subject did not learn to insert the stick on the correct side in an insightful way; half of the time she inserted it on the wrong side, pushed it gently until she saw that the food was about to go in the trap, and then withdrew the stick and went to the other end of the tube. In a follow-up study, Visalberghi and Limongelli (1994) rotated the tube 180° so that the trap was on top of the tube where it presented no danger of losing the food. The successful subject did not change her strategy, however, and avoided pushing the food to the trap side nonetheless. Apparently the subject had learned over trials to simply push the food away from the trap side without understanding ahead of time (with foresight) the causal relations involved.

In a different approach to monkeys' understanding of causality in a different behavioral domain, Cheney, Seyfarth, and Silk (1995a) presented evidence that Chacma baboons (*Papio cynocephalus ursinus*) are able to understand something of the causal structure of the social interactions that must be taking place outside their visual field on the basis of the vocalizations they hear. These investigators used a tape recorder to play for individual subjects the vocalizations of some females interacting, presumably out of their sight. When the interaction was what the investigators called causally inconsistent (subjects heard the vocalizations of two known individuals, with the dominant giving a fear bark to the subordinate) subjects responded vigorously. When the same vocalizations were made causally consistent by adding in the vocalizations of an individual dominant to both of the original females (as a possible cause of the fear barks), subjects took much less notice. The investigators concluded that baboons are able to understand something of the causal structure of social interactions on the basis of vocal information alone. On the other hand, these same investigators reported observations suggesting that vervet monkeys may not always understand causal relations in the natural world around them. For example, when vervets see the visible tracks left by dangerous predators, they do not seem to infer that those tracks were caused by a predator who might still be lurking nearby (Cheney & Seyfarth, 1990b).

3.3.2 Apes

Recently Visalberghi and colleagues presented both versions of their tube tasks to great apes. Visalberghi, Fragaszy, and Savage-Rumbaugh (1995) presented parts of the original Visalberghi and Trinca task to five chimpanzees, four bonobos, and one orangutan, some of whom had been brought up in a humanlike cultural environment. (They also administered the same tasks to six capuchin monkeys and basically replicated previous results.) Eight of the ten apes solved the basic tube task on the first trial, and the other two were successful later. When given a bundle of sticks, all subjects immediately disassembled the bundle; no subject attempted to insert the bundle, as capuchins sometimes did. When the stick had two transverse sticks blocking insertion on each end, the apes had more difficulty in the 10 trials administered (2 blocks of 5). All subjects on at least one occasion attempted to insert the stick without modification, or else removed the transverse stick from one end and attempted to insert the other end. The majority of subjects made this error at least once, even in the later trials, and two subjects made the error on every trial. Five subjects tried to use the broken-off transverse stick (which was much too small and too short) at least once; one subject made this mistake on several trials. For five of the seven subjects who completed both blocks of trials in this condition (three subjects made many errors in the first block and were not given a second block), the number of errors decreased across blocks, while for two it increased.

The investigators concluded that the apes showed more foresight in this task than did capuchins because they were successful in the basic task so quickly, and they made almost no errors in the bundle condition. However, three considerations make this conclusion less than convincing. First, the performance of the apes in the transverse-stick problem contained many errors of the type made by the capuchins, and indeed a statistical comparison of the two species in this condition revealed no significant differences (see also Khrustov, 1970, whose adult chimpanzee subjects made many of these same kinds of error on a similar task). Second, these ape subjects were not systematically administered the version of the problem requiring subjects to insert three small sticks in the same end of the tube, the most difficult variation for the capuchins. Third, in an administration of the same task to somewhat younger chimpanzees (2–4 years old), Bard, Fragaszy, and Visalberghi (1995) found that in the two most difficult versions of the task (the short sticks and the transverse stick), the performance of the majority of subjects actually deteriorated over trials (more errors in later trials), arguing that they may not have come to understand the causal relations involved.

Limongelli, Boysen, and Visalberghi (1995) presented the second version of this task, the tube with the trap from Visalberghi and Limongelli (1994), to five chimpanzees in 14 blocks of 10 trials each. All subjects behaved at chance levels for the first 70 trials. Two subjects learned during the second 70 trials to avoid the trap by using a stick to push the food out the opposite end. In an attempt to assess the understanding of these two subjects, a follow-up study was administered. Recall that the capuchins in this task (Visalberghi & Limongelli, 1994) seemed to be operating with the rule "push the food out the side to which it is closest," which, because the trap was always in the center of the tube, always meant away from the trap. The investigators therefore varied the placement of the trap in the tube for the chimpanzees. In some cases the trap was located very close to one end with the food just beyond it, so that subjects actually had to push the food out the end from which it was farthest. In other cases, the opposite arrangement was used. Both subjects solved these variations easily, with almost no errors, and so the investigators concluded that these two chimpanzees understood the causal relations in this task better than the capuchin monkeys. It should be noted, however, that the variations used in this experiment could still be solved by the performance rule "push the food away from the trap," which could have been learned in the preceding 140 trials of the basic trap task. It is also important that they did not administer the "inverted tube" version of the task in which the trap was rotated so that it was on top of the tube (and no longer a threat to the food) to probe the depth of subjects' understanding of the functioning of the trap. When Reaux (1995) gave this task variation to one 6-year-old chimpanzee (the only one of four subjects who was successful in the basic tube task), she continued to avoid the inverted trap, in the same manner as Visalberghi's capuchin.

Also relevant in assessing the type of causal understanding that apes may possess is a series of tasks presented by Mathieu, Daudelin, Dagenais and Decarie (1980). The subjects were two 2-year-old chimpanzees. In a number of tasks subjects showed an understanding of causality in the sense of doing something to make an external event occur based on some knowledge of the causal relations involved. For example, subjects pulled in an object tied to a string or placed on a cloth, and they used a rake tool to rake in food. Of particular interest to issues of mental representation is their responses in tasks involving a "hidden cause." For example, subjects were presented with a familiar toy missing one of its key parts. Both of them looked for the missing part. Also, an experimenter situated behind the subject threw a paper airplane over the subject's head. Both subjects turned around to look where it came from. Finally, after the experimenter operated a toy using a device under the table (a cord went from the toy to under the table), subjects were given the toy. One of the two subjects looked under the table. The investigators concluded that the subjects were operating with a Piagetian stage 6 level of causal understanding in which events external to themselves were connected by causal links and displayed a "closed" sense of causality: every action must have a cause even if it is not immediately observable. Without more knowledge of the histories of these individuals with these materials, however, it is not possible to know if this conclusion is correct.

In a different approach to the causal understanding of apes, Premack (1976) presented the language-trained chimpanzee Sarah with pictures depicting, for example, an apple and a cut apple. Sarah's job was to choose from several alternatives the picture of a knife as the possible instrument of the cutting that presumably had taken place. Sarah performed quite well on these tasks, although it is not totally clear that all of the possible associative strategies that could be used (e.g., knifes are often associated with objects split into parts) were ruled out effectively. Premack and Premack (1994) reported some variations on this theme with the same basic findings, interpretations, and limitations. In addition, they reported a new study that they considered to involve causal reasoning. Each of four chimpanzees saw a human place an apple in an opaque container and then a banana in another opaque container some 10 m away. After having been distracted for 2 minutes, the subject then saw the human eating either an apple or a banana, and then the subject was released. Ideally, subjects would infer that if the human was eating one of the fruits, only the other would be left for their own consumption. Only one of the four chimpanzees consistently chose the correct container. In a variation on this test, subjects observed a fruit being eaten by a human but the time interval between hiding and eating was so brief that it could not have come from one of the containers (because the food was now made difficult to access from the container, and the subjects were made aware of this). The previously successful chimpanzee again chose the other food, indicating a lack of appreciation of the temporal contingencies involved. Human 3- and 4-year-old children did well on both versions of the task.

3.3.3 Comparative Studies

Two other studies have directly compared monkey and ape understanding of causality in tool-use situations using fairly simple tasks. Spinozzi and Potì (1989) presented subjects with an out-of-reach object resting on a cloth support that they could reach. On some trials they placed the object on the ground next to the cloth but not on it. Of interest was whether subjects would differentiate between the two situations by pulling the cloth to them when the object was on it and by refraining from pulling when the object was not on it. Two longtail macaques, one Japanese macaque, two capuchin monkeys, and one gorilla all differentiated between these two situations. The investigators concluded that subjects of all four species understood that the causal relation by means of which the cloth transported the object required spatial contact. Spinozzi and Potì (1993) recently administered the support problem to two chimpanzees at two points in their second year of life and found that one was successful in differentiating the object-on-the-cloth and object-off-the-cloth conditions. Mathieu et al. (1980) found that two chimpanzees in their third year of life both solved the support problem, although it is not clear that they presented the object-off-the-cloth condition to check for differentiation.

Natale (1989a) presented eight subjects from these same four species (five of whom were the same subjects as those of Spinozzi and Potì, 1989) with an out-of-reach object and a stick, placed in different positions relative to the object in different experimental conditions. Because the stick was presented removed from the object, a successful subject in this task would need to do more than just recognize the required spatial relation (as in the cloth support problem), but would also need to produce a behavior that brought the tool and food into a specific type of contact. None of the macaques was successful in this task. On the other hand, three of the four capuchin monkeys and the gorilla were at least moderately successful, even when the tool and object were presented to them in different arrangements. They were on average several months older than when they solved the cloth support problem. Natale argued that successful performance in this task is evidence that subjects understand a complex of causal relations, for example, the understanding that the stick must be of the appropriate size and material (e.g., long and rigid) and that only certain kinds of contact (e.g., with a certain force and directionality) would likely be successful.

3.3.4 Comparison of Species

In Visalberghi and colleagues' studies of primate understanding of tool use, there is some evidence that apes may have a deeper causal understanding of tool-use tasks than monkeys. However, the apes are far from perfect, require many trials for learning in some variations, and in many aspects of the problems behave just like monkeys. Moreover, the only monkey species studied in the various experimental manipulations is the capuchin monkey, who is well-known to be a rapid-fire ("leap before looking") manipulator. It is possible that other monkey species who have shown some tool-use skills, such as the liontail macaques who seemed to behave in a foresightful manner in Westergaard's (1988) study, or those who have received more extensive human training, would behave more insightfully in these experimental tasks.

It is important also to point out that Visalberghi's tasks were fairly difficult conceptually—depending on some fairly specific physical and causal knowledge—and thus that human children are not proficient at all these task variations until after their second birthdays (Visalberghi & Limongelli, 1996; they are not proficient at the short-stick variation until 30 months of age), well beyond Piaget's sixth stage of sensory–motor development when the understanding of causality as an external independent process is first understood (18 months). On a simpler task assessing causal understanding (the support), two macaque species and a species of capuchin monkeys were as proficient as a gorilla, and in another fairly simple problem (the stick), the capuchin monkeys were as proficient as the gorilla. The overall idea, then, is that causal understanding may be of a fairly simple or a more complex variety. The simpler varieties in which straightforward spatial relations predominate may be attainable by many primate species. The varieties that rest on the use of tools may only be understood deeply by those animals that use tools naturally and habitually, especially apes, and even then the basis of this understanding may be fairly simple spatial relations. How these forms of causal understanding relate to the causal understandings of human children in which the notion of force plays an essential role (Frye, Zelazo & Palfai, 1995) is a question in need of systematic investigation.

Causality is a difficult concept to specify, and there is much disagreement over its proper domain of application by philosophers and theorists of all persuasions. The prototype of causal understanding involves two objects interacting forcefully with one another independently of the observing subject (e.g., two billiard balls), and so the case of tool use is less than prototypical, since it involves the subjects active manipulation of one object relative to another. Also an issue is whether it is legitimate to include under the purview of causality such things as chimpanzees looking over their shoulders for the source of a flying plane, or baboons reacting more strongly to anomalous than to expected social events. The issue of causal understanding and its role in primate cognition will be discussed more fully in chapter 12.

Figure 3.8. Young chimpanzee using a stick to bring an object within reach. (Photo courtesy of F. de Waal.)

3.4 What Primates Know about Tools and Causality

There seems to be a basic difference in the evolutionary significance of animals' understanding of space, as reviewed in chapter 2, and their understanding of object relations and causality. The difference is that, whereas all animals have to know where food and other important items are located in their environments, not all animal species have to know about object relations and causality to forage and perform other subsistence activities effectively. It should not be surprising, then, to see differences in the skill and motivation to manipulate objects among different animal species, and this should be related both to their phyletic status and their foraging ecologies. In the current context, the most general point is that as hand-feeders, most primates, especially simian primates, need to be more adept at handling objects and processing food than most other species. Other mammals such as raccoons and otters that use their hands to process food do not use their hands in a wide variety of other ways to manipulate or investigate objects.

When primates are presented with objects outside of any clear-cut functional context, all species typically orient visually and show at least some interest (Glickman & Sroges, 1966). Within primates, however, some species show more interest than others in exploring and manipulating them. Prosimians and some leaf-eating monkeys (both Old World and New World) show the least interest, while the great apes and several species of monkeys, especially capuchins and baboons, show the most interest. This greater interest is manifest in a stronger tendency to observe the effects that various self-produced actions have on objects and in relating objects to one another in systematic ways. Interestingly, the species that are least interested in manipulating objects tend to be insectivores and leaf-eaters, presumably because they have become relatively narrowly adapted to foraging in their own specialized ways. The species that are more interested in objects have wider diets, and in some cases need to engage in extractive foraging techniques; these are all the same species that seem most proficient at tool use as well. Nevertheless, it must be recognized that nonhuman primates are not nearly as interested in manipulating objects as are human children. The differences are diminished, however, when nonhuman primates are raised by humans with exposure to human objects and artifacts and encouragement in their use. This accords with the arguments of Vauclair (1984), who believes that humans become especially interested in manipulating objects early in ontogeny at least partly because they live in an encouraging social environment in which many important social interactions with others revolve around objects.

The use of tools involves not only object manipulation, but also the systematic relating of the object being manipulated (the tool) in relation to another (the goal). Only one species of nonhuman primate habitually uses a variety of tools on a species-wide basis in its natural habitat: all groups of chimpanzees who have been observed extensively use tools on a regular basis. One group of orangutans uses two tools on a regular basis in their natural environments, and several monkey individuals in the wild have been observed to use tools on a sporadic basis. In captivity, however, many primate species use tools with great facility, the most adept being orangutans, chimpanzees, and various species of baboons, macaques, and capuchins. This might imply that they can understand the dynamic tertiary relations between objects, as the use of most tools requires that they monitor the effect of tool on goal as they manipulate it, but there are ways to be skillful with tools that do not require a deep understanding of the causal relations involved (Visalberghi, 1993). Again, human children seem to be especially proficient tool users as they reach early childhood, and there are some influences of the human environment on the tool-using propensities and skills of nonhuman primates (Call & Tomasello, 1996). This effect of the human environment is presumably due to the fact that in captivity monkeys and apes see many objects and how they work (their affordances), experience social encouragement and rewards for tool use, and are forced to use objects in some contexts if they desire certain materials controlled by humans.

From the point of view of cognition, those primates that manipulate objects and/or use tools do so in very flexible ways that depend on individuals adapting to their particular ecological circumstances—seemingly in ways that are more flexible than those of other animal species that manipulate objects and tools in some foraging circumstances. But the recent discovery of some especially flexible uses of tools in natural environments by nonprimate animal species (Hunt, 1996) suggests caution in making general comparisons of primates with other large taxonomic groups. In terms of mental representation, not enough is known about the "symbolic play" of human-raised apes to come to definitive conclusions. The domain in which mental representations seems to be most evident is in the use of tools in new and creative ways without overt trial and error; that is, the use of tools with insight and foresight, especially the modification or making of tools to fit particular task demands. The supposition by many researchers is that in these cases the individual mentally represents the needs of the situation and then chooses or fashions a tool accordingly. Chimpanzees, orangutans, olive baboons, tufted capuchins, and liontail macaques have all been observed making tools or otherwise showing foresight in anticipating the needs of a problem. In experimentally controlled studies investigating the ability of great apes and tufted capuchins to show foresight in their understanding of causal relations in Visalberghi's tool task, there is some evidence that apes understand causal relations more deeply than do capuchins, but the evidence is not overwhelming, and other species of monkeys (especially Old World monkeys) need to be tested in this situation. This is necessary especially since several species of monkeys do quite well in other situations involving the understanding of somewhat simpler causal relations, such as the cloth-support task which only requires the recognition of a simple spatial relation (i.e., support). Chimpanzees also show some evidence of understanding invisible causes, but there has been only one, somewhat informal, study, and monkeys have not been tested for this type of understanding (and baboons can make causal inferences about social interactions on the basis of vocal information alone). What forms of causal understanding, as opposed to more straightforward forms of spatial understanding not involving notions of force, might be involved in primate tool use is discussed more fully in chapter 12.

As alluded to above, the context for the evolution of these skills of object manipulation and tool use was presumably foraging. The primate hand may have received a head start based on adaptations for arboreal life in general (i.e., climbing), but further adaptations for hand feeding were presumably responsible for the different manipulative propensities and skills of the various primate species. Most primates do not regularly engage in tool use in their natural foraging activities, and so the widespread use of tools in captivity by many primate species is still something of a mystery. It is possible that the underlying cognitive skills involved derive from adaptations for some other form of intentional activity in which one sensory–motor procedure on an object is used in the service of another, leading to general sensory–motor skills that the individual may then co-opt for the use of tools if forced to do so. It might even be that some intentional activity from another cognitive domain, for example, intentional communication in which one individual attempts to manipulate another by means of some intentional behavior, is somehow co-opted for the use of tools. If this were true, it might mean that even the more flexible forms of tool use employed by nonprimate species would be different from primates in that they would be more specialized adaptations for the flexible use of one particular tool in one particular situation.

Of all primates, there is no question that humans are the most interested in objects and their interrelations, and they are the most creative users of tools as well. From a very early age human infants explore and manipulate objects outside of functional contexts, and from sometime after their first birthdays they begin using tools in various ways as well, with many adultlike forms of tool use appearing in the early childhood period. The cultural dimensions of the process are obvious—children do not routinely invent the tools they use but only learn how to manipulate tools invented by others—and so how much beyond other apes human children might go if left to their own individual devices is not clear. Investigations of the causal understanding of human infants have focused on tool use, in which infants show some foresight from a very early age (e.g., Piaget, 1954), and on the understanding of the interactions of independently moving objects with older children, in which they show some understanding of the notion of force as essential to the causal analysis of the physical interactions of independent objects (Frye et al., 1995; Piaget, 1974). Primate understanding of different forms of causality and their relation to the understanding of intentionality, which involves a notion of psychological "force," is discussed more fully in chapters 12 and 13.

4

Features and Categories

Underlying the ability to locate and manipulate objects is the ability to identify particular objects or particular types of objects. Much object identification is done by means of perceptual mechanisms that are highly similar throughout the order Primates, and probably throughout many other orders of mammal as well (see Fobes & King, 1982; Passingham, 1982; Prestrude, 1970; Riesen, 1970). These perceptual mechanisms are not our concern here. But animals also identify objects in some cases on the basis of conceptual categories that go beyond direct perception, and these are of obvious importance in any study of cognition. As all texts of cognitive psychology emphasize, organisms make sense of the virtually infinite amount of information available to their senses at any given moment by categorizing experiential items based on their similarities and differences, both perceptual and otherwise. We are concerned in this chapter, therefore, with research aimed at revealing processes of conceptual mediation and categorization in nonhuman primates. Because the detection of such processes requires comparing the responses of animals to a systematic array of stimuli, almost all the relevant research has been conducted in the laboratory.

In terms of theory, it should be stressed at the outset that most of the research on nonhuman primate categorization has been carried out in the context of associationistic theories of learning. This makes it difficult to integrate the research with much of the other work reported in this book for two reasons. First and most obviously, all varieties of learning theory derive historically from one or another form of behaviorism, which has traditionally eschewed cognitive phenomena altogether. Indeed, the entire theoretical vocabulary of behaviorism was set up to describe only "stimuli," "responses," and their association, all from the point of view of the external observer. When behaviorists finally turned their attention to cognitive phenomena, they did so with a theoretical vocabulary that consisted of little more than "discrimination," "generalization," and, for some researchers, "conceptual mediation," a far cry from the rich set of epistemological categories that are the focus of research from the Piagetian point of view that structured our investigations of space, objects, tools, and causality. Adding in a few selected concepts such as "memory," "attention," and the like from traditional cognitive psychology does not improve the situation significantly. In any case, at least partly because of this impoverished set of theoretical tools, the study of animal cognition in the learning theory tradition has focused on only a small set of cognitive phenomena from one fairly narrow theoretical point of view.

The second reason that the learning theory approach to cognition is difficult to integrate with other approaches to cognition is that behaviorism has traditionally not taken an ecological approach (nor has traditional cognitive psychology for that matter; Neisser, 1984). From its inception, behaviorism searched for "the laws of learning" that applied generally across all species and all domains of cognition. This has had two consequences. First, since the laws of learning are supposed to be the same for all species, the particular species that behaviorists choose to study are few in number. Specifically, there has been an inordinate amount of work done on those species that live robustly in the laboratory, especially rats, pigeons, and rhesus macaques. Second, following traditional cognitive psychology in focusing on "general processing mechanisms" such as attention, memory, learning, and conceptual representation, the natural competencies of the species involved have basically been ignored. The consequence has been that learning theorists have not often looked for qualitative differences in the types of cognitive mechanisms employed by different species in different cognitive domains, but rather have looked for quantitative differences in such things as speed of learning or number of concurrent discriminations that can be made. In its turn to the study of cognitive processes, learning theory has attempted to some extent to change its nonecological ways by recognizing that different species are "prepared" to learn in different ways. But the change has not been fundamental. Even fairly recent texts of comparative psychology (or comparative cognition) from a learning theory perspective have retained the central focus on the general processing mechanisms displayed by a few animal species (e.g., Roitblat, 1987).

There are four areas in the large body of research on learning theory approaches to comparative psychology that are relevant to primate cognition. First, there are several phenomena of simple discrimination learning that have led behaviorist researchers to invoke some type of cognitive mechanisms. The three major instances are learning sets, cross-modal recognition, and delayed response, all of which have demanded that theorists go beyond perceptual stimuli and motor responses and deal with conceptual mediation of some sort. Second, in contrast to these studies of discrimination learning that employ as experimental stimuli such things as squares and circles, there are some studies of primate discrimination learning that employ more ecologically meaningful materials, seeking to investigate processes of primate categorization with stimuli such as pictures of human beings or other animals. In these tasks, individuals must learn to respond similarly to natural objects (or pictures of natural objects) that humans would say belong to the same cognitive category. Third, nonhuman primates have also been asked to learn categories in which the appropriate discrimination is made on the basis of a comparison among stimuli, and the resulting category is not of entities but of the relations among entities. Fourth, some primates have been asked by researchers (including some from outside the learning theory tradition) to physically sort objects into spatially distinct groups on the basis of their perceptual similarities and differences. In addition to sensory–motor manipulative skills, object sorting of this type requires a simultaneous understanding of both the similarities and differences among objects so as to create a set of cognitive categories that relate to one another in specific ways.

4.1 Discrimination Learning

Basic discrimination learning is exemplified by studies in which an individual animal is rewarded for choosing one stimulus over another; if it learns to do this successfully, it is said to have learned the discrimination. The experimenter arranges ahead of time the choices to be presented and the stimuli to be rewarded, with successful learning on the part of the animal demonstrating its ability to make the appropriate discrimination. For example, many animal species, including primates, can learn to respond to an object presented to the right side over one presented to the left side, or to a green object over a yellow object, or to a round object over a square object. As the stimuli to be rewarded become more complex or subtle, learning takes longer. Thus, animals may be asked to respond on the basis of multiple cues simultaneously, or they may be asked to respond to one stimulus in one context and to another stimulus in another context ("conditional discrimination"). Nonhuman primates take from hundreds to thousands of experimental trials to master these more complex tasks, and there seem to be few if any differences in how quickly the different species succeed (and, indeed, few if any differences between primates and other mammals such as rats, sheep, and cows) (for a classic review of basic phenomena of primate discrimination learning, see Harlow, 1951; for a more recent review see Meador, Rumbaugh, Pate, & Bard, 1987).

The discrimination learning phenomena of most interest to the study of primate cognition are those that have been studied in some depth in nonhuman primates and have led researchers to posit cognitive processes. There are three areas of research in simple discrimination learning that meet these two criteria. The first is problems of discrimination learning that require animals to identify dimensions of similarity across sets of problems ("learning sets"), implying in some cases what behaviorist researchers have called conceptual mediation. The second is tasks of cross-modal perception, which imply a flexibility of object recognition that goes beyond what is needed for simple stimulus-response learning (perhaps even requiring some notion of conceptual representation). The third is studies of delayed response in which the animal must remember (cognitively represent) some stimulus or set of stimuli over time before responding.

4.1.1 Learning Sets

The basic phenomenon of learning sets, also called "learning to learn," is that an individual will become better at discrimination problems as it solves different sets of similar problems over time; it has learned to deal with a particular type of problem in general. Harlow (1949) argued that the formation of a learning set was a cognitive phenomenon because it indicated that the animal was learning something beyond particular stimulus-response associations; it was also learning something more abstract. Over the years it has been found that rhesus macaques and chimpanzees, but also rats, dogs, and some birds, show the learning-to-learn pattern (see Fobes & King, 1982; Passingham, 1981; Rumbaugh & Pate, 1984a,b, for reviews). There is even recent evidence for this type of learning in fairly natural social settings (Drea & Wallen, 1995; Lepoivre & Pallaud, 1986). There are two basic paradigms in this domain of research: (1) object discrimination learning set and (2) reversal learning set.

Object discrimination learning sets were first studied extensively by Harlow (1949). He presented rhesus macaques (and some mangabeys) with two real objects in a Wisconsin General Test Apparatus (WGTA). Choosing one of the objects was rewarded, while choosing the other was not. The same object pair was presented for six trials in a row. The performance of subjects on trial one was, of course, random, as they had no way of knowing which object was correct, but they learned over trials to choose correctly. After the six trials with the first pair of objects (the first problem), two new objects were presented in each of a series of later problems. Performance on trial one remained random for the new problems, but the monkeys did learn to choose correctly on these later problems eventually. Indeed, the learning-to-learn phenomenon was that over problems the monkeys learned to choose correctly in fewer trials. Of special importance is performance on trial two across problems. In this paradigm, after choosing randomly on trial one and either being rewarded or not, a subject has all the information it needs to respond correctly on trial two of that problem. Harlow (1949) found that the monkeys' performance on trial two was around chance for the first eight problems he presented to them, but it was nearly perfect after several hundred problems. He argued that the transfer of learning across problems suggested that some kind of conceptual mediation was at work.

Even though Harlow (1958) cautioned against it, many studies of learning set formation were subsequently performed with an eye to comparing "the intelligence" of species. Both Warren (1974) and Passingham (1981, 1982) reviewed these studies and concluded that primates were somewhat quicker to form learning sets than nonprimate animals. However, in their review, Fobes and King (1982) found no differences between primates and nonprimates (see especially their figure 9.10), although there are difficult issues of equating problems for different species with their different perceptual and behavioral capabilities. With regard to differences within primates, Passingham (1981) implies that apes are quicker to form learning sets than monkeys, but he cites only one study in favor of this conclusion (Hayes, Thompson & Hayes, 1953). In that study, several chimpanzees were tested and then the data from the best single subject was compared to Harlow's data for rhesus macaques; overall that study found no difference in the speed of learning set formation among rhesus macaques, chimpanzees, and human children. Also, Passingham cites but dismisses the study of Rumbaugh and McCormack (1967) in which macaques did as well as great apes. In their review, Fobes and King (1982) found some differences among species, but no differences that corresponded to broad-based taxa such as monkeys and apes (see their figures 9.8 and 9.9).

Figure 4.1 summarizes all the major studies done on object discrimination learning set formation with primates that meet certain criteria of comparability. As can be seen in this figure, although there are species differences in learning set formation, there are no differences corresponding to large taxonomic groups such as prosimians, monkeys, and apes. After 200 problems, approximately 80% correct performance is achieved by species as different from one another as black lemurs (*Lemur macaco*), chimpanzees, rhesus macaques, and gorillas. It is of interest to note, in this regard, that some species are represented by only one or two studies, sometimes each with only one or a few indi-

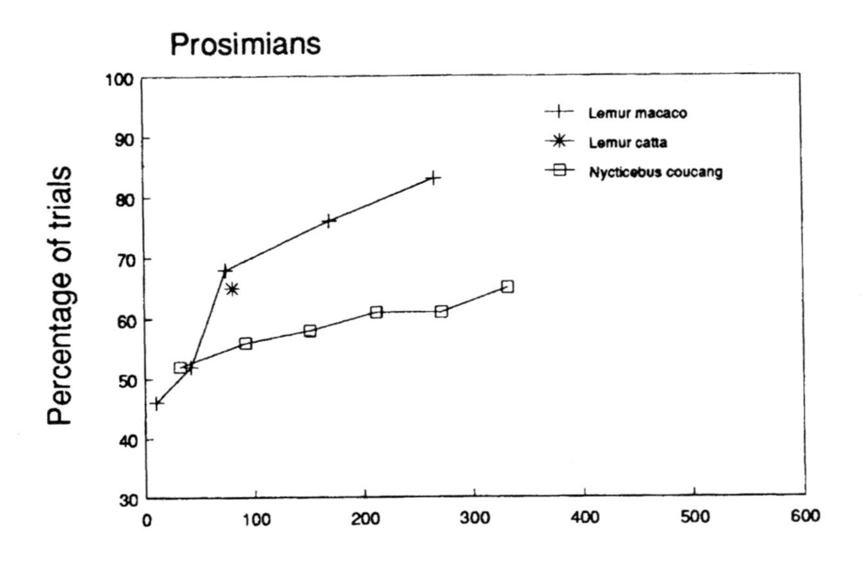

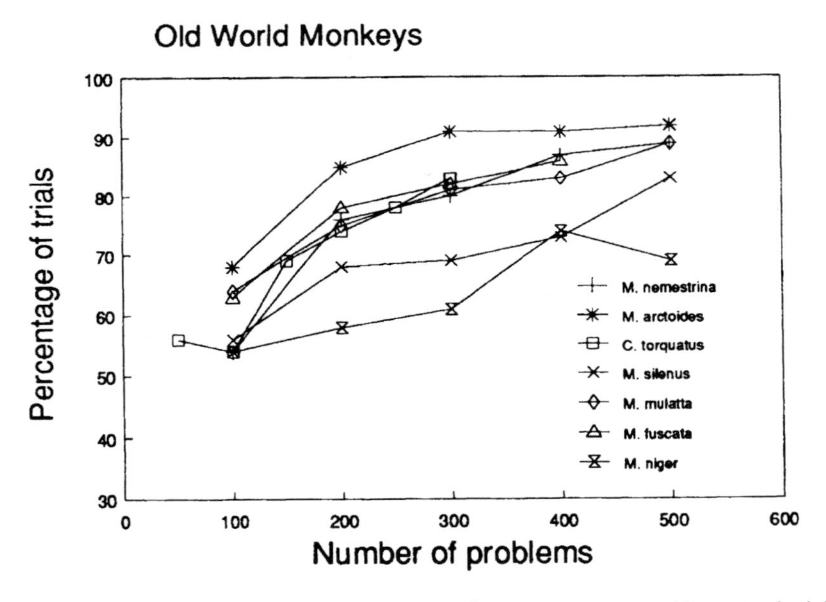

Figure 4.1. Object discrimination learning set in primates as represented by second-trial performance. Average values have been used for species represented by more than one study. SOURCES: *Ateles geoffroyi* (Shell & Riopelle, 1958), *Callithrix jacchus* (Miles & Meyer, 1956), *Cebus albifrons* (Shell & Riopelle, 1958), *Cebus apella* (de Lillo & Visalberghi, 1994), *Cercocebas torquatus* (Behar, 1962), *Gorilla gorilla* (Fischer, 1962; Rumbaugh & Rice, 1962; Rumbaugh & McCormack, 1967), *Hylobates* (Rumbaugh & McCormack, 1967), *Lemur catta* (Ehrlich et al., 1976), *Lemur macaco* (Cooper, 1974), *Macaca arctoides* (Riopella & Moon, 1968), *Macaca fuscata* (Yagi & Furusaka, 1973), *Macaca mulatta* (Harlow & Warren, 1952;

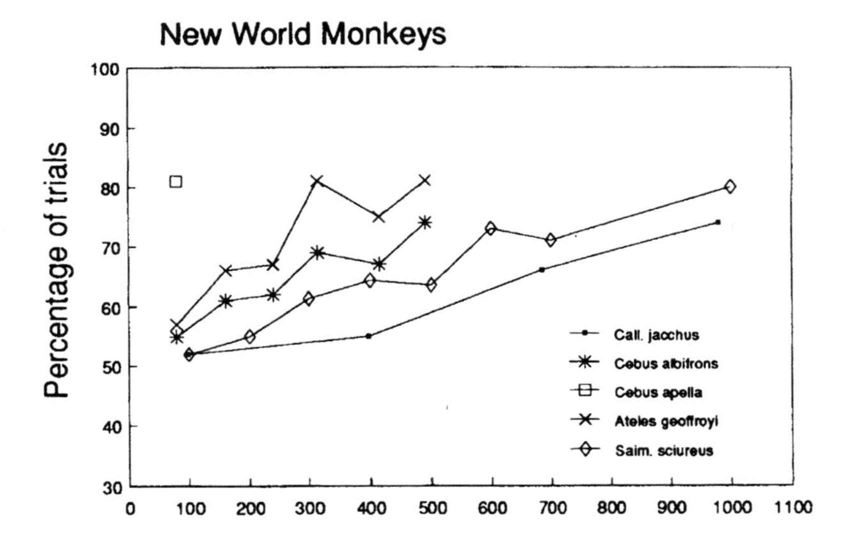

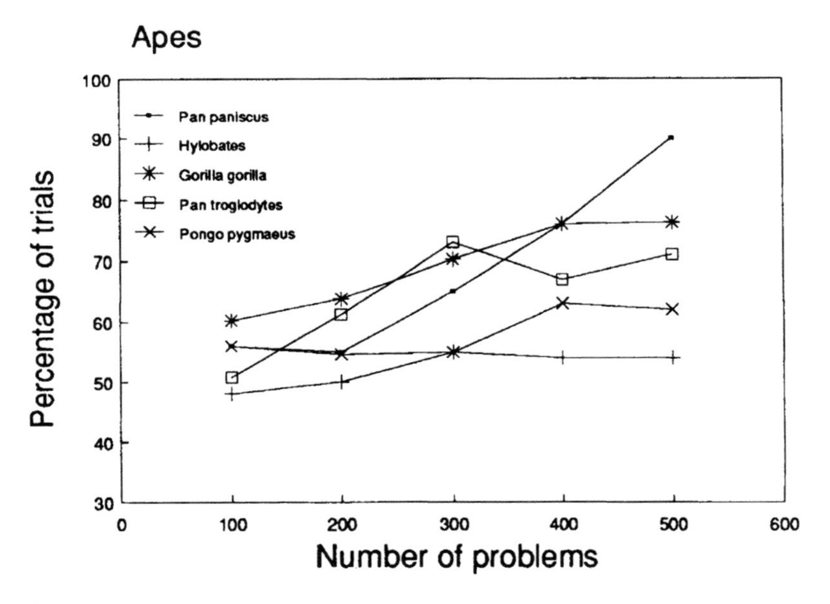

Figure 4.1. *(continued)*
Darby & Riopelle, 1955; Riopelle & Francisco, 1955; Behar, 1962; Sterrit et al., 1963 Harlow et al., 1969), *Macaca nemestrina*, *Macaca niger*, *Macaca silenus* (Rumbaugh & McCormack, 1967), *Nycticebus coucang* (Ohta, 1983), *Pan paniscus* (Rumbaugh & McCormack, 1967), *Pan troglodytes* (Davenport et al., 1969; Rumbaugh & McCormack, 1967; Rumbaugh & Rice, 1962), *Pongo pygmaeus* (Rumbaugh & McCormack, 1967; Rumbaugh & Rice, 1962), *Saimiri sciureus* (Miles, 1957; Rumbaugh et al., 1965; Shell & Riopelle, 1958).

vidual subjects. The problem with small numbers is illustrated by the fact that rhesus macaques have been investigated in several dozen studies, and much variability in their performance across studies is found; for example, after 200 problems, the mean scores of groups of rhesus from different studies ranges from 63% to 87% correct (with the scores of individuals ranging much more widely).

The other discrimination learning problem that has been widely used with primates is reversal learning. As a way of comparing different species on reversal learning, Rumbaugh (1970) developed what he called the "transfer index." The transfer index is basically how quickly an animal will learn in a situation in which responses to particular stimuli from previous trials are pitted against learning to learn. For example, in reversal problems, one stimulus of a pair is rewarded for many trials (for instance, the red one is always rewarded and the green one is never rewarded); after some trials this contingency is reversed (so that, for instance, the green one is now rewarded and the red one is not). If an individual is learning by conditioning, it will take many trials to learn the new contingency because the original response in the prereversal trials has to be extinguished and the new one learned. If it learns the new contingency quickly (faster than the prereversal discrimination), there has been some "transfer" across the discrimination problems. Rumbaugh argues that transfer in such situations must be accomplished by mediational (cognitive) mechanisms because a strict learning theory explanation would predict at least as much time for the postreversal discrimination as for the prereversal discrimination. In more explicitly cognitive terminology, if there is transfer, the animal is doing something like understanding that in the new trials the "opposite" stimulus will be rewarded. Rumbaugh's transfer index is higher the more quickly an animal learns the postreversal discrimination.

In reviewing studies using the reversal learning paradigm for 10 different primate species, Rumbaugh and Pate (1984a,b) found that there is a general increase in the transfer index from prosimians to monkeys to great apes. Moreover, and importantly, differences in species may be observed as a function of what criterion of learning was attained before the reversal was administered. In many studies prereversal trials were learned to either a 67% correct criterion or an 84% correct criterion. For the three prosimians studied, learning the prereversal discrimination to an 84% criterion led to *less* transfer than when the prereversal discrimination was learned to a 67% criterion. This implies that the process was one of conditioning; the stronger the prereversal learning, the more it interfered with the postreversal learning—presumably because it took some trials to extinguish the earlier learning. For all three great ape species, on the other hand, learning the prereversal discrimination to an 84% criterion led to *more* transfer than when the prereversal discrimination was learned to a 67% criterion. This implies that subjects were mediating their postreversal learning by means of their prereversal learning. Far from having to unlearn their responses from the prereversal trials, they were using their prereversal training to help them learn the postreversal discrimination; they learned something like: "pick the other one." However, although the performance of the four monkey species as a group (along with one species of gibbon) was in between prosimians and apes in the extent to which the level of prereversal training influenced their postreversal learning, this was not true for all species. Some species more closely resembled the prosimians (e.g., talapoin monkeys; *Cercopithecus talapoin*), but others performed as well as the great apes (e.g., rhesus, which outperformed many of the apes). Figure 4.2 summarizes these results (Rumbaugh, 1995, has added data but with the same pattern of results).

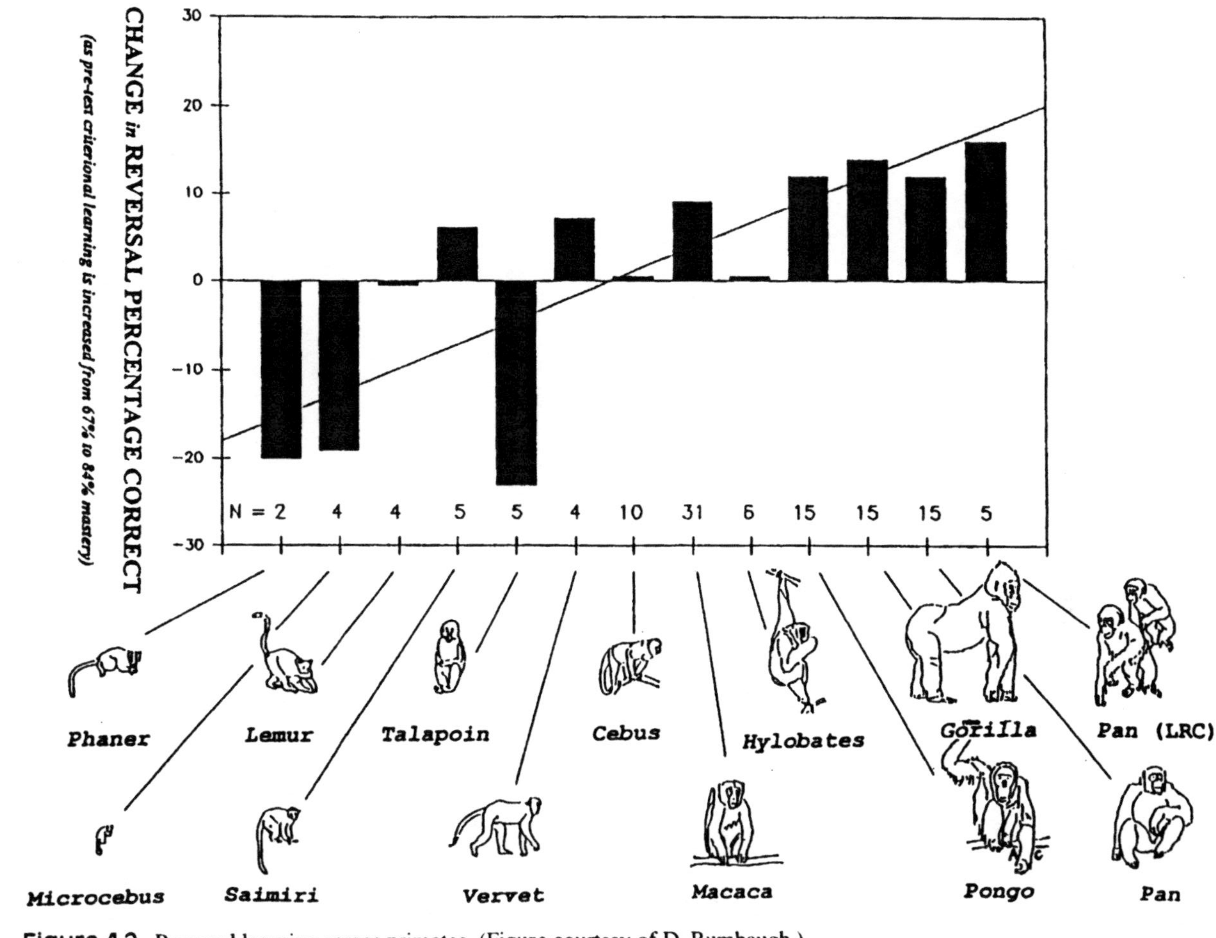

Figure 4.2. Reversal learning across primates. (Figure courtesy of D. Rumbaugh.)

Rumbaugh and Pate (1984a,b) also administered variants of the reversal paradigm to a number of primate species, such that a simple conditioning explanation would predict one outcome and a mediational (hypothesis-testing) account would predict another (the comparison also includes the data of Essock-Vitale, 1978). Their overall conclusion was that, in addition to traditional stimulus-response learning, great apes and some monkeys (i.e., rhesus) also engage in conceptually mediated learning (see their figure 31.5). This was also the original conclusion of Essock-Vitale (1978), who found that both rhesus macaques and apes engaged in conceptually mediated hypothesis testing, albeit in slightly different ways (see also Washburn & Rumbaugh, 1991b, for evidence that rhesus macaques engage in conceptually mediated learning in a similar task). A number of monkey species do not seem to engage in such mediated hypothesis testing (e.g., tufted capuchins; De Lillo & Visalberghi, 1994).

4.1.2 Cross-Modal Perception

A second phenomenon of simple discrimination learning that has led learning theorists to posit mediation processes is cross-modal transfer, again because such transfer seems to involve going beyond specific stimulus-response associations. Many species of primates show cross-modal transfer. One example is the study of Davenport and Rogers (1970), in which chimpanzees and orangutans were given a pair of objects to explore haptically while being prevented from perceiving them visually. They were then shown a visual stimulus that matched one or the other of the objects they had felt, and they were rewarded for choosing the felt object that matched the one they had seen. Both species performed at a very high level, even on trials in which they were presented with objects they had never before encountered. The authors note that the subjects' performance seemed in every way to be a process of active matching. For example, if by chance a subject was feeling the incorrect object as the visual stimulus appeared, it moved on to the second object and correctly chose it. If, on the other hand, the subject was by chance feeling the object that matched the visual sample when it appeared, it often chose that object without ever feeling the other one. Similar types of performance have also been observed for rhesus macaques (Jarvis & Ettlinger, 1978; Weiskrantz & Cowey, 1975) and pigtail macaques (Gunderson, 1983).

In his review of this research, Elliot (1977) concluded that all primate species investigated show essentially equivalent performance in tasks requiring cross-modal transfer between the haptic and visual modalities. It should be noted, however, that chimpanzees have also shown a number of other modes of transfer as well; for example, auditory to visual (Davenport, 1977), gustatory to visual (Premack, 1976), and even some transfer involving symbolic stimuli (Savage-Rumbaugh, Sevcik, & Hopkins, 1988; including one bonobo). No other species has been tested in these ways. Interestingly, human infants show similar forms of cross-modal recognition (both visual–haptic and visual–auditory) in the first few months of life (Meltzoff & Borton, 1979; Spelke, 1979).

We should note that results of cross-modal transfer studies employing learning (rather than recognition) paradigms have produced some curious results. For example, some studies had animals learn a discriminative stimulus over many trials in one modality and then presented the "same" stimulus in another modality (e.g., the same temporal pattern of light flashes in training trials and of auditory flashes in transfer trials).

Prosimians such as bushbabies (*Galago senegalensis*) have been found to transfer learned responses from vision to audition and vice versa (Ward, Silver, & Frank, 1976; Ward, Yehle & Doerflein, 1970). Surprisingly, however, most studies of cross-modal transfer in monkeys (both New World and Old World) have found negative results (see Meador et al., 1987, for a review). Some positive evidence has been found for one individual capuchin monkey (*Cebus albifrons*) (Blakeslee & Gunter, 1966), for some vervet monkeys (*Cercopithecus aethiops*) (Stepien & Cordeau, 1960), and for some rhesus macaques using some experimental procedures (Ettlinger, 1977). Chimpanzees also have not shown impressive and reliable abilities to transfer learning across modalities (Ettlinger & Jarvis, 1976). Although it seems likely that problems in defining what is "the same" stimulus across modalities may underlie these curious results, the basis for the difference in cross-modal functioning in the two different experimental paradigms is not known at this time. (See section 4.3.1 for some further results on primates failing to transfer some types of learning across perceptual modalities.)

4.1.3 Delayed Response

The third area of research that has led behaviorist researchers to invoke cognitive explanations is delayed response. We have already seen that both monkeys and apes have some skills in delayed responding when they must remember a spatial location (Tinklepaugh, 1928, 1932). In a systematic comparison across a number of primate genera, Harlow and colleagues (Harlow, 1932; Harlow, Uehling, & Maslow, 1932; Maslow & Harlow, 1932) presented delayed spatial choice problems to a number of primate species. In these problems, the subject sees a food reward hidden in one of two locations and then after some delay may search for it; across trials experimenters measure the length of delay subjects can tolerate and still remain accurate in their choices. Harlow concluded from his findings that great apes (one each of orangutan, gorilla, and chimpanzee) were best at remembering the food's location after long delays. Next were various Old World monkeys, including various species of baboons, mandrills, guenons, and macaques (and a few others), along with gibbons (a lesser ape). Next were various New World monkeys (mostly capuchins), whose performance was disrupted by moderately short delays. Finally, the species who could least well remember the location of the food were all prosimians, all from one or another species of lemur. Figure 4.3 summarizes Harlow's results.

In general, subsequent research has continued to find that prosimians are not skillful at delayed-response tasks (Jolly, 1964a, for lemurs and galagos). Also, despite the findings of Harlow and Bromer (1939) that under some conditions capuchin monkeys and spider monkeys may perform reasonably well on delay tasks, subsequent research has also found that New World monkeys are generally not as skillful as Old World monkeys (see Miles, 1957, who compared marmosets and rhesus, and French, 1959, who found poor performance in squirrel monkeys). The finding that apes are better than monkeys at delayed-response tasks is not as clear-cut, however. The graphs in Figure 4.3 show no large differences across these taxa. For example, at 60-second delays, apes ranged from 51% to 92% correct and monkeys ranged from 55% to 84% correct. At 120-second delays, two of the ape species are at the lower end of the monkey range; the only ape that performs especially well is the orangutan. Furthermore, in the study of Fischer and Kitchener (1965), which is sometimes cited in support of ape–monkey dif-

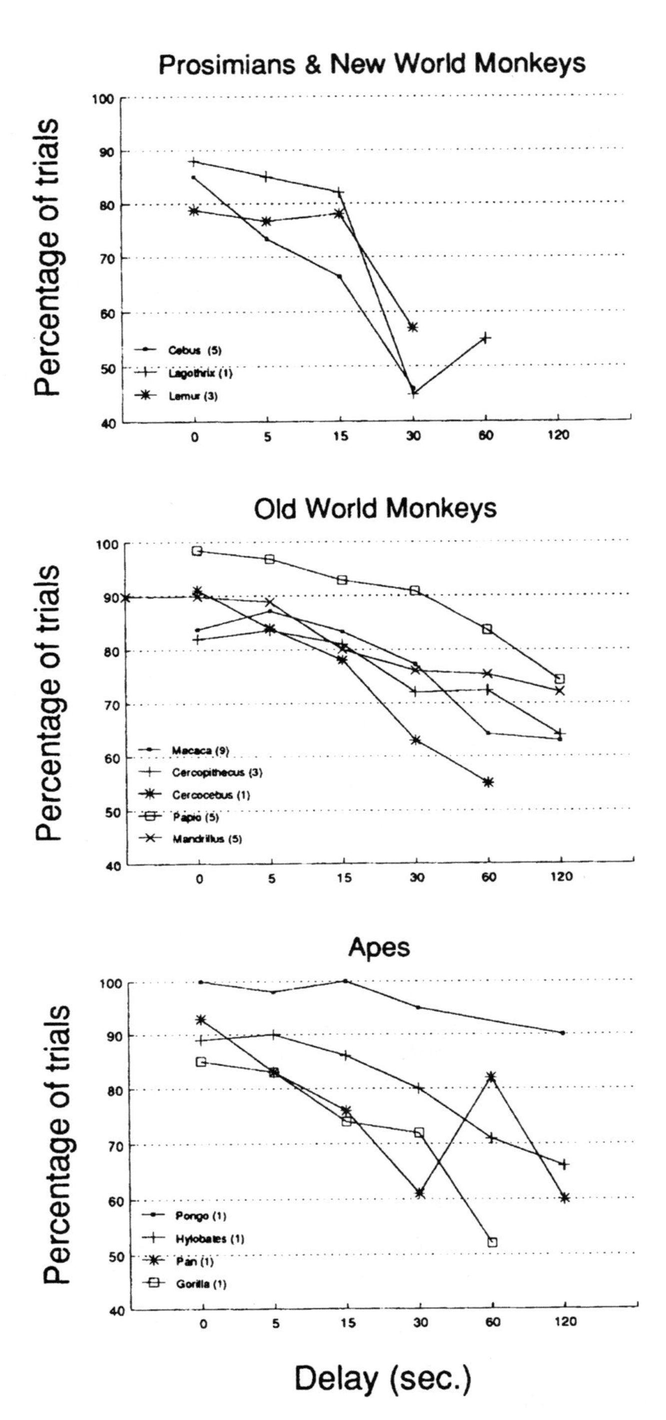

Figure 4.3. Spatial delayed response in primates. Bracketed numbers next to the species name indicate the number of subjects tested. SOURCE: Harlow et al. (1932), Maslow & Harlow (1932).

ferences, the longest delay period at which monkeys and apes may be compared is 16 seconds. The results at this delay were orangutans 91%, rhesus 85%, gorilla 82%, and marmoset 65%. The one conclusion to be drawn from the monkey–ape comparisons on delayed response is thus that orangutans as a species seem especially skillful.

As further tests of delayed responding, Harlow and colleagues, in these same studies, administered a series of three follow-up tasks (in which no great apes were tested). They found that the Old World monkeys displayed virtually no decrement in performance if (1) they were distracted by a human to another location during the delay; (2) they were presented with two delayed-response problems simultaneously; or (3) the containers were moved about (in sight and returned to their original locations) during the delay. The New World monkeys, on the other hand, experienced some decrements of performance in these complicating circumstances. The performance of the lemurs was completely disrupted by these variations of the task.

Many researchers believe that spatial memory may be a specialized skill, and so they have also investigated delayed responding in a variety of nonspatial tasks. For example, Yerkes and Yerkes (1928) had chimpanzees watch experimenters hide some food in one of four easily identifiable boxes. While the subject was out of the room, they moved all four boxes about spatially, and then brought the subject back into the room after delays of up to 30 minutes. Although they showed an initial location bias, once this was overcome, chimpanzees had little trouble with such delayed responding. Similar findings were reported by Berkson (1962) for gibbons and by Fischer and Kitchener (1965) for orangutans and gorillas. In a delayed match-to-sample paradigm, Weinstein (1941) had rhesus macaques observe a sample stimulus, which was then immediately removed. The subjects' task was to remember the sample so that they could pick out a match from a pair of presented alternatives some 15 seconds later. After many trials they were able to do this at an accuracy rate of 75%. Chimpanzees have been found to reach a similar level of accuracy in somewhat fewer trials (Finch, 1942), and capuchin monkeys have been found to be capable of delaying their response to the sample presentation by as long as four minutes without losing accuracy, even when the sample was presented for only 60 milliseconds (D'Amato & Worsham, 1972). Finally, Vauclair, Rollins, and Nadler (1983) had chimpanzees view a display panel consisting of a 3 × 3 or 4 × 4 matrix of lights. A sample pattern was presented by lighting some lights (e.g., a cross or a diagonal), which was then turned off. The subjects' task was to produce the same pattern by touching, and thus activating, the appropriate lights. They were able to do this quite well for most patterns (they had some trouble with diagonals), and a 5-second delay between presentation of the sample and opportunity to respond led to basically no decrements in performance. A similar study with rhesus macaques was reported by Medin (1969).

4.1.4 Comparison of Species

These three skills of simple discrimination learning—learning set formation, cross-modal transfer, and delayed response—are all skills possessed by some nonprimate animals. Learning set formation (of both the object discrimination and reversal varieties) is quite widespread among mammals and even pigeons (see Fobes & King, 1982; Passingham, 1981; Sutherland & MacKintosh, 1971 for reviews), and there is even some recent evidence that human infants as young as 9 months form learning sets as well

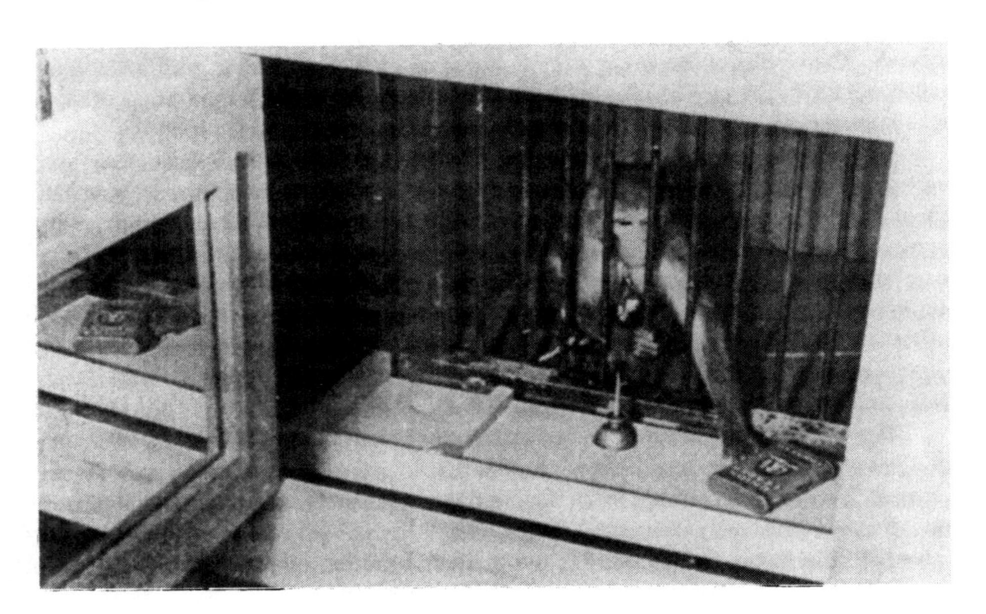

Figure 4.4. Monkey in Wisconsin General Testing Apparatus. (Photo courtesy of American Psychological Association.)

(Coldren & Colombo, 1994). Although solid data are lacking, there is at least some evidence for cross-modal transfer in nonprimates such as some species of rodents (see Ettlinger, 1977); human infants show cross-modal transfer in early infancy (Spelke, 1979). And delayed response has been demonstrated in pigeons, rats, and a number of other nonprimate animals as well (see Roitblat, 1987, for a review), with one study even showing that a species of squirrel was superior to a species of New World monkey (King, Flaningam & Rees, 1968); human infants show delayed response in the object permanence task at approximately 8 months of age (Piaget, 1954).

Within primates, there are species differences in the different tasks, but differences among larger taxa such as monkeys and apes are not robust. In object discrimination learning set formation, there are no apparent differences of performance across large taxonomic groups (i.e., prosimians, monkeys, and apes). In reversal learning, apes, on average, sometimes have better performance than monkeys, on average, but there are in all cases one or more monkey species who perform well within the ape range, and in ways indicative of conceptual mediation. Prosimians as a group do not do as well on reversal learning sets. In cross-modal recognition, there is no evidence of robust species differences among primates, including humans. In delayed-response tasks, there is some evidence that Old World monkeys and apes are better able to remember locations after long delays than are New World monkeys and prosimians; however, there is no evidence of a difference in this skill between monkeys and apes as taxa (although orangutans seem especially skillful as a species).

4.2 Natural Categories

In an attempt to provide animals with materials that are more ecologically valid, researchers have performed discrimination studies using more natural stimuli. The first study of this type was by Herrnstein and Loveland (1964), who employed a straightforward discrimination learning paradigm to demonstrate that pigeons could discriminate between pictures that had humans in them and pictures that did not. Because the images of humans varied widely from one picture to the next (individuals of different ages, genders, and dress, pictured in many different settings) the investigators argued that the pigeons had a "natural concept" of humans. Subsequent studies have found similar results for pigeons forming natural concepts such as bodies of water, trees, and fish (e.g., Herrnstein, 1979), and recent studies have shown even more complex skills of this type (Bhatt, Wasserman, Reynolds, & Knauss, 1988; Wasserman, Kiedinger & Bhatt, 1988). There are even some similar findings for bees categorizing flowers (Dukas & Waser, 1994). The central theoretical issue in this research is whether successful recognition of novel pictures of natural objects requires concept formation or may be accomplished by simpler processes of stimulus generalization (e.g., Chater & Heyes, 1994). As a result, the central methodological issue is how many training exemplars to use and how many training trials to conduct, given that the more training trials given and exemplars used, the greater the probability that animals could be generalizing to new exemplars in transfer tests on the basis of stimulus generalization rather than concept formation.

In the case of primates, there is reason to believe that in their natural environments they operate with some form of natural categories. First, in their social behavior there is some evidence that they categorize conspecifics (reviewed in section 7.3) and predators (reviewed in section 8.1) in a variety of ways. Second, in their foraging activities, they seem to understand something about types of food. For example, Milton (1981) reported that after the spider monkeys she was observing discovered a tree with fruit that had just become ripe, they often proceeded immediately to other trees with the same type of fruit, implying that the fruit trees in their environment were categorized on the basis of the particular type of fruit they bore. More systematic observations of this type were made by Menzel (1991), who found that when Japanese macaques found an experimentally provided akebi fruit they immediately stared upward from the provisioning site into the trees where akebi vines might be growing, they went to locations where akebi vines grew, and actually inspected or touched akebi vines, in all cases statistically more frequently than in two other experimental conditions (see section 2.1.1).

4.2.1 Concept Formation

Experimental studies with nonhuman primates have also found natural concepts. Following the lead of Herrnstein and colleagues, Schrier, Angarella, and Povar (1984) presented stumptail macaques with various pictures of humans in conjunction with pictures with no humans and trained them to choose the picture with the human in it; they did the same thing with pictures of monkeys. Within several hundred trials, subjects learned to respond correctly to humans and to monkeys, even to totally novel pictures, thus demonstrating the acquisition of these two concepts. The performance of the monkeys was not as impressive as that of the pigeons in Herrnstein's previous studies, however. A subsequent study with rhesus macaques (Schrier & Brady, 1987) found better evidence for the concept of humans, but this study used large sets of training exemplars, thus increasing the possibility that subjects might transfer by means of simple stimulus generalizations. Neiworth and Wright (1994) studied rhesus macaques using a different method, in which subjects were taught to respond either "same" or "different" to pairs of stimuli, and found categorization of a number of natural stimuli after several thousand trials.

D'Amato and van Sant (1988) questioned the "natural concepts" explanation for these results in both pigeons and monkeys. The problem is that subjects usually viewed several hundred different exemplars of the to-be-learned concept in a training phase before the relevant discrimination was demonstrated in transfer tests, which typically took several hundred more trials. D'Amato and van Sant pointed out that a subject could be generalizing to new instances in the transfer test not on the basis of an already existing natural concept, but rather on the basis of stimulus generalization from specific exemplars and their features shown in the training phase. Even something simple like the presence of a flesh-colored patch within the picture might be the controlling stimulus in these studies, or, perhaps, the presence of one or more of several features that are common in pictures of humans (e.g., eyes, limbs, flesh color, etc.). In a series of four experiments, D'Amato and van Sant (1988) attempted to address this problem by presenting capuchin monkeys (*C. apella*) with relatively few training trials containing few exemplars of humans. After several hundred training trials using from 20 to 100 exemplars, subjects showed transfer to new exemplars in less than 100 trials, demonstrating acquisition of the concept of human. The authors argue that with so few exemplars and training trials relative to previous studies and the relatively rapid transfer to new exemplars, there is much less chance that their subjects were responding in the transfer phase on the basis of simple stimulus generalization from the training phase.

Nevertheless, D'Amato and van Sant expressed skepticism that a "natural concepts" explanation was appropriate, either in their study or in others. Most important was their analysis of the errors their subjects made. In the Schrier and Brady (1987) study, stimuli were classified into "good" examples of humans (the human form occupied more than 50% of the surface area of the slide) and "poor" examples (the human form occupied less than 25% of the surface area of the slide). Subjects were correct on only 66% of the poor examples, but on 93% of the good examples, showing a fairly restricted concept of humans, if indeed they had a concept at all. In analyzing their own study in this way, D'Amato and van Sant found that subjects had particular difficulty with close-up photographs of human faces, and with some frequency misidentified pictures with reddish patches in them as human images (e.g., an open piece of watermelon). Both of these types of errors are ones that would be expected if the animals were learning the task by means of stimulus generalization.

In discussing their results, D'Amato and van Sant (1988) express the opinion that monkeys very likely do have a natural concept of humans in their everyday lives, but they also are worried that this will be difficult to demonstrate via the study of concepts using two-dimensional pictures and artificial training procedures. However, at least two recent studies have shown categorization by nonhuman primates without extensive training. Irle and Markowitsch (1987) studied three squirrel monkeys that had been previously trained to choose one member of a pair of pictures that did not match a sample ("non-match-to-sample"). Without any further training, the investigators changed the task so that the member of the choice pair that matched the sample no longer matched

Figure 4.5. Capuchin monkey. (Photo by J. Call.)

it identically, but only matched it in terms of conceptual class; for example, the sample might be a picture of one cat and one member of the choice pair might be a picture of a different cat. The subjects performed well from the beginning on this new task (as well as three adult human subjects).

The second study was reported by Fujita and Matsuzawa (1986) and centered on the categorization skills of a single language-trained chimpanzee named Ai. Ai was trained to press a key in order to see a picture. The picture stayed on as long as the key was pressed and went off as soon as the key was released. After the key was released, pressing it again within 10 seconds led to the same picture again, but after 10 seconds the picture was different. Some of the slides had humans in them and some did not. From the beginning and without specific training in the task, Ai showed a strong preference for the slides with humans in them, as evidenced by her choosing to view those slides for a longer period of time than the slides without humans in them. It should be pointed out that although there were no training trials in either of these two studies (thus satisfying one of D'Amato and Van Sant's 1988 concerns), the slides within categories shared many perceptual features and thus could have been grouped on the basis of some fairly simple form of stimulus generalization.

There is also some evidence that nonhuman primates operate with categories in the auditory modality, although these may be species-specific perceptual categories evolved for dealing with various biologically significant sounds. For example, in the wild a variety of primate species recognize such things as a particular class of alarm call, even though the calls come from different individuals and are made in different circumstances, so that they sound very different to the human ear (Cheney & Seyfarth, 1990b). In laboratory studies, Owren (1990a,b) found that vervet monkeys were able to differ-

Kingfisher	Bird	Animal
Kingfisher (+)	Robin (+)	Tiger (+)
Robin (-)	Tiger (-)	Lake (-)
Cardinal (-)	Horse (-)	Tree (-)
Kingfisher (+)	Gazelle (-)	Gazelle (+)
Bluejay (-)	Kingfisher (+)	House (-)
Pigeon (-)	Pigeon (-)	Mountain (-)
Kingfisher (+)	Elephant (-)	Bluejay (+)
Crow (-)	Sheep (-)	Landscape (-)

Figure 4.6. Samples of Roberts and Mazmanian stimuli in the three tasks. For each task, stimuli were always presented in pairs with one positive and one negative exemplar.

entiate between two types of alarm calls that were a part of their species-typical repertoire, even though each had multiple exemplars recorded from a variety of different vervet monkeys in the wild. Similar findings, using different methodologies, have been reported by Peterson, Wheeler, and Armstrong (1978) and Zoloth, Petersen, Beecher, Green, Marler, Moody and Stebbins (1979) for Japanese macaques, by Snowdon (1987) for pygmy marmosets, and by Masataka (1983) for Goeldi's monkeys (*Callimico goeldi*). The problem in all of these cases is simply that it is very likely that we are dealing here with categorical perception, not with conceptual categories that are learned and used in a flexible manner by individuals.

4.2.2 Levels of Abstraction

Roberts and Mazmanian (1988) pointed out a different problem with studies of natural categorization. The pictures used as negative exemplars would seem to be an important factor that none of the researchers has addressed; for example, the subject's task would seem to be different in the studies of the person concept if the negative exemplars were chimpanzees (fairly difficult to discriminate from humans) or a blank screen (very easy to discriminate). In all previous studies with both pigeons and monkeys, the negative exemplars were always pictures completely devoid of humans or whatever was the relevant concept. The fact that subjects can pick out pictures of humans from a variety of other pictures in such a study does not necessarily mean that they have a concept of person; it might mean that they have a concept of primate, or mammal, or animal. The abstraction level of the concept (if indeed it is a concept) is thus uncertain. In a series of four studies, therefore, Roberts and Mazmanian presented their subjects with different pictures of animals and arranged the training regime to specify concepts at different levels of abstraction (see figure 4.6). At the most concrete level, subjects were rewarded for choosing pictures of one species of bird (the common kingfisher) and not rewarded for choosing pictures of all other birds. At an intermediate level of abstraction, subjects were rewarded for choosing pictures of birds and not rewarded for choosing pictures of all other animals. At the most abstract level, they were rewarded for choosing pictures of animals and not rewarded for choosing pictures without animals. Subjects for the studies were pigeons, squirrel monkeys, and human adults.

Roberts and Mazmanian found that humans easily formed categories at all three levels of abstraction, but they were fastest with the middle-level concept of bird (and about 10 times faster than both nonhuman species). This is in accord with Rosch's (1978) theory that people form concepts easiest at the "basic level," the typically intermediate level of abstraction at which objects have the most number of "correlated attributes" (in this case, wings, beaks, feathers, etc.). Despite repeated testing in a number of different ways, however, pigeons and monkeys never exceeded chance performance on novel transfer probes at the level of bird. In a variety of different tests, the investigators showed that the easiest level for the concept formation of pigeons and monkeys was the most concrete level of kingfisher, with pigeons being particularly good at this level. The generality of these findings on the species level, with special reference to the subject's own species, was demonstrated by Yoshikubo (1985), who found that rhesus macaques were able to discriminate between different pictures of rhesus and Japanese macaques. Similarly, Fujita (1987) found that subjects from four different macaque species could discriminate pictures of individuals of their own species from pictures of individuals from other monkey species. After some methodological modifications, both species of animals in the Roberts and Mazmanian (1988) studies also learned to form the more abstract concept of animal, with monkeys being slightly faster than pigeons at this level.

On the species level, the basis for the animals' concept of kingfisher was quite easily identified, as this concrete level was composed of pictures with more common perceptual features than the more abstract levels. But further analysis of the animal pictures failed to identify any common set of features (e.g., eyes, limbs) that were consistently chosen by subjects. Whatever its featural bases, however, it is possible that the evolutionary foundation for the finding that monkeys and pigeons form categories at the level of species and animal may be simply that it is important for them to identify in some circumstances (1) particular species of animal that are prey or predator, and (2) that in general it is useful to be attentive whenever other animals are around. The intermediate, basic level has no very obvious functional value, except perhaps for humans for purposes of linguistic communication. It must be pointed out, however, that the artificiality of the stimuli used present a problem in drawing conclusions about animals' natural categories. Perhaps exemplars of real birds that fly and sing, for example, would make a difference in how subjects categorize animals.

4.2.3 Comparison of Species

Research on natural categories is clearly an important line of research on primate cognition. No research has been done with any prosimian species (and there is only one study with apes), but there is little reason to believe that they would behave differently from pigeons and monkeys. Humans form concepts relatively rapidly, presumably with the aid of linguistic symbols. Although there is some question about the artificiality of the stimuli used, the Roberts and Mazmanian study points to a potentially important difference between the natural categories of humans and nonhuman primates. It is possible that the basic level organization that seems to be so important in human cognition derives from the workings of language. It is likely that humans form many of their concepts around linguistic symbols, which have been created for purposes of communication, and that the intermediate level of abstraction characteristic of basic level concepts is particularly useful for that purpose. In this view of concept formation, it is possible that nonhuman animals group objects on the basis of particular perceptual or behavioral (rather than communicatory) functions, and thus that the basic level is not of great significance for them.

4.3 Relational Categories

In simple discrimination problems, including those that use natural stimuli (or pictures of natural stimuli), subjects learn to respond to a single stimulus or to a stimulus category at some level of abstraction. Discrimination learning of relational categories, on the other hand, involves concepts that can be learned only by comparing stimuli to one another and inducing some relational stimulus (e.g., "same as," "larger than"). The three most studied instances of relational concepts are (1) the identity relation as manifest in generalized match-to-sample problems, (2) the oddity relation as manifest in generalized oddity problems, and (3) the sameness–difference relation as manifest in generalized relation-matching problems. In all three cases, the subject is given some problems in training that can be solved by attending to a relation and is then given transfer tests that use completely different objects that can be seen as representing that same relation. If learning is relatively fast in the transfer phase, the inference is that the subject acquired a relational concept in the training phase and is now applying it in the transfer phase. If the learning is at the same basic rate in training and transfer (with some allowance for the formation of a learning set), the inference is that the subject has not learned a relational concept but is treating each new problem as a separate entity with its own particular stimulus characteristics.

Relational concepts are especially important in the study of primate cognition because they appear to be more complex than object-based concepts, relying as they do on comparisons among stimuli. Moreover, there have been some claims that only primates are capable of forming relational concepts (Thomas, 1986), and that only language-trained apes and humans are capable of forming certain kinds of relational concepts (Premack, 1983). In their natural environments, nonhuman primates seem to form relational categories quite readily in the social domain; this evidence will be reviewed in section 7.1. In the physical domain, the evidence comes almost totally from the laboratory.

4.3.1 Identity

In traditional match-to-sample problems, an individual is presented with a target visual stimulus (the sample) and two other stimuli, one of which is the same (matches it). The subject is rewarded for choosing the matching alternative. For example, the sample might be a circle and the two choices might be a circle and a square, with the circle always being rewarded. In repetitions across trials with these same stimuli, or closely related stimuli, traditional learning processes of stimulus–response learning and stimulus generalization may operate. However, in a generalized match-to-sample procedure, completely new stimuli are used in the transfer phase (e.g., a cross as sample and a cross and triangle as choices). If the animal immediately "knows what to do" and matches this new sample, the implication is that it learned initially that the task is to choose "the same" stimulus as the sample (see figure 4.7). Such behavior goes beyond any straightforward learning of stimulus characteristics, at whatever level of abstraction, and has thus been designated as an example of a relational concept (Thomas, 1980). Although there are claims that some birds and dolphins can acquire the relational concept of

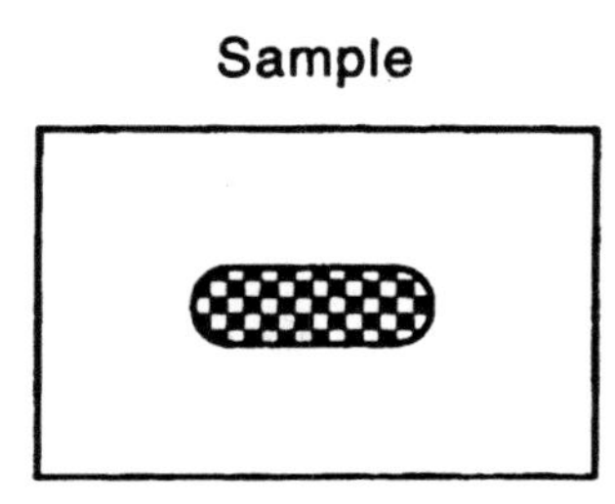

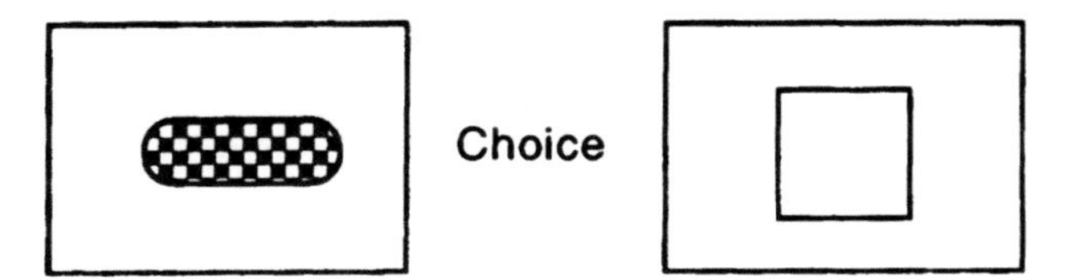

Figure 4.7. The identity problem (match to sample): choose one object from the choices available.

identity (e.g., Herman & Gordon, 1974; Wilson, Mackintosh, & Boakes, 1985; Zentall, Edwards, Moore, & Hogan, 1981), D'Amato, Salmon, and Colombo (1985) argue and present evidence that no nonprimate species has yet been shown to have acquired the relational concept of identity.

Acquisition of the concept of identity in generalized match-to-sample tasks has been reported in rhesus macaques and Japanese macaques (see D'Amato & Salmon, 1984, for a review). Of most interest in the current context is the fact that D'Amato, Salmon, Loukas, and Tomie (1986) directly compared capuchin monkeys with pigeons on the same generalized match-to-sample tasks. Although the monkeys and the pigeons both learned the initial task in the same number of training trials (more than 1000 trials), when the generalization test trials involving novel stimuli were presented, capuchins learned much more rapidly than pigeons. The conclusion was that pigeons were learning to respond to the particular perceptual dimensions of the stimuli presented, and so they did not generalize "identity" to new match-to-sample problems. The capuchin monkeys, on the other hand, were learning the concept of identity, and so they saw the new problems as the same as the previous problem even though the particular stimuli involved were different. In attempting to characterize the nature of the concept acquired, however, it should be noted that the capuchin monkeys of D'Amato et al. (1986) did not extend the identity relation to further transfer trials in which the match-to-sample problems were presented auditorially (see also D'Amato & Colombo, 1985). Similar findings involving a lack of transfer from form to color were reported by Jackson and Pegram (1970a,b) for rhesus macaques, and some other failures to transfer with facility to new types of stimuli were reported by Kojima (1979) and Fujita (1982) for Japanese macaques (although see Wright, Shyan & Jitsumori, 1990, for a different finding with rhesus macaques).

Studies with apes also provide evidence for an identity concept. Nissen, Blum, and Blum (1948) reported that all four of their chimpanzee subjects generalized their match-to-sample responses to new stimuli (forms) almost immediately, with only two sets of stimuli used during training (see also Finch, 1942; Riesen & Nissen, 1942). Oden,

Thompson, and Premack (1988) noted that the success-based reinforcement provided in the transfer phase of all three of these generalized match-to-sample studies with chimpanzees may have accounted for the fast transfer reported. They therefore used a similar procedure but did not differentially reward responses in the transfer phase. They trained four chimpanzees to above-chance responding to two different match-to-sample problems (using common objects such as cups and locks), taking an average of 800 trials to do so. All four of the subjects generalized immediately to new problems involving other objects, maintaining the same level of performance in the transfer trials as had been attained in the final trials of training. They also transferred successfully to very different types of objects (namely, swatches of cloth and bits of food). To our knowledge, there are no studies in which apes were asked to transfer across perceptual modalities.

In the only study we are aware of in which human children were tested in this same type of procedure, they, like the chimpanzees, generalized immediately to new match-to-sample problems using only two sets of stimuli in training (Weinstein, 1941). To our knowledge, no prosimian species has been tested in this paradigm.

4.3.2 Oddity

Testing for the oddity concept is in many ways similar to testing for the identity concept. In training, subjects are presented with trios of objects, two being identical and one being different in some way. The subject is rewarded for choosing the odd item of the trio for multiple trials involving several different types of trios with placement of items randomized (e.g., square-square-triangle and red-blue-red). After meeting some criterion in training, the subject is then given some completely new trios in a transfer phase (e.g., line-circle-line). If it chooses correctly in the earliest transfer trials with a particular trio (especially trial one), the inference is that it has understood the relational concept of "odd" (see figure 4.8). If, on the other hand, learning in the transfer trials takes some time, the inference is that the subject does not have a concept of "odd" but is learning each new problem anew based on the particular stimuli presented (with some improvement during the experiment due to learning set effects; see Meyer & Harlow, 1949). Although there are some claims that pigeons and rats can acquire the relational concept of identity, Thomas and Noble (1988) argue and present evidence that no nonprimate species has yet been shown to have acquired the relational concept of oddity, since in all cases learning set explanations are more probable (see also Roitblat, 1987).

There is good evidence that many primate species can acquire the concept of oddity as evidenced by their generalizing to new oddity problems successfully the first time they are presented (Meyer & Harlow, 1949; see King & Fobes, 1982, for a review of the classic literature). For example, Thomas and Boyd (1973) showed that after some initial experience with oddity problems, both capuchin monkeys and squirrel monkeys were able to solve novel oddity problems on the first transfer trial (see also Thomas & Frost, 1983). In a comparative study, Rumbaugh and McCormack (1967) found equivalent and good generalized oddity performance in gorillas, orangutans, chimpanzees, and macaques, but basically no generalization in the transfer phase for gibbons. Davis, Leary, Stevens and Thompson (1967) found some generalized oddity learning in the one prosimian species they tested (*Lemur catta*). Some primate species have also been able to solve dimension-abstracted oddity problems in which the odd object must be distinguished from four other alternatives that are not identical to one another (as in tradi-

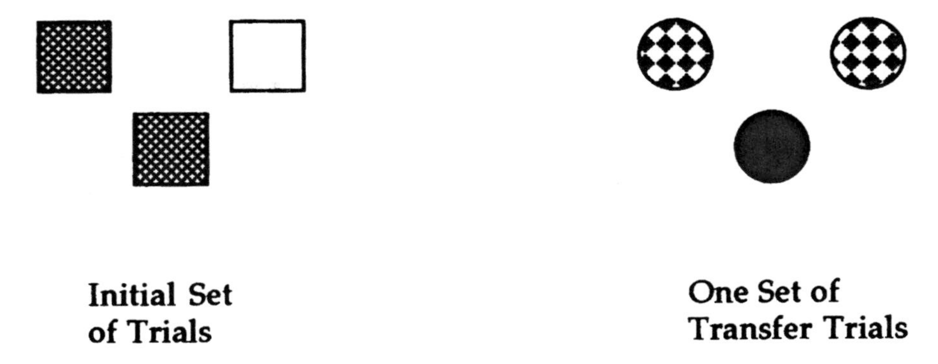

Figure 4.8. The oddity problem: choose odd object (out of three) on each trial.

tional oddity problems) but only resemble one another in some dimensions (e.g., objects of different shapes which are all red, as opposed to the odd object which is blue). Bernstein (1961) found that rhesus macaques, pigtail macaques, chimpanzees, and orangutans all learned to become proficient in this task (with methodological differences preventing a comparison among species). Thomas and Frost (1983) found that squirrel monkeys were able to solve this type of problem as well.

In a direct comparison of rhesus macaques, chimpanzees, raccoons, and cats, Strong and Hedges (1966) found that the raccoons and cats did not generalize oddity in the number of trials given, whereas both species of primate did, with the chimpanzees generalizing a bit more quickly than the monkeys. Human children have been presented with oddity problems in a number of studies and generally perform well in the earliest trials of transfer (e.g., Lipsett & Serunian, 1963). Premack (1978) and Thomas and Frost (1983) argued, based on these and other findings, that no nonprimate animals have acquired a concept of oddity.

4.3.3 Sameness–Difference

In contrast to the studies of identity in which an individual simply learns to match to sample with novel materials over time, some studies have investigated primate concepts of sameness–difference by presenting pairs of objects simultaneously and asking subjects to indicate whether they are the "same" or "different" (see figure 4.9). Wright, Santiago, and Sands (1984) employed such a procedure with rhesus macaques, training them to move a lever to the right when the pair of pictures of objects were the same and to the left when they were different. Subjects were then presented with novel pairs of items in 100 trials of a transfer test. No subject was different from chance. After more training, transfer was found in an additional set of 100 transfer trials. In a similar study, Fujita (1983) rewarded Japanese macaques for pressing a lever when a pair of stimuli matched in color and did not reward them for pressing the lever when they did not match in color. Only two colors were used in these training trials. Three of four subjects successfully transferred the correct discriminative responding to two new colors. King and Fobes (1975) found similar results with capuchin monkeys. Using similar procedures, Robinson (1955, 1960) found acquisition of a concept of sameness–difference in

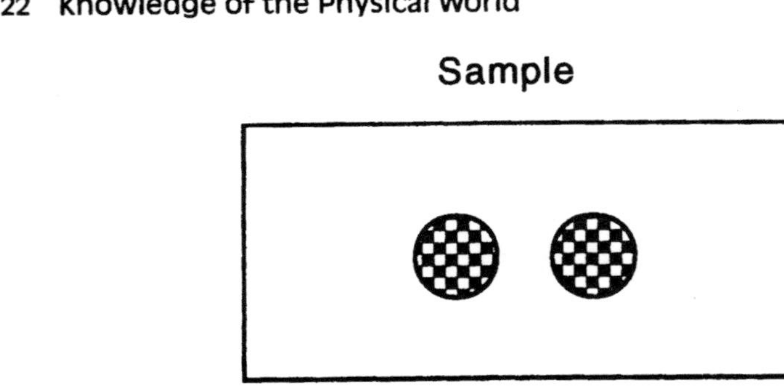

Figure 4.9. Relation matching task (match to sample): Choose one pair from the choices available.

chimpanzees (see also McClure & Helland, 1979), and King (1973) found this concept in both chimpanzees and orangutans. Dasser (1988b) found that longtail macaques were able to correctly identify pictures of conspecific mother–offspring relations when the comparison was other pairs of conspecifics, based on training with one exemplar of this relation; Dasser (1988a) found something similar for the sibling relation (see section 7.1.4 for more details of these experiments).

Premack (1983) argued that identity, oddity, and sameness–difference tasks as traditionally administered do not require the kind of relational concepts that investigators have claimed. Because the matching takes place across trials in all of these tasks, he claimed that "the animal simply reacts to whether it has experienced the item before. Old/new or familiar/unfamiliar would be better tags for this case than same/different" (p. 354). Instead, he advocated use of a generalized match-to-sample procedure in which the matching involves the relations between items. Solving this type of problem, which may be called a relation-matching problem, requires the subject to understand what Premack calls second-order relations; for example, noting that a sameness relation holds between the relation of X to X' and Y to Y'. Following this line of reasoning, Premack (1983) presented chimpanzees with a sample pair of stimuli that either matched (AA pairs, such as two apples) or that did not match (CD pairs, such as a pear and an orange). Their task was to pick which of two alternatives matched the relation exemplified in the sample: either a pair of new items that matched (BB pair, such as two bananas) or a pair of new items that did not match (an EF pair, such as a plum and a grape). Thus, when the sample was AA, the subject was to choose BB (rather than EF) because the relation among items in both cases is one of "sameness"; if the sample was CD, the subject should choose EF (rather than BB) because the relation among items in the two cases was one of "difference." What Premack reported was that the chimpanzees who had undergone several years of training with a languagelike system of communication (in which paying attention to the matching and nonmatching visual relations among plastic tokens was an integral part, and in which they had learned plastic tokens standing for "same" and "different") solved such tasks, whereas chimpanzees who had not undergone such training did not solve the task. He argued that the learning of a symbolic code such as language fundamentally changed the nature of the cognitive representations used by chimpanzees by providing them with an abstract propositional code (rather than a concrete imaginal code) in terms of which they might interpret their experience.

However, two sets of studies make Premack's interpretation of his results problematic. First, Premack apparently overlooked a previous study of a generalized relation-matching in chimpanzees with no language training in which very good performance was found (Smith, King, Witt, & Rickel, 1975). Furthermore, in a subsequent study, Oden, Thompson, and Premack (1990) used a different procedure and found that four infant chimpanzees (around 1 year of age with no language training) also engaged in the matching of relations. What they did was to simply present the subject with a sample stimulus that consisted of a pair of objects mounted on a small board; the pair could either match (AA) or not match (CD). Subjects could play with this sample as desired, and their play time was recorded. They were then presented with two test pairs of objects, also mounted on board, that they might play with; one was a matching pair (BB) and one was a nonmatching pair (EF). What was found was that subjects' initial play with the sample affected their handling time with the new test pairs. If subjects had played with the sample pair that matched (AA), they were no longer interested in the matching relation and so played more with the nonmatching test pair (EF); if they had played with the nonmatching sample (CD), they played more with the matching test pair (BB). The conclusion of these investigators was that chimpanzees can understand relations among relations, even if they do not always show this competence in tasks in which they must actively choose stimuli. The modified conclusion of these authors was that while chimpanzees understand second-order relations, language training helps them to incorporate this into their instrumental behavior. This modified conclusion has recently been modified further by Thompson, Oden, & Boysen (in press), who found that chimpanzees who had experience with the use of tokens for representing abstract quantitative relations were able to actively match realtions in a match-to-sample task.

The other conflicting study, also overlooked by Premack, was reported by Burdyn and Thomas (1984), in which a species of monkey with no language training performed skillfully on a generalized relation-matching problem (see figure 4.10). Adult squirrel monkeys were given a choice of a pair of two identical objects (AA) or a pair of two different objects (CD). The investigators trained the subjects to choose the pair of identical objects (AA) in the presence of one arbitrary symbol (a triangle), but to choose the pair of different objects (CD) in the presence of a different arbitrary symbol (a heptagon). In subsequent generalization tests using novel pairs of objects, the squirrel monkeys showed that they understood the significance of the arbitrary symbols for "same" and "different" [in much the same manner that Premack's (1983), language-trained chimpanzees had understood them in their language training]. The monkeys were even able to match the relation indicated by the symbol when there was a significant delay (16 seconds) between presentation of the symbol and the choices. This finding illus-

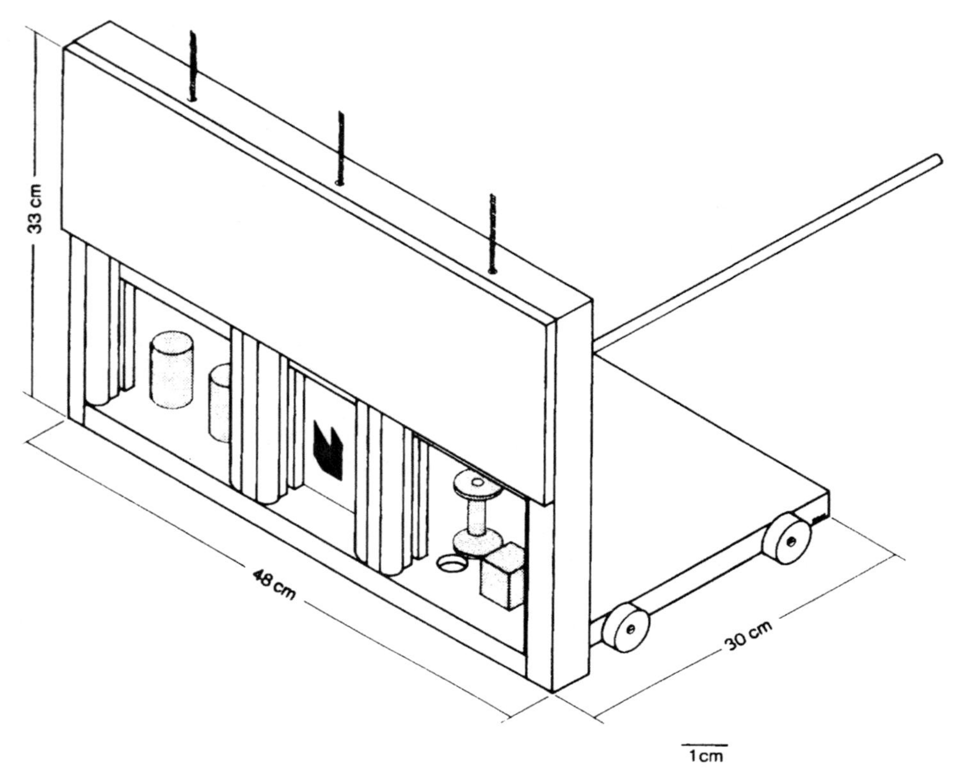

Figure 4.10. Experimental apparatus used by Burdyn and Thomas (1984) to test conditional discrimination in squirrel monkeys. (Figure courtesy of R.K. Thomas with permission from the American Psychological Association.)

trates not only that abstract relational concepts (second-order concepts) are not the exclusive province of language-trained chimpanzees, but that they are not the exclusive province of chimpanzees at all.

As a different example of a chimpanzee forming relational concepts, Gillan, Premack, and Woodruff (1981) presented the language-trained chimpanzee Sarah (Premack, 1976) with pairs of objects that had various relations; Sarah's job was to identify another pair that had an analogous relation. For example, in "figural" problems, Sarah was presented with an odd shape with a dot on it and that same shape without the dot; she was then presented with another shape with a dot and had to choose from a pair of alternatives that same shape without the dot (i.e., the analogous relation of two shapes with and without a dot). In so-called "conceptual" problems, Sarah was presented with household items with which she was familiar and asked to draw analogies, for example, between a key and lock as compared with a can opener and can. On figural items Sarah performed correctly about three-quarters of the time, and on conceptual items she was correct at a slightly higher rate. Having ruled out various possible alternative explanations, the investigators concluded that Sarah was able to understand the relation in the first pair of stimuli at a level of abstraction sufficient to allow her to identify it in subsequent stimulus pairs, both perceptually and conceptually. In this interpretation, these problems require the understanding of second-order relations among a wider array of first-order relations than previously tested.

4.3.4 Comparison of Species

There is currently controversy over which animal species are capable of understanding relational categories. This controversy is based partly on disagreements over the definition of relational category and partly on disagreements over the criteria that should be used to identify the use of relational categories in particular types of experiments. Thus, some researchers believe that some birds and rats have demonstrated relational concepts in generalized match-to-sample and oddity training studies (e.g., Nakagawa, 1993; Wilson et al., 1985; Zentall et al., 1981; see also Pepperberg, 1987). However, other researchers have argued that the performance of these nonprimates is best explained in terms of learning set formation applied to transfer trials, since transfer for these species is never on trial one (e.g., D'Amato & Salmon, 1984; Thomas, 1986) or else all of the appropriate controls have not been employed (Thomas, 1994). These more critical investigators have thus argued that the understanding and formation of relational categories may be an exclusively primate cognitive achievement (e.g., Thomas, 1992a). The major conflicting evidence was reported by Pepperberg (1987), who found that her language-trained African Grey parrot performed well on a language-based relational categorization task (essentially, learning labels for relations). The task was somewhat differently structured from those administered to primates, however, making direct comparison difficult (not to mention the difficulties in comparing such widely divergent animal species). In all, given the difficulties of interpreting the performance of individuals over time in traditional identity and oddity tasks, it would be useful to present nonprimate mammals with relation-matching problems to see if they are able to learn and generalize appropriately in these tasks.

Within primates, there do not seem to be any major differences among species in the formation of relational concepts (although the data for prosimians are mostly lacking). Many species of monkeys (mostly rhesus macaques, squirrel monkeys, and capuchin monkeys) and apes (mostly chimpanzees) have performed skillfully in generalized identity, oddity, and relation-matching tasks. Human children are also skillful at forming relational concepts, although the process has not been studied with methods precisely comparable to those employed with nonhuman primates (see Gentner, Rattermann, Markman and Kotovsky, 1995; Goswami, 1991).

4.4 Classification

Primates in their natural environments exhibit a number of behaviors that indicate a kind of natural classification of objects; for example, they gather particular types of food items or particular plant materials for making nests. How one tests the nature of these skills is a difficult issue. In the studies of natural and relational categories already reviewed, subjects were only required to respond yes or no to the question of whether a single stimulus array was similar to those that had been rewarded in the past—after extensive training in how to do this, with categorization being inferred later by experimenters examining responses over trials. Another way of testing the ability of organisms to categorize objects is to present them with many objects simultaneously and ask them to physically sort the objects into groups on the basis of their similarities and differences: what might be called classification. This may be a more demanding test of categorization skills because in object sorting, subjects are required to coordinate both the similarities and differences of multiple objects simultaneously and then to actually manipulate the objects in line with that coordinated understanding.

The paradigms in which sorting tasks have been presented to primates vary mainly in the training animals have been given, varying from intensive discrimination training to no training, and the constraints that have been placed on how the objects must be sorted. Almost all of the tasks use human objects and artifacts, and sometimes even abstract shapes that have no functional significance for primates. It must also be noted that some animals that have been asked to sort objects were previously raised in many humanlike ways around human objects, and some were even trained in humanlike linguistic symbols, whose appropriate use requires the categorization of referents.

4.4.1 Sorting after Formal Training

A number of studies have used the discrimination learning paradigm to actively train primate subjects to place objects in particular locations for food rewards. For example, Weinstein (1945) trained a rhesus macaque to designate objects of specific colors when presented with a sample of that color (e.g., to push red objects off a tray when presented with a red triangle as a sample). He also taught the monkey to pick out red objects in response to one discriminative stimulus, to pick out blue objects in response to another discriminative stimulus, and not responding when the choice array contained no objects of the appropriate color. Weinstein did not have the subject attempt to sort objects into more than one group.

Garcha and Ettlinger (1979) rewarded animals for sorting objects into three groups. Subjects were five chimpanzees, six rhesus macaques, and three capuchin monkeys. Each subject was seated in front of an apparatus consisting of three clear jars (each with a reward tube emerging from under it). The nine wooden objects to be classified varied systematically in shape, color, and size. For training, subjects were given the objects successively in three sets of three. The first object placed in a jar determined what was "correct" for that jar, that is, to be counted as correct (and so rewarded from the appropriate tube), subsequent objects placed in that jar had to resemble the first object in one of the three dimensions. Results were that none of the monkeys sorted objects at an above-chance level within 100 trials. One of the rhesus macaques and two of the capuchin monkeys did perform at above-chance levels after 880–2418 trials. On the other hand, four of the five chimpanzees performed sorting at above-chance levels, 2 within the first 100 trials and 2 within the first 650 trials. In a subsequent test, two of the chimpanzees learned to sort a new set of objects either by size or by color on different occasions (although they had great trouble learning to sort by shape).

Braggio, Hall, Buchanan, and Nadler (1979) tested six juvenile chimpanzees on a multiple classification task involving a 2 × 2 matrix of objects with one cell missing; for example, the top row of the matrix might contain a large circle on the left and a small circle on the right and the bottom row a large square on the left and a blank on the right. Success at this task is supposed to measure the ability to classify objects on the basis of two dimensions simultaneously (e.g., shape and size). The chimpanzees learned to perform the task accurately and reasonably quickly, but, as Doré and Dumas (1987) point out, the choices from which the subjects chose their response were three objects identical to those already in the matrix (e.g., a large and small circle and a large square) plus the one that correctly completed the matrix. Subjects might have been choosing their response, therefore, based only on a simple strategy in which they chose from the response array the item that was not in the sample matrix.

Savage-Rumbaugh, Rumbaugh, Smith, & Lawson (1980) taught two language-trained chimpanzees, Sherman and Austin, to sort various objects into what humans would identify as food and tools by placing them in predesignated locations (see figure 4.11). These two categories are generally considered to be superordinate categories, defined functionally by what can be done with them, and thus the members within each category were different from one another in terms of perceptual form (although it is possible that the materials of which food and tools are composed differ from one another systematically). The two subjects mastered this task fairly quickly, classifying novel food and tool items correctly almost all the time. The investigators then had the two subjects sort in this same way plastic tokens (lexigrams) that had previously been learned as "words" for these same individual food and tool items. The crucial test for the issue of conceptual representation was when the investigators presented Sherman and Austin with novel lexigrams that had never before been sorted, nor had their referent food or tool items ever before been sorted. They had no trouble sorting the novel lexigrams. Savage-Rumbaugh et al. argued that the chimpanzees in this study were mentally representing the referents of the novel lexigrams and categorizing those referents as a basis for their choice of where to place the plastic tokens.

4.4.2 Sorting after Informal Training

Other chimpanzees with more informal training have also learned to become proficient sorters of objects. Savage-Rumbaugh (1986) found that Sherman and Austin, after some informal training with demonstrations and social praise, learned to sort both natural objects and more artificial stimuli such as those used by Garcha and Ettlinger (see also Yerkes & Petrunkevitch, 1925). Premack (1976) reported similar results for his chimpanzees Peony and Elizabeth, with Peony even changing criteria across trials with the same set of objects. Spontaneous classification of different items according to certain morphological features such as material, color, shape, or size has also been reported in a human-raised bonobo (Savage-Rumbaugh, 1986), a human-raised gorilla (Patterson & Linden, 1981), and a human-raised orangutan (Miles, 1990), all of whom had only informal instruction in classification.

Matsuzawa (1990) presented the language-trained chimpanzee Sarah and some human children with several objects such as small bells and blocks and two plates to sort them onto (see Figure 4.12). For example, in some trials subjects were presented with two identical bells and two identical toy cars. In other trials, however, subjects were presented with three bells and only one car. In some trials, subjects were presented with plates already containing one object each, and they were to place a third object with the one it was like. Overall, in all of these and similar tasks, Sarah sorted the objects correctly the overwhelming majority of the time, and indeed her performance was better than the performance of the 1- and 2-year-old children and equal to

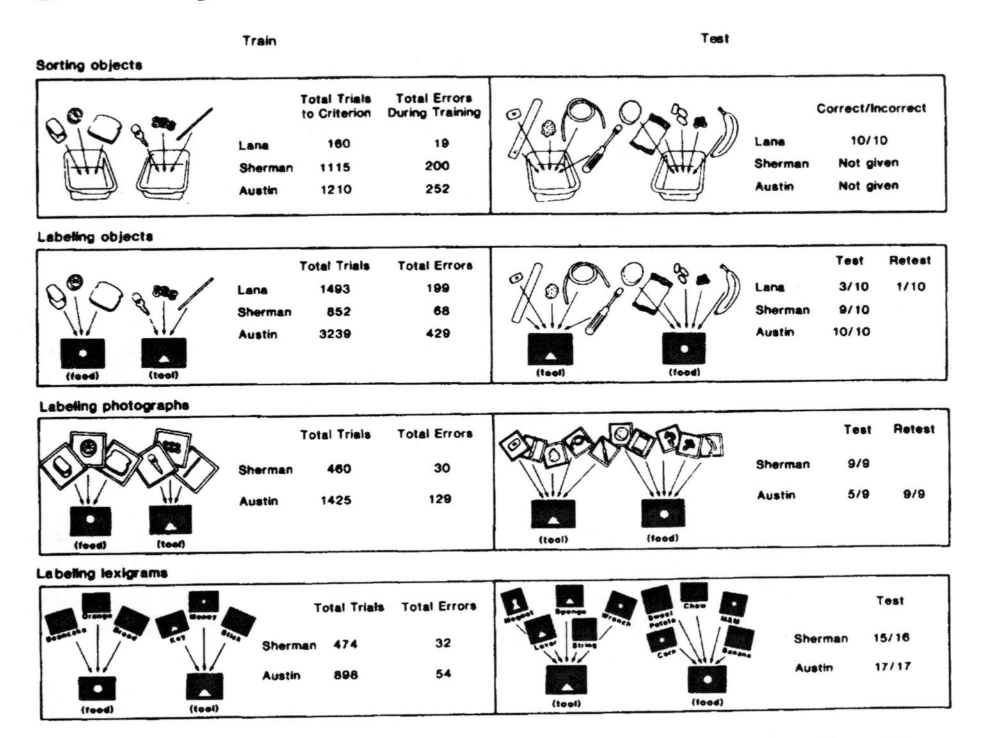

Figure 4.11. Diagram of the Savage-Rumbaugh et al. (1980) study. (Reprinted from *Science* with permission from the American Association for the Advancement of Science.)

that of the 3- and 4-year-old children. Tanaka (1995b) used a similar methodology with five chimpanzees and replicated many of these results. In addition he found that chimpanzees would spontaneously group objects that had been used together in their previous manipulative experience, for example, bottles and their tops. In some cases subjects also grouped objects with which they were familiar versus those that were novel to them.

Undoubtedly the most impressive demonstration of classification behavior by a nonhuman animal was reported by Hayes and Nissen (1971). They reported that their home-raised chimpanzee Viki began to spontaneously sort toys on her own at 3 years of age. They encouraged this behavior and began supplying her with two plates on which to place the objects belonging to the different classes. At age 5 they gave her a series of more formal tests. In these tests Viki was given a total of 13 different sets of objects, each of which could be divided into two groups. In some cases the two potential groups differed widely from one another, but the objects within a group were identical (e.g., matches versus safety pins). In other cases the two potential groups differed widely from one another, but there was some variability within groups (e.g., different kinds of buttons versus different kinds of screws). In still other cases the two potential groups differed from one another in one dimension only (e.g., short versus long nails, black versus white buttons). Overall, in a total of 200 object placements, with no samples given by humans to initiate the process, Viki made only 20 errors. In a final test

Figure 4.12. The chimpanzee Sarah sorting objects. (Photos courtesy of T. Matsuzawa.)

of this type, Viki sorted a pile of objects into nuts, bolts, nails, screws, washers, and paper clips into the six sections of a muffin pan without error. It should be noted that when presented with objects that could be classified either on the basis of form or function (wood and metal eating and writing instruments), Viki invariably classified them on the basis of their perceptual form (often on the basis of the material of which they were composed).

Hayes and Nissen (1971) also presented Viki with sets of objects that could be grouped in more than one way. On a number of occasions they gave her a pile of objects (from 12 to 80) and two bowls to sort them into. After she sorted in one way, she was simply given the objects again to see if she would sort them in a different way. If she did not, the experimenter would put one sample object in each bowl illustrating the new classification desired. Of eight object sets presented, Viki sorted six of them in more than one way: two in response to the samples only, three spontaneously but on the trial immediately following one in which she had done so in response to a sample, and one completely spontaneously. The one totally spontaneous sorting was Viki's classification of her button collection in three different ways: white–black, large–small, and round–square. These investigators argued that this flexibility of sorting showed that Viki could adopt when needed the "abstract attitude" characteristic of human classification skills.

4.4.3 Sorting without Training

Given the sorting skills of apes with some type of training, it is surprising that Mathieu (1982) found that when her nursery-raised chimpanzees were simply given human objects to sort freely, with no training or rewards or preestablished sorting locations, they did not spontaneously create identifiable groups. When she then trained her four subjects to sort, they did so with some skill, however. This study thus raises the question of whether nonhuman primates that have not been trained to sort objects (either formally in a discrimination learning paradigm or more informally through various play and communication activities) have the ability to classify them on the basis of their perceptual or functional features.

In an effort to answer this question, Spinozzi and Natale (1989a) adopted the method used by Langer (1980) to study the spontaneous classification behavior of human infants (see also McClure & Culbertson, 1977; Rensch, 1973). Following this method, they simply gave subjects a set of logically structured objects and observed their free play with them, including both the sequence in which they manipulated objects and any spatial arrangements of objects they made (see figure 4.13). Subjects were two longtail macaques and three capuchin monkeys. On any one trial, a subject was presented with 6 objects from a set of 16 possibilities: 4 different forms (cups, rings, crosses, and sticks) each realized in 4 different materials (wood, plexiglas, rubber, and wood covered by plastic quartz). In the additive condition, subjects were presented with two groups of identical objects (three of each) that differed in only one way (e.g., three wood cups and three wood crosses, or three plexiglas sticks and three rubber sticks). In the disjoint condition, subjects were presented with two groups of three objects that differed in both dimensions (e.g., three wood cups and three rubber sticks). In the multiplicative condition, no two objects were identical but rather they varied systematically from one another on the two dimensions in a multiplicative fashion (e.g., a wood and rubber cross,

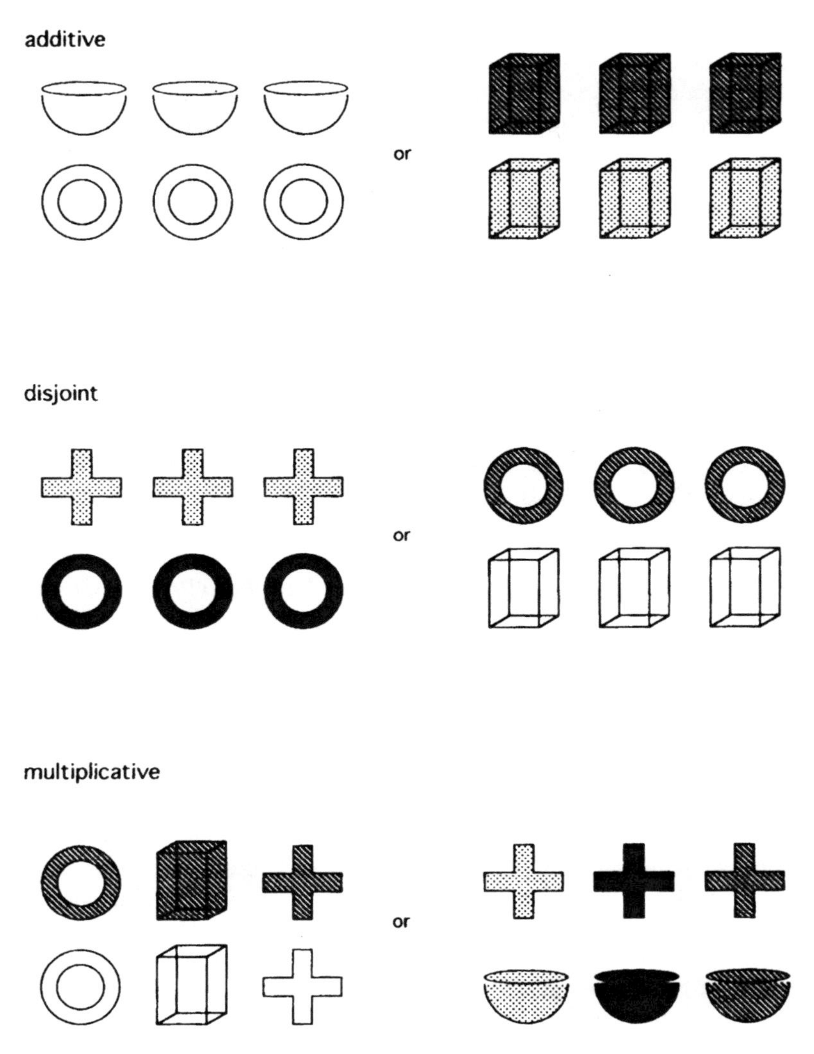

Figure 4.13. Sample of possible combinations of stimuli employed by Spinozzi and Natale (1989) representing three different levels of classification: additive, disjoint, and multiplicative.

a wood and rubber ring, and a wood and rubber stick). Subjects were given credit for composing a group if they placed objects in spatial proximity to one another, (within 10 cm of each other and 20 cm away from other objects), including when they picked them up and held them together in their arms. The order in which they manipulated objects that composed a group was also noted.

Spinozzi and Natale found that the two longtail macaques composed identifiable groups in all three conditions, but almost all of them were composed of only two objects. In the multiplicative condition, where there was a choice of criteria on which to classify objects, the macaques invariably chose the material (not the form). The capuchin monkeys also composed groups in all three conditions, and by the oldest age tested (48 months) had as many three-object groupings as two-object groupings. In the multiplicative condition, some individuals sorted by form and some by material. None of the subjects was reported to form two identifiable groups of objects simultaneously. None of the subjects was reported to classify the objects in the multiplicative condition in more than one way.

Spinozzi (1993) replicated this experimental procedure with five chimpanzees (the one difference being that the chimpanzees got four different colors instead of four different materials—a potentially important difference). The chimpanzees ranged in age from 1 to 4 years, and one chimpanzee was tested at three ages. No classificatory behavior was found at the age of one year. At 2 years of age, there was classification in all three conditions, but only with two objects at a time. At ages 3 and 4, subjects began forming consistent three-object groups. Of particular note is the fact that one subject at age 4 also engaged in what these investigators called second-order classification; that is, composing two multiple-object groupings, each of which was consistently composed (this same subject had not done this when tested earlier at age 3). This is behavior characteristic of human children at 18 months of age (Langer, 1980, 1986). None of the chimpanzees classified the objects in the multiplicative condition in more than one way. In terms of the actual types of spatial groupings involved, Potì (1996) reported that one chimpanzee in these sessions formed vertical stacks of objects (see also Köhler, 1925; Matsuzawa, 1990); however, no chimpanzees constructed the other types of multiple-object constructions characteristic of the object manipulations of human children (e.g., containment relations and various spatial alignments such as "bridges").

4.4.4 Comparison of Species

Beyond such specific activities as food gathering and nest building, physically sorting objects is not something that most animals have a need to do. Indeed, there are no studies to our knowledge of the object sorting skills of any nonprimate animal. Human children, on the other hand, begin showing some basic skills of classification by spontaneously sorting objects around 1 to 2 years of age, and they actively sort objects to adult specifications at 3 to 4 years of age (Langer, 1980, 1986; Sugarman, 1983).

With regard to nonhuman primates, classification in the form of object sorting is a skill that may differ between monkeys and apes, although the evidence is far from overwhelming. When monkeys (capuchin and longtail) are given human objects to physically manipulate, they often make groups of identical or similar objects (often by picking them up in order). They do this only for single groups, mostly composed of a single pair or trio of objects, however, and it is difficult to train them to sort objects into places designated by humans. Chimpanzees in their spontaneous manipulations also handle pairs and sometimes trios of similar objects in succession and place them in spatial proximity to one another (often by picking them up in order). Unlike any monkey, however, one chimpanzee on several occasions composed two coherent groups simultaneously. Without doubt, however, the most impressive classification performances have come from human-raised apes that have been exposed to many human artifacts and to humans manipulating and classifying the objects in various ways. These apes seem to take to object sorting quite readily, with one chimpanzee composing six coherent groups of objects on one occasion, and on another occasion sorting the same group of objects in three different ways within a single session. What monkeys would do if they were raised and trained with human artifacts in some of these same ways is not known at this time. It is also important to note in making a general monkey–ape comparison that chimpanzees, the only natural tool users, are the only apes that have been observed in classification tasks outside of human influence and training, and that some monkey species that might be expected to perform well (e.g., baboons) have never been tested for classification skills.

4.5 What Primates Know about Features and Categories

The ability to make sense of the perceptual world by discriminating features and categorizing phenomena is a basic cognitive capacity of many animal species. A wide variety of animal species demonstrate this capacity both in their daily lives in the wild and in discrimination learning experiments in which they are trained to make human-specified discriminations and to form human-specified categories. Understanding not only "where" things are, but also "what" they are categorically is of clear evolutionary significance to most animal species in numerous ways.

In discrimination learning experiments, primates and other animals learn not only particular features of stimulus materials, but they also learn something about the "rules" governing the kinds of problems with which they are presented across trials (learning sets). This implies to many investigators some processes of conceptual mediation. Primates and other animals also can recognize objects cross-modally and deal with discrimination learning problems when they must delay their responses. These skills also imply something more than the ability to react to particular stimuli and lead once again to issues of cognitive representation. Although there have been some claims that apes engage in conceptual mediation, whereas monkeys rely on perception only, recent research has shown that when monkeys are given the chance, they too perform in ways consistent with conceptual mediation (e.g., Essock-Vitale, 1978; Washburn & Rumbaugh, 1991b). Many of these same skills are shown by human infants (e.g., Coldren & Colombo, 1994; Piaget, 1954; Spelke, 1979).

Primates and other animals also seem to employ cognitive processes in discrimination learning experiments in which the task is to form categories of objects. The idea here is that if organisms react not just to particular stimuli but to categories of stimuli, once more they are going beyond the information given to perception. There are issues about whether the categories formed in these experiments are perceptual categories, based on the reliable presence of particular perceptual features across category exemplars, or whether the categories are indeed conceptually based (see Tomonaga, 1995, for the argument that some primate categories may be perceptually based). In either case, primates do not seem to have special skills relative to other animals in this domain. It is noteworthy that the natural categories of primates seem to be, like those of pigeons, on the concrete species level (e.g., crow) and the generic level (e.g., predator), not on the basic level (e.g., bird) which seems to be so important in human linguistically based cognitive processes. Humans show evidence of natural categories of this type in infancy (e.g., Resnick, 1989).

In addition to these basic processes of object categorization, primates also categorize relations. They come to understand abstract relations among stimuli over time as new stimuli are presented to them in generalized identity and oddity experiments, and

Figure 4.14. The chimpanzee Ai engaged in a categorization task. (Photo courtesy of T. Matsuzawa.)

they also identify and match the relations between objects when different pairs of objects with different relations are presented. Some theorists have argued that only primates are able to form relational categories (Thomas, 1992a), and indeed there is some fairly strong evidence that this is the case. The most impressive performance of a nonprimate species is undoubtedly an African Grey parrot, Alex, who has used his human-trained "language" skills to identify relational categories for human experimenters (Pepperberg, 1987). The phyletic distance of this bird species from primates and the different methodology used to test his skills makes it difficult to determine if the same process is at work. It is important in this regard that experience with humans and their material and symbolic artifacts also may play some role in the skills of apes in this domain, as human-raised apes can learn to actively match relations and even form analogies across object relations. It is unclear at what age human infants acquire skills in forming relational categories of these various types (although infants show some relevant perceptual skills; Tyrell, Stauffer & Snowman, 1991). Of what use such relational categories might be to nonhuman primates in their natural environments is not known at this time, but one possibility is that the ability to recognize and categorize relations is an important aspect of the social cognition of many primate species (see chapters 7 and 11).

Although their skills are not so impressive overall, nonhuman primates also show some abilities to physically sort objects into coherent groups. Chimpanzees seem to be a bit more skillful at this than several monkey species that have been tested in various ways. Thus, chimpanzees may be trained more easily than rhesus macaques to group objects to human specifications, chimpanzees make multiple groupings of objects in their spontaneous manipulations, whereas macaques and capuchin monkeys make only single groupings, and the only known instance of a nonhuman primate classifying the same group of objects on the basis of more than one criterion on different occasions was by a human-raised chimpanzee, Viki (Hayes & Nissen, 1971). It is unclear whether these differences between chimpanzees and several monkey species indicate a general difference between monkeys and apes, especially since the chimpanzee is the one primate species that uses tools naturally in the wild. Moreover, human influence is an important factor in the object-sorting skills of chimpanzees, and it is typically the case in classification studies that chimpanzee subjects have had more experience with humans and their artifacts than monkey subjects in these same studies. Human infants show evidence of categories through the sequential grouping method during their second year of life, and they more clearly place objects into groups in their third year of life and beyond (Langer, 1980, 1986).

In all, it is noteworthy that even learning theorists, whose roots are mostly in behaviorism, have been led to posit cognitive processes to explain the behavior of nonhuman primates. Primates and other animals do not just react to particular physical stimuli. They remember things, they learn about categories of things, and they acquire general strategies to use across situations. Although language may play an important role in facilitating human categorization skills and the formation of certain kinds of problem-solving strategies, and it may make possible certain kinds of unique categorization processes (e.g., hierarchies of inclusion relations), language is clearly not responsible for the initial appearance of such skills phylogenetically.

5

Quantities

To forage efficiently, primates not only must be able to locate, manipulate, and identify food, they must also be able to estimate its quantity. Individuals sometimes must decide such things as which of several food sources they should return to on a given morning based on an assessment of the quantities observed at each location the day before, or even whether to travel a short distance for a small amount of food or a long distance for a larger amount of food. Other situations in which estimating quantity may be important include estimating the number of potential opponents one might have in a fight or the amount of foliage available for making a nest. In all of these situations there are different ways of making such estimates, and the skills needed to do so do not necessarily include a humanlike concept of number or complex arithmetic operations. Again, because teasing apart the different cognitive processes involved in primate estimation of quantities is a somewhat delicate affair, most of the research has been done in the laboratory, much of it in the form of discrimination learning experiments.

There are basically two types of quantitative skills that prefigure a humanlike concept of number, both of which are possessed by nonprimate animals as well as by human children before they have adultlike numerical competence. These skills fall into the general categories of cardination and ordination. The cardinal aspect of number refers to quantity as a simple property of some set of objects. Thus the property "twelveness" characterizes such things as the months of the year, the Christian apostles, and the collection of eggs in most egg cartons. What is missing in this aspect of number is the notion that this property of "twelveness" occupies a particular place in an ordered series of quantities, that is, ordinality. So, for example, a foraging primate can recognize that it should go out on a limb with three pieces of fruit rather than on a limb with only two pieces of fruit because it recognizes that in the past trios have provided more sustenance than duos. In the human literature this skill is often called subitizing, and it works because duos and trios and other small quantities can be identified on sight as instances of prototypical perceptual patterns (or on the basis of "rhythm" in the case of auditory or other sequential stimuli; von Glasersfeld, 1982). Many animal species display abilities to discriminate small quantities in this way; the evidence is most compelling for various species of rats and birds (see Davis & Perusse, 1988, for a review). Because subitizing (what Thomas, 1992b, calls "prototype matching") is basically a skill of perceptual pattern recognition, the role of cognitive processes is unclear.

The ordinal aspect of number concerns the ability to place objects in a series on the basis of some quantitative property (e.g., from short to tall) and, perhaps, to identify certain places in the series (e.g., the third place, including such things as placing group mates in dominance rankings). What is missing in this case is cardinality, that is, some notion of the quantity designated by a particular place in a series. So, for example, a prenumerical human child may memorize a sequence such as the English words for the Arabic numerals without being able to match any item in that series with a particular quantity of objects. Several species of nonprimate animals have been trained to quit responding after the "third" response or to go to the "fourth" burrow in a row of burrows for a reward, for example, again with the evidence being most compelling for various species of rats and birds (see, e.g., Boysen & Capaldi, 1993; Davis & Perusse, 1988). Most importantly for the study of primate cognition, some investigators have stressed that if individuals can learn a fully ordered series, they should be able to make transitive inferences about the relative values of the different positions in the series, even in cases in which they have not had the opportunity to compare particular values directly. Such inferences would seem to be prototypical cognitive processes because they imply a cognitively represented ordinal scale—since to make a transitive inference requires mentally evoking the "middle term"—on the basis of which the individual makes informed quantitative choices.

In the theory of Piaget (1952), a humanlike concept of number is a "synthesis" of the cardinal and ordinal aspects. At some point children know about the number three both the quantity of items in the world to which it may be correctly applied and the relative value of three in an ordered series of quantities. The basic test for children's concept of number in Piaget's theory is their ability to pass a conservation of number task in which they show an ability to conserve the number of objects in a display across a transformation that changes the spatial arrangement of the objects in misleading ways (see also Gelman & Gallistel's, 1978, criteria for adultlike counting skills). Number of objects is just one type of conservation judgment; however, others may involve continuous quantities such as liquids, areas, lengths, volumes, and weights. Conservation of any of these various quantities is a prototypical cognitive skill because it clearly requires subjects to go beyond the information given to perception (indeed, in the classic test the perceptual information is actually misleading) and instead use some type of conceptually-based information to make quantitative judgments. A major problem is that, as with many Piagetian tasks past the sensory–motor stage, an important part of the assessment procedure is children's verbal explanations of their reasoning, and this obviously presents a problem for the study of nonhuman animals.

In this chapter, we evaluate the skills of nonhuman primates with quantities and number. We do this first for tasks that focus on cardinal aspects of number, that is, on animals' judgments of the absolute or relative numerousness of a set of objects. Second, we look at nonhuman primates' skills with the ordinal aspects of number in tasks in which they learn the ordering of a set of objects and even learn, on some occasions, to make transitive inferences about the members of the series. Third, we review studies of nonhuman primates that investigate their skills at the counting and "summing" of quantities. Finally, we look at studies involving tasks of conservation, including number as well as other quantitative concepts.

5.1 Estimating Numerousness

The ability of animals to judge, or to estimate, the numerousness of a set of objects is typically assessed in one of two ways. (Note that the somewhat awkward term "numerousness" is typically used instead of "numerosity," as this latter term is thought to imply an underlying concept of number, which prejudges the issue.) The simplest way is to use a discrimination learning paradigm to train the animal to use the number of items presented as a discriminative stimulus. Thus, just as an animal may be trained to choose a red over a blue stimulus for a reward, it may similarly be trained to choose a set of four objects over sets composed of three or two objects. These are usually referred to as judgments of absolute numerousness. The second type of assessment involves judgments of relative numerousness. In this paradigm animals are trained to pick the one of two sets of objects that contains a larger number of objects. These judgments may be relatively easy, such as discriminating 2 objects from 10 objects, or they may be relatively difficult, such as discriminating 8 objects from 9 objects. In the best of these studies, the spatial arrangement of objects in the two sets is randomized, and other variables such as the overall circumference of the set, the density of the set, the average size of the shapes, and the average distance between shapes are experimentally controlled. Although it is impossible to construct two displays that differ in no other way than in the number of items they contain, if the confounding stimulus properties are randomized across trials, this problem is presumably eliminated (i.e., on one trial the more numerous set covers the same area and has the same-size dots but is denser, on another trial the two displays have equal-sized dots and are equally dense but the more numerous set covers more area, and so forth).

5.1.1 Absolute Numerousness

In the classic study of this type, Hicks (1956) presented rhesus macaques with either two or three stimulus cards that depicted one to five geometric shapes. Subjects were always rewarded for choosing the card with three geometric shapes, no matter what was on the other card. They learned to do this after several thousand trials, and they did so even though the outline configuration and space covered by the shapes were experimentally controlled.

In a complication of this procedure of training subjects to recognize a single numerical stimulus, Ferster (1964) trained two chimpanzees to associate numerals in binary form with arrays of one to seven objects. The binary numerals were embodied in a series of three lights on a panel (e.g., off-off-on represented "one," off-on-on represented "three," etc.). The chimpanzee was then trained to produce the appropriate binary numeral when presented with a set of objects by operating a series of three levers that activated the lights on the panel. Several hundred thousand trials were required to train this performance to a high criterion for all seven quantities, but eventually the chimpanzees were performing consistently at greater than 95% correct. More recently, Matsuzawa (1985) trained his language-trained chimpanzee Ai in a similar manner, only in

Figure 5.1. Skills of quantification are used by a number of species such as chimpanzees. (Photo by J. Call.)

this case the associated stimulus was not a panel of lights but rather a visual display of the appropriate Arabic numeral. Ai had to both choose the appropriate set of objects (out of several) when presented with an Arabic numeral and choose the appropriate Arabic numeral (out of several) when presented with a set of objects. Perhaps because of Ai's background of human rearing and training, the training took about one-tenth the trials of Ferster's conditioning technique. The highest number recognized or produced by Ai in this procedure was six.

Hayes and Nissen (1971) presented their home-raised chimpanzee Viki with a match-to-sample version of the absolute numerousness task. They presented her with a sample card containing a small number of dots and then two cards from which to choose, one of which matched the sample in terms of number of dots. The size and spacing of the dots was randomized on all cards. Viki was reliable at matching the numerousness of a sample of three objects when the choice given her was between sets of three and four. She was not able to perform accurately when she had to match a sample of four objects and her choice was between sets of four and five. Viki was also not able to match a sample presented in the auditory mode; that is, she was not able to reproduce the same number of taps on a table that a human produced, even when very small numbers (two and three) were used.

5.1.2 Relative Numerousness

The other procedure in this same vein involves assessing judgments of relative numerousness (see figure 5.2). In this procedure an animal is presented with two stimulus arrays and rewarded for choosing the one with the greatest (or least) relative numerousness. Because numerousness covaries with other features, such as the size or density of the stimulus array or the objects involved, in order to say that what is being responded to is the relative number of items presented, these variables must be controlled in some way.

Thomas, Fowlkes, and Vickery (1980) trained two squirrel monkeys to respond to the member of a pair of cards that had fewer dots, with area, brightness, and spatial pattern cues controlled. They began with "easy" discriminations such as two versus seven, and then narrowed the gap until the difference was only two versus three. Using this method they found that both animals learned to reliably discriminate seven versus eight, and one of them learned to discriminate eight versus nine. In a related study, Terrell and Thomas (1990) taught four squirrel monkeys to discriminate irregular polygons with different numbers of sides. This method has the advantage that more factors than number could be controlled in the two stimuli to be compared; for example, in this method there is no worry about the tradeoff between area covered by the dots and the average size of the dots in the two stimulus sets. Two monkeys learned to discriminate seven- and eight-sided figures, while the other two discriminated six from seven and five from seven. It should be noted, however, that in this procedure the problem is that differences of shape must result from the way the two stimuli are constructed, and there may be some commonalties of shape among all polygons with the same number of sides.

Thomas (1992b) reports that he and his colleagues replicated these two experiments with human adults. In these cases, however, the two choices were presented for only 200 milliseconds (and there was post-stimulus masking) so as to preclude the possibility of subjects using their counting skills. Following pretraining on three versus four, subjects were presented with more difficult pairs. Eight of 20 subjects were successful at discriminating 7 from 8 dots, and 3 of 20 subjects were successful at discriminating 7- from 8-sided polygons. Thomas (1992b) points out that the superior performance of the monkeys relative to the humans is likely attributable to the longer exposure times they had to the stimulus materials.

Chimpanzees have also been tested on relative numerousness judgments. Dooley and Gill (1977a) presented their language-trained chimpanzee Lana with two sets of washers of various sizes in the context of a lexigram for either "more" or "less," which she had learned previously. Lana was rewarded for choosing the set that was consistent with the lexigram presented. The investigators report reliable discrimination for the pair three versus four (see Thomas & Chase, 1980, for a similar finding with squirrel monkeys, including judgements involving intermediate values). Using a slightly different procedure, Dooley and Gill (1977b) presented Lana with two different sets of food items (pieces of sweet cereal) that differed in number. They assumed she would always choose the set she perceived as containing more food items. Lana reliably chose the set with more pieces of cereal when the two sets were 9 and 10. It must be pointed out, however, that since all of the pieces of cereal were the same size, the surface area of the array of food was confounded with number in this study.

Mathieu (1982) attempted to deal with the issue of stimulus variables that are confounded with number by changing the arrangements of arrays systematically across trials. She presented three nursery-raised chimpanzees with two rows of food containing different numbers of food items. She presented these rows in three conditions: (1) the length condition, in which density was held constant (so that the longer row always had

Trial 1

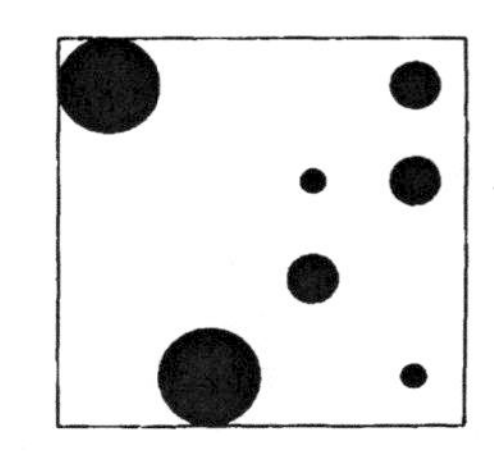

Trial 2

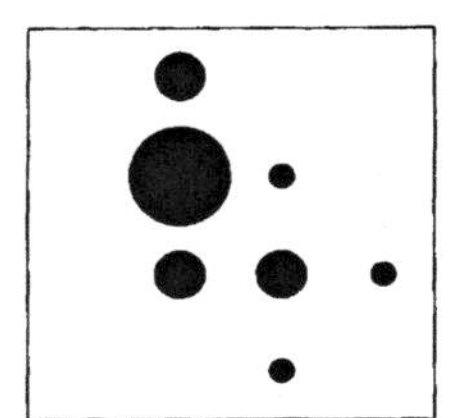

Figure 5.2. Sample of numerosity stimuli analogous to those used by Thomas, Fowlkes, and Vickery (1980).

a greater number); (2) the density condition, in which length was held constant (so that the denser row always had the greater number); and (3) the number condition, in which the row with the greatest number was either shorter and denser or longer and less dense or longer and denser. Only one of the three subjects succeeded in discriminating numbers up to six in all three conditions.

Finally, the issue of the stimulus characteristics of the two quantities to be compared was partially resolved in a study of two laboratory-trained chimpanzees (Boysen & Berntson, 1995). In this study, two chimpanzees (Sheba and Sarah) were taught to play a game in which they were presented with two dishes of candies and asked to indicate one of them. The dish they indicated went to the other chimpanzee, and they were left with the dish of candies not indicated. Both subjects had great difficulty with this task, presumably because of difficulties in inhibiting their tendency to reach for the greater amount of food for themselves. However, one of the subjects (Sheba) had previously learned to associate Arabic numerals with different numbers of candies, and so she was next asked to do the same task but with Arabic numerals. In this case she chose the smaller Arabic numeral (so that the other subject would get the smaller amount of candies) more than 75% of the time. The important point for current purposes is that the reliable choice of the smaller of two Arabic numerals was made, and in this case there are no confounding stimulus characteristics associated with a larger number of stimulus items. Boysen, Berntson, Hannan, & Cacioppo (1996) have recently replicated these findings with five chimpanzees, including some additional interesting task variations.

In terms of auditory stimuli, it should be noted that in the classic studies of Woodrow (1929) with rhesus macaques and Kuroda (1931) with longtail macaques, subjects successfully discriminated between one and two sounds, or between two and three sounds, whose intersound intervals were varied in different ways. In this same vein, Douglas and Whitty (1941) found that a Guinea baboon could be taught to discriminate two from three flashes of light.

5.1.3 Comparison of Species

The skills primates display in judging either the relative or absolute numerousness of sets of objects are not unique to their order. With regard to absolute judgments, Davis (1984) found that a raccoon could be trained to discriminate the number three. With regard to relative judgments, Koehler (1950) found that, after training, various species of birds were able to choose a nonpatterned array of dots that matched in number a differently patterned array of dots, up to six or seven. Pepperberg (1983, 1987) taught an African Grey parrot to vocalize the appropriate Arabic numeral words in English to arrays with varying (but always small) numerosities, including arrays with novel types of objects for which it had not been trained. Human infants in the first few months of life can be habituated to stimulus arrays containing a particular number of items (always small) such that they are surprised when the number in that array changes (Starkey, Spelke & Gelman, 1983). Within the order of primates, although there are not enough studies with enough species to draw definitive conclusions, there is no evidence of systematic species differences in making numerousness judgments (with no data on prosimians to compare).

In the view of most investigators, the skills displayed in numerousness judgments are mainly perceptual pattern recognition skills—direct perception of numerousness, either absolute or relative—not a conceptual understanding of number. It is interesting that this pattern recognition ability can operate to extract numerosity as a salient dimension even when the overall shape of the set and size of the objects involved is systematically varied, but there is no reason in any of these studies to believe that anything like counting underlies animals' skills in these tasks (Thomas, 1992b). The relative commonness of these skills of quantitative judgment across species suggests that this skill is useful in the lives of many different species, presumably for purposes of estimating the amount of food in foraging.

5.2 Ordinality and Transitivity

The other foundational aspect of number that has been extensively investigated in nonhuman primates is ordinality, including the ability to make transitive inferences. In studies of human children, the stimuli used to test for ordinality and transitivity are objects that vary systematically and naturally along a quantitative dimension; for example, height (e.g., Piaget, 1952). In one measure of their understanding, children are given the opportunity to manipulate objects freely (or in response to an experimenter's directions), and their constructions are assessed for evidence of ordinality. To test for transitive inference, subjects are trained with pairs of objects of different values (e.g., A taller than B, B taller than C), and then they are tested with novel pairs of objects that can only be related transitively; that is, through a common object to which they have both been compared previously (e.g., A taller than C). Transitive inference is in many ways a prototypical cognitive skill in that it involves the individual learning and cognitively representing a series of values and then choosing flexibly among those values in an adaptive manner.

Primates and other animals have not been tested in this form of transitivity (with one minor exception; see section 5.2.2). Instead, they have been tested in "associative transitivity," in which completely arbitrary stimuli are used (e.g., boxes of different colors). In this case, the relative "values" are created experimentally through differential reinforcement. For example, if an individual is reinforced for choosing a black box over a white box and a white box over a purple box, the series of values black–white–purple may be created (see figure 5.3). Again, transitive inference involves choosing the higher-valued box of a pair of boxes even when they have never before been presented together (e.g., the black box is to be chosen over the purple box in the above example). The adaptive value of skills of associative transitivity would seem to lie in being able to estimate the relative payoffs of the items in an array of possible choices in terms of various amounts of reinforcement.

5.2.1 Monkeys

D'Amato, Salmon, Loukas, and Tomie (1985) reinforced capuchin monkeys for choosing A over B (e.g., a red over a blue square), and then they taught them to choose B over C, despite novel spatial arrangements of the items on each trial. When subjects were presented with A and C, which had never before been presented together, subjects chose A over C, presumably on the basis of their respective relations to B, which was not perceptually available at the time of the A–C comparison. They called this ability associative transitivity, since there was no natural physical basis for ordering the stimuli, only the training history. It is interesting to note that pigeons tested in this same manner did not choose A over C; they did not show evidence of learning the concept of associative transitivity (D'Amato, Salmon, Loukas, & Tomie, 1986). The investigators argue that to solve this problem subjects had to employ representational processes; that is, in order to make the A–C comparison, they had to mentally evoke the absent B to which A and C could each be compared.

One problem with this study is that with only three items, the A item was never the non-reinforced item of a pair and the C item was never the reinforced item of a pair. D'Amato and Colombo (1988) therefore trained capuchin monkeys to touch five arbitrary items in a specified order (labeled A, B, C, D, and E), again with a novel spatial arrangement of items from trial to trial. After they had mastered this task (which pigeons master in roughly the same number of trials; Straub & Terrace, 1981), they were presented with pairs to order correctly. Because the two stimuli on either end of the five-item series (A and E) were always associated with first or last position, of particular importance were the internal pairs BC, CD, and especially BD. The BD comparison was especially important because these two items were both internal to the series and they were nonadjacent to one another in the previous training. It was found that subjects ordered these three internal pairs correctly 81–88% of the time, well above chance. When presented with triplets from which they were to choose the highest item, they ordered the internal triplet BCD correctly 94% of the time, also well above chance. This finding essentially replicates, with even stronger results, the findings of McGonigle and Chalmers (1977) with squirrel monkeys. It is interesting to note that when pigeons in the Straub and Terrace (1981) study were presented with internal pairs that had not previously been seen together, they responded in all cases at chance levels. Human children at 6 years old are as successful as the squirrel monkeys (Chalmers & McGonigle, 1984).

D'Amato and Colombo (1988) also presented data on response latencies in the pair tests to see if this might shed some light on the cognitive processes involved. They

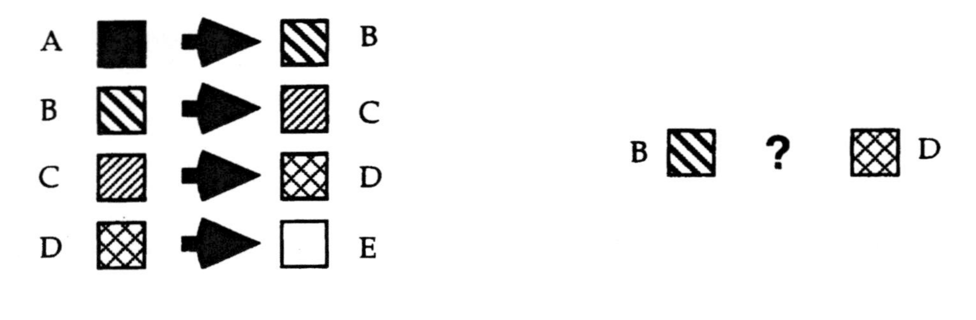

Figure 5.3. Associative transitivity task: choose higher valued item of pair. (All possible pairs presented in training except b and d.)

found two interesting patterns. First, subjects' latency to respond to the first item of a pair increased as that item moved down the series; that is, they responded most quickly to pairs in which the first item was A, then for pairs in which the first item was B, then C, then D. The implication is that each time they are presented with a pair, the subjects are mentally reconstructing the entire five-item series. Second, animals responded most quickly to the second item of a pair for pairs with adjacent items (e.g., AB, CD, etc.), then for pairs separated by one gap (e.g., AC, CE. etc.), then for pairs separated by two gaps (i.e., AD, BE), and they were slowest on the second item when the gap was three (AE). Again, the implication is that subjects are going through the entire series mentally on every trial. Swartz, Chen, and Terrace (1991) essentially replicated these results, both in terms of ordinal judgments and in terms of reaction times, for rhesus macaques. Interestingly, however, D'Amato and Colombo (1988) also reported that data supplied to them by Terrace revealed no such patterns in the response latencies of the pigeons tested by Straub and Terrace (1981) (see also Terrace, 1993, and Terrace & McGonigle, 1994, for confirmation of this monkey–pigeon difference). They therefore argued that the most reasonable interpretation of the data is that the capuchin monkeys have a mental representation of the series as presented them, whereas pigeons rely on some form of simple discrimination learning.

D'Amato and Colombo (1989), however, pointed out that the results of this study are compatible with an associative chain interpretation in which each item in the sequence simply serves as a discriminative stimulus evoking the next item, with no inherent notion of the serial order of items necessary for perfect performance. In the transitive inference tasks, a mentally evoked item must enter the chain, but the process is still one in which a series of conditional discriminations are made. To investigate whether capuchin monkeys were also associating a specific serial position with each item in the associative chain, D'Amato and Colombo (1989) used a procedure that essentially broke the chain. Using monkeys that had already learned the ABCDE sequence, on some trials they introduced a "wild card" item at a particular point in the sequence (e.g., ABCXE). This was a novel item that had never been used as part of the training, and thus it had no associations with any other items. What these investigators found was that no matter what the position of the wild card item, subjects treated it in a manner similar to the item it replaced at above-chance levels, touching it at the appropriate place in the sequence approximately 60% of the time. They performed just as well with sequences containing two wild card items. Consequently, D'Amato and Colombo argued that the monkeys in this study, and presumably in previous studies, were operating with something more than an associative chain, that is, they were operating with some mentally represented sequence of items in which the ordinal position of each item was essential information.

A final piece of evidence for this more cognitively rich interpretation of the monkeys' performance on tasks of associative transitivity was presented by D'Amato and Colombo (1990). They reasoned that if monkeys had a truly ordinal representation of the ABCDE series, their judgments on items presented in pairs should show what in the human literature has been called the "symbolic distance effect." If monkeys know the ordinal position of each item in a series without having to go through the entire chain associatively, then their decision times to touch the first item of a pair should be shorter the farther apart the items are in the series (since the ordinal positions of items farther apart on the series are more easily distinguished than the ordinal positions of items closer together in the series). Analyzing newly collected data on their capuchin subjects after completion of the D'Amato and Colombo (1989) wild card study, these investigators indeed found the symbolic distance effect. McGonigle and Chalmers (1992) also found the symbolic distance effect in squirrel monkeys (see also Harris & McGonigle, 1994). It is important to note that this finding concerns the time to make a decision about the first item of a pair, whereas the result reported above for D'Amato and Colombo (1988) concerned how long it takes a subject to respond to the second item of a pair, which increases as the items are farther apart in the series. The overall conclusion of these investigators is that capuchin monkeys operate in associative transitivity tasks with both associative chain information and with information about the particular serial position of items.

Finally, in a variation on this associative transitivity task, Washburn and Rumbaugh (1991a) taught two rhesus macaques to associate visually presented Arabic numerals with the reception of a corresponding number of food pellets (see figure 5.4). Relying on the fact that they naturally prefer receiving more food, monkeys were presented with experimental trials in which there was a forced choice between the members of a pair of numerals. Within fewer than 900 trials, these macaques learned to reliably choose the numeral associated with the larger number of food pellets. In addition, some pairs of numerals were held out from the initial training so that when they were presented they had never before been seen together by the subjects, and thus their relative values could only be determined by transitive inference using a mentally evoked "middle term" to which they had both been compared. One of the two subjects was above chance in choosing the larger member of the novel pair in the first set of trials. In a later variation on this procedure, Washburn and Rumbaugh presented the subjects with five numerals simultaneously, and they both chose the largest one at above-chance rates. These investigators interpreted their results as indicating that the monkeys formed a representation of a "matrix of values" corresponding to the numerals. Interestingly, Washburn (1994) later found with these same two subjects (and four other conspecifics) that when the numerosities compared consisted of Arabic numerals presented in different numbers of instances, subjects showed stroop-

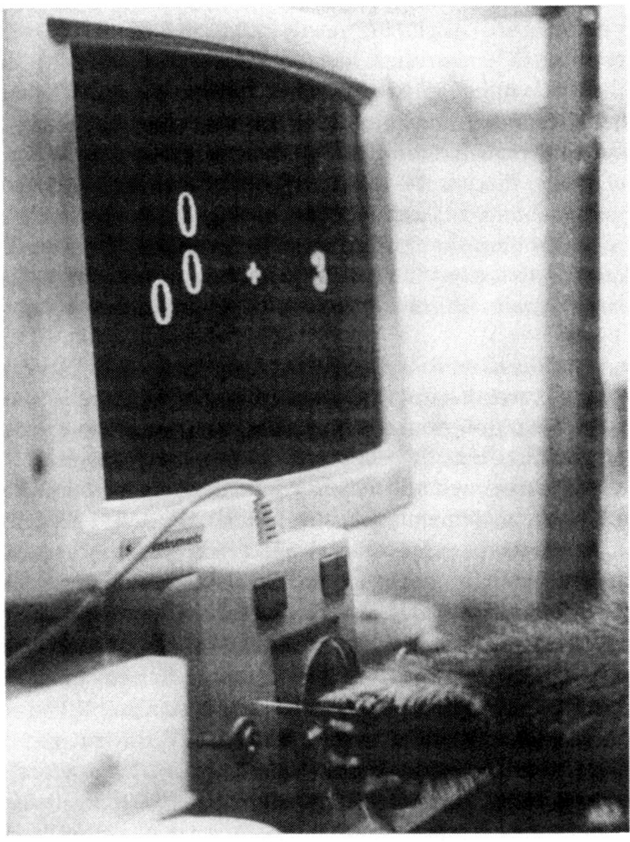

Figure 5.4. Rhesus monkey working on a numeric task with the Language Research Center's Computerized Test System. (Photos courtesy of D. Rumbaugh.)

like interference effects. That is, subjects experienced interference effects when they were presented with three instances of the Arabic numeral "two" and asked to make some judgment with respect to its quantity, for example. Washburn argues that this finding further indicates rhesus macaques' appreciation of the significance of Arabic numerals.

5.2.2 Apes

The only apes that have been tested for skills of serial order and transitivity are chimpanzees. Gillan (1981) presented three young chimpanzees with pairs of colored boxes. Food was placed in one box so as to designate it as the preferred choice. Food was presented in a pattern that formed the series ABCDE; that is, the four adjacent pairs (AB, BC, CD, DE) were presented with the "higher" box holding the food in each case. Gillan then tested individuals on the novel BD pair (internal and never before experienced as a pair). One animal was accurate from the beginning, one learned to be accurate over some trials, and one responded randomly throughout. (Menzel, 1969, also reported some evidence of transitive ordering of stimulus objects in seven adult chimpanzees.)

In a replication of this study, Boysen, Berntson, Shreyer, and Quigley (1993) taught three chimpanzees (two nursery-reared and one human-raised) an ABCDE series of colored boxes in the same pairwise manner (i.e., only adjacent pairs were presented). When presented with the novel BD pair in subsequent testing, all three animals chose D at a greater-than-chance rate. These investigators followed up this study by generalizing it to Arabic numerals. Because these animals had already acquired some skills at associating Arabic numerals with corresponding numbers of items, they were given the same ordinality task using Arabic numerals instead of colored boxes. After being trained with pairs of adjacent numerals in the same pairwise manner as in the original study, they were presented with the novel pair 2–4. In this study it was found that one individual (Sheba) learned the series of numerals much more rapidly than the other two and successfully chose the 4 over the 2 in the novel pairwise test, whereas the other two chimpanzees did not. Since Sheba had had much more varied and intense training with Arabic numerals prior to the study, the other two animals were given additional training with Arabic numerals and then retested. In this retesting, they too showed generalization to the novel nonadjacent pair. The investigators concluded that with appropriate training, chimpanzees can learn the serial order of symbolic stimuli. Tomonaga, Matsuzawa, and Itakura (1993) reported a similar performance by their language-trained chimpanzee Ai, and an analysis of Ai's reaction time data also showed the symbolic distance effect (see previous subsection).

One interesting study of the ability of chimpanzees to order objects naturally (i.e., on the basis of a real physical dimension without training rather than on the basis of a trained series) was reported by Perrault (1982; cited by Doré & Dumas, 1987). Perrault presented one chimpanzee with a series of sticks that went in small incremental steps from short to long, with one missing step. The subject was then given a choice of sticks to fill in the missing step. Although there was no evidence that it could solve this task even after hundreds of trials, Perrault noted that the subject almost invariably attempted to fit the alternatives into the missing step in order from the shortest to the longest of the alternatives, showing at least some ability to discriminate short and long sticks and perhaps to order those discriminations.

5.2.3 Comparison of Species

Both primates and pigeons can be trained to construct and order a series of items on the basis of the reinforcements associated with them. However, primates are able to do this in such a manner that accurate transitive inferences may be made about pairs of items in the series that have never before been seen together ("associative transitivity"), whereas pigeons seemingly cannot. Moreover, primates can order either a series of colored boxes or a series of Arabic numerals, and their performance shows a number of reaction-time phenomena consistent with the hypothesis that they are constructing a mental representation of a series of items, each of which has an ordinal position associated with it independently of the total chain. Pigeons show no such reaction-time phenomena, presumably because the making of transitive inferences across a series of items involves the use of relational categories such as "larger than" and "smaller than." Thus, although to our knowledge no nonprimate mammals have been tested in this paradigm, these findings of a difference between primates and pigeons are consistent with the hypothesis, developed in chapter 4, that primates can form relational categories in a manner that nonprimates cannot. The major possible exception to this generalization is reported by Roberts and Phelps (1994), who found that rats can be trained to perform well in the basic associative transitivity task, although only under certain conditions (if trained linearly) and without testing for any of the related phenomena (such as reaction time, symbolic distance effect, and wildcards) that make the performance of primates so impressive.

Within the order of primates, there is no evidence of any species differences in this ability, with a number of monkey and ape species performing in much the same manner across an array of tasks (once again, there are no data on any prosimian species to compare). Human children typically perform at the level of the nonhuman primates in these studies at 4–6 years of age (Bryant & Trabasso, 1971; Chalmers & McGonigle, 1984).

5.3 Counting, Summation, and Proportions

Several number-related phenomena have been studied in nonhuman primates that seem to depend on some kind of mathematical, or quasi-mathematical, operations. First are studies of counting, which differ from those of ordination and associative transitivity mainly in the fact that there is a stopping point in the sequence that corresponds to some external criterion (and there are other items still present that could potentially be counted). There is still much debate, however, about whether the "counting" done by nonhuman animals relies on a concept of number involving both cardinality and ordinality (Davis & Perusse, 1988). Second, the only truly arithmetic operation that has been studied in nonhuman primates is summation. Again, there is some question as to whether this is the same operation engaged in by humans in which true numbers are added together in a totally general operation, or whether something more concrete has been learned (i.e., some variation on subitizing). Finally, there is one study with one chimpanzee in which some notion of proportion that is general across different contents seemed to be evident, although in this case there is no claim that truly mathematical operations are involved. All three of these phenomena have been studied only in apes, mostly chimpanzees.

5.3.1 Counting

The one report of a nonchimpanzee primate counting items is that of MacDonald (1994) in her study of gorilla spatial memory, as reported in chapter 2. In that study the gorilla subject foraged eight food sites for four food rewards randomly placed across trials. The gorilla quickly learned to stop searching after the fourth food reward was found, even when there were unsearched food sites remaining. This same behavior was not observed in a similar study with marmosets and yellow-nosed monkeys (MacDonald & Wilkie, 1990; MacDonald, Pang & Gibeault, 1994). MacDonald argues that the gorilla was using counting as a foraging strategy to minimize energy expenditure. Because the food sites were visited sequentially and the food at each one was consumed as it was encountered, there can be no question of simple visual subitizing, as in the studies of estimating numerousness (see section 5.1). However, there are some claims that subitizing can be done sequentially in terms of patterns of rhythms (von Glasersfeld, 1982), and so it is unclear to what extent counting is involved in this gorilla behavior.

Using a design first employed by Yerkes (1916, 1934a) and then modified by Spence (1939), Rohles and Devine (1966, 1967) presented a female chimpanzee with rows of items from which she was to choose the numerically "middle" item; for example, the third item in a series of five. The spatial arrangement of items was varied from trial to trial so that the numerically middle item was not in the same absolute spatial position over trials, and in a second study the spatial arrangement was irregular so that the numerically middle item was not necessarily the spatially middle item. Especially in this latter task, the investigators thought some operation like counting was needed. The investigators first taught this subject (via discrimination training) to choose the middle of 3 items, then 5, and so on up to 17 items with constant spacing between items. Although the subject took more time to learn the middle item in larger sets, she learned to successfully identify the numerically middle item in all of the sets presented up to and including 17 (2600 trials to reach a 95% correct criterion for a single day's testing). Choice of one of the two stimulus objects on either side of the numerically middle item accounted for 97% of all errors. When the items were irregularly spaced, the subject was successful with series of up to 11 items. Rohles and Devine (1967) reported informally that they were unable to train this behavior in rats and that they were able to train some monkeys with three- and five-item series but not seven-item series [a result that Yerkes and Coburn (1915) also found with pigs].

Premack (1976) attempted to teach four juvenile chimpanzees to match objects in one-to-one sequences, an important skill, as true counting clearly depends of the establishment of some kind of one-to-one correspondence between series (e.g., in human counting between number names and objects). The subjects generally were not proficient at this skill. Although they matched items, they often added extra items to one row or the other, spoiling the one-to-one correspondence. The language-trained Sarah did learn to make more accurate correspondences, but only after much training and the addition of a marker to signal that she was done. Woodruff and Premack (1981) also found that these chimpanzees behaved similarly in a match-to-sample task in which their task was to see a sample number of objects and then to choose from two sets of other objects the one that matched the sample in numerosity. Only Sarah responded at above-chance levels. (It should also be noted that in this same series of studies, Sarah later failed a number conservation task; see next section.)

In a computer-based counting task, Rumbaugh, Hopkins, Washburn, and Savage-Rumbaugh (1989) taught the language-trained chimpanzee Lana to use a joystick in a video-game format. At the top of the video screen they presented her with an Arabic numeral for each trial. Her task was to use the joystick-cursor to "remove" that same number of squares from the bottom of the screen (from a large array) by touching squares one by one, designating with the cursor when she was through counting. After several different types of training with different types of feedback after each touch, Lana learned to perform correctly at a very high rate with the numerals 1, 2, and 3. In a final test she learned to do this even if the squares disappeared as she touched them (with a tone sounding as they disappeared), implying that she could keep track mentally of how many she had already touched. Rumbaugh, Hopkins, Washburn, and Savage-Rumbaugh (1993) reported informally that Lana, as well as two other language-trained chimpanzees (Sherman and Austin), had learned to master this task through the number six. These investigators argued that these subjects were indeed counting in a way similar to human children in that they knew not only the ordinality but also the cardinality of the Arabic numerals involved. Rumbaugh et al. did not report, however, whether the training on larger numbers went more quickly as a result of generalization from the earlier training on smaller numbers.

Finally, Boysen and Berntson (1989) taught the home-raised, laboratory-trained chimpanzee Sheba to count objects using Arabic numerals. They began by teaching her a one-to-one correspondence task in which she had to place only one object in each of six compartments of a divided tray. She was then taught a match-to-sample task in which she had to pick a card with the same number of dots as there were pieces of food in a simultaneously presented sample for the numbers one, two, and three. They then replaced the cards with dots by cards with Arabic numerals and continued training so that Sheba had to pick the Arabic numeral that corresponded to the number of dots on a card. They then trained Sheba to comprehend the Arabic numerals; they presented her with an Arabic numeral, and she had to pick the card with the corresponding number of dots. Sheba learned both of these tasks with Arabic numerals (production and comprehension) within several hundred trials. Boysen and Berntson then tested her for generalization in two ways. First, they presented her with one, two, or three common household items and asked her to pick the corresponding Arabic numeral, which she readily did. Second, they introduced the Arabic numerals 4, 5, and 0 directly (without first using cards with dots), and Sheba readily learned to associate these with the correct number of objects as well.

During this training of Sheba, Boysen and Berntson noticed that she often engaged in "indicating acts" as she counted, that is, she touched, displaced, or "pointed to" objects serially in attempting to determine the appropriate Arabic numeral, much as human children touch or otherwise indicate objects as they count them. In a follow-up study, therefore, Boysen et al. (1995) looked to see if the number of indicating acts Sheba used as she engaged in these tasks correlated with the number of items in the array (by the time of this study Sheba knew the numerals 0–7). They gave her some counting tasks, using the numerals 0–7, and found that she correctly counted 54% of the time (with errors distributed equally across the 1–7 range). They found further that, whereas the absolute number of Sheba's indicating acts did not correspond to the number of items in the array precisely, typically being about twice as large, these did correlate significantly ($R = .74$). It is unclear if this correlation is due to counting or to the

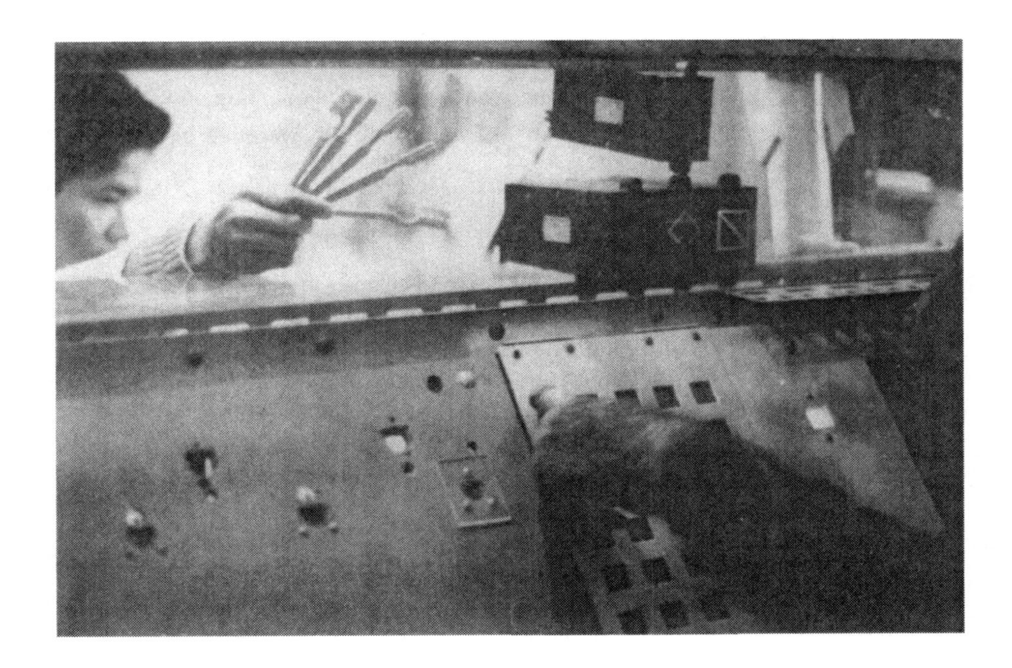

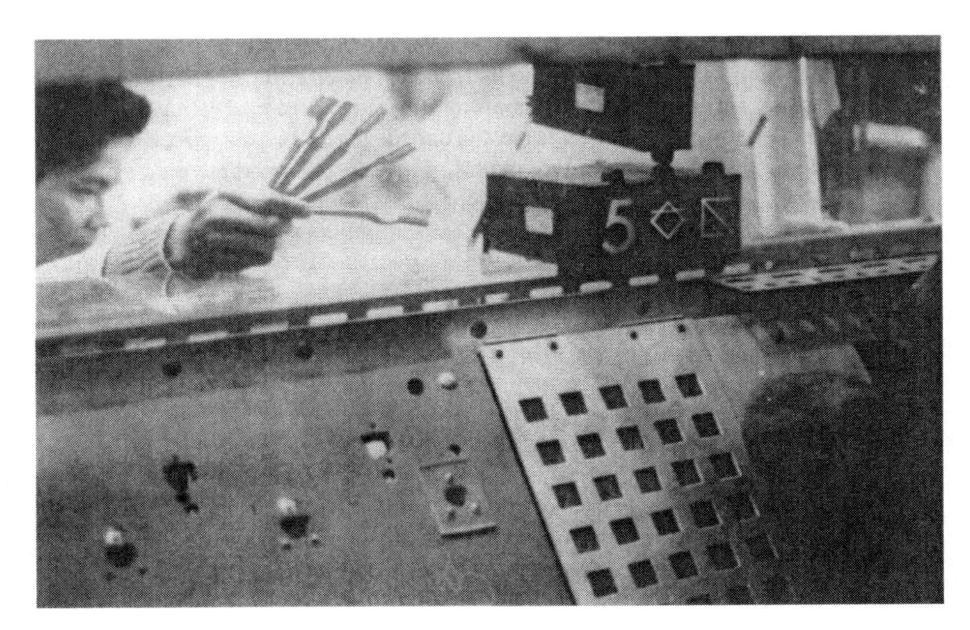

Figure 5.5. The chimpanzee Ai selecting the numeral 5 to represent the quantity of toothbrushes presented to her in Matsuzawa's (1985) study. (Photo courtesy of T. Matsuzawa.)

fact that determining the numerosity of the larger numerals requires more time, so that a constant rate of indicating acts across all numerals would lead to the correlation. In any case, the investigators concluded that Sheba was indeed counting objects in much the same way as human children, and that her indicating acts were serving a mediating function in the process.

5.3.2 Summation

Three sets of investigators have claimed that chimpanzees can learn to summate separate sets of objects. First, building on the work of Menzel (1969), Rumbaugh, Savage-Rumbaugh, and Hegel (1987) and Rumbaugh, Savage-Rumbaugh, and Pate (1988) presented two language-trained chimpanzees (Sherman and Austin) with two sets of chocolates, one of which had more individual pieces (see also Perusse & Rumbaugh, 1990). They were allowed to consume the set that was chosen, and so they quickly learned to choose the set with the largest number of pieces. The investigators then presented the subjects with two sets of chocolate pieces, but in this case each set had its items distributed in various ways in a pair of food wells (see figure 5.6). For example, an individual might be presented with 4 versus 3 pieces of chocolates, with the 4 distributed between its pair of wells as either 2–2 or 3–1 or 4–0 and the three distributed as either 1–2 or 3–0. In more than 90% of trials, subjects chose the pair of food wells with the greater number of chocolate pieces (up to 7), and this was even true when the correct pair did not have a single well with a greater number than either of the single wells on the incorrect side (e.g., they chose 4–3 over 5–1). The investigators do not claim that their subjects are "adding" numbers in the human way, especially because the spatial volume of the chocolates was confounded with their number; that is, because each piece of chocolate was the same size, the chocolates in the pair of wells containing the greater number took up a greater amount of space in a way that could be directly perceived without addition, counting, or other sophisticated numerical skills. Nevertheless, they do claim that the skills required for this task go beyond simple subitizing, since the items of each of the two quantities to be compared are separated into two spatially distinct subsets.

Second, in a recent study by Hauser, McNeilage, and Ware (in press), rhesus macaques observed pieces of food being hidden behind a screen. When they saw one piece being hidden, followed by another, they were more "surprised" when the experimenter removed the screen to reveal either one or three pieces of food, and less "surprised" when removal of the screen revealed two pieces of food. As in the research with human infants on which this study was modeled (Wynn, 1992), surprise was measured by the amount of looking time: longer looking times indicate greater surprise. Similar results were obtained in a reversal of this task in which pieces of food were systematically removed. Although the human findings have been interpreted by some researchers to indicate the arithmetical operations of addition and subtraction, further research has indicated that these results may be interpreted more simply as the subject keeping track of objects behind occluders, as in object permanence tasks (Simon, Hespos & Rochat, 1995). This same interpretation is possible for rhesus macaques, as Hauser et al. (in press) note.

Finally, the most impressive arithmetic performance by a nonhuman animal is that of the chimpanzee Sheba. After training her to count objects, Boysen and Berntson

Figure 5.6. Chimpanzee choosing one of the two sets of trays presented by Rumbaugh et al. (1987). (Reprinted from *Journal of Experimental Psychology: Animal Behavior Processes* with permission from the American Psychological Association.)

(1989) tested Sheba's understanding of summation in two additional counting tasks. Sheba sat at a home base, and there were three "foraging" sites in other parts of the room. At two of the three sites a small number of food items were hidden (or possibly zero). Sheba's task was to move among the three sites, observe the food at each, return to the home base, and pick a card with the Arabic numeral corresponding to the total number of pieces of food at the three sites combined. There were never more than 4 pieces total, and the cards from which she was to choose at the home base were the numerals 1–4 arranged in ordinal sequence (see figure 5.7). Sheba's performance was above chance very quickly, and averaged 75% correct over many hundreds of trials (chance = 25%). The final assessment of Sheba's skills was to repeat the foraging task but with Arabic numerals (0, 1, 2, or 3) instead of food items hidden at each site. Sheba's performance did not drop at all from the functional counting task; she was 76% correct on the first 21 trials. It is important to note that despite Sheba's extensive training in foundational tasks, until these two foraging tasks she had never before been asked to sum two quantities using either real items or numerals.

Of all the performances of primates in number-relevant tasks, Sheba's performance is undoubtedly the one that inspires the most humanlike interpretations. Of special note is the final task with Arabic numerals, which would seem to preclude the possibility that she was doing anything perceptually based. That is, in this task Sheba

Figure 5.7. Sheba's experimental setting for summation from Boysen and Berntson (1989). (Diagram courtesy of S.T. Boysen with permission from the American Psychological Association.)

could not simply use visual subitizing, since the items that needed to be quantified were never perceptually present. Nor could she have used some kind of sequential subitizing based on rhythm because all that was present at each location was a single Arabic numeral. At the very least, she must have been sequentially subitizing mental representations of the items signified by the Arabic numerals, which would seem to stretch the concept of subitizing beyond recognition. Given that Sheba also can make transitive inferences with an ordered series of items (Boysen et al., 1993; see section 5.2) and uses indicating acts as she attempts to determine the numerosity of sets of objects (see previous subsection), the investigators hypothesize that she is actually counting, in a humanlike way, in these foraging tasks, and that her number concept is very much like that of a young human child.

5.3.3 Proportions

Another interesting arithmetic operation that chimpanzees have shown is the ability to understand (in some sense) proportions. In a match-to-sample paradigm, Woodruff and Premack (1981) taught language-trained Sarah and four non–language-trained conspecifics to match one-quarter of an apple to one-quarter of another apple, as opposed to three-quarters of another apple. They did the same with liquid in a jar, filled either one-quarter or three-quarters full. They then combined these two types of materials to present, for example, one-quarter of an apple as sample and a jar one-quarter full and another three-quarters full as alternative choices. Control procedures were instituted to rule out simple perceptual strategies. Woodruff and Premack found that all four of the non–language-trained chimpanzees failed at this task, but Sarah passed easily, choosing the correct alternative on the first eight trials in which the two types of material were used in the same task (see also Premack & Premack, 1983). The investigators argued that, as in the case of the understanding of abstract relations (see section 4.3), the language training that Sarah received allowed her to use an abstract code to make the analogical comparison. Whether a task measuring analogical skills but not relying on match-to-sample performance would yield better results for the non–language-trained chimpanzees, as it did in the case of the understanding of abstract relations, is not known.

5.3.4 Comparison of Species

A number of nonprimate animals show some skills that might seem to rely on counting. Many birds can be trained to eat only a certain number of food items serially and then to quit (e.g., Koehler, 1950), which is reminiscent of the behavior of the gorillas of MacDonald (1994). Moreover, both rats and canaries can be trained to enter one of a number of tunnels based solely on its serial position relative to its starting position (Davis & Bradford, 1986; Pastore, 1961), which is similar in some respects (i.e., items are present sequentially and then disappear) to the "counting" of chimpanzees. The African Grey parrot of Pepperberg (1983, 1987) can work accurately with small-valued Arabic numerals. However, no nonprimate animal has been observed to do anything like the subject of Boysen and Berntson (1989) did in summing objects and numerals, although to our knowledge no other subject of any species has been tested in this paradigm.

The problem with comparing different primate species in the set of skills that are of current concern is that only apes, mostly chimpanzees, have been explicitly tested for the understanding of counting, summation, and proportions. It should be noted for a point of comparison that human children first begin to count accurately and reliably at 4–6 years of age. What they would do at earlier ages if given the training and testing regimen of the chimpanzee subjects in the experimental paradigms reported in this section is unknown.

5.4 Conservation of Quantities

In Piaget's (1952) theory of number, the clearest test of a concept of number is the ability to pass a number conservation task. In this task, children are presented with two rows containing an identical number of objects. They are then asked to judge that they have the same number. Then, as the children watch, one row is pushed together so that its length is shorter (or sometimes the objects are stacked), and they are asked if they still have the same number or if one row now has more or less than the other. If they give the correct answer, that there is still the same number in each row, the children are asked how they know this. Correct explanations are responses such as: "you could just

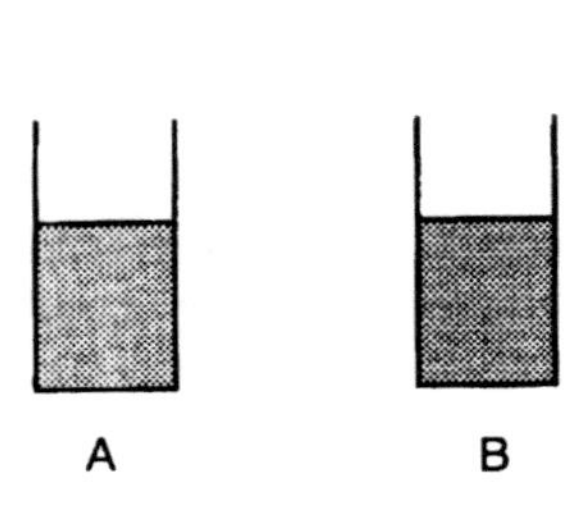

Step 1: Amount of liquid in two jars is equal.

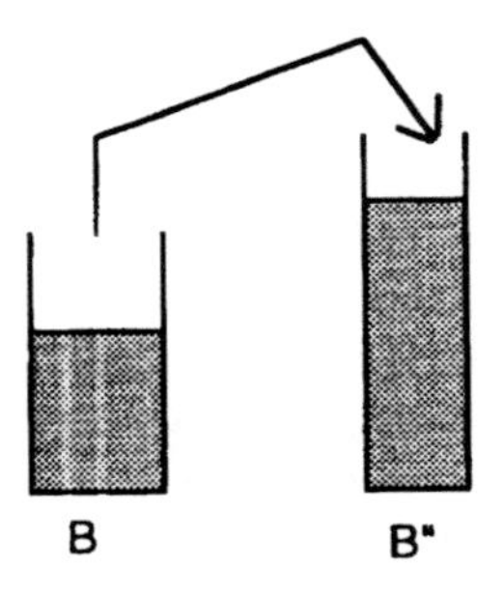

Step 2: Liquid in one jar poured into jar of different shape.

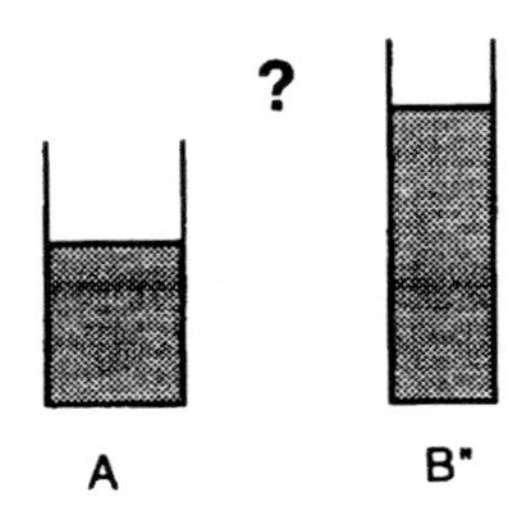

Step 3: Subject judges amounts in two jars after transformation.

Figure 5.8. Diagram of a liquid conservation task as presented to human children.

put them back the way they were and see that they were still the same" (reversibility) or "this row looks like it has more because it's longer but in this row the objects are closer together" (compensation). It should be noted that children's verbal comments in this situation often reveal that they consider the conservation of numerical quantity despite the spatial transformation of objects to be a logically necessary conclusion: they say "there *have* to be the same number still; just moving them *cannot* change the number," for example. Other quantitative concepts, for example, liquid quantity, solid quantity (substance), length, area, weight, and volume, may be tested in this same way (see figure 5.8 for an illustration of conservation of liquid quantity). At some point in the 5- to 10-year age range, human children come to use these quantitative concepts as stable parts of their reasoning about the physical world.

5.4.1 Monkeys

Two studies have tested conservationlike skills in monkeys. First, Czerny and Thomas (1975) trained squirrel monkeys to respond to the left food-well when two stimuli contained the same volume and to the right food-well when they contained different volumes. The subjects learned to do this and generalized it to new stimuli. However, as the investigators point out, such generalized judgments of volume are relevant to conservation, but they are not the same thing. The whole point of conservation judgments is that they establish that a quantity stays the same despite a spatial transformation of some sort (see also Pasnak, 1979, for a similar study with macaques).

Second, using these same squirrel monkeys, Thomas and Peay (1976) trained subjects to respond to the right food-well when two objects were of the same length and to the left food-well when they differed in length. They then had subjects judge the relative lengths of two wooden blocks. One of these was then moved in some way (for example, so that one of its ends reached spatially beyond the end of the other), and subjects were asked to judge again the relative lengths of the two blocks. After several hundred trials, two subjects learned to be successful in their post-transformational judgments. Although several control procedures were used, one essential control procedure was not employed. Because animals were not given the chance to judge the relative length of the two blocks without a transformation, it is possible that subjects had learned to ignore the transformation and simply to judge the relative lengths of the two blocks in the configurations they were left in after the various transformations (and indeed they had been trained to make the pre-transformational judgments using many of these configurations, as Thomas and Peay point out). The overall conclusion from these two studies is thus that we do not know whether any monkey species can perform in humanlike ways in conservation problems.

5.4.2 Apes

Two studies with chimpanzees employed more controls and thus demonstrated more humanlike performance in conservation tasks. First, Woodruff, Premack, and Kennel (1978) tested language-trained Sarah on tasks of the conservation of liquid quantity, solid quantity, and number. Before the experiment Sarah had learned to use plastic symbols indicating that two items were the "same" or were "different." Therefore, Sarah observed two quantities that were the same in both quantity and overall shape (e.g., two identical beakers of colored water with the same amount of liquid in each) and the transformation of one of these so that it now took on a different shape (e.g., the water reached a higher level because the water was poured into a narrower beaker). She was then asked to judge the post-transformational quantities, one of which always differed in appearance from its pre-transformational state. Sarah was able to accurately judge the conservation of both liquid and solid quantity, but she was not given the number test due to a failure in a pretest to make correct judgments of equally and unequally numbered rows.

Three important control procedures were employed in this study. First, Sarah was also presented on some trials with two quantities that were unequal both pre- and post-transformationally (this prevented a simple response perseverance on "same"). Second, there were some trials with transformations that did change quantity, for example, extra liquid was added to one beaker (this prevented Sarah from simply judging pre-transformationally and then ignoring whatever the experimenter did). Third, Sarah was presented with a sample of the post-transformational pairs to judge—to rule out the possibility that she was simply ignoring everything the experimenter was doing and simply making a perceptual judgment of the amount of liquid or solid in the post-transformational options. Sarah's performance in these control tests (especially her failure on control trials of the third type) suggested to the investigators that she was indeed using her knowledge of the initial quantities and the transformational process to infer the relative quantities of the post-transformational choices, that is, she was conserving a continuous quantity.

Muncer (1983) replicated Woodruff et al.'s (1978) result for liquid quantity with two language-trained chimpanzees using a different procedure. He relied on the chimpanzee's tendency to choose the larger of two amounts of a preferred liquid to drink. Using a number of control procedures to ensure that the subjects were not simply judging the post-transformational volumes or employing some perseverance strategies, Muncer found that one of his two language-trained chimpanzees succeeded in conserving the unequal amounts of liquids across a spatial transformation. Interestingly, this same subject was also successful in a number conservation task of the same type using bits of preferred food as the stimulus items (with controls for always choosing the denser row, and so forth). Muncer concluded that number conservation was also within the grasp of chimpanzees, and that language training was not necessary for chimpanzees to develop skills of judging and conserving relative quantities in a number of different domains.

Recently, however, some findings with orangutans have provided an alternative explanation for these findings. Call and Rochat (1996) presented four orangutan subjects with a series of choices between two glasses of juice (see figure 5.9). In the first series of tests, three of four subjects consistently chose the container with the larger quantity of juice, regardless of the shape of the container, and chose randomly when the two containers had equal amounts of juice, again independent of the shape of the containers. In the second series of tasks, all four subjects chose the container with more juice after transformations into new containers, which sometimes resulted in misleading perceptual cues (e.g., the glass in which the level was lower actually had more juice, because it was also wider). The interpretive problem is that in all of these tasks, as in all of the previous conservation tasks with nonhuman primates, the subject could be making a judgment about the quantity of the two containers of juice and then just tracking the greater quantity of liquid spatially, basically ignoring the perceptual transformations that occurred. In a final series of tasks, therefore, Call and Rochat broke the integrity of one of the quantities by pouring it (sometimes the greater quantity, sometimes the smaller quantity) into a collection of smaller cups. Under this transformation, none of the subjects systematically chose the larger of the two quantities across the various trials of different types, although some of the 6- to 8-year-old children they tested did so. These investigators interpreted their results as indicating that orangutans are very good at estimating quantities and at tracking the quantity they prefer across various spatial displacements, but they do not conserve quantities across perceptual transformations in a humanlike manner.

5.4.3 Comparison of Species

No nonprimate animals have to our knowledge been given these conservation tasks, which are normally mastered by human children during the 5- to 10-year age period. The problem is that, whereas young children can be presented with this task by means of language such as "Which glass has more water? Do they still have the same *amount* of water?" the chimpanzees and monkeys cannot. Moreover, the children are

Figure 5.9. The orangutan Chantek pointing to the container with the most liquid despite the fact that the liquid in the other container reaches a higher level. (Photo by J. Call.)

asked to justify their judgments verbally, again by using language talking about *amount*. Nonverbal assessments thus always have the problem that we can never be certain that subjects are indeed focused on what we wish them to focus on. Indeed, in a nonverbal version of a conservation of weight problem, preverbal human infants showed signs of knowing that the weight of an object is not changed by a transformation of its shape; their grasping motions that were normally calibrated for weight did not change when they reached for an object pre- and post-transformationally (Mounoud & Bower, 1974).

While not gainsaying the ability of chimpanzees and monkeys to make sensitive judgments of relative quantities, it is difficult to conclude that they are really conserving the quantities across transformations (nor are the preverbal infants). Thus, it is possible that they are making an initial judgment of the relative quantities of the two choices pre-transformationally, which they are perfectly capable of doing (see section 5.1), and then simply tracking the object/liquid/row with the greater quantity across the transformation, sticking with their initial judgment. What is being "conserved" across the transformation, therefore, is only the identity, not the quantity, of the liquid, substance, or objects involved. The control employed by Woodruff et al. (1978) in which an additional quantity was added and the quantity thereby changed, does not solve the problem because adding material to existing material is well-known by the subjects to change the entity involved. What is needed is some kind of control in which perceptually tracking the pre-transformational materials across the transformation will not work, and there is no adding or subtracting in the process.

5.5 What Primates Know about Quantities

Many animal species have some quantitative skills. In particular, some bird and rat species can estimate quantities, identify specific places in a series, and be trained to perform an act a specific number of times. Primates can do all of these things as well, although we should point out that there are no studies that investigate any of these skills, or any other quantitative skills, in any prosimian species. Primates (i.e., a few monkey and great ape species) have been shown to perform some additional behaviors in this cognitive domain, but in many cases no nonprimate species have been tested.

Both monkeys and apes can make accurate quantitative judgments of small quantities of items, although whether this skill is perceptual or conceptual is a difficult issue. Moreover, both monkeys and apes can learn to order stimuli and make transitive judgments about the place of these stimuli in an ordered series. They can do this even about items in the series that have never been seen or trained together, and examination of the reaction times in these tasks suggests humanlike decision processes. The primates' behavior in these situations would seem to indicate an ability to construct a complex sequential representation, of a type that pigeons appear not to be capable of, with the question of rats still outstanding, as their performance has some curious limitations (Roberts & Phelps, 1994). One problem is that there are still some associative explanations of performance in associative transitivity tasks, since these tasks have to be extensively trained with many food rewards, and these undermine the conceptually based interpretation of these tasks (De Lillo, in press).

Skills of counting, summation, and proportionality have only been systematically investigated in chimpanzees. Chimpanzees have some skills in these domains. But it should be noted that all of the subjects involved in these studies were home-raised or language-trained by humans, so it is unclear whether any monkey species, if given sufficient training of the appropriate kind, would also learn to be skillful in these types of tasks. A similar, though slightly different, story may be told of conservation. The one study with monkeys involved only one New World species and did not include at least one necessary control procedure, so the results are inconclusive. The ape studies, taken together, are more suggestive, but still far from definitive given the difficulty of ruling out the use of alternative strategies in the absence of verbal reports from subjects. Again, the apes in these studies came to task with backgrounds of training by humans; it is unknown how monkeys (especially some Old World species that have never been tested) would perform if they were given this kind of training.

Perhaps the most difficult question is how the performance of the apes in tasks of counting, summation, and conservation compares to the performance of human children. While the comparison of humans and nonhumans is in all cases difficult and fraught with methodological difficulties, the current case is particularly problematic. The major problem is that children's skills of counting, summation, and conservation rely on their prior acquisition of linguistic symbols; they learn number words as other words and use them productively well before they proceed to mathematics per se. They are taught many things in these domains, including explicit counting procedures in many cases, but they are never explicitly trained in the manner of the apes. Moreover, human children can help clear up ambiguities in their behavior in experiments by verbally explaining their reasoning process, which is especially important in the conservation tasks, whereas even the language-trained apes are not in a position to help experimenters this way. The two best tests of a humanlike understanding of number are Gelman and Gallistel's (1978) five prerequisites for human adultlike counting and performance in number conservation tasks.

With respect to Gelman and Gallistel's (1978) criteria for human adultlike counting, the picture is as follows. (1) Several of the apes (especially Lana and Sheba) have shown that they can use the principle of one-to-one correspondence to consistently align each item in a set of numerals with one item in a set of objects. (2) Lana and Sheba have also demonstrated the stable-order principle in which they use a reliably ordered series of numerals in making the one-to-one correspondence. (3) It is difficult to know the notion of cardinality underlying Lana's procedure, but Sheba certainly seems to know that a set of items that has been counted is characterized by the last numeral counted, which is its quantity. (4) No research has specifically addressed whether these counting procedures are abstract enough to be applied to any set of objects (the abstraction principle), although Sheba and Ai count different types of food. (5) No research has addressed the order-irrelevance principle in which an individual knows that the order in which items are counted is irrelevant to their cardinality. Overall, the performance of at least some highly trained apes would seem to present good evidence of something like a human procedure for counting, at least for small quantities.

With respect to number conservation, we have already expressed many of our reservations with the experimental procedures as adapted for apes. But, in addition, we must reiterate that one of the essential characteristics of conservation is its logically necessary quality; that is, human children say that the amount of water *must* still be the same because all you did was change the shape and that *cannot* change quantity. It is difficult to know how to address such a question with nonverbal animals, but we need to in order to decide whether the nature of the skills apes are showing in conservation tests is similar to that shown by human children in the early and middle childhood periods. The precise nature of the concept of number that might underlie nonhuman primates' various skills of quantification, therefore, has still to be unequivocally determined.

6

Theories of Primate Physical Cognition

Theories of primate physical cognition may be classified into two broad groups depending on their aims. The first group of theories has to do with proximate mechanisms, that is, with how cognition is structured. This has been our major concern in the preceding four chapters, and all the data on how primates cognize the world in all the various behavioral domains are relevant to theories of proximate causation. The second group of theories has to do with ultimate causation, that is, with the ecological conditions that provided the adaptive context within which particular cognitive mechanisms might have evolved. Here the data are mostly lacking because we have no direct access to the behavior of our primate ancestors, and the fossil record is woefully incomplete (Martin, 1990). Nevertheless, there are interesting and testable hypotheses, and even some data, about the relation between particular ecological conditions and cognitive mechanisms as they may have arisen in primate evolution.

In this chapter we do three things. First, we review the empirical facts that we have established in the previous four chapters. Second, we provide accounts of existing theories of primate physical cognition, first with respect to proximate mechanisms and then with respect to ultimate causation. We then provide suggestions for further research on primate physical cognition from an ecological point of view.

6.1 Summary of Primate Physical Cognition

Table 6.1 provides a compendium of the primate species that have been investigated in each of the domains of physical cognition reviewed in the previous four chapters. To summarize the data, we relate very briefly the major characteristics of the cognitive processes employed in each of these cognitive domains by the various species in each of the three major groups of primates (in the discussion we combine New World and Old World monkeys). This is just a method of convenience; each species within these groups likely has some unique cognitive skills.

Table 6.1. Catalog of species studied in each cognitive domain (see appropriate section for details of findings).

	Prosimians	New World Monkeys	Old World Monkeys	Apes
I. Space & Objects				
Cognitive maps	*Daubentonia madagascariensis* *Microcebus murinus*	*Alouatta palliata* *Callicebus moloch* *Callithrix jacchus* *Cebus apella* *Saguinus fuscicollis, mistax* *Saimiri sciureus*	*Cercocebus torquatus* *Cercopithecus aethiops, ascanius* *Colobus badius* *Macaca fascicularis, fuscata, mulatta* *Papio cynocephalus, hamadryas, papio*	*Gorilla gorilla* *Pan troglodytes, paniscus* *Pongo pygmaeus*
Hidden objects	*Loris tardigradus*	*Cebus apella, capucinus* *Lagothrix flavicauda* *Saimiri sciureus*	*Macaca arctoides, fascicularis, fuscata, mulatta*	*Gorilla gorilla* *Hylobates lar* *Pan paniscus, troglodytes* *Pongo pygmaeus*
Invisible displacements		*Cebus apella, capucinus* *Saimiri sciureus*	*Macaca arctoides, fascicularis, fuscata, mulatta*	*Gorilla gorilla* *Pan troglodytes*
Other spatial skills	*Lemur catta*	*Ateles geoffroyi* *Cebus apella, capucinus* *Lagothrix humboldti* *Saguinus fuscicollis* *Saimiri sciureus*	*Cercocebus sp.* *Cercopithecus nictitans* *Macaca arctoides, fascicularis, mulatta, nemestrina* *Mandrillus sphinx* *Papio anubis, cynocephalus, papio*	*Gorilla gorilla* *Hylobates lar* *Pan paniscus, troglodytes* *Pongo pygmaeus*

(continued)

Table 6.1. *(continued)*

	Prosimians	New World Monkeys	Old World Monkeys	Apes
II. Tools & Causality				
Object manipulation	*Galago crassicaudatus, garnettii, senegalensis* *Lemur catta, fulvus, variegatus* *Loris tardigradus* *Nycticebus coucang* *Perodicticus potto*	*Aotus trivirgatus* *Ateles belzebuth, fusciceps, geoffroyi, paniscus* *Callicebus moloch* *Callithrix argentata, jacchus, penicillata* *Cebus albifrons, apella, capucinus, nigrivittatus* *Lagothrix lagothricha* *Saguinus labiatus, midas, mistax, weddelli* *Saimiri sciureus*	*Cercocebus albigena, atys, galeritus, torquatus* *Cercopithecus aethiops, albogularis, ascanius, cephus, hamlyni, l'hoesti, mona, neglectus, nigroviridis talapoin* *Colobus guereza* *Erythrocebus patas* *Macaca arctoides, assamensis, cyclopis, fascicularis, fuscata, maura, mulatta, nemestrina, radiata, silenus, sinica, sylvanus, thibetana* *Mandrillus leucophaeus, sphinx* *Nasalis larvatus* *Papio anubis, cynocephalus, hamadryas, ursinus* *Presbytis cristatus entellus, francoisi, obscurus, pileatus* *Theropithecus gelada*	*Gorilla gorilla* *Hylobates lar, moloch, klosii* *Pan paniscus, troglodytes* *Pongo pygmaeus* *Symphalangus syndactilus*

Tool use		*Cebus apella, capucinus*	*Erythrocebus patas* *Macaca fascicularis, fuscata, mulatta, nemestrina, radiata, silenus tonkeana* *Mandrillus sphinx* *Papio anubis, hamadryas papio, ursinus*	*Gorilla gorilla* *Hylobates lar* *Pan paniscus, troglodytes* *Pongo pygmaeus*
Causal understanding		*Cebus apella*	*Cercopithecus aethiops* *Macaca fascicularis, fuscata* *Papio ursinus*	*Gorilla gorilla* *Pan paniscus, troglodytes* *Pongo pygmaeus*
III. Features and Categories				
Discrimination learning	*Galago senegalensis* *Lemur catta, macaco* *Microcebus sp.* *Nicticebus coucang* *Phaner sp.*	*Ateles geoffroyi* *Callithrix jacchus* *Cebus albifrons, apella* *Saimiri sciureus*	*Cercocebus torquatus* *Cercopithecus aethiops* *Macaca arctoides, fascicularis, fuscata. mulatta, nemestrina, niger, silenus* *Miopithecus talapoin* *Papio papio*	*Gorilla gorilla* *Hylobates sp.* *Pan paniscus, troglodytes* *Pongo pygmaeus*
Natural categories		*Alouatta palliata* *Callimico goeldi* *Cebuella pygmaea* *Cebus apella* *Saimiri sciureus*	*Cercopithecus aethiops* *Macaca arctoides, fascicularis, fuscata, mulatta, nemestrina, radiata*	*Pan troglodytes*

(*continued*)

Table 6.1. *(continued)*

	Prosimians	New World Monkeys	Old World Monkeys	Apes
Relational categories	*Lemur catta*	*Cebus apella* *Saimiri sciureus*	*Macaca fascicularis, fuscata, mulatta, nemestrina, niger, silenus*	*Gorilla gorilla* *Hylobates lar, moloch, pileatus* *Pan paniscus, troglodytes* *Pongo pygmaeus*
Classification		*Cebus apella*	*Macaca fascicularis* *Macaca mulatta*	*Gorilla gorilla* *Pan paniscus, troglodytes* *Pongo pygmaeus*
IV. Quantities				
Estimating numerousness		*Saimiri sciureus*	*Macaca fascicularis, mulatta* *Papio papio*	*Pan troglodytes*
Ordinality & transitivity		*Cebus apella* *Saimiri sciureus*	*Macaca mulatta*	*Pan troglodytes*
Counting, summation & proportion		*Callithrix jacchus* *Macaca mulatta*	*Cercopithecus ascanius* *Pan troglodytes*	*Gorilla gorilla*
Conservation		*Saimiri sciureus*	*Macaca mulatta*	*Pan troglodytes* *Pongo pygmaeus*

6.1.1 Prosimians

Overall, we know very little about the cognitive mechanisms that prosimians might employ in any domain of their activity (see Wilkerson & Rumbaugh, 1979, for a review of learning skills). Of the 15 domains listed in table 6.1, we have information on prosimians in only 5 (as compared with 14 for monkeys and 15 for apes). One species of loris did not search for an object after it saw the object hidden; however, this was likely due to the interference of their typical foraging behaviors, which involve poking and snatching at foods. Removing a covering obstacle is thus likely not a part of their object manipulation skills. Indeed, in the domain of object manipulation, the only domain in which a reasonable number of prosimian species have been observed, prosimians do not seem to manipulate objects in as many or in as complex ways as other primate species. In particular, they do not engage in actively relating objects to one another or the use of tools. Prosimians also seem to be a bit slower and more error-prone in discrimination learning (including delayed responding) than other primate species, but they do seem to have some cross-modal skills. In terms of cognitive mapping, ayes-ayes are skillful in mapping out the tunnels of their insect prey, but the flexibility of their skills in this activity is not known. Basically nothing is known about the categorization or quantitative skills of any prosimian species, although the skills of some nonprimate species might suggest that they have skills as yet undiscovered in these domains (since this might imply a homology in these skills among all mammals).

6.1.2 Monkeys

The available data for monkeys in many cognitive domains represents only a very few species: in 11 of the 15 domains of table 6.1, we have data for 5 or fewer species. Generalizations across species are thus risky and ignore some demonstrable differences. However, it seems safe to say that all monkeys have cognitive mapping skills of some sort, including the ability to remember what is at various locations (although there may be some exceptions among the New World monkeys). All monkeys look for hidden objects in small-scale space even if this involves removing obstacles and following visible displacements. One individual monkey tested in a rigorous manner has followed the invisible displacements of objects (*Cebus apella*), but only about a dozen individuals from four species (three macaques) have been appropriately tested. One baboon species has shown the ability to mentally rotate objects, and other monkey species have shown various abilities to solve detours and mazes, both bodily and by manipulating objects, in one case by manipulating a cursor on a video screen.

Different monkey species have different tendencies to manipulate objects; baboons and capuchin monkeys are the most highly manipulative. No monkey species uses tools extensively in the wild, although in captivity some baboons, capuchins, and macaques use tools quite skillfully. Capuchins and macaques have shown an understanding of causality in some situations (the support problem), but capuchins have not in other situations (complex forms of tool use). Various monkey species have been shown to learn discriminations, including those based on natural categories and relational categories, and to form learning sets of various kinds. Monkeys are not particularly proficient at physically sorting objects into groups; however, this may be due partly to their modest manipulative skills relative to chimpanzees. At least three monkey species are quite proficient at learning to estimate the numerousness of groups of objects and at learning to create a mental representation of an ordered series of stimuli that may support transitive inferences.

6.1.3 Apes

The data for apes are also very spotty in some domains and almost nonexistent for the lesser apes. However, we have more data for the great apes than any other group of primates, with data for chimpanzees available in all 15 domains of table 6.1 (data are available for 8 domains for gorillas, 7 for orangutans, and 4 for bonobos). In all, it seems safe to say that all great apes have cognitive mapping skills of some sort, including the ability to remember what is at various locations, and all great apes look for hidden objects in small-scale space even if this involves removing obstacles or following invisible displacements (although the data on invisible displacement using rigorous testing procedures is only for one gorilla individual). Great apes are as skillful as monkeys in solving detours and mazes.

Different great ape species have different tendencies to manipulate objects, but they all seem to be more manipulative than the various monkey species (with the possible exception of capuchins). Chimpanzees are the only apes that use tools extensively in the wild (although one population of a subspecies of orangutan has recently been observed in two tool-use practices), although the other great ape species use tools in captivity with some skill. Several great ape individuals have shown some understanding of causality in some complex tool-use situations and have also shown an understanding of hidden causes in some instances. All great ape species can learn to make various discriminations, including those based on relational categories (they have not been tested for natural categories), and can learn to form learning sets. Chimpanzees (the only great apes tested) are proficient at physically sorting objects into groups, especially with training. Chimpanzees also can estimate the numerousness of groups of objects and learn to create an ordered series of stimuli that will support transitive inferences. Chimpanzees have been taught to "count" objects and "sum" quantities, and chimpanzees and orangutans have shown some abilities to "conserve" quantities across transformations.

6.1.4 Comparison of Species

It may be useful to make some summary comparative statements about the physical cognition of the different primate species. Because there are so few data available on the prosimians and lesser apes, it is best to simply compare the cognitive skills of monkeys and great apes, which is in any case the comparison of most interest theoretically to many researchers in the study of primate cognition. In general, based on our review, these differences have been grossly exaggerated by some researchers (see chapter 12 for further discussion).

Of most importance, many learning theorists have tried to maintain the position that there is a qualitative difference between the learning and cognitive skills of monkeys and apes and that this is correlated with their relative brain sizes. Many of the older studies claim monkey–ape differences with little or no support from their data other than a mean difference in performance between monkeys and apes as groups, even though some monkeys perform better than some apes (e.g., Harlow and colleagues). More recently, Rumbaugh and colleagues have claimed in some reports that monkeys approach problems associationistically, whereas apes approach them representationally. But their own data belie that difference, as both Rumbaugh and Pate (1984a,b) and Washburn and Rumbaugh (1991b) found representational strategies in rhesus macaques (see also Essock-Vitale, 1978). It should also be noted that several monkey species have shown clear skills of cognitive representation in tasks of spatial memory and transitivity. And so, while it is possible that apes as a group may be a bit quicker to learn in some tasks than monkeys are, we find basically no data to support the contention that there is a qualitative difference in the learning or cognitive skills of these two primate taxa.

Visalberghi, Antinucci, and colleagues have also claimed that the crucial difference between great apes and monkeys (specifically, capuchin monkeys for Visalberghi, and both capuchins and macaques for Antinucci) is that only great apes employ cognitive representation in their problem-solving strategies. But, as noted above, monkeys and many other animals cognitively represent all kinds of absent realities in foraging and other contexts, and monkeys can form mental representations of hidden objects and even transitive series. In the domain of causality, which has been Visalberghi's major concern, she finds that some great apes are more skillful in some tasks of causal understanding (involving tools) than are capuchin monkeys. But until other monkey species are tested, the question is still open whether this difference characterizes apes versus monkeys as taxa. In any case, it would not seem that the difference is that apes but not monkeys can form mental representations of the world.

It should be noted, finally, that apes have been administered a number of tasks that monkeys have not, for example, in classification and in the quantitative domain. And in many cases the subjects of these studies are apes that have received much human training of one sort or another. Since there is some evidence that the human rearing and training of nonhuman primates can lead to some new cognitive skills (Call & Tomasello, 1996), a fair comparison in these cases would be to raise and train some monkey species (preferably large-brained Old World monkeys such as baboons), and then give them some of these same tasks.

6.2 Theories of Proximate Mechanism

There are basically two systematic theories of comparative primate cognition in the physical domain, one based on behavioristic learning theory (Thomas, 1980, 1986) and one based on Piaget's theory of human cognitive development (Parker & Gibson, 1977, 1979). What they both share is the positing of a common metric along which "the intelligence" of species may be measured and compared. We are not convinced that such a common metric is useful, because we do not believe in a single unidimensional entity called "intelligence," although elements of both theories are extremely useful in particular cognitive domains in which we find behavioral adaptations showing flexibility, complexity, and representation. We are much more partial to the views of theorists such as Schneirla (1966), who view the evolution of behavior as involving changes in the way organism–environment behavioral interactions are organized in various domains.

6.2.1 A Learning Theory Approach

Thomas (1980, 1986; Thomas & Walden, 1985) is interested in "intelligence" and explicitly equates intelligence with the ability to learn. He therefore adopts an approach to primate cognition that comes directly from the learning theory of Gagné (1970). As can be seen in Table 6.2, there are eight levels in the hierarchy of learning types, some including the learning of conceptual material. The lowest levels are (1) habituation, (2) signal learning (classical conditioning), (3) stimulus–response learning (operant conditioning), (4) the chaining of operants, and (5) the concurrent discrimination of multiple stimuli. Thomas notes that these are levels of learning that have been shown, in one form or another, in many different vertebrate species, and so they are clearly not unique to primates. There may be quantitative differences in how quickly particular species learn at each of these levels; however, there is no compelling evidence that primates are particularly quick discrimination learners as compared with other mammals. Also, at the fourth and fifth levels there may be differences in the number of chains or concurrent discriminations a species may make, although there is no compelling evidence for special primate skills at these levels either.

Level 6 is the first level that involves concepts that in some way go beyond the perceptual information given. Level 6 involves class concepts, of which there are two types: absolute and relative. Absolute class concepts involve such things as natural categories and may be characteristic of a wide variety of vertebrate species, although they have only been extensively studied in pigeons and monkeys. However, according to Thomas (1986), there is currently fairly reliable evidence that the ability to learn relative class concepts (e.g., identity, oddity, sameness–difference) may be an exclusively primate skill. There is no evidence of differences within primates. Levels 7 and 8 deal with discrimination learning based on the logical operations of conjunction (*and*), disjunction (*or*), conditional (*if*), and biconditional (*if and only if*), given that the stimuli being related in these ways are class concepts. An animal might thus learn to respond in a discrimination task in conditions where both A and B, or either A or B, are present, where A and B are themselves class concepts (e.g., some type of red object, or a natural category). Although the interpretation of the study is less than straightforward, some monkeys and chimpanzees may have the ability to learn discriminations based on such conjunctions, disjunctions, and conditionals (e.g., Wells & Deffenbacher, 1967; no animals were tested in the biconditional). Whatever the interpretation of this single study, there are currently no demonstrated differences within primates at this level of discrimination learning either.

While this approach to primate cognition has much to offer (and indeed its distinctions among different types of concepts is very useful), it also has some limitations. The first limitation is that it views intelligence (or the ability to learn) as a unidimensional construct. But more and more evidence from behavioral ecology is showing that an animal may learn in one way in one behavioral domain and in another way in another behavioral domain, depending on how it is biologically prepared. For example, many seed-caching birds can learn many places where food is hidden, but it is unclear if they can learn to remember as many stimuli in other domains. It seems possible also that an animal could form concepts in one domain (e.g., species of trees bearing fruit), but not in other domains (e.g., species of predator animals). The ecological approach proposes to examine each domain separately and to examine the particular skills used to solve particular problems within domains.

Table 6.2. A hierarchy of intellective (learning) abilities (from Thomas, 1980).

Concept type	Level	Description
Relational	8	Biconditional concepts[a]
	7	Conditional concepts[a] Conjunctive concepts[a] Disjunctive concepts[a]
Class	6	Affirmative concepts[a] (Absolute / Relative)
	5	Concurrent discriminations
	4	Chaining
	3	Stimulus–response learning
	2	Signal learning
	1	Habituation

[a] The complements of these logical operations also apply at their corresponding levels (namely, affirmation/negation, disjunction/joint denial, conjunction/alternative denial, conditional/exclusion, and biconditional/exclusive disjunction (*Millward*, 1971: p. 940) and appropriate tests of the complements might be used.

A second and related problem is that learning theorists tend to study animal cognitive skills in highly artificial situations for which there can be no biological preparation. This means that in most learning experiments the animal is actively prevented from deploying its cognitive skills to try to understand what is happening in a situation. In most discrimination learning experiments there is no question of solving a problem by causal analysis of how objects relate to one another or the like, but rather the organism must discern what a human experimenter wants it to do (as communicated by the way the human dispenses food reinforcements). In Thomas's approach, therefore, even though there is a supplementing of traditional behavioristic learning processes with so-called conceptual processes, the focus remains on discrimination learning, and so there is no account of intelligent behavior involving concrete epistemological categories such as space, objects, and causality. It is operating with these kinds of concepts that allows for insightful problem solving and other varieties of intelligent behavior.

6.2.2 A Piagetian Approach

Parker and Gibson (1977, 1979) propose a theory in which the ontogeny of a primate species repeats, to some extent, its phylogeny. Drawing on the analysis of Gould (1977) of the different types of relationships that may obtain between ontogeny and phylogeny, they concluded that in the order primates, the *scala naturae* from prosimians to monkeys to apes to humans is characterized by "terminal additions." Terminal additions, which may take place in conjunction with neotony of various types, means that a "higher" species repeats the ontogeny of the common ancestor it had with "lower" species, and then continues on with some added complexities that have evolved subsequently, for example, when the "higher" species radiates into a new and more complex niche while the"lower" form remains in the niche to which the common ancestor was adapted. Terminal additions in the domain of cognition represent added complexity and so may be said to be "more intelligent."

The theoretical framework within which Parker and Gibson make the comparisons among primate species is Piaget's (1970) theory of cognitive development, created for the analysis of human ontogeny. As can be seen in Table 6.3, Parker and Gibson (1979) believe that the different primate species may be ordered with respect to the Piagetian stages and substages. They posit in the evolution of primate intelligence a prosimian stage, an Old World monkey stage, a great ape stage, and a hominid stage, each of which corresponds, respectively, with a Piagetian stage characteristic of older and older human children. Of special importance in this scheme is "tertiary sensory–motor intelligence," that is, object manipulation skills involving relating objects to one another (including, possibly, tool use) characteristic of Piaget's stage 5 of sensory–motor intelligence. Also important is stage 6 of sensory–motor intelligence (and the later preoperational period) in which organisms deal with absent realities of various complex sorts. New World capuchin monkeys would seem to present a problem for this account, as they manipulate objects and use tools in complex ways (in captivity, at least). Parker and Gibson deal with this apparent discrepancy by hypothesizing that capuchin monkeys underwent a parallel evolution with apes in some of their intellectual skills. The ecological conditions that might have led to this parallel evolution of tertiary sensory–motor intelligence in apes and capuchin monkeys are discussed below.

The Piagetian approach to primate cognition has much to recommend it. The main problem with the Parker and Gibson (1979) theory specifically, however, is that, like learning theory, it views intelligence as a unidimensional construct of which species may have more or less (and in this theory remnants of the *scala naturae* are even more apparent as current day species are taken as representative of ancient primate ancestors) . This view derives from the Piagetian assumption that there are stages of cognition that may be described generally [actually, in terms of a special kind of mathematical group theory that was the formal basis of his theory (Piaget, 1970)] so that all the different domains of cognition may be accounted for within the same set of stages. It turns out, however, that these stages do not emerge together across domains in human cognitive development as Piaget thought they should (Gelman & Baillargeon, 1983). As Parker and Gibson so usefully reveal, some primate species show stages of development in some domains (e.g., object permanence) that, using Piaget's numbered stages, they do not show in other domains (e.g., object manipulation).

Indeed, one general point that comes out clearly in Parker and Gibson's analysis is that many nonhuman species, primate and otherwise, may have cognitive skills that simply do not manifest themselves in the manipulation of objects. Since manipulating objects is the behavioral expression that Piaget used almost exclusively as an index of cognitive functioning in human infants (even in object permanence), the cognitive skills of some species will be seriously underestimated. As one instance, we find it hard to believe, as Parker and Gibson (1979) claim, that prosimians do not have an understanding that objects continue to exist when they are not perceived; the Jolly (1964a) study only showed that prosimians will not remove a cover from a piece of food. If prosimians were tested in the Baillargeon et al. (1985) procedure in which observers measure their habituation to various possible and impossible events relying on an understanding of the existence of absent objects, for example, it is likely that they would show stage 4 object permanence, especially since this level of understanding is

Table 6.3. Parker and Gibson's theory of primate physical cognition. (Reprinted from *Behavioral and Brain Sciences*, 1979.)

Stage of sensory–motor intelligence	Prosimian	Old World monkey	Great ape	Early hominid
Stages 1 & 2				
Simple prehension, hand–mouth coordination	Manual prey catching, branch-clinging, climbing by grasping			
Stage 3				
Hand–eye coordination		Manual foraging		
Secondary circular reactions			Object play for tool use	Same
Stage 4				
Coordination and application of manual schemes on single objects		Manual food preparation and cleaning, manual grooming		
Stage 5				
Object permanence		Food location memory (?)	Same	Same
Object–object coordinations, trial-and-error investigation of object properties (tertiary circular reactions), discovery of new means (tool use)			Trial-and-error discovery of tool use for extractive foraging on embedded foods	Same
Stage 6				
Deferred imitation of novel schemes			Imitative learning of tool-use traditions	Same
Mental representation of images of actions			Search for new embedded foods, insightful tool use	Same

shown by many nonprimate species whose manipulative skills are more adapted to the task. Overall, then, while the series of stages that Piaget proposed may be extremely useful for analyzing primate cognition within particular domains, their use in constructing a single unidimensional scale of intelligence is suspect on both theoretical and empirical grounds. A more ecological version of the theory, sensitive to the possibility of different cognitive mechanisms in different behavioral domains, is needed.

6.3 Theories of Ultimate Causation

Along with most other theorists, we assume that the evolution of primate physical cognition has taken place mainly in the context of foraging for food (with possible contributions from other factors such as escaping predators). The two main theories along these lines have focused on the complexities involved in primates: (1) locating and identifying food so as to become efficient foragers in their particular niches, and (2) extracting and processing food (involving object manipulations) so as to exploit new food resources. The first of these complexities is best represented by Milton (1981, 1988) and the second by Parker and Gibson (1977, 1979). There are almost no fossil remains of primates that speak directly to either of these theories.

6.3.1 Foraging

Milton (1981, 1988) proposed the hypothesis that the diets of various primates species determine certain kinds of foraging behaviors and that foraging demands are the central adaptive factor in determining primate cognitive skills. She begins by noting that all extant primates, and probably most if not all extinct primates, have plant foods as a relatively major part of their diets—this is especially true of those species with relatively large bodies who live in tropical forests. The problem is that particular types of plant foods are patchily distributed in the space of the forest. Particular types of plant foods are also patchily distributed in the forest temporally: flowers, new leaves, and ripe fruits are only available for several months of the year, and these times may differ for different plant species. Interestingly, however, Milton (1981) also observed that although many plant foods are patchily distributed in both space and time, they are also somewhat predictable. A given tree does not move in the lifetime of an individual, trees of the same species tend to occur in clumps in the forest, and the growth of new leaves, flowers, and fruits occurs at predictable times during the year for a given plant species.

The hypothesis is that the intelligence of primates, especially monkeys and apes, evolved to deal with foraging in which there is a complex set of problems that certain kinds of cognitive skills could help to solve (since the problems have some predictable solutions). Patchiness in space and time is equivalent, in this theory, to complexity. That is, because day-range size is a limiting factor (animals can only cover so much ground in a day), spatial patchiness determines that a given species has to have a relatively diverse diet of plant foods. Similarly, because energetic requirements are also a limiting factor (animals can only go so long without eating), temporal patchiness also determines diversity of diet. There is thus a correlation of patchiness with diversity of diet, and diversity of diet with the spatial–temporal complexities needed for efficient foraging. The hypothesis is thus that the greater the complexity of foraging demands on a species (the

Figure 6.1. Howler monkey. (Photo courtesy of M. Seres.)

patchier the food resources it exploits), the greater the complexity of its cognitive skills, given that the patchy resources have predictable spatial and temporal parameters.

The outcome is that many primates evolved complex cognitive skills to locate and identify (and possibly quantify) food sources. Because conditions in tropical forests may change rapidly, it is best that primate youngsters be born not with knowledge of particular kinds of foods or locations but with flexible and adaptable learning skills so that they can learn to map out their particular environments and forage efficiently. But there are also possible differences in this basic foraging paradigm within different primate species, and indeed, Milton (1981) proposes a scale along which plant foods that primates eat may be measured for patchiness. In general, mature leaves are least patchily distributed in space and time, then new leaves and flowers, and then ripe fruits.

As evidence for her hypothesis, Milton (1981) reviews her studies of two species of New World monkeys that are of about the same body size, live in social groups of similar size, and inhabit roughly comparable (indeed often sympatric) parts of the forests of Central and South America: mantled howler monkeys (*Aloutta palliata*) and black-handed spider monkeys (*Ateles geoffroyi*). Howler monkeys are mainly leaf

Figure 6.2. Spider monkey. (Photo courtesy of M. Seres.)

eaters, frequenting over 150 species of plants, but avoiding many others (and indeed they often refuse novel foods in captivity). Spider monkeys, on the other hand, are mainly fruit eaters. Because there are fewer species of plants that bear fruit than leaves and because fruit is more patchily distributed temporally, the foraging range of spider monkeys is approximately 25 times the size of the foraging range of howlers (at least for her island-living study groups). It is noteworthy that ontogenetically spider monkeys have a much longer period of dependence on their mothers, which is often associated with more that needs to be learned before independence. Milton has also observed that when spider monkeys find a particular species of tree with ripe fruit, they will often feed and then go directly to nearby trees of this same species, presumably on the expectation that because they are trees of a similar type they will be fruiting also (see also Menzel, 1991, for further evidence of this type).

The correlation that Milton uses to test her hypothesis is essentially between the patchiness of the food resources exploited by a species and its brain size (corrected for body size if necessary). She finds that spider monkeys have brains that are roughly twice as large as those of howler monkeys. In this connection, she also cites the work of Clutton-Brock and Harvey (1980), who found a correlation between patchiness of food sources (as measured by the amount of territory a species has to cover each day; day-range size) and relative brain size among a large number of primate species (see also Milton, 1988). (A similar correlation has also been found among species of the order Chiroptera by Eisenberg & Wilson, 1978.) Milton speculates that when hominids moved onto the savanna, their history of adaptation to patchy food sources (relying on cognition) acted as a preadaptation when they needed to compensate for the insufficient plant foods available in that environment (given the extreme patchiness of edible plants on the savanna and new concerns for water and terrestrial predators) by such things as hunting for meat. Given that their previous vegetarian lives had given them no "jaws or claws" of consequence, they had to hunt in groups, which provided a context for the evolution of many important social–cognitive and communicative skills as well.

Recently a number of problems with this hypothesis have been raised. First, the hypothesis assumes, as do all the other theories we have reviewed, that intelligence is one entity that can be assessed with one measure. Moreover, some have questioned the one measure of intelligence Milton uses, that is, relative brain size (Byrne, 1994); the real measure should be encephalization, since body size (which Milton and others use as a correction factor) is irrelevant in this context. Dunbar (1992, 1995) also questioned the use of day-range size as a measure of foraging complexity, pointing out that "the same" environment is differentially complex for different organisms, with size being an important determiner (he thus corrects day-range size for body size). When these two changes in the operationalization to measures of intelligence and day-range size were made, Dunbar (1992, 1995) did not find a correlation between intelligence (as measured by neocortical expansion) and complexity of foraging requirements (as measured by day-range size corrected for body size) (see chapter 12 for further discussion). Finally, it is also not entirely clear that some nonprimate species do not face foraging demands that are just as challenging as those faced by primates (especially some species of birds; King, 1986a).

6.3.2 Extractive Foraging

Parker and Gibson (1977, 1979) provide an evolutionary dimension for their Piagetian theory of primate cognition. Like Milton, their focus is on foraging, but they focus more particularly on extractive foraging. "Extractive foraging" refers to the process by which an animal extracts food that is not directly perceptible from a substrate or shell. Examples include opening nuts and hard-shelled fruits, digging in the ground for tubers, scrounging in leaves or under bark for insects, turning over rocks for insects, and using tools to probe inside holes for insects or honey. Extractive foraging usually is associated with omnivorous animals that feed on a variety of foods that are available only seasonally. Parker and Gibson's idea is that extractive foraging poses two unique foraging problems: the food is not directly perceptible and extracting it often requires complex object manipulation skills (sometimes including tool use). For an animal to be a skillful extractive forager, therefore, the hypothesis is that it should have skills of object manipulation and mental representation at or beyond the level of Piaget's stage 5 of sensory–motor development (tertiary object relations and the search for hidden objects after visible displacements).

Figure 6.3. Stumptail macaque engaged in extractive foraging. (Photo by J. Call.)

The data Parker and Gibson use to support their hypothesis are observations and studies showing that chimpanzees, who score high on Piaget's sensory–motor scales, are omnivorous extractive foragers that use tools (as are humans). Gorillas and orangutans, who also score quite high on Piagetian scales of sensory–motor intelligence, are not species-wide tool users in the wild, but they are extractive foragers and skillful tool users in captivity. The hypothesis is thus that the common ancestor of humans and great apes was a tool-using, extractive-foraging omnivore. The reason gorillas and orangutans do not use tools extensively to forage in the wild (but do use tools when necessary in captivity) is that they have mostly radiated into feeding niches that do not require the use of tools. The other primates that are a tool-using, extractive-foraging omnivores are all species of capuchin monkey, and they also score high on the Piagetian scales. Their phylogenetic distance from great apes, and the supposed absence of this same foraging pattern in other primates, leads to the hypothesis that the tertiary sensory–motor adaptation arose a second time in capuchin monkeys in response to extractive foraging problems similar to those faced by great apes. Gibson (1986) provides support for the hypothesis in a general way by finding a relationship within primates between omnivory and relative brain size (corrected for body size).

The major problem with the theory, in addition to its conceptualization of intelligence as a unidimensional construct, is that extractive foraging is a more widespread feeding strategy than Parker and Gibson recognize, and it is not always correlated with high-level performance on Piagetian scales of intelligence. King (1986a) notes that many prosimians are classified by Jolly (1985) as "pokers and pryers" as they unroll leaves, chew open dead trunks, pry bark off trees, and turn over stones in search of insects. But prosimians are classified by Parker and Gibson as possessing only stages

1 and 2 of Piagetian sensory–motor intelligence, and indeed there are no theories of primate intelligence that rank prosimians highly (see Parker and Gibson, 1979: table 3). Many monkey species also engage in extractive foraging as well, sometimes in ways that seem as complex as that of great apes. For example, King (1986a) notes that Chacma baboons (*Papio ursinus*) and yellow baboons (*P. cynocephalus*) each have three or four types of extractive foraging, including digging earth, stripping bark, overturning stones, unfolding plants, and so forth (and other species of baboons have been observed to use tools in captivity; see chapter 3). The problem in this case is that Old World monkeys are not included in the specific proposal of Parker and Gibson of the two evolutionary origins of tertiary sensory–motor intelligence in great apes and capuchin monkeys (see also Westergaard, 1988, for further evidence of monkey skills in this domain). Perhaps most importantly, when Dunbar (1995) attempted a specific correlation of extractive foraging and relative neocortex size among a number or primate species, no correlation was found. King (1986a) also pointed out that extractive foraging is not correlated with any known measures of intelligence in taxa other than primates.

6.4 Directions for Future Research

As we have stressed repeatedly, the main problem with all of these approaches to primate cognition is their concern with providing a metric for species' overall "intelligence." With regard to proximate mechanisms, Thomas (1980, 1986) does this with an amalgamation of learning theory concepts and concepts based on truth-conditional class logic. Parker and Gibson's (1977, 1979) Piagetian theory is based on a wider array of epistemological categories and a more broad-based logic of relations and so is preferable—if we can modify it to eliminate reliance on broad-based stages that cut across all domains of functioning and all animal species. The approaches from the point of view of ultimate causation (e.g., Gibson, 1986; Milton, 1981) suffer from this same malady as well, as evidenced by their continued reliance on overall brain size (measured in various ways) to assess "the intelligence" of a species. In an ecological approach to cognition, on the other hand, we look first at the ecological problem to be solved and then at how a particular species solves it, with comparisons among species taking place only within this domain. This does not preclude the possibility that in some cases solutions in one domain may in some way be useful in other behavioral domains.

How to clearly delineate different behavioral domains is a difficult problem. But one approach is to focus on adaptive functions, for example, locating food, manipulating food, identifying food, and quantifying food (corresponding, roughly, to the four content chapters that make up part I of this book). We then look for the use of cognitive mechanisms in these domains, that is, behavioral adaptations that display some degree of flexibility, complexity, and cognitive representation, and how they relate to the particular demands faced by a species. In doing this across species, phylogentic relatedness must always be kept in mind because a particular environmental demand will not lead to the same outcome in different animal taxa; a demand always occurs in the context of the species' existing morphology and behavioral competencies. Thus, the same foraging demand may lead to different solutions in different species; for example, extracting insects from holes may be solved by evolving an elongated beak/

Figure 6.4. Gorilla processing food. (Photo by J. Call.)

tongue and narrowly adapted behaviors to use it, or it may be solved via more flexible behavioral adaptations involving tool using skills. Moreover, sometimes even "the same" behavioral solution may be supported by different cognitive mechanisms, for example, the use of a stick to obtain insects from their holes may be accomplished in a context-specific and inflexible manner, as it seems to be by some birds, or it may involve flexible use of the stick and even the modification or creation of new stick-tools through an understanding of the spatial or causal relations involved, as it seems to be by chimpanzees.

With these considerations in mind, we would like to close our account of primate physical cognition with some suggestions for future research. These will be placed within the context of what we know about each of the four basic domains of primate physical cognition as we have reviewed them. It must be acknowledged, once again, that for most of the general statements we can make, we know next to nothing about prosimians and lesser apes.

6.4.1 Finding Objects and the Knowledge of Space

The foraging ecology of primates is, in broad outline, similar to that of many mammalian species. Like other mammals, primates employ cognitive mapping skills, enabling shortcuts and detours, for efficient travel during foraging. But primates have an especially challenging set of resources to exploit, as the fruits and plants on which most primates survive are patchily distributed in space and time. Presumably because of cognitive adaptations to this feeding niche, at least some primates are able to remember not just where food is (as do many food-caching animal species), but they actually remember what kinds of foods are at particular locations (even when they did not hide the food themselves). In terms of temporal parameters, some primate species are also able to make categorical inferences about the availability of particular types of food; for example, finding one instance of a ripe fruit implies that other trees of the same species should also be bearing ripe fruit at the same time. An important question for future research into the ecology of primate cognition, therefore, is whether other mammals have these same skills to remember "what" is "where" and to make categorical inferences about food availability. Knowing this would help answer the question of whether these skills emerged in response to the particular feeding ecologies of primates, or some species of primates, or from more general ecological demands as faced by some mammalian ancestor before primates emerged. Also important, of course, is to know how widely these skills are distributed among primates, as only a few species have been studied in each case, and whether they are easily applied to different types of objects, for example, social objects and their locations. Also of interest would be quantitative comparisons of how many different objects at how many different locations can be remembered and how this relates to the particular feeding ecologies of different species.

Basic skills of object permanence in which animals know that food and other objects exist and remain at that location when they are not being perceived (stages 4 and 5), are widespread among mammals as well and are necessary for the efficient foraging of many species. (However, new methods are needed for assessing these skills in prosimians, for whom the traditional task is so poorly adapted.) Because appropriate methodologies have only recently been developed, it is not clear which species are able to track the invisible displacements of objects (stage 6). In general, this skill would not seem useful in the foraging of folivores because plant foods generally stay in one place; it might seem more useful in tracking moving prey as they attempt to escape, tracking moving predators as they approach, or keeping track of group mates. Consistent with this view, at least some nonprimate mammals (e.g., dogs) seem to understand the invisible displacement of objects in a way that at least some primate species (e.g., several species of monkeys) do not. Although the data are far from definitive, gorillas also seem to be able to track invisible displacements. Since they mostly feed on completely sessile plant foods and do not have many stalking predators, it is possible that the tracking of the invisible displacements of objects evolved for apes (if indeed other apes have this same skill) in the context of keeping track of social partners who move about continuously. Another important question for future research, therefore, is just which species, primate and nonprimate, are able to track the invisible movements of objects and whether this bears any relation to their particular feeding practices or to some other aspect of their ecological niches.

6.4.2 Manipulating Objects and the Knowledge of Causality

The object manipulation skills of primates are, on the whole, especially flexible and complex as compared with those of other mammalian species, presumably because primates have evolved as arboreal hand feeders. Within primates, however, different species use their hands with respect to objects in many different ways. Because an organism's perception and understanding of an object is intimately tied to the affordances it has for its action (Gibson, 1979), an important line of future research would involve systematic comparisons of the object manipulation skills of different primate species, using more complex and more meaningful sets of objects than have previously been used in comparative studies. One good example is a small study by Milton (1984), who gave an unusual type of nut to several different species of captive primates. Howler monkeys and spider monkeys sniffed and/or bit and/or licked the nut shell for a second or two and then dropped it and lost interest; they did not seem to recognize it as food. Woolly monkeys cracked it open with their teeth. Capuchin monkeys banged it against the ground and pried open the resulting cracks in the shell with their teeth. Studies such as these tell us much more about the manipulative proclivities of particular primates and how this relates to their individual needs in nature, than do studies of primates manipulating human objects whose functions are irrelevant to their lives. Studies in which multiple objects may be manipulated with respect to one another are especially needed.

One of the great mysteries in the study of primate cognition is why so many species are such skillful tool users in captivity, yet they do not use tools extensively in the wild. Presumably they have some general skills of sensory–motor cognition that develop in their natural habitats in contexts other than tool use that are then somehow exploited in tool-use tasks. More research is clearly needed into the different tool-use skills of primates, especially in species other than chimpanzees and capuchin monkeys. Of particular promise are studies of the type pioneered by Visalberghi and colleagues in which not just the manipulative skills employed, but also the causal understanding involved, is identified. So far, however, only a few species have been tested (great apes and capuchin monkeys), and these are phylogenetically quite far removed from one another. Investigating the causal understanding of some Old World monkeys would therefore seem to be an especially important direction for future research. It is also possible that manipulating objects outside of foraging situations might also be important for causal understanding, for example, in nest building, grooming, and other activities, and these have not been widely studied. Also needed are a wider range of tasks to assess causal knowledge more broadly, perhaps even in situations in which subjects do not manipulate objects or use tools (e.g., some kind of habituation procedure), in order to assess their understanding of causal events in which their own action is not the initiating cause. In all cases the skills of tool use and causal understanding that different species display should be related to their manipulations of objects and their experiences of the interrelations of objects in their natural environments.

6.4.3 Identifying Objects and the Knowledge of Categories

Discriminating and categorizing objects depends on the perception of object features and the similarities and differences among features. These skills might be useful in a number of domains including foraging and social behavior. The perception of basic features and categories of objects seem to be the same among many mammalian species. Where primates have skills that other mammals do not is in relational categories. Thus,

Figure 6.5. Liontail macaque. (Photo courtesy of M. Seres.)

in discrimination learning experiments, nonprimate animals do not seem to comprehend cognitive categories that depend on a comparison of objects (e.g., identity, oddity, sameness–difference), with the possible exception of one individual, highly trained bird. More research is clearly needed to see if this cognitive skill is indeed uniquely primate. What might be the evolutionary basis of such categories in foraging or other manipulative activities, and why they should be uniquely primate, is not immediately clear, and indeed these categories are only in evidence after repeated experimental trials in artificial settings. Overall, foraging does not seem a likely place for the evolution of relational categories, and indeed we argue in chapter 12 that they evolved in primates for understanding social relationships, a domain in which they would seem to be especially useful and in which primates may have some special ecological demands. Discrimination learning experiments in the laboratory simply train individual primates to apply these categories to inanimate objects. In line with Premack's (1983) analysis, of special importance in this domain of research with new and different species would be studies of relation matching in which the relation to be identified is simultaneously present in two pairs of objects that must be somehow related to one another.

Few animal species have any skills at sorting objects into groups the way that human infants and children do with such facility during the preschool years, and the ecological relevance of such skills for most species is not easy to determine. Primates have some relevant skills (combining their skills of categorization and manipulation), but they would seem to be extremely limited compared with humans. Since categorization tasks, at which primates are skillful, simply ask individuals to identify objects from the same category across trials, it is possible that one of the problems that primates and other animals have with sorting objects is the need to coordinate simultaneously both similarities and differences. New methodologies need to be developed to assess the skills of various primate species at categorizing and physically sorting objects, both when the process is carried out sequentially and when it must be carried out with many objects simultaneously. Also important is the fact that the one individual chimpanzee that showed some humanlike skills of object sorting (e.g., sorting the same objects in more than one way, a skill demonstrated by no other individuals) was raised by humans in the presence of much object manipulation with human objects and artifacts. It is thus possible that other species would develop more sophisticated sorting skills if they had extensive experience with many different objects throughout their early ontogenies. This possibility needs to be further investigated, perhaps especially with Old World monkeys.

6.4.4 Quantifying Objects and the Knowledge of Number

Primates do not seem special relative to other animals in their skills at estimating quantities (subitizing), which would be an obvious advantage to all animals in increasing foraging efficiency. Summation skills have only been tested in apes, and they might just be an elaboration of estimating quantities and could be present in other species if tested. It is likely that skills of "counting" related to estimating the number of reinforcers to be given or the number of trials to be finished before a reinforcer is given are not true counting based on a concept of number, and, in any case, primates do not have special skills in this regard. Capaldi and Miller (1988) believe that numerical skills of one sort or another are fundamentally important in the foraging of many species.

Primates may have an understanding of ordinal relations not possessed by other animals, however, as evidenced by the failure of pigeons and the success of capuchin monkeys and chimpanzees at comprehending transitive inferences based on mental representations. Much needed in this regard are more studies on nonprimate mammals (beyond the one study with rats, which differed in methodology from the primate studies), and of special importance would be studies of transitivity based on real physical dimensions, rather than just experimentally constructed series based on reinforcement contingencies (De Lillo, in press). Understanding real ordinal relations and transitivity would be of some value in foraging (tree A has more fruit that tree B, which has more fruit than tree C; therefore A has more than C). In this case, however, we would not expect to see differences between primates and other foraging species. It is thus also possible that constructing ordinal relations among objects is most directly related to the understanding of dominance hierarchies in the social domain and that, with training, some primates can learn to apply this skill in the physical domain as well. In this case, we might expect to see special skills in primates and, perhaps, other animals phyletically close to primates for whom dominance is an important part of their social lives. It is important to note, in this regard, that ordinal relations are a special type of relational category, and so it may be that only primates (and perhaps only primates with certain kinds of social lives) have a need to employ them. More research is needed into the different species that do and do not possess the relevant skills. Also, the special "counting" skills displayed by one human-raised and trained chimpanzee should be investigated as well in other primate species who are given the same kinds of training.

Primate skills at understanding some of the basic logic of quantities, namely, that quantity is conserved across certain spatial transformations such as change of location and change of shape, have not been extensively investigated. The studies that exist are not conclusive and concern only three species (squirrel monkeys, orangutans, and chimpanzees). Conservation skills would seem to have some relevance to foraging, although it is not clear that the subtleties involved would be important to an animal's survival, nor is it clear that there is any ecological reason why primates should have special skills at conservation. But conservation skills do seem to involve some relating of quantities to one another, and so they could conceivably derive from other relational skills that primates alone possess (as kind of epiphenomena). There is once again a need for more research, especially with Old World monkeys and a variety of conservation tasks. Although it is difficult to test for quantitative concepts of this nature with nonverbal organisms, if it could be determined that nonhuman primates have conservation skills it would be a very important fact because these are skills that human beings do not develop until 5–7 years of age, an age well beyond that at which the cognitive skills of human and nonhuman primates normally compare.

6.5 Conclusion

The cognitive revolution is just now making itself felt in the study of primate behavior. There is much to be done, and in some cases done better. Research from a Piagetian point of view has often lacked the kind of rigorous controls needed to make definitive scientific statements. Research from the learning theory tradition has often studied skills in highly artificial circumstances that are only displayed after many hundreds, or even thousands, of experimental trials. In neither theoretical paradigm has there been sufficient appreciation for the domain-specific dimensions of cognition and how these might relate to the particular ecological niches of different primate species. In neither paradigm have a wide range of primate species been investigated and compared both with each other and with nonprimate species.

We have advocated throughout our account that primate cognition should be investigated, to the extent possible, in the context of the natural ecology of the species involved. We have also inveighed against investigations into "the intelligence" of species and have instead advocated investigations into particular cognitive domains. But this point of view does not mean that there are no cognitive skills that underlie behavior in a wide range of circumstances—either within the various domains of physical cognition or even across the physical to the social domain. Indeed, in several places we have argued that a skill such as cognitive mapping might have evolved for finding food but then, because conspecifics are physical objects as well, might have been applied in locating group mates (or the skill might have evolved for its usefulness in both contexts). Conversely, we have suggested the possibility that experiments in which primates relate objects to one another might be tapping skills that evolved so that primates could relate group mates to one another. Thus, we do not believe that domain specificity is an inviolable principle of cognitive architecture, as some evolutionary psychologists seem to think (e.g., Tooby & Cosmides, 1991), but only that it is the most appropriate point of view from which to begin an investigation.

We now proceed to an investigation of primate social cognition, also organized into four functional domains (with divisions among domains made along different lines than those of physical cognition), and then to an attempt to construct an overall theory of primate cognition that encompasses both of these behavioral domains.

PART II

KNOWLEDGE OF THE SOCIAL WORLD

Primates acquire and use cognitive skills not just in dealing with their physical worlds, but also in dealing with their social worlds. For various reasons, primatologists and other behavioral scientists were somewhat slow to recognize this fact. Although there were a few prescient theorists before this time (Chance & Mead, 1953; Jolly, 1966; Kummer, 1967), it is only in the last two decades that primate social cognition has emerged as its own field of study. Indeed, many primatologists today believe that the most complex cognitive problems faced by primates arise mostly in the social domain. It may even be that those aspects of primate cognition that are unique to the order all arose evolutionarily as responses to the complexities of the social environment (Humphrey, 1976). The basic point is that although conspecifics are physical objects that must be located, identified, and in some cases quantified—just as inanimate objects—they also create additional cognitive problems that simply are not present in the world of inanimate objects. In our view there are three basic ways in which social life adds cognitive complexity to the world of primates, and this second part of our investigation is organized around them.

The first and most basic fact of primate social life, indeed of social life in general, is that other organisms behave spontaneously on their own, raising the difficult problem of prediction. Because primates individually recognize many of the members of their social groups, they come to know much about the specific behavioral tendencies of specific individuals, both toward themselves and toward one another making for a highly complex "social field" in which virtually every decision made must take account of the behavioral tendencies and social relationships of virtually all the individuals present. This is especially true as individuals compete with one another by forming various types of coalitions and alliances: cooperation for purposes of competition. These coalitions and alliances also may create a set of "obligations" (if I help you now, I expect you to help me in the future), which constitutes another dimension of complexity in the primate social field. Overall, primates' individualized knowledge of the social fields in which they operate is the central fact of their social lives (as it is for many social species), and so it is with this knowledge that we begin our investigation of social cognition in chapter 7.

The second basic fact of primate social life is that conspecifics cannot be manipulated like physical objects, but rather their behavior can only be influenced or controlled by means of various types of social and communicative strategies. Based on experience within their own particular social fields, primate individuals may create such strategies during their individual ontogenies, or perhaps flexibly modify their use of phylogenetically ritualized communicative signals. These strategies and communicative signals occur in both the gestural and vocal modalities, and even include interactions that some researchers have labeled as deception. The acquisition and use of the most flexible social and communicative strategies by means of which primates attempt to actively manipulate and influence their individual social fields are our central concern in chapter 8. Also considered in this chapter are the types of communicative signaling that may result when primates, especially great apes, are raised in humanlike cultural environments in which they are taught humanlike skills of linguistic communication.

The third fact of primate social life that presents special cognitive problems involves social learning. There are any number of ways that the behavior of conspecifics may be useful to primates and other animals in their adaptations to both the physical and social worlds. For example, as youngsters follow their mothers in the typically mammalian pattern, they may be exposed to novel objects and social interactants, novel transformations of the environment that result from the environmental manipulations of adults, and even the behavioral strategies of adults in problem-solving situations. It is also possible that primate and other species of mothers may actively teach their youngsters how to accomplish certain tasks. These different processes of social learning and teaching and the social understanding that may underlie them are our central focus in chapter 9. If the result of these processes is something like "cultural transmission," then the outcome will be some qualitatively new processes of cognitive evolution involving the "transmission of acquired characteristics" that simply do not take place on the level of purely organic evolution.

Social life thus constitutes an especially important, and perhaps unique, arena for the operation of primate cognition: individuals operate within a complex social field that requires the use of social and communicative strategies and provides all kinds of opportunities for learning from groupmates socially. Underlying all three of these domains of social interaction and cognition is the way in which individuals understand the behavior, and perhaps the psychological states, of the conspecifics with whom they are interacting. Chapter 10, "Theory of Mind," thus attempts to discern what primates know about how their groupmates and other conspecifics work as psychological beings. Evidence is taken from analyses of what individuals would need to know in order to behave as they do in each of the three domains of social interaction reviewed in chapters 7–9, as well as from a number of laboratory experiments aimed at this question directly.

It should be noted that all of the important features of primate social cognition may be characteristic of some nonprimate animal species as well. Perhaps because many primatologists are especially attuned to looking for complex cognitive processes (and perhaps because primatologists are primates), many of the most humanlike cognitive phenomena are reported for primates. But this may not reflect the true state of affairs, as many other animal species also lead cognitively complex social lives (see Rowell & Rowell, 1993, for an interesting discussion). This is a continuing concern in all of the chapters that follow, but it is a special concern of chapter 11, "Theories of Primate Social Cognition," which deals with issues of both proximate mechanism and ultimate causation. It will also be documented in this chapter that all of the theories used to account for primate social cognition have so far been very crude, invoking only very general notions such as "theory of mind," "metarepresentation," and the like, with nothing quite like the six Piagetian stages or the eight qualitatively distinct levels of discrimination learning that have been proposed in the domain of physical cognition. It is possible to discern a rich continuum of possibilities in this domain as well, however, and we present our modest attempt in this direction in this chapter as well.

Before beginning our investigation of primate social cognition, two preliminaries are necessary, one methodological and one theoretical. First, methodologically, readers must be forewarned that much of the research reviewed in this part of the book is based on observational evidence, sometimes from natural habitats. This increases its value as information about the state of affairs in nature, which is, of course, our ultimate concern, but in many cases it also creates problems of interpretation for researchers trying to identify cognitive processes—since, as outlined in chapter 1, making such determinations with confidence often requires experimental methods. With this situation in mind, we try to be especially careful in this part of our investigation to report the precise nature of the observations on which our conclusions are based. The data used to make comparisons to other animal species are also more problematic than the many experimental data available for the physical domain.

Second, because the research on primate social cognition is still fairly new, the theoretical terminology of different scientists is not always consistent. We attempt to be consistent by using the following terminology to designate primate understanding of the social world (see Tomasello, Kruger & Ratner, 1993, for more details):

1. Understanding other organisms as *animate agents* means understanding that they generate their own behavior spontaneously, which distinguishes them from inanimate objects. Understanding that a behavior is *directed* means being able to predict particular events that will follow from it, based on past experience with similar behavior in similar circumstances in the past.

2. Understanding other organisms as *intentional agents* means understanding that they engage in flexible decision-making processes. These processes manifest themselves behaviorally as organisms choose among several possible means or *strategies* toward an independent goal, and perceptually as organisms choose to pay *attention* to some things in their perceptual fields to the active neglect of others.

3. Understanding other organisms as *mental agents* means understanding that they have *thoughts* and *beliefs* about external situations that may conflict with one's own and that may even differ from the true state of affairs as one knows it.

When we wish to remain agnostic about the distinction between items 2 and 3, we speak of organisms understanding others as *psychological* agents, as opposed to animate agents only.

To be absolutely clear about what is at issue, we must emphasize at the outset that there is a fundamental difference between the claim that an organism behaves intentionally or has mental states—which is part and parcel of the cognitive approach we have adopted here—and the claim that an organism understands the intentionality or mental states of others—which is one of the questions we are investigating in this part of the book. Much complex social interaction, requiring complex cognitive processes and skills, may concern the behavioral level only, as organisms make informed decisions about the likely behavior of conspecifics in various situations. If those conspecifics are conceived as intentional or mental agents, it only adds to the complexities involved.

7

Social Knowledge and Interaction

Most primates live in some form of social group. These groups range from the relatively simple social organizations of some prosimian species in which individuals forage alone but then congregate to sleep, to the monogamous pairings of the several species of lesser ape, to the complex societies of many macaque species dominated by matrilineal kinship groupings, to the fission-fusion societies of chimpanzees in which there is a constant flux of smaller groupings within an overall community structure (Smuts et al., 1987). Each of these forms of social life, as well as those of many other animal species, creates difficult cognitive problems that individual primates must learn to deal with adequately if they are to survive and procreate.

The problems of social life are especially complex for species whose cognitive skills serve to create a complex "social field" (Kummer, 1971, 1982). Across virtually all types of social organization, the primate social field is based on several fundamental components. Most importantly, primates recognize individual groupmates and are able to make use of a variety of behavioral and contextual cues to predict some aspects of their behavior. Along with general cognitive processes of learning and memory, these skills enable individuals to remember the social interactions they have had with particular groupmates in the past and so to form social relationships with them (direct relationships). These skills also may enable primates to remember and learn about the previous social interactions that particular groupmates have had with one another and so to understand something of *their* social relationships (third-party relationships). Obviously, the specific individuals of a group and their various interrelations are things that primates must learn for themselves during their ontogenies, and this learning allows them to formulate predictions about what will happen next—what particular groupmates are likely to do—in a variety of social situations. It is possible that in some cases individuals may even construct cognitive categories of groupmates and their relationships and so make predictions about what is likely to happen in a given circumstance based on experience with other social relationships and circumstances of the same type.

The importance of this social field for primate social life and social cognition cannot be overestimated. It creates a whole new dimension of complexity for various competitive interactions, transforming what would otherwise be relatively fixed and inflexible social interactions, such as mating, into problems that require complex decision making involving such things as who can mate with whom in the presence of whom. One especially important aspect of the social field for many primate species is dominance status as individuals compete with one another for valued resources—to the extent that there may even be competition over dominance when no tangible resources are currently at issue (see de Waal, 1987a, on primate "status striving"). Developing individuals therefore must acquire skills either for establishing dominance or for finding other ways to obtain needed resources. Toward this end, in many species individual primates form with one another either relatively transient and situation-specific "coalitions" or more long-term and generalized "alliances." These coalitions and alliances, which are also formed in many other mammalian species, in many instances interact with knowledge of the social field in complex ways, for instance, in an individual's choice of a particular coalition partner in a particular social circumstance.

In addition to these immediate cooperative–competitive interactions, the formation of coalitions and alliances among individuals may also lead to other cognitive complexities. Most importantly, if an individual helps an unrelated individual at some risk to itself (e.g., in a fight with a dominant), there needs to be some long-range benefit to the helper or else such altruistic behavior could not evolve into an evolutionarily stable strategy. One possibility is that there could be some reciprocation of aid in the future, so-called reciprocal altruism (Trivers, 1971). Over time, this might lead to the need for individuals to keep some type of cognitive tally sheet of who has done what for whom in what situations in the past, so as to balance out the reciprocity over time, including in some cases balancing out different kinds of favors with one another. These "obligations" add another dimension of complexity to the primate social field.

Finally, individual primates also cooperate with one another in situations that do not involve overt competition with conspecifics (although the "competitive exclusion" of other individuals may result as a by-product) and that involve physical rather than social obstacles to goal attainment. In many cases such cooperation, which is also found in other mammalian species, helps both individuals to obtain immediate goals that neither of them could obtain on their own. One example might be several groupmates cooperating to capture a prey who can locomote more quickly than any of them individually, or possibly mobbing a potential predator so as to render it ineffectual against any one individual. Cooperative problem solving of this type may in some cases result from each individual pursuing its own immediate self-interest and ignoring the others, but in other cases it might involve the use of more complex social knowledge and strategies.

In this chapter, we explore primate knowledge of the social field and some basic types of social interaction in which individuals must make informed behavioral decisions. We look first at the general nature of the primate social field, including their knowledge of individuals, their use of a variety of behavioral and contextual cues to understand and predict the behavior of others, and their understanding of direct and third-party social relationships. We then look at the nature of the coalitions and alliances that individuals form to obtain valued resources within the context of this social field, including the possibility that there is some form of mental record keeping of credits and debits among participants over time (reciprocity and interchange). Finally, we look at skills of cooperation in problem-solving contexts in which being able to understand the role of the other participant is of immediate benefit.

7.1 The Social Field

Just as most primates live in a complex spatial field, most of them also live in a complex social field. The primate social field has four main components: (1) individuals specifically recognize other individuals in their groups; (2) individuals understand and can predict at least some of the behavior of other individuals; (3) individuals remember something of the previous behavioral interactions they have had with other individuals, and so form direct relationships with them; and (4) individuals remember something of the interactions other individuals have had with one another, and so understand their third-party relationships. The most important kinds of relationships that primates recognize involve "kinship" (based not on genetics but on certain patterns of association), "friendship" (based on relatively recent affiliative or aggressive interactions), and dominance rank, all of which are reducible to a basic understanding of the behavioral interactions that specific individuals have with one another. Many other avian and mammalian species also seem to recognize individual groupmates, to predict their behavior, and to remember the interactions they have had with those individuals in the past (Green & Marler, 1979; Hinde, 1987), but it is unknown whether nonprimate animals understand the relationships that third parties have with one another.

7.1.1 Knowledge of Individuals

The evidence that individual primates recognize one another is overwhelming. Informally, it is clear to virtually all fieldworkers that primate individuals preferentially associate with or avoid certain individuals. Most of the more systematic evidence for primates' individual recognition of one another concerns monkeys and focuses on the vocal modality. For example, there is evidence, mostly from playback experiments in which experimenters play for a group the prerecorded vocalizations of individuals, that (1) squirrel monkey and vervet monkey mothers recognize the individual cries for help of their own offspring (Cheney & Seyfarth, 1980; Kaplan, Winship-Ball, & Sim, 1978); (2) rhesus monkey juveniles recognize the vocalizations of their own mothers (Hansen, 1976); (3) grey-cheeked mangabeys (*Cercocebus albigena*) distinguish the calls of individual groupmates from those of strangers, as well as distinguish certain adult individuals within their own groups (Waser, 1977); (4) pygmy marmosets, spider monkeys, and squirrel monkeys can identify the contact calls of individual groupmates (Biben & Symmes, 1991; Masataka, 1986; Snowdon & Cleveland, 1980); (5) vervet monkeys expect the calls of individuals that do not belong to their group to originate from particular places (Cheney & Seyfarth, 1982b); and (6) savannah baboons recognize that particular threatening screams and reconciliation vocalizations come from the same individual (Cheney, Seyfarth & Silk, 1995b).

Although there are no systematic data on apes' individual recognition of the vocalizations of others, few primatologists would doubt that they can recognize at least some of their important groupmates by voice. Experienced primatologists can recognize the long distance calls of individuals with relative ease, especially chimpanzee pant-hoots

(Marler, 1976; van Lawick-Goodall, 1968b). Moreover, Boesch (1991a) reports that the vocalizations of the dominant male chimpanzee in a group in Western Africa induce others to travel, whereas the same types of call from other individuals do not induce travel. Other chimpanzee grouping behaviors (including maternal responses to infant distress cries and the recruiting of allies) seem to rely on individual recognition as well (Goodall, 1986; Mitani and Nishida, 1993). In fact, on the basis of his review of the literature, Snowdon (1986) concludes it is likely that all primate species are able to recognize individuals on the basis of their vocalizations.

In the visual modality, Boysen and Berntson (1986) found that a captive chimpanzee's heart rate decelerated at a greater rate when it viewed pictures of familiar, as opposed to unfamiliar, human individuals, suggesting individual recognition. In a laboratory matching task, Bauer and Philip (1983) demonstrated the ability of three other human-raised chimpanzees to match the pictures and voices of familiar conspecifics and humans (see also Boysen, 1994), and Itakura and Matsuzawa (1993) found very rapid matching of letters with pictures of individual humans by a human-raised chimpanzee. The only studies of the ability of monkeys to identify individuals visually are those of Dasser (1987), who found that longtail macaques discriminated between pictures of individuals from inside and outside their group, and Humphrey (1974), who found that rhesus macaques who were habituated to pictures of one individual dishabituated when shown a picture of a new individual (and they did not do the same when pictures of individual pigs were used). Again, despite this scant information, few primatologists would doubt monkey abilities in this domain given the nature of their social interactions, which seem in many cases to rely on individual recognition in the visual modality, and their experimentally demonstrated skills of face recognition (rhesus macaques and squirrel monkeys: Dittrich, 1990; Mendelson, Haith, & Goldman-Rakic, 1982; Phelps & Roberts, 1994; Rosenfeld & van Hoesen, 1979).

7.1.2 Knowledge of Behavior

The coordinated social interaction of individual primates, like that of most mammals, depends on the ability of individuals to predict what others are likely to do at any given moment—whether, for example, they are likely to fight or flee. There are many sources of information that aid prediction. These sources include the recognition of ritualized displays, the comprehension of learned communicative signals, and the knowledge of various recurrent behavioral sequences that take place in particular contexts. The cues by means of which individuals learn to predict the behavior of others in particular situations may be thought of as another dimension of the primate social field.

Species-specific display behaviors typically make public the motivational state of the displayer, which is highly correlated with what the displayer will likely do next in the face of certain contingencies. For instance, if an adult male chimpanzee is stomping the ground, scowling, and the hair on its shoulders is erect, there is a high likelihood that a playful initiation by a juvenile will be rebuffed or punished. If a female chimpanzee is in estrus and approaches a male with head bowed and a submissive facial expression as she presents her rump to the male, there is a high likelihood that sexual advances will be accepted. Each primate species has a large number of such ritualized displays, many of which are in the repertoire of individuals raised in conditions of social isolation and are tied in an invariant fashion to relatively involuntary motivational

Figure 7.1. Group of baboons in a complex social field. (Photo by J. Call.)

states (although some learning may be required to use these displays in appropriate social contexts; Menzel, 1964). Because of the relatively fixed manner in which they are tied to eliciting conditions, the production of such ritualized displays is not of direct concern to an investigation of primate cognition. However, comprehending these displays in the context of various social and ecological conditions and using them to predict the likely behavior of others in these contexts does require some flexibility of understanding and directly influences all kinds of behavioral decisions. Also, individuals may learn certain behavioral sequences by simply observing the behavior of others in particular contexts, and so learn to use the initial part of the sequence to predict its final part in that context. For example, an individual might learn that a groupmate approaching a tree containing desirable food is likely to climb that tree and attempt to obtain and eat the food, or it might learn that when a particular male approaches a particular female when she is in estrus, another particular male will react aggressively. The combination of all these sources of information—knowledge of ritualized displays and learned behavioral sequences, all in the context of specific ecological and social conditions—allows individual primates to understand and predict the behavior of groupmates in many important situations and so to make informed behavioral choices.

A particularly instructive illustration of how this process might work comes from the well-known studies of Menzel (e.g., 1971a, 1973c, 1974, 1979) with a captive colony of chimpanzees. In these studies, experimenters hid food in one of various places in an outdoor compound while the chimpanzees were locked inside an indoor area. The chimpanzees were then released to the outdoor compound to search for the food. In some cases the experimenters allowed one individual to observe the hiding process. He was then put back with the others in the indoor enclosure. Over time, the chimpanzees learned how to identify the one who had observed the food being hidden as soon as they were released into the outdoor compound, presumably on the basis of the informed individual's involuntary display behaviors indicating anticipation of food, given the knowledge of the chimpanzees that in this game there was food hidden somewhere. The chimpanzees then discerned where the food was located by the nature of the purposeful locomotion of the informed individual, again in the context of the task: informed individuals proceeded with a definite direction of locomotion and moved "confidently" toward the food, looking back to see if others were following, speeding up as they got close to the goal. Because the uninformed chimpanzees knew that there was food somewhere, knew how to identify the informed individual, and knew how to predict the direction of locomotion of that individual, they learned very quickly in this context how to find the food themselves, sometimes even anticipating the informed individual's destination and arriving there first. Coussi-Korbel (1994) replicated many of these observations with white-collared mangabeys (*Cercocebus torquatus*).

In variations on this theme, experimenters then informed one chimpanzee individual of the type and amount of food hidden at one location and another individual of the type and amount of food hidden at another location across a large compound (neither was informed of the contents of the other location). When the other members of the group, who had not witnessed the hiding, were released, they most often went to the pile with the most preferred type or amount of food. The information that specified the desirability of the food might have been the relative level of excitement of the informed individuals as manifest in various ways, one of which may have been that excited individuals would go farther toward the food initially before they looked back to the group (these juveniles were so attached to one another that they would not go far alone). In fact, in a small experimental test, Menzel himself attempted to inform individuals of which food was in which location by walking a long way toward the location with more food before looking back but only a short way toward the location with less food. The chimpanzees who witnessed these behaviors most often went to the location that Menzel had walked the farthest distance toward. In a similar experiment to test the cues chimpanzee's used for direction, an experimenter indicated a single food's location by simply walking and pointing himself in the direction of the food (versus spinning around in place). The chimpanzees succeeded quite readily in following his direction and finding the food (though they were not successful when the experimenter simply spun around).

There are many other situations in which primates employ this same kind of "reading" and anticipating of the behavior of others in particular situations. For example, an infant may anticipate that its mother is preparing to leave, a mother may anticipate that its infant is coming toward her to nurse, a male may anticipate that two other males are about to attack him, or a female may anticipate that grooming is about to cease. In an experimental investigation, Koyama and Dunbar (1996) found that captive chimpanzees learned to anticipate the different social situations that would arise under different feeding regimes, one in which food was given in dispersed fashion and one in which it was presented in clumped fashion (leading to more intergroup competition and conflict), on the basis of various contextual cues (see also de Waal, 1982, and Mayagoitia et al., 1993). Such knowledge is not unlimited, of course, as recently Premack and Premack

(1994) reported that when one chimpanzee was led down a path to a scary snake and then reunited with a conspecific, that conspecific, when it was his turn, did not take a cue from the cagemate and anticipate that something scary awaited him down the path. Presumably, the simple information that another was afraid when he entered the cage did not signal to the other that she also would encounter something frightening when she left the cage; some other information might have made this inference more likely.

In general, we may refer to this kind of understanding of antecedent–consequent relations in the behavior of others, based on whatever kinds of cues, as an understanding of the animacy and directedness of behavior. Understanding the animacy of behavior means understanding that it is self-produced; unlike sticks and rocks, animate beings generate behavior on their own. Understanding the directedness of behavior consists essentially in being able to learn from experience which conspecific behaviors typically lead to (temporally precede) which other behaviors. Understanding of this type is clearly crucial in the way individuals of all primate species learn to predict the behavior of conspecifics and so adjust their own behavioral responses and strategies to the situation at hand. Thus, as we have described it so far, the primate social field comprises both a knowledge of individual groupmates and at least some knowledge of their likely behavior in particular situations. The combination of these two aspects of the social field means that primates potentially may understand and remember something about how particular individuals interact with other particular individuals. This is raw material for the understanding of social relationships.

7.1.3 Knowledge of Direct Relationships

Recognizing individuals and knowing something of their behavioral proclivities toward one another opens up the possibility of social relationships. Although in some sense each particular relationship may be reduced to those individuals' past interactions with one another (Heyes, 1994a), it seems most parsimonious to simply say that after a certain number of interactions individuals, begin to form and to perceive a relationship that may allow them to reliably predict the other's behavior in a variety of novel situations (Hinde, 1987). The major categories of social relationship in which primates are involved are "kinship" (based on long-term affiliations), relative dominance, and "friendship" (based on more transient affiliative and agonistic encounters). Most of the systematic data concern apes and cercopithecine Old World monkeys.

A number of monkey species direct affiliative behaviors such as grooming or huddling preferentially toward individuals who are their kin (Gouzoules & Gouzoules, 1987). Kinship also plays an important role in aggressive encounters, as kin are the most likely perpetrators of aggression in many monkey species (Bernstein, 1988a) as well as the most likely supporters of victims (Walters and Seyfarth, 1987). Moreover, when rhesus macaque juveniles scream for help in aggressive incidents, the acoustic structure of their calls contains specific information about their kinship to the aggressor (Gouzoules et al., 1984). Great apes show many of these same patterns, although kinship in the social interactions of adult apes is generally less prominent than it is in the social interactions of many monkey species. For example, bonobo and chimpanzee mothers and infants groom one another preferentially (Furuichi, 1989; Nishida, 1987a), and orangutan, chimpanzee, and bonobo mothers share food with their offspring more than with other juveniles (Bard, 1990; Kano, 1992; Silk, 1978). There is no definitive

Figure 7.2. Rhesus macaques engaged in the common activity of grooming matrilineal kin. (Photo courtesy of F. de Waal.)

evidence on how primates recognize their "kin," but most data point to individual learning during ontogeny based on familiarity. For example, adults of many primate species form close affiliative bonds (friendships) irrespective of their genetic relatedness, and as a result their interactions begin to resemble those of kin (Cheney & Seyfarth, 1990b; Galdikas & Vasey, 1992; Goodall, 1986; Smuts, 1985). This interpretation is corroborated by the one experimental test of this issue in which individual longtail macaques were raised in artificially composed groups, with some kin in the same groups and some kin in different groups (Welker et al., 1987). When the groups were later merged, the factor that most influenced the nature of the social interactions of individuals in this new situation was their past history with one another in their artificial groups (familiarity), not genetic relatedness.

Dominance is another type of social relationship that influences primate interactions. A variety of studies have found that monkeys approach or retreat from other individuals predictably based on their dominance relation to them (e.g., Rowell, 1966; Takahata, 1991), and in some species dominant individuals are clearly preferred as grooming partners (Seyfarth, 1980). In many species the dominant individual even has a dominance posture or vocalization and the submissive individual has a submissive posture or vocalization (de Waal, 1987a; van Hooff, 1967). As in the case of kinship, Gouzoules et al. (1984) found that rhesus macaque juveniles encode in their recruitment screams their dominance relation to their attacker (see Gouzoules & Gouzoules, 1989a, for a similar finding in pigtail macaques). Apes display many of these same patterns.

Individuals of all four great ape species retreat from dominants (Galdikas & Vasey, 1992; Goodall, 1986; Kano, 1992; Schaller, 1963); male chimpanzees show asymmetric grooming patterns based on dominance (Simpson, 1973; Takahata, 1990); and certain greeting vocalizations in chimpanzees and bonobos are directed exclusively from subordinates to dominants (de Waal, 1988; Hayaki, 1990), with greeting among males ceasing altogether in periods of rank instability (de Waal, 1982).

Finally, the differential allocation of behavior by individuals to groupmates is affected not only by relatively long-term and stable parameters such as kinship and dominance, but in some cases it can also be affected by more transient factors such as the immediately preceding interactions of individuals. For instance, individuals of many monkey species are more likely to direct affiliative behavior toward an immediately preceding opponent (i.e., to reconcile) than toward others in the group (see de Waal, 1989b, 1993, for reviews). Seyfarth and Cheney (1984) observed that vervet monkeys that had been groomed by an unrelated individual were more likely to attend to that individual when it was attacked by a third party than in the absence of such grooming. Chimpanzees, gorillas, and bonobos have also been observed to reconcile with agonists after conflicts (de Waal, 1989b; de Waal and van Roosmalen, 1979; Watts, 1995), indicating their recognition of their immediately past interactions with that individual.

7.1.4 Knowledge of Third-Party Relationships

Perhaps of special importance from the point of view of cognition are the relationships that primates seem to perceive and understand in the interactions of others, so-called third-party relationships. These relationships require special observational skills to learn, as the observer gains understanding by watching social interactions in which it is not directly participating. They also make the multiparty interactions of primates especially complex, as each individual now must keep track not only of its own relations to each of the other participants but also of their relationships to one another, which increases the amount of information to be processed geometrically (Whiten & Byrne, 1988c). Although little research has been done with nonprimate species, it is possible that the understanding of third-party relationships is either unique to, or especially complex and important to, the order Primates. Again, the major categories of relationship involved are kinship, dominance rank, and immediately past interactions, and the database is composed exclusively of apes and cercopithecine Old World monkeys.

With respect to the kinship relations of third parties, the most widespread observation is that of "dependent rank" in various monkey and ape species involving the third-party relation of mother–offspring. Kawai (1958) observed that individual Japanese macaques sometimes act aggressively toward a juvenile when that juvenile is alone, but they do not behave aggressively when its mother, who is dominant to them, is present. Since the presence of other dominant adults does not have this same inhibiting effect, this differentiation presumably indicates a recognition that the adult is indeed the mother of (a protector of) the potential victim. [Chapais (1992) reported a number of studies in which macaques were experimentally placed into and removed from a group, providing experimental evidence for the phenomenon of dependent rank; see also Keddy-Hector, Seyfarth & Raleigh (1989) for vervets.] It is possible, of course, that the inhibition apparent in the dependent rank is a phenomenon based on past experiences in which an individual aggressed toward a juvenile and received a thrashing from its mother

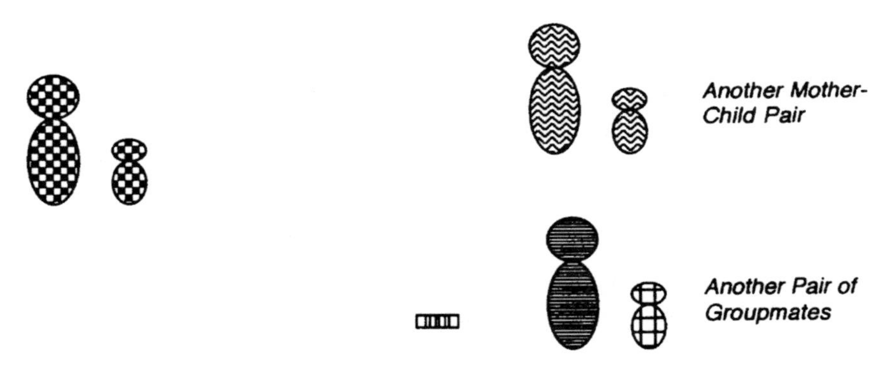

Figure 7.3. Dasser (1988) study with longtail macaques. The question is which pair at step 2 the subject will dishabituate to—indicating that it is not another instance of a mother-child dyad.

as a result, so that her presence becomes a simple predictive cue. But it is also possible that the potential aggressor understands something of the relationships of mothers and their offspring and what kinds of things they might do in particular circumstances.

This more cognitively generous interpretation of dependent rank receives support from two experimental studies. First, Dasser (1988b) rewarded longtail macaques for choosing various pictures of one mother–offspring pair over various pictures of other pairs of unrelated groupmates (see figure 7.3). All of the individuals pictured were familiar to subjects from their past social interactions in a single group setting. Subjects were then asked to (1) choose between pictures of other mother–offspring pairs over other unrelated pairs and (2) choose the appropriate offspring when presented with a picture of its mother (with again all individuals being familiar to subjects from past experience). Dasser found that subjects correctly identified the mother–infant pairs in the first procedure and chose the mother's offspring in the second procedure (see Dasser, 1988a, for a similar finding with respect to this species' understanding of the sibling relationship). This finding thus demonstrates that longtail macaques see at least some similarity in different instances of familiar mother–child dyads, although the basis for this perceived familiarity is not known.

In the second experiment, Cheney and Seyfarth (1980) played a previously recorded vocalization of a juvenile vervet monkey to its mother and to two other adult females (who also had offspring). The females responded to their own infant's calls. But, in addition, as soon as the call was played, the other females looked to the appropriate mother (more than to the other female), often before that mother made any overt movement or sign of recognition. This anticipatory behavior seems to indicate that individuals recognize at the very least particular mothers and their particular offspring (see figure 7.4). Overall, in concert with the naturally occurring phenomenon of dependent rank, these two experimental studies provide strong evidence that monkeys understand the third-party relationship of mothers and their offspring, although the generality of

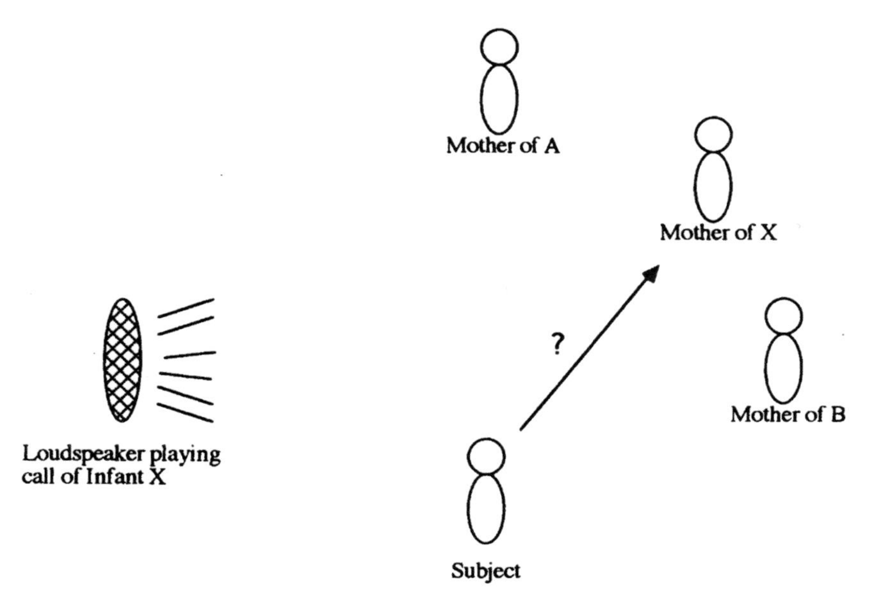

Figure 7.4. Cheney and Seyfarth (1980) study with vervet monkeys. The question is whether the subject, upon hearing the call of an infant, will look to that infant's mother.

this understanding and the features on which it is based are still open questions. There are no comparable experimental studies with any ape species.

There are several other sets of observations that indicate primate understanding of third-party kinship relations other than mother–offspring. Most important are various phenomena of "redirected aggression." For many primate species, individuals that have been the victims of aggression sometimes retaliate against the aggressor's kin or close associates, which implies an understanding of the kinship/friendship relation of those parties. This has been observed in pigtail macaques (Judge, 1982), longtail macaques (Aureli & van Schaik, 1991), Japanese macaques (Aureli, Cozzolino, Cordischi, & Scucchi, 1992), vervet monkeys (Cheney & Seyfarth, 1989), orangutans (Galdikas & Vasey, 1992), and chimpanzees (van Lawick-Goodall, 1968a). Furthermore, Aureli et al. (1992) and Cheney and Seyfarth (1986, 1989) also found that the victim's kin (not involved in previous aggression) were more likely to attack the aggressor's kin (also not involved in previous aggression), suggesting an even more complex understanding of kin relations. Relatedly, in the context of reconciliation after fights, Judge (1991) observed increased levels of postconflict affiliation between aggressors and the kin of their opponents in pigtail macaques, and Cheney and Seyfarth (1989) observed the same phenomenon in vervet monkeys.

Primates also understand third-party relationships concerning dominance rank. Kummer (1967) reported the phenomenon of "protected threat" in which one female baboon threatens a dominant female while simultaneously sexually presenting herself to a male dominant to both. This implies that the threatening animal has the ability to compare the relative ranks of the individual it is threatening and the one it is soliciting. Along these same lines, Seyfarth (1976, 1980) observed that grooming was preferentially directed to high ranking partners in vervet monkeys, with all individuals' preferences observing basically the same linear sequence, suggesting the possibility that individuals compare the relative ranks of third parties to choose the highest ranking partner. Cheney and Seyfarth (1990b) found further that when a dominant female vervet approaches two subordinate females in search of grooming, the lower ranking of the groomers almost always retreats, while the higher ranking groomer almost always stays put; the fact that the higher ranking groomer does not retreat suggests that she is able to judge that, although she is subordinate to the approaching animal, the other groomer is "more subordinate" (although other interpretations are possible in terms of these individuals' experiences with one another in the past in similar circumstances). Cheney, Seyfarth, and Silk (1995b) found that savannah baboons responded less vigorously to playbacks of call sequences that were consistent with their past experience of the rank relationships of the individuals involved (i.e., the more dominant individual gave a "dominant grunt" and the less dominant individual gave a "fear bark") than to sequences that were dissonant with their experience of those same individuals (i.e., the less dominant individual gave a "dominant grunt" and the more dominant individual gave a "fear bark"). It is also relevant in this context that when juvenile rhesus and pigtail macaques scream for help, their mothers respond differentially on the basis of the rank relation between the infant and its attacker as encoded in the juvenile's scream (Gouzoules & Gouzoules, 1989a; Gouzoules et al., 1984).

In terms of apes, chimpanzee recognition of third-party rank relationships is reported by Goodall (1990), who found that adolescent males enter the adult hierarchy by first challenging low-ranking females, then moving up to challenge higher ranking females, and only then challenging the males, again in order of dominance, perhaps indicating some recognition of the relative ranks of the different individuals challenged. Galdikas and Vasey (1992) reported that when ex-captive orangutans are attacked by other orangutans, they seek refuge with humans that were dominant to those aggressors, rather than submissive to them, in the past, again suggesting an ability to compare the ranks of the third parties involved. In both of these cases, however, other interpretations in terms of individuals' past interactions with other individuals are possible.

Monkeys and apes also understand something of the "friendship" relationships among third parties. For instance, Kummer, Goetz, and Angst (1974) observed that male hamadryas baboons (*Papio hamadryas*) formally recognize the "ownership" of certain females by other males in their natural habitats by staying away from those females. Further, in a series of experimental studies, Bachman and Kummer (1980) found that male baboons did not compete for females that had been observed to show a preference for another male by interacting with him (see also Sigg & Falett, 1985, and Kummer & Cords, 1991, for further studies of respect for "ownership" in hamadryas baboons and longtail macaques, respectively). In agonistic contexts, Smuts (1985) observed redirected aggression against the friends of a previous aggressor in savannah baboons. Also, Kummer (1971) found that male gelada baboons (*Theropithecus gelada*) in a newly formed group engaged in separating interventions toward the affiliative interactions of third parties whose relationship posed a threat. De Waal (1982) and Nishida and Hiraiwa-Hasegawa (1987) have described similar behavior for chimpanzees, and Chapais, Gauthier, and Prud'Homme (1995) have seen the same phenomenon in Japanese macaques. In a similar phenomenon, Maestripieri (1995a) found that rhesus macaque mothers assessed potential danger from groupmates differentially for themselves and for their infants, indicating an appreciation for how third parties interacted with their offspring. On occasion, chimpanzees (but not two species of macaques) also attempt to mediate between two antagonistic parties in an attempt to get them to reconcile (de Waal & Aureli, 1996; de Waal & van Roosmalen, 1979). Finally, Evans and Tomasello (1986) found that infant chimpanzees were more likely to associate with their mother's adult friends than with other adults in the group (independent of spatial proximity), implying the possibility that they understood her relationship to them.

7.1.5 Comparison of Species

It is likely that all primate species recognize individual groupmates, as do most mammalian species (although there is no systematic evidence for any prosimian). Primates are able to predict the behavior of conspecifics in various ways, based on their ability to read certain behavioral cues and derive predictions from them in specific social and ecological contexts—that is, to discern in a given situation the *animacy* and *directedness* of another individual's actions in terms of the outcome to which they are likely to lead. Among these cues are the species-typical displays that serve at least in part to index the emotional state or the impending behavior of the producer, the predictable behavioral and communicative sequences that individuals learn from their observations of others, and the ecological context in terms of such things as the current location of food or predators. Knowledge of this type is clearly crucial to the manner in which members of all primate species understand and predict the behavior of conspecifics and so adjust their own behavioral responses and strategies to different situations. Although each species has its own peculiarities in this regard, it is unlikely that primates as an order have any special cognitive skills in this dimension of their social fields.

Based on this knowledge of individuals and their behavior, primates come to form and understand direct relationships with others. Most simian primates probably recognize some third-party relationships among groupmates as well, although all of the evidence comes from great apes and cercopithecine Old World monkeys. The understanding of third-party relationships adds a whole new level of complexity to primate social interaction as it requires individuals to monitor many different social relationships each time they act. Although we do not have all the data necessary, perusal of the current literature on the social interactions of many nonprimate mammals suggests the hypothesis that the understanding of third-party social relationships may be unique to the order Primates, and thus that it may be responsible for much of the special complexity that characterizes primate social life. Although most scientists studying the social behavior of nonprimate mammals have not specifically looked for evidence of an understanding of third-party relationships, it is nevertheless the case that, despite extensive and intensive observations, none of the requisite behaviors (as listed above) has been reported for such well-studied mammals as elephants (Moss, 1988), dolphins (Connor, Smolker & Richards, 1992), or lions (Packer, 1994).

A major lacuna in the existing evidence for primates' understanding of social relationships, both direct and third party, is that we know little of their abilities to categorize and so to generalize these relationships. On the one hand, it is possible that all of their knowledge of relationships is based on direct observation. For example, they may

Figure 7.5. Conflict and postconflict behavior between siblings in stumptail macaques. The older sibling (top) threatens (open mouth display) the younger one, (bottom) they slap each other, and (facing page) they embrace to reconcile. (Photos by J. Call.)

know that several mother–infant pairs stay close together, the infant nurses from the mother, the mother protects the infant, and so forth, because they have observed all of these behaviors for all pairs. On the other hand, it is also possible that upon encountering a new mother–infant pair for the first time and only observing the two staying close together and nursing, they infer that the mother will protect the infant because this is another instance of a mother–infant pair and that is what other mother–infant pairs have done in the past. We simply do not know which of these alternatives is correct because, even in Dasser's (1988a,b) experiments, we do not know precisely what experiences subjects previously with their groupmates depicted in the stimulus materials.

Figure 7.5. *(continued)*

The findings of Oden et al. (1990) that young captive chimpanzees show some evidence of categorizing relationships of inanimate objects (i.e., in terms of identical versus different object pairs) might lend some credence to the more generous interpretation, but the materials and relations in this experiment are very different from kinship and dominance relations among conspecifics.

In any case, primates' knowledge of individual groupmates and their various social relationships with one another, combined with a more generalized ability to comprehend the directedness of the behavior of others in particular situations, makes for a highly complex social field (see Cords, in press, for a review). The combination of these two types of social knowledge is sufficient to enable an individual to determine such things as who it can and cannot attempt to mate with in the presence of which other individuals; who knows where food is; who one can attempt to take food from in the presence of which other individuals; who is about to leave the area; who will retaliate if a juvenile is attacked; who is likely to be a strong ally in a fight; where a frightening object or predator might be located; and who is likely to form an alliance against whom in the future. Following the general lines of de Waal (1982), the social field of primates might even be thought of as a "political field," as we shall see most clearly in the section that follows.

There are no compelling species differences within primates in any of these abilities, although studies with prosimians are mostly lacking. Only chimpanzees have been observed to seek to mediate reconciliations between third parties, but it is likely that this has more to do with various constraints on monkey social interaction (e.g., the likelihood that mediators will be attacked) than with any cognitive differences among the taxa (Tomasello & Call, 1994). It should also be noted in this context that all the experimental evidence for the understanding of third-party relationships concerns monkeys, with no experiment we are aware of making the same determination for any ape species.

7.2 Coalitions and Alliances

One arena in which primate social knowledge plays a particularly important role is in directly competitive interactions over food and mates. In such situations individuals may use a variety of strategies, for example, attempting to defeat the mating rival in a fight or attempting to race ahead to capture the desirable food first. But because competitive interactions over many years produce dominance relationships that are difficult to circumvent directly, individuals of some species regularly form coalitions and alliances with one another—so-called cooperation for competition, which integrates the vertical dimension of primate social structure (dominance) with its horizontal dimension (affiliation) (de Waal, 1987a). One excellent example is the phenomenon of dependent rank (Kawai, 1958), in which kin provide protection for one another, especially mothers for their offspring, and this allows infants and juveniles with powerful mothers to have free access to resources even in the presence of much bigger and stronger adults from lower ranking matrilines (Bernstein & Ehardt, 1986).

But in many primate species, spontaneous cooperation for competition takes place among unrelated individuals as well, and these involve less directly processes of inclusive fitness in which individuals derive direct evolutionary benefits for cooperating with close kin. Consequently, these interactions often involve more difficult cost–benefit decisions for individuals than those involving offspring. For example, in some primate species adult males are often in a situation in which they desire to mate with a female, but there is a rival male nearby. This situation forces some kind of cost–benefit analysis, based on some prediction of what is likely to happen in the situation given certain actions. This prediction and analysis can only be effective in most primate species if it includes information about both the affiliative and dominance rank relationships, of both the direct and third-party kind, of all the individuals present (Chance & Mead, 1953). Because we are concerned with cognitive processes, it is these more open-choice situations that are our focus.

Open-choice coalitions and alliances among unrelated individuals are engaged in by many mammalian species, including hyenas (Zabel, Glickman, Frank, Woodmansee & Keppel, 1992), sheep (Rowell & Rowell, 1993), lions (Packer, 1994), and dolphins (Connor et al., 1992), and involve the acquisition of food, mates, and dominance rank. They are also engaged in by all four great ape species (chimpanzees: Goodall, 1986; bonobos: Kano, 1992; gorillas: Harcourt & Stewart, 1989; orangutans: Galdikas & Vasey, 1992) and by virtually all the monkey species that have been observed for extended lengths of time in the field (almost all are cercopithecine Old World monkeys characterized by much "contest competition"). There is no information on prosimians. An excellent overview of the relevant research may be found in Harcourt and de Waal (1992).

7.2.1 Functions

Primate coalitions and alliances may serve one of three functions. First, as just mentioned, many coalitions and alliances of primates find their evolutionary explanation in straightforward kin selection terms (individuals cooperate most with those who share the most genes with them), raising the possibility that the proximate mechanism is relatively simple and direct (e.g., a genetic predisposition to cooperate with kin). A second possibility is some form of mutualism in which each individual perceives that it is to its own immediate self-interest to form a coalition, for example, in increasing the probability that it will gain access to a mate or some food. This may involve some complex decision making in estimating the relative costs and risks involved, given the nature of the interactants and their relationships to one another. The third possibility is reciprocal altruism, in which an individual deliberately incurs an otherwise unacceptable risk now for the possibility of some appropriate compensation later. This adds another dimension of cognitive complexity to the decision-making process in the form of issues of reciprocity and record keeping over time (discussed more fully in section 7.3).

In terms of proximate functions, individual primates mostly join forces against third parties for one of a few predictable reasons. In summarizing four monkey species (rhesus macaques, vervet monkeys, chacma baboons, and Barbary macaques), Walters and Seyfarth (1987) estimate that roughly 10% of all disputes arise over food, roughly 20% over access to social partners (including mates), and fully 70% occur over no perceptible resource other than perhaps space or freedom of movement. It is assumed in this last case that the conflict concerns the establishment of dominance relations pure and simple.

In each of these types of conflict, primate coalitions may be formed in more reactive or more proactive ways. On the most reactive end of the continuum, a dominant individual is in the process of winning a fight against a groupmate and others join in on the winning side. This is often called "winner support" and involves neither active planning on the part of the helper nor active soliciting on the part of the dominant individual, although it does involve individuals actively monitoring a number of different social interactions simultaneously. In many other cases a coalition is formed because an individual is in the process of being attacked by a superior groupmate, solicits help from others, and a groupmate (often kin) reacts appropriately to the solicitation (Bernstein & Ehardt, 1986; Gouzoules et al., 1984). This type of interaction is often called "loser support." It involves no planning on the part of the helper, and the solicitations of the loser take place only after the conflict has begun. Over time, when certain individuals reliably help one another in both loser support and winner support situations, we may speak of an alliance (such as the alliance of three adult male baboons in the classic study of Hall and DeVore, 1965).

At the most proactive end of the continuum, individuals sometimes join forces first and only then seek to obtain access to a resource or to assert their dominance (this has also been reported for dolphins; Connor et al., 1992). The best known example of this is female hamadryas baboons who simultaneously threaten another female and present themselves sexually to a dominant male—the "protected threat" (Kummer, 1967; see also Colmenares, 1991). Male chimpanzees also sometimes "swing by" a potential ally to solicit help for an upcoming dominance conflict (de Waal, 1982). Although there are to our knowledge no data available on the relative frequency of such proactive coalitions for any primate species, they would seem to be relatively infrequent as compared with the more reactive types of coalition formation in all species studied extensively (de Waal, personal communication). From the point of view of cognition, these types of coalition are of special importance because they seem to involve the kind of insight and foresight that is so important for problem solving in the domain of physical objects.

Figure 7.6. Chimpanzee coalition. (Photo courtesy of F. de Waal.)

7.2.2 Tactics

In coalitions and alliances that do not involve close kin but that do involve some measure of planning, there are four basic ways in which primates demonstrate their social-cognitive skills. They demonstrate foresight in knowing who to solicit, how to solicit them, who to retaliate against, and how to intervene to prevent future conflicts.

First, in many primate species, individuals appear to select their allies carefully, with high-ranking individuals being solicited for help more often than low-ranking individuals. Indeed, Harcourt (1992) argues and presents evidence that the recruitment of the specific individuals based on characteristics such as their dominance relative to potential opponents is the one characteristic that most clearly differentiates primate coalitions from the coalitions of nonprimate species. In addition, in some primate species there seems to be active competition for more powerful allies that takes place outside of any immediately competitive situation. For example, in a number of observational studies researchers have found that relatively low-ranking individuals more often groom high-ranking individuals than the other way around, with some researchers interpreting this as low-ranking individuals competing with one another for grooming access to high-ranking groupmates (e.g., baboons: Cheney, 1977; rhesus macaques: Chapais, 1983; chimpanzees: de Waal, 1982; vervets: Seyfarth, 1980). It is thus possible that this behavior is being used to curry favor with potential allies in future conflicts. It has also been observed that individuals often join a higher ranking individual in its attack on a low-ranking individual when that individual needs no help, possibly as a way of building up a low-risk "debit" to be repaid in the future (Walters & Seyfarth, 1987). The interpretation of these types of behavior as the recruitment of allies for the distant future depends on a reciprocal altruism interpretation of the interactions individuals have with one another and is discussed more fully in section 7.3.

Second, most primate species that have been studied extensively show some identifiable communicative signals used reliably when they are attempting to recruit an ally. When this occurs in the immediacy of a competitive encounter, de Waal and van Hooff (1981) call it "side-directed" behavior. For example, savannah baboons often look emphatically back and forth between the current opponent and the potential ally and sometimes engage in 'staccato grunting' (Noë, 1992). Rhesus monkey juveniles have a special set of recruitment screams in agonistic contexts, and the alternating of gaze between opponent and potential helper often accompanies the screams (Gouzoules et al., 1984). In more proactive recruiting, hamadryas baboons often engage in various sociosexual behaviors (e.g., presenting or mounting) with the recruitee before the conflict has begun or as it is beginning (Kummer, 1967). When in danger of being attacked, chimpanzees sometimes give the submissive begging gesture of an outstretched hand to a potential ally (de Waal & van Hooff, 1981), and, as mentioned previously, in more proactive recruiting situations they will "swing by" a recruitee in a characteristic way (de Waal, 1982). It is important to note that in many of these cases the communicative signal used is not one that has been phylogenetically ritualized specifically for a recruitment function, but rather is some gesture that occurs regularly in other contexts as well, indicating a flexible adaptation for current purposes.

Third, although the nature of this system is a matter of some debate, it seems that some primate species have a "revenge system" in which they retaliate against others who have opposed them in the past (either singly or in a coalition) and sometimes even the kin of those who have opposed them in the past ("redirected aggression"). De Waal and Luttrell (1988) report such a system for chimpanzees but none for rhesus and stumptail macaques. More recently, however, Silk (1992) found paybacks against former opponents in bonnet macaques (see also Smuts, 1985, for savannah baboons), suggesting the possibility that their absence in rhesus and stumptails may have more to do with social structure (i.e., a relatively inflexible system of dominance that might inhibit subordinates attempting to retaliate against dominants) than with cognitive factors (Aureli et al., 1992, report similar findings for Japanese macaques). If such a revenge system does exist, it would rely on the individual recognition of past opponents, the understanding of third-party relationships such as kinship and friendship, and on some kind of cognitive tally sheet over time.

Finally, some primate species have been observed to engage in behaviors that are especially proactive in the sense that they are preventive. For several primate species, researchers have observed repeated instances of a dominant individual spying on two individuals, grooming or otherwise engaged in affiliative interactions, who together would make a dangerous alliance. The dominant individual then approaches the affiliative interaction and breaks it up, presumably so that the groomers will not form the alliance. Such separating interventions have been reported for male chimpanzees (de Waal, 1982; Nishida & Hiraiwa-Hasegawa, 1987), gelada baboons (Kummer, 1971), and Japanese macaques (Chapais et al., 1995). De Waal and van Roosmalen (1979) and de Waal and Aureli (1996) also report some instances of third-party mediation by chimpanzees in which an individual attempts to help others who have just had a fight to reconcile with one another, possibly in an attempt to preempt future conflicts. These researchers have not observed this behavior in rhesus and stumptail macaques, although once again this may be due to the different social structures of the species rather than to differences of cognition (e.g., it may be riskier for an outsider to approach a just-defeated monkey than a just-defeated chimpanzee).

It should be noted, at least briefly, that there are few reports of any significant "teamwork" in the coalitions and alliances of nonhuman primates in the sense of proactive strategies in which different individuals play different roles that require an understanding of the role of the other cooperating individuals. One could imagine, for example, strategies in which a female used sexual solicitations to entice a male away from food, which could then be obtained by another female (who might then share it with the first female). In general, these kinds of complex cooperative strategies have not been specifically looked for by field researchers, although de Waal (1982) made some intriguing observations of this type with captive chimpanzees. This has been a much more important issue in the domain of cooperative problem solving and so will be discussed more fully in section 7.4.

7.2.3 Comparison of Species

Few of the studies of primate coalitions and alliances are experimental, and so appropriate caution must be exercised in interpreting the many naturalistic observations of these phenomena. Overall, however, it seems to be fairly well established that when faced with competitive social situations, many species of monkeys and apes sometimes recruit and compete for allies selectively, employ communicative signals in the recruitment process, use a revenge system to pay back others (and their kin) who have opposed them in the past, and engage in separating interventions and other proactive mediations. The social cognition that is manifest in these behaviors mainly concerns the ability of individuals to assess ahead of time the social problem at hand and the strategy needed to deal with it effectively, often involving some assessment of the third-party relationships of others. In all of these behaviors, primates do not seem so different from many other mammalian species that also form coalitions and alliances (Harcourt & de Waal, 1992), except insofar as the understanding of third-party social relationships is involved, as in the selective recruitment of allies (Harcourt, 1992).

Overall, the combination of individuals' knowledge of how coalitions work and their individualized social knowledge of the group makes for much of the cognitive complexity that is assumed to underlie primate cooperation in the social domain. The only difference between monkeys and apes that has been suggested and empirically corroborated in this domain is that only apes engage in third-party mediation in which they attempt to bring together previous combatants; however, plausible explanations for this difference in terms of social structure are possible as well. There is basically no information on coalitions and alliances in prosimians, New World monkeys, or colobine Old World monkeys.

7.3 Reciprocity and Interchange

Added to the complexity of coalitions and alliances in their immediate contexts are complexities added by the possibility that individuals keep mental "score sheets" on their debits and credits with groupmates based on their past encounters; they may even

Figure 7.7. Stumptail macaques engaged in reciprocal grooming. (Photo by J. Call.)

attempt to affect the score sheet prospectively by doing favors for individuals whose help they will need in the future. When these score sheets all concern the same "currency" (e.g., trading help in fights for help in fights), we speak of reciprocity. When there is more than one currency involved (e.g., trading grooming for help in fights), we speak of interchange. The evidence for and against true reciprocity and interchange in primate coalitions and alliances is complex and different for different species, involving both correlational and experimental analyses. But a relatively clear picture has recently emerged of fairly widespread skills of reciprocity and interchange in a number of species (although their interpretation remains controversial). The important point for current purposes is that if reciprocity and interchange are involved in primate social interactions, another layer of cognitive complexity may be added in the form of skills of memory and quantification, and perhaps even strategic "bargaining."

7.3.1 Reciprocity

In attempting to establish whether primates engage in what de Waal (1992) calls "calculated reciprocity" (i.e., reciprocity relying on some form of cognition), two alternatives must always be kept in mind. First, if a coalition or alliance is in the selfish best interest of all cooperating parties, then costs and benefits do not need to be calculated and no paybacks need to be foreseen. There is no reciprocity, only mutualism. Second, any time researchers observe a correlation that may be indicative of reciprocity (e.g., two unrelated individuals consistently support one another in fights), the role of other factors must always be considered. For example, what looks like reciprocity might exist if individuals have a tendency to help any time a fight is nearby spatially, and some individuals, for whatever reason (including kinship), tend to stay in close proximity to one another. Or perhaps individuals only help others whose dominance ranks are high because there is little risk involved. In none of these cases do calculations need to be done or plans for the future made.

The first claim for reciprocal altruism necessitating a mental score sheet of credits and debits was reported by Packer (1977). He reported that unrelated male savannah baboons (*Papio cynocephalus anubis*) sometimes solicited a partner to help them displace another male who was consorting with an estrus female. In the six relevant observations, the initiating male in the coalition obtained access to the female, while the solicited male continued to confront the displaced male for a short time. The argument was thus made that the solicited male put himself at serious risk for no tangible benefit, and thus there would presumably be some reciprocation by the initiating male at some time in the future (or else males who had a tendency to help would soon die out). And, indeed, Packer (1977) found that there was reciprocation: particular individuals preferred coalitions partners whom they both solicited and were solicited by more often than other potential partners. Moreover, individuals often refused the solicitations of individuals who had refused their solicitations on earlier occasions.

Bercovitch (1988) observed a different group of this same species of baboon. His findings were very different from Packer's, however, and so his interpretation of baboon coalition formation was different as well. First, using a different analytic technique from Packer, Bercovitch did not find reciprocation in coalition formation: males did not preferentially solicit those who had solicited them in the past (with the exception of one pair that clearly preferred each other), and, indeed, individuals who were partners on one occasion were often combatants on other occasions. Individuals did not differentially refuse the solicitations of those who had accepted or rejected their solicitations in the past. Most importantly for the reciprocal altruism interpretation, Bercovitch did not find that initiators obtained sexual access to the female any more often than the solicited males. The key factor in actual success in obtaining access to females in Bercovitch's study was simply the number of attempts made, regardless of whether the male was the initiator or solicited partner. (Noë, 1992, and Rasmussen, 1980, also found no differential success rate as a function of which of the coalition partners was initiator.) Because the costs associated with participating in a coalition were judged by various measures to be low, Bercovitch's interpretation was that male baboon coalitions for access to estrous females is a mutualistic (cooperative) effort in which neither partner risks much and both have close-to-equal probability of gaining access to the female. In one other study of animals in their natural habitats, Hunte and Horrocks (1987) found reciprocation in the agonistic aiding of nonkin, adult vervet monkeys.

The other line of research relevant to primate reciprocity concerns captive animals and has been more focused on establishing the existence of reciprocity in coalition formation than on the outcome of those coalitions. De Waal and Luttrell (1988) observed groups of rhesus macaques, stumptail macaques, and chimpanzees over extended periods to determine if there was reciprocity in agonistic aiding (whatever its goal). They found significant correlations for all three species: individuals solicited most often individuals whose solicitations they had agreed to most often in the past and were solicited most often by individuals whom they had solicited previously. Unlike previous field studies, de Waal and Luttrell statistically controlled for kinship and the spatial proximity of individuals so that these factors could not be accounting for the correlations observed. Their findings thus provide strong support for the reciprocal altruism interpretation. De Waal and Luttrell (1988) also looked at "revenge" in these species, that is, reciprocation for past opposition. They found that chimpanzees, but not macaques, joined coalitions more against those who had attacked them in the past than against others in the group. In a more recent study, however, Silk (1992) found that bonnet macaques do seem to reciprocate oppositional interactions, with kinship statistically controlled.

Hemelrijk and Ek (1991) argued that there is another variable, dominance, that might account for observed correlations in the agonistic aiding of individual primates. The problem is that individuals might tend to aid only individuals who are fairly close in rank to themselves, or even that high-ranking individuals aid more absolutely than others (to assert their rank with little risk) and all individuals tend more often to aid higher ranking individuals (because they are strong allies). If this were the case, it would still be true that individuals helped one another in kind, but the reciprocal altruism explanation (and the consequent mental record keeping) would be gratuitous. In further analyses of the captive chimpanzee group previously observed by de Waal and Luttrell (1988), Hemelrijk and Ek found that reciprocity of agonistic aiding in males could be observed even with dominance rank statistically controlled. (They also controlled for the total number of initiations by individuals, although it is unclear whether they controlled for kinship or spatial proximity.) It is also significant that they found this correlation only in periods of rank instability, suggesting in another way that the reciprocity observed was independent of dominance factors. In contrast to de Waal and Luttrell (1988), however, these investigators found no evidence of reciprocity of oppositional interactions in any kind of revenge (although they worked with a smaller sample of subjects).

Two other behaviors in which reciprocity has been investigated are grooming and food sharing. With regard to grooming, even from the initial reports, the role of kin and rank have been apparent: savannah baboons tend to groom kin and those close in rank to themselves, which results in reciprocity (Seyfarth, 1976, 1977). However, Hemelrijk (1990a,b) reanalyzed Seyfarth's data and still found reciprocity of grooming even when rank was statistically controlled. Other studies have found grooming reciprocity while controlling for kinship, for example, Stammbach (1978) for hamadryas baboons, Fairbanks (1980) for vervet monkeys, Muroyama (1991) for Japanese macaques, and Hemelrijk and Ek (1991) for chimpanzees. The issue in the case of grooming, however, is not whether there is reciprocity, but whether grooming is costly to either party and thus truly a case of reciprocal altruism. There are arguments on both sides of this issue (e.g., compare Dunbar, 1988, with Cords, 1995, and Maestripieri, 1993). With respect to food sharing, de Waal (1989a) observed the reciprocal exchange of food among members of a captive chimpanzee group (in terms of allowing others access to a clump of food that could be individually monopolized, not in terms of active offering or giving), even when both proximity and dominance were statistically controlled (see figure 7.8). Although sharing food is clearly costly in nature, how costly it might be in a captive environment (and de Waal's observations were made during a supplemental feeding) is unclear (see Lefebvre, 1982, Lefebvre & Hewitt, 1986, and Paquette, 1992b, for further observations of chimpanzees exchanging food with humans and objects among themselves).

Figure 7.8. Chimpanzee food sharing. (Photo by J. Call.)

7.3.2 Interchange

Interchange occurs when two individuals exchange acts or objects of different types for one another, for example, grooming for future or past support in fights. As in the case of reciprocity, if a mutualistic (cooperative) explanation can be placed on the acts exchanged, then the question of reciprocal altruism and the keeping of mental tally sheets is moot. Also as in the case of reciprocity, a simple correlation between the number of acts of a particular type that individuals direct toward one another is not easily interpretable unless a number of other possibly confounding variables (e.g., kinship, proximity, dominance) are in some way taken into account. The two types of interchange that have been studied are the interchange of grooming for agonistic aiding and sometimes other things and the interchange of food for agonistic aiding, grooming, and sometimes other things.

First, Seyfarth (1980) found a correlation in vervet monkeys between the amount of grooming engaged in by an individual and the support that individual received in fights (see also Cheney & Seyfarth, 1990b). He interpreted this correlation as reciprocal altruism. Hemelrijk (1990a,b), however, reanalyzed the 1980 data statistically, controlling for rank relations, and found no evidence of interchange. Also, Fairbanks (1980), Silk (1982), and de Waal and Luttrell (1988) did not find an interchange between grooming and agonistic aiding in vervet monkeys, bonnet macaques, and rhesus macaques, respectively. However, Hemelrijk (1990a,b) did find an interchange between grooming and agonistic support in female baboons (data from Seyfarth, 1976), and Hemelrijk and Ek (1991) found a similar correlation in female chimpanzees, in both cases with dominance rank and kinship statistically controlled. Hemelrijk, van Laere, and van Hooff (1992) found that chimpanzee males groomed estrous females more than anestrous females, especially those with whom they mated more often, but there was no relationship between mating and agonistic aiding. Muroyama (1994) found evidence for reciprocity between grooming and allomothering in nonkin, female patas monkeys.

Unlike the case of reciprocity, in the case of interchange three controlled experiments have been conducted that help in interpreting this conflicting set of correlations. This is important because in correlational studies extraneous variables can only be controlled statistically, not experimentally (raising the possibility that some third, unobserved variable is an explanatory factor). Furthermore, in correlational studies there is no assessment of the sequence of events and thus no ability to assign with any confidence a causal relationship between the acts interchanged. The first experimental study is that of Seyfarth and Cheney (1984). In a field experiment of vervet monkeys, two unrelated individuals were selected for observation. The investigators then waited for one to groom the other. Thirty to 90 minutes after the grooming bout had ended, they played back from a loudspeaker the "threat grunt" of the animal that had been the groomer at a distance not too far from the animal previously groomed. (The "threat grunt" of vervets is used when they are threatening a groupmate and sometimes to solicit aid from others.) Some days later they also played back the same "threat grunt" for the same individual but after there had been no grooming between the target pair. They repeated this procedure for a total of nine pairs. What Seyfarth and Cheney found was that when an individual had recently been the recipient of grooming from another individual, as opposed to when it had not, they looked longer in the direction of the speaker (by several orders of magnitude). They interpreted this increased responsiveness as a readiness to help in the supposedly impending fight. They did not find this same effect, however, when pairs of kin served as subjects, presumably because the kin were already predisposed to help one another without any preceding grooming.

The second experiment was by Hemelrijk (1994), and was similar, but it used captive longtail macaques. The study was designed to make two improvements on the Seyfarth and Cheney design. First, the response studied was not just looking in a direction, but actual agonistic aiding in an real fight (induced with food). Second, because the grooming bouts observed by Seyfarth and Cheney were mostly reciprocal, it is possible that any grooming interaction at all would have led to the same result; that is, perhaps when individuals groom others they are led to provide them with further aid in the future (even more so than when they themselves are groomed), which would undermine the explanation in terms of an interchange of favors. Hemelrijk thus gave individual longtail macaques the opportunity to provide agonistic support for others under three conditions: after being groomed by the individual needing help, after grooming that same individual, and after no grooming (see figure 7.9). In agreement with Seyfarth and Cheney, Hemelrijk found that individuals were more likely to provide support to a groupmate if they had recently been groomed than if they had performed the grooming or if there had been no grooming at all, with the latter two conditions not leading to different levels of support. In interpreting these results, Hemelrijk pointed out that it is still possible that being groomed puts monkeys in the mood to be helpful to any individual, not just the one who groomed them; to test the specificity of their tendency to aid, groomed subjects would also need to be presented with the opportunity to help an individual who had not groomed them previously. It is relevant to this point, however, that

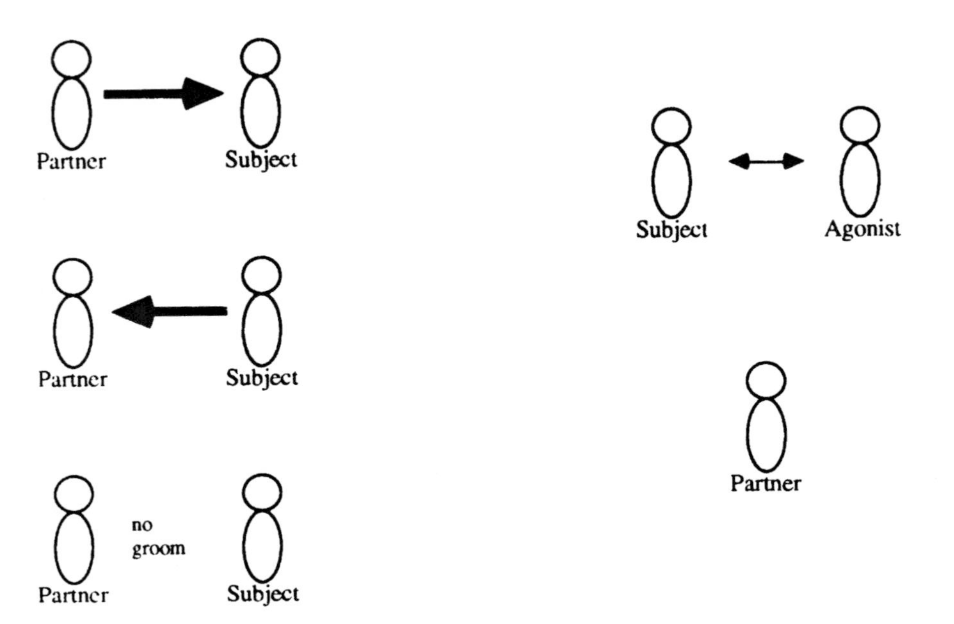

Figure 7.9. Hemelrijk (1994) study with longtail macaques. The question is whether the partner will aid the subject in a fight at step 2 as a function of their previous interactive experience at step 1.

Seyfarth and Cheney (1984) did not find the effect with kin pairs, demonstrating that if there is some mood-altering effect of being groomed on one's disposition to be helpful, it does not always hold.

The third experimental study was that of Chapais et al. (1995). The logic of their design was to experimentally alter the dominance relationships in a group of Japanese macaques and to observe changes in the social interactions, including agonistic aiding, as a result. Beginning with a group composed of three matrilines, ranked as A-B-C, experimenters artificially constructed a new group (by carefully choosing selected members of each matriline) so that the ranking was changed to C-A-B. Of most relevance in the current context, it was found that the individual from the C matriline most in need of help in disputes (because she was young and small) dramatically increased her grooming of the newly dominant members of the group from her matriline relative to the preexperimental period. Conversely, those dominants assisted her in fights much more than they had previously (and she joined them in fights against others more than previously as well). Also important, the highest ranking individual from the A matriline changed her allegiance in agonistic support from the B matriline (just below her top rank preexperimentally) to the C matriline (occupying the top rank postexperimentally), even preventing on many occasions affiliative advances of B toward C individuals in the postexperimental period. Nevertheless, Chapais et al. (1995) did not interpret their results as requiring reciprocal altruism with mental record keeping. They hypothesized that physical proximity to dominants is associated with protection from attacks by others, and so all individuals prefer to groom and engage in other affiliative behaviors toward dominants for that immediate benefit. Conversely, dominants support their affiliates either because their source of grooming is being threatened or, in some cases, because their affiliates outrank the opponent and supporting winners in fights is a way of reinforcing their own rank. Either way, both individuals derive immediate benefits, and so the reciprocal altruism explanation involving mental record keeping is superfluous.

The second type of interchange that has been investigated involves food. In addition to his finding, reported above, that chimpanzees reliquinsh food to one another in a reciprocal fashion, de Waal (1989b) also found that individuals who had groomed another individual in the hours immediately preceding introduction of food were more likely to receive food from the individual they had groomed than were other individuals. This correlation held even when the effects of proximity and dominance were statistically controlled. A negative correlation held in the other direction: an individual who had performed grooming on another individual previously was less likely to share food with that individual. (There was also some evidence of aggression for not sharing.) Boesch's (1994) observations of the hunting of wild chimpanzees are consistent with this finding: hunters shared meat with co-hunters more than with nonhunters regardless of rank. These findings may be especially significant in the current context because in all of the other cases of interchange the received good is agonistic support or grooming, and it is not totally clear that these are not mutualistic in the sense that both participants benefit (Bercovitch, 1988). Giving up food, however, would not seem to be mutualistic in any respect (although it must be kept in mind that in this study the food was supplemental to the animals' normal diet). It should also be noted that Nishida, Hasegawa, Hayaki, Takahata and Uehara (1992) reported a correlation between the sharing of meat by the alpha male of a chimpanzee group in the wild and the grooming he received from particular individuals, although dominance and proximity were also factors and were not statistically controlled in this correlation. Hemelrijk et al. (1992) found no interchange of food and sex by male and female chimpanzees, although de Waal (1987b) found some evidence for such an interchange in bonobos.

There is one experiment involving the interchange of food that may provide partial corroboration for the positive correlational findings in this domain. Stammbach (1988a,b) trained individual members of a group of longtail macaques to expertly operate a food-dispensing apparatus, with experts changing over trial blocks. The experts were always relatively low-ranking so that they could not monopolize the food after it was dispensed, but they nevertheless maintained control over when the apparatus was operated. Stammbach found that during the block of trials in which an individual was the expert, many of the others began to treat her differently. Of 15 individuals whose amount of grooming activity differed depending on whether an individual was expert, 14 increased their grooming for the expert. Of the 12 significant correlations between grooming of the expert and the expert's allowing access to the apparatus, 11 were positive correlations. The clear implication is that these longtail macaques curried favor with those who controlled the food (via grooming) so that they would allow them access to it at some time in the near future. Given the relatively free access of the expert to the apparatus, it is not clear what costs the expert may have incurred for allowing access to some individuals.

7.3.3 Comparison of Species

Many animal species employ some type of group defense against predators and other groups of conspecifics, and many mammalian mothers protect offspring from potential harm. Both of these cooperative activities are important for most social species and may be identified as the evolutionary roots of other forms of cooperation (Harcourt, 1992). Presumably, these two forms of cooperation are so fundamental and widespread because they depend in fairly straightforward ways on evolutionary processes of inclusive fitness and mutualism. When inclusive fitness and mutualism are not directly involved, however, cooperation becomes a more problematic phenomenon evolutionarily and requires some form of reciprocal altruism if it is to become an evolutionarily stable strategy.

The only widely cited case of reciprocal altruism in nonprimate animals concerns the food-sharing of vampire bats (Wilkinson, 1984) and is based on a small sample of observations. In primates, many of the cases of reciprocity and interchange that have been widely cited (e.g., male baboon coalitions for mate acquisition) may be more properly interpreted as cases of mutualism or cooperation in which both parties benefit in one way or another—and there may be more of these cases than is generally acknowledged (Chapais et al., 1995). Other cases in which reciprocal correlations are reported for behaviors such as grooming or agonistic aiding may reflect the simultaneous influence of extraneous variables such as kinship, spatial proximity, or rank. In either case, the implication is that, although a number of other complex social-cognitive processes might be involved, the mental tallying of credits and debits need not necessarily be involved.

Although the data are not entirely consistent, other cases of primate reciprocity and interchange appear to involve reciprocal altruism. This conclusion is supported by both controlled correlational and experimental studies in which at least one party incurs some specifiable cost in terms of risk of harm or the relinquishing of food (especially the experimental studies of Hemelrijk, 1994, and Stammbach, 1988a,b). At least some primate social interactions would thus seem to involve some form of mental record keeping. The form that record keeping might take is unknown at this time, but there is no evidence that the tallying is quantitatively precise. The experimental studies are especially helpful in this context because failures to find reciprocity or interchange by using correlations of behavior in relatively natural settings could easily occur because the investigator does not know the "exchange rate" that holds in those domains, for example, how many times one individual must groom another before agonistic aid is "deserved" (Seyfarth & Cheney, 1988).

There is no strong evidence for differences among monkeys and apes in either reciprocity or interchange; and, again, there are no data for prosimians, New World monkeys, or colobine Old World monkeys. In the two main cases of reciprocity (grooming and agonistic support, including revenge), there are similar types of evidence for both monkeys and apes. Food sharing in a reciprocal pattern has only been observed in chimpanzees, but there have been no systematic studies of any monkey species. In terms of interchange, there is similar correlational evidence for monkeys and apes in the two major types: grooming and agonistic support, and grooming and food sharing. The only experimental evidence of interchange in both of these cases is with macaque monkeys, and in all cases it is positive.

7.4 Cooperative Problem-Solving

In reciprocal altruism, only one organism benefits from a given act on a given occasion, and it is only when there are repeated occasions, with individuals changing roles over time, that both may be said to benefit. Mutualism or more immediate forms of cooperation, on the other hand, involve two or more individuals in a situation in which neither can benefit alone, or at least not to the same degree, as when they act in concert. It should also be noted that most of the research on reciprocal altruism in primates has focused on situations in which the barriers to an individual's deriving maximum benefit from a situation are all social barriers (i.e., competition with conspecifics). In the research on cooperative problem solving, however, individuals typically must act together to overcome nonsocial barriers. The interest of cooperative problem solving from a social-cognitive point of view is the extent to which, and manner in which, individuals might understand the role of the others in the interaction and/or use various social and communicative strategies to help coordinate their behavior.

7.4.1 Chimpanzee Cooperative Hunting

The most interesting and complex case of primate cooperation in the wild is chimpanzee cooperative hunting. Although some species of monkeys and other mammals prey on other animals in groups (see Boesch & Boesch, 1989, for a review), the extent to which and the manner in which they actively cooperate mostly cannot be determined from field reports. Boesch and Boesch (1989), however, sought to discriminate the different ways in which chimpanzees coordinate their behavior during hunting, thus making possible inferences about the nature of the social cognition involved.

The chimpanzees of the Tai forest hunt colobus monkeys, mostly red colobus monkeys (*Colobus badius*), relatively high in the forest canopy. The hunting is done by small groups (two to six members) and exclusively by males. Hunting occurs 5–10 times per month, lasts on average just under 20 minutes per episode, and is successful just more than half the time. Many occasions of hunting are opportunistic in the sense that individuals are engaged in other activities, see or hear some monkeys nearby, and respond. Approximately half of the hunts, however, appear to begin spontaneously as a small group moves off in relative silence, stops periodically to peer up in the trees, and proceeds until monkeys are found. When monkeys are found individuals often give a "hunting bark" that alerts groupmates, and when prey is captured, they give a very distinct "capture call" that also alerts others. Meat is usually shared by the primary captor with other hunters, and sometimes with other group members who appear only after the hunt.

Approximately two-thirds of the hunts involved what Boesch and Boesch call "collaboration." Collaboration is the most complex form of cooperative hunting and involves different individuals playing different roles as they coordinate their movements in time and space. For example, one male might climb the tree the monkey is in, while others climb other trees it might be tempted to flee to. Most hunts also have some bystanders who wait on the ground, becoming actively involved only if the monkey falls to the ground. It should be noted that this pattern is somewhat different from that observed in the Gombe chimpanzees by Busse (1977), Goodall (1986), Stanford, Wallis, Mpongo and Goodall (1994), and Teleki (1973). Gombe chimpanzees more often

Figure 7.10. Chimpanzee eating a colobus monkey (Photo courtesy of C.B. Stanford).

hunt alone. When they do hunt together, the cooperation is almost always opportunistic, not planned, and it is not collaborative in the sense of differentiated roles (all individuals simply chase the prey simultaneously). Boesch (1994) explains these differences as due to the greater chance of success of single individuals hunting at Gombe, given the less densely spaced trees.

The seeming planfulness of these hunting activities might indicate foresight of the same type that many primate species use in their coalitions and alliances and physical problem solving, although it is also possible that the chimpanzees are simply setting off in a "trap-line" to find resources as do many animal species. From the point of view of social cognition specifically, the main question is the extent to which individuals in these hunting parties understand the role being played by others in the cooperative enterprise. Boesch and Boesch (1989) point out that although it is rare, lions and wolves sometimes hunt collaboratively with some individuals "hiding in ambush" while another drives the prey toward them. One can easily envision in all of these cases, including the chimpanzee case, that each individual is pursuing its own best interests with little understanding of others. For example, one chimpanzee may climb a tree in pursuit of a monkey, which leads the others, based on their past experiences, to move directly to the places they predict the monkey will flee. Such behavior clearly requires learning, and possibly insight, but it does not necessarily require individuals to understand the different roles or perspectives of others. This is as opposed to a form of collaboration in which individuals actively monitor both the prey and their collaborators, predicting the behavior of the collaborators based on a knowledge of their goals (or even attempting to influence their collaborators' behavior). To choose among these alternatives, experimental evidence is needed.

7.4.2 Chimpanzee Cooperative Problem-Solving

In captivity there are occasionally observations of individuals cooperating in ways that seem to involve different roles and perhaps even the understanding of the goals of the other. For instance, Menzel (1972) observed one juvenile chimpanzee holding a pole that another juvenile was climbing, seemingly to steady it so that the climber could reach its destination (see also de Waal, 1982). In terms of truly experimental observations, there are three studies of ape cooperation for material ends (the first is actually a series of three).

The first study is the classic investigations reported by Crawford (1937, 1941), in which several dyads of chimpanzees were observed. In the first of these experiments Crawford (1937) first trained each of two juvenile chimpanzees to use a rope to pull a box containing food within reach. The juveniles were then placed in the cage together and presented with an out-of-reach box containing visible and very desirable food. There were two ropes extending from the box to the cage, and the box was so heavy that both individuals had to pull simultaneously if they were to succeed. For some trials, there was almost no coordinated behavior. Each individual was then given further training by humans, with not much better results. Finally, Crawford simply taught each individual to pull singly in response to a verbal cue ("Pull!"), and they were then put back together and the cue was given. This resulted in successful coordination of behavior, which continued even after the cue was no longer given. After the cue was no longer given, one member of the pair became very careful to wait at the ropes until the other approached, and only pulled when the other began pulling. When that same individual was later paired with another experienced individual, after more trials she even began to solicit the other when necessary by touching, using the "tandem-walk" gesture, or vocalizing. After some failures, this new pair was also successful. In general, most of the dyads went through this same basic sequence: uncoordinated behavior, coordination after human assistance, active coordination (e.g., waiting for another), solicitation of another. Surprisingly, however, when the same subjects were given a new task with much the same structure as the first (two ropes that had to be pulled down from the ceiling for food, or another small variant), all subjects reverted to uncooperative behavior. This was also true of subjects that learned the two tasks in reverse order, and even though all individuals in all studies were capable of performing their individual roles with ease.

Several years after this investigation, Crawford (1941) presented some of these same subjects and some new juveniles with a new cooperative task. Each was first trained to press four differently colored shapes on a panel in an invariant sequence, at which time food was delivered. Pairs were then tested in a situation in which two of the colored shapes were on a panel in one cage with one subject, and the other two colored shapes were on a panel in an adjoining cage with the other subject (with the first and second shapes to be pressed always in different cages). One pair of subjects, experienced from the previous cooperation studies, coordinated their behavior successfully rather quickly, although when one subject was slow to respond (sometimes due to human intervention), there were initially no attempts by the other at solicitation. Later such solicitations occurred, and consisted mainly of attempts of one subject to push or lead the other (through the cage) toward the color that needed to be pushed. The second pair of subjects (4-year-old twins with less experimental experience) never were very successful in this situation, and there were never any solicitations.

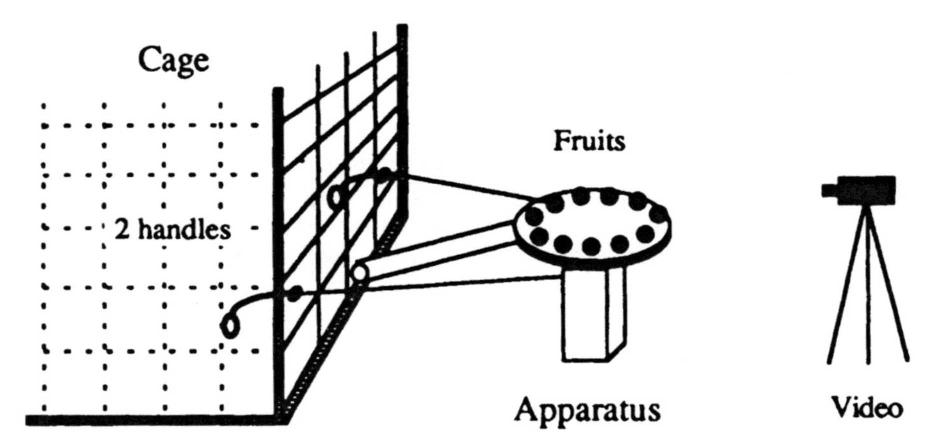

Figure 7.11. Chalmeau's experimental setup to investigate chimpanzee cooperation. (Reprinted from *Primates* with permission from the Japan Primate Centre.)

The final investigation of this type was reported by Chalmeau (1994), who presented a group of six chimpanzees with a food-dispensing apparatus attached to the side of their cage (see figure 7.11). To obtain a piece of fruit, two chimpanzees had to pull simultaneously on two different handles, at which time a cherry was dispensed. The apparatus was presented on 30 occasions of several hours each. In general, the dominant male dominated the apparatus. Of particular interest, this individual, like several of the individuals Crawford studied, learned to wait by one handle until another individual approached and pulled on the other handle. He obtained the fruit on most of the successful trials. On some occasions he attempted to catch and physically lead a female to the apparatus (succeeding on five occasions), presumably so that she would pull the other rope appropriately. Analysis of the pattern of gazing that took place between individuals in these sessions further confirmed the interpretation that at least some individuals knew that others must be at their appropriate stations for the apparatus to work (Chalmeau & Gallo, 1996).

A second type of experimental study of chimpanzee cooperation was reported by Savage-Rumbaugh et al. (1978a). They worked with two chimpanzees trained in the use of a lexigram keyboard intended to simulate the use of human linguistic symbols (see figure 7.12). Each subject was first trained by humans in two roles: (1) to identify the tool needed to open a particular kind of locked box (containing food) and press the appropriate lexigram to request it; and (2) to respond to another's pressing of a particular lexigram by retrieving the appropriate tool and giving it to them. After this training, the two chimpanzees were placed in a two-cage situation in which one of them had sole visual and physical access to the locked box and the other had sole visual and physical access to the tools. From the second day of testing, the subjects played their previously trained roles efficiently together: one identified the needed tool and communicated that via the keyboard to the other, who then retrieved the appropriate tool and gave it to him, at which time the first subject used the tool to open the box and obtain the food - sometimes sharing it with his partner. When the lexigram keyboard was removed and the animals were left to their own devices of gesturing and vocalizing, they were not able to cooperate successfully.

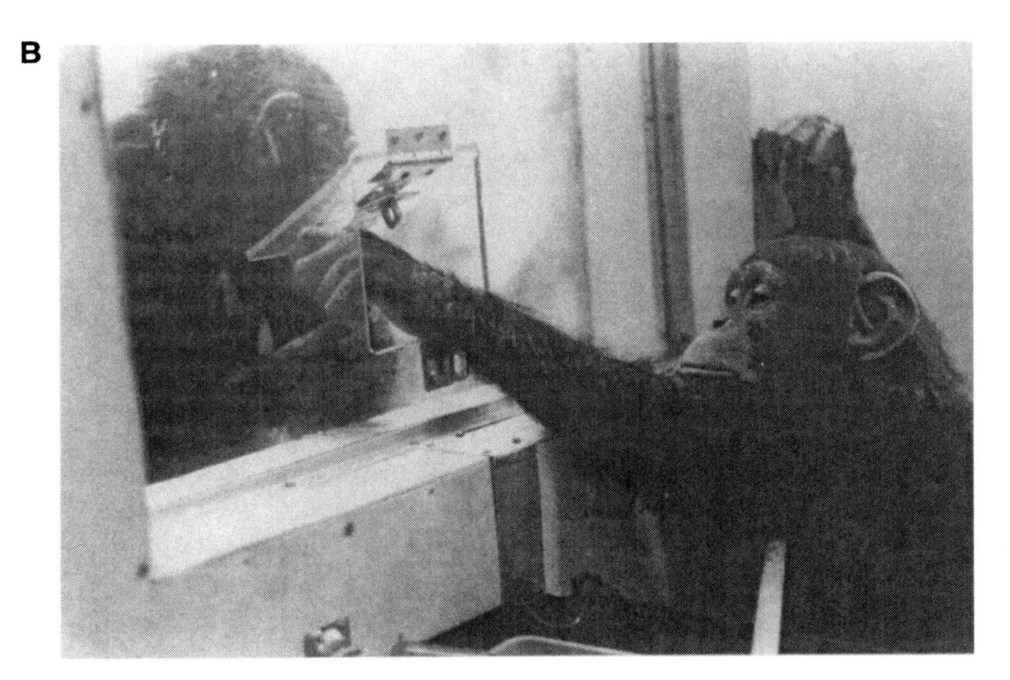

Figure 7.12. The chimpanzees Sherman and Austin in a cooperative task: (a) one chimpanzee requests via the computer keyboard a tool that is located in an adjacent room, (b) the other watches and then retrieves the requested tool, (c) hands the tool to the first chimpanzee, (d) who uses the tool to get the food. (Photos courtesy of E.S. Savage-Rumbaugh with permission from Cambridge University Press.)

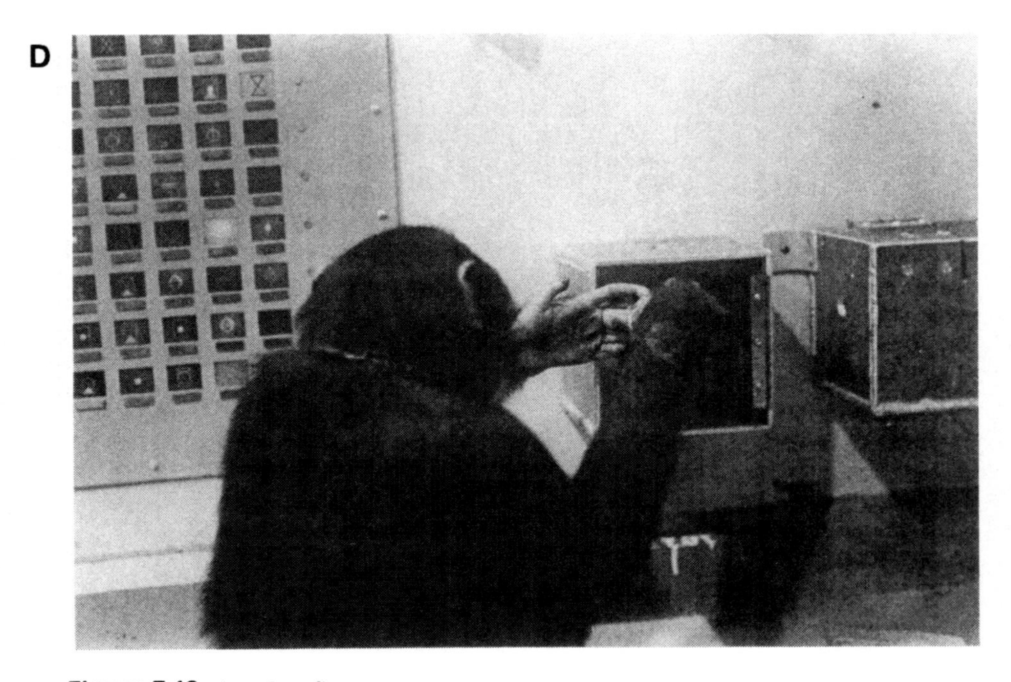

Figure 7.12. *(continued)*

The third type of study was reported by Povinelli, Nelson, and Boysen (1992), who were interested in how chimpanzees understand the different roles in a cooperative task. In this case the cooperation was with humans. They presented a chimpanzee–human pair with an apparatus first used by Mason and Hollis (1962) (see figure 7.13). The apparatus consisted of a platform supporting four pairs of food-baited cups on runners extending the length of the platform, only one pair of which contained food on any given trial. Which pair contained the food could only be seen from one side of the platform. On the other side of the platform were four handles, one per pair of food wells on runners. When a handle was pulled, the corresponding pair of food wells proceeded down the runners, one in each direction, so that both participants got food. Thus, the cooperative task is for one member of a pair (the informant, who could see the food) to indicate to the other member of the pair by "pointing" or looking which one of the four food well pairs contained the food. That other individual (the operator) then pulled the appropriate handle and both got food. Pretraining was given to all four chimpanzee subjects in which they learned how to pull the appropriate handle to obtain the food when it was fully in sight.

The basic idea of the experiment was that each subject first learned to play one role (through human training), and then was asked to play the other role with no further training. Two chimpanzees were first trained by a human in each role: informant or operator. It took each subject some time to learn its role. The same human experimenter then rotated the apparatus so that the subject had to play the opposite role, with no specific training. Note that what was required of the subjects who had first been trained as informants but were now operators was not that they had to learn to operate the apparatus; they had learned to do that in pretraining. Instead, they had to learn to recognize the signs given by the human as to where the food was located so that they would know which handle to pull. On the other hand, the subjects who had been first trained as operators but were now informants now had to learn to give a sign to the human to designate the location of the food (which the human did not know). Results were that one of the two chimpanzees that had first been trained as informant played its new role as operator efficiently from the beginning. It should be kept in mind, however, that this subject was Sheba (the mathematics wizard from chapter 5) who had extensive human experience before this study, and it is possible that she came to the study knowing how to comprehend a human pointing gesture (this was not pretested for her individually). Both of the two individuals who had initially been trained as operators and then were switched to informants learned to designate for the human the location of the food fairly quickly. However, it is not clear to what extent these subjects actually pointed manually for the human and to what degree the human was simply able to discern the location of the food from the chimpanzees' involuntary behaviors such as looking (no criterion for the scoring of a pointing response was given in the study, and it is possible that the subjects simply "gave away" the location of food by approaching it and holding onto the cage in front of it, with no specific gesture such as pointing).

According to Povinelli et al. (1992), the results of this study were consistent with the hypothesis that three of the four chimpanzees were taking the perspective of the other subject in solving the problem, as evidenced by their rapid transfer. They did note that other interpretations were possible, however, especially since there was no pretest to see to what degree specific individuals comprehended human pointing before the experiment. The lack of an independent coder to characterize the "designating" behaviors of the chimpanzees was also a problem. For now, therefore, this study presents only intriguing results in need of replication (see also Tomasello & Call, 1994; Heyes, 1993a). It is also important to keep in mind that these chimpanzees all had received a large amount of human contact in their earlier lives, and this sometimes leads to different skills of communication with humans (see section 8.4, "Communication with Humans").

7.4.3 Monkey Cooperative Problem Solving

Savannah baboons hunt in groups for animal prey (Strum, 1981). The behavior of the baboons does not appear especially coordinated, however, as each individual seems simply to be chasing the prey on its own. Defense against predators is another form of monkey cooperation in the wild (see Cheney & Wrangham, 1987, for a review), as it is for many other mammalian species. For instance, Stanford (1995) observed red colobus monkeys collaborating in their defense against the attacks of chimpanzees in Gombe National Park. The colobus males regularly group together and block the chimpanzees' way to the females and infants, sometimes jointly attacking them as they approach. Similarly, Boinski (1988) observed three capuchin monkeys mobbing a poisonous snake.

Reports of a lack of cooperation in various problem-solving situations based on experimental manipulations have come from Burton (1977) with Japanese macaques, Fady (1972) with Guinea baboons, and Petit, Desportes, and Thierry (1992) with rhesus macaques. In these studies, experimenters placed pieces of food under stones that were too heavy to be moved by a single individual. All species failed to coordinate their efforts to remove stones and uncover the food. In contrast, Petit et al. (1992) observed several instances of successful coordinated activity aimed at removing the stones in Tonkean macaques. Using a different task in which subjects had to pull a string to put some food within their cagemate's reach, Wolfle and Wolfle (1939) also observed cooperation in eight monkeys of four different species (spider monkey, capuchin monkey, hamadryas baboon, rhesus and pigtail macaques). Beck (1973a) observed a female hamadryas baboon learn, without human training, to bring to a male in the adjoining cage a tool that he needed to retrieve some food. Experimentally, Mason and Hollis (1962) presented rhesus macaques with the task described in the previous subsection in the Povinelli et al. (1992) study (see figure 7.13). In a series of experiments, they found that the monkeys working with monkey partners (not human partners as in the Povinelli et al. study) learned to cooperate with one another to obtain the food (the informants did not point but simply oriented to the food). They did not reverse roles when the apparatus was rotated, however. One interesting behavior that emerged in the cooperative process was that the operator learned to "read" the behavior of the informant such that it would approach one of the handles and touch it, wait to see the informant's response (either excited or not), and then pull the handle or not depending on that response.

Still using this same apparatus (see Figure 7.13), Povinelli, Parks, and Novak (1992) replicated the Mason and Hollis (1962) study with rhesus macaques, but made it more comparable to their chimpanzee study. Again, rhesus macaques learned to cooperate effectively in the task, but in this case it was with a human partner. However, when the roles were reversed, the subjects' performance fell to chance levels. This was clearly true when they went from being informants to operators (i.e., when they had to

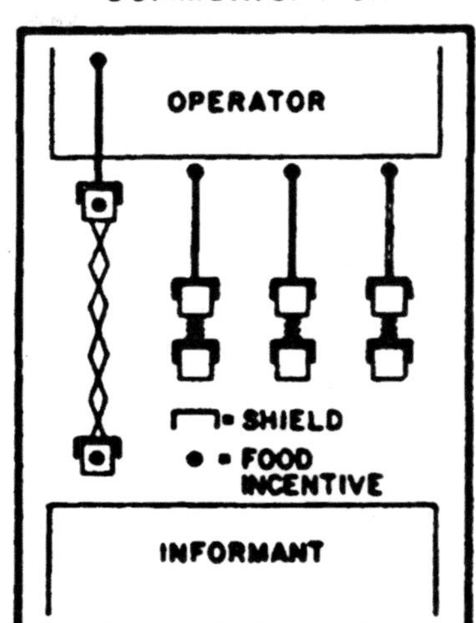

Figure 7.13. Mason and Hollis (1962) experimental apparatus, also used by Povinelli et al. (1991). (Reprinted from *Animal Behaviour* with permission from Academic Press.)

go from indicating to comprehending indications). In the opposite direction (i.e., from comprehending indications to indicating), they did not transfer immediately, but they learned to indicate the food's location very quickly (a kind of pointing ritualized from reaching for the food). It should be noted that although these subjects had had experience with humans in previous experiments, it is not clear whether they had had the kinds of communicative experiences of the chimpanzees of Povinelli et al. (1992). Povinelli et al. interpret their results as indicating that rhesus macaques lack an understanding of the role of their cooperative partner.

7.4.4 Comparison of Species

Cooperation in the sense of multiple conspecifics engaged in pursuing the same goal at the same time is a general phenomenon of mammalian behavior. Cooperation in the sense that two or more individuals adopt different roles toward a common end and adjust their behavior depending on the behavior of their partner(s) would not seem to be a common activity, either of mammals in general or of primates in particular. The only clear example of primates in the wild is the cooperative hunting of chimpanzees in the Tai forest. Other primates, even other chimpanzee communities, do not seem to cooperate in this same way in their natural habitats or in other activities that require differentiated cooperative efforts. In the somewhat conservative opinion of Cheney and Seyfarth (1990b: p. 292), "Chimpanzee hunting behavior still appears to be no more cooperative than the hunting behavior of wolves, wild dogs, or even some spiders."

In captivity, when a situation is arranged so that food can be obtained only through cooperative activity, chimpanzees show some limited skills. In all cases except one, human intervention and training was an important part of the process. The exception is the Chalmeau (1994) study, in which one individual of a group of chimpanzees learned to inhibit its own response with a handle until another individual was performing the required activity on the other handle, and on some occasions even attempted to drag another individual to the apparatus. Monkeys learned to cooperate with one another in the Mason and Hollis (1962) study, but without much understanding of the role of the other. Human children learn to cooperate with one another in complex ways involving an understanding of the role of the other from sometime after their third birthdays (Ashley & Tomasello, in press; Brownell & Carriger, 1990).

7.5 What Primates Know about Others in Social Interaction

Many mammals live in complex and dynamic social groups. Many mammals recognize individual members of their groups, understand something of their behavior, and remember past interactions with them (i.e., form individual relationships with them). This means that many mammals live in a complex social field in which they must monitor many social relationships simultaneously. However, there is no evidence that mammalian species other than primates understand third-party social relationships—although the problem may be that nonprimate species have not been observed with these questions in mind. Because there is clear evidence of this ability in primates, both from experiments and naturalistic observations of their social relationships based on kinship, friendship, and dominance, we hypothesize that this ability is unique to primates. If the understanding of third-party social relationships is indeed a uniquely primate ability, the implication is that the social fields in which they operate are more complex by several orders of magnitude than those of other social mammals. In combination with a wide-ranging and flexible set of skills for predicting the behavior of conspecifics, the understanding of third-party relationships makes for extreme cognitive complexity in all of the everyday interactions that comprise primate social life. It is also interesting in this regard that the understanding of the tertiary relations among objects that is so important to some primate problem solving in the physical domain may also be unique to the order (see chapters 4 and 6). The relation between the physical and social forms of tertiary understanding is a central theme of our overall theoretical account of primate cognition in chapter 12.

All primate social interactions take place within this complex social field, even relatively stereotyped species-specific behaviors such as mating (e.g., the mating of certain individuals cannot take place in the presence of other individuals). Coalitions and alliances are an especially interesting type of interaction that take place within this field because they require coordination of the individual's cooperative and competitive skills. The individualized social knowledge that constitutes the social field of primates leads to selectivity in the choosing partners and targets that nonprimate mammals may not show (Harcourt, 1992). Primates also show some skills of foresight and planning in their coalitions and alliances, as well as skills of learning and memory. There is scant evidence for differences among primate species in any of these skills (and in some cases good evidence for equality of skills across large taxa such as Old World monkeys and great apes), although in some cases only a few species have been systematically studied.

In forming coalitions and alliances and in engaging in a number of other social behaviors such as grooming, many primates behave reciprocally. This reciprocity may be based on kinship or mutualism, or it may emerge as a by-product of proximity or dominance relations. Despite any other cognitive complexities that might be involved, these interactions do not necessarily involve mental record keeping of interpersonal credits and debits. In addition to these cases, however, some behaviors of primates would seem to qualify as true reciprocity (and at least one behavior of one nonprimate species). The nature of the mental record keeping that mediates this reciprocity—for example, in what sense it is explicitly quantitative—is not known at this time, but in whatever form it exists, it contributes additional complexities to the primate social field. Again, there is no convincing evidence of broad-based differences in skills of reciprocity among the primate species studied (mostly cercopithecine Old World monkeys and great apes).

Cooperative problem solving in which different individuals play different roles (rather than, for example, several individuals simultaneously chasing a prey) is not common in nonhuman animal species. The only well-documented example of such cooperation in primates is the group hunting of one community of chimpanzees. The nature of the cooperation involved in this case is not totally clear, but it is likely that the individuals do coordinate their behavior to take account of the location and perhaps behavior of others in the hunt. How they understand the role of the others, in the sense of understanding their goals so that their behavior can be predicted, is much less certain. There is little evidence from the laboratory on cooperative problem solving to help answer this question, and the studies that do exist do not show any great skills that have not been observed in the wild, especially if we exclude cases in which there is explicit human training. Chimpanzees can learn that they must coordinate their behavior in time with that of another, but the extent to which they understand the role being played by the other in the cooperative enterprise is not clear. The one study supposedly showing role reversal involved indicating the location of food for humans (Povinelli et al., 1992), not two chimpanzees coordinating their problem-solving behaviors, and even here there are other interpretations of the supposed role reversal, as the investigators noted.

A crucially important question is whether and in what ways primates understand the psychology, over and above the behavior, of conspecifics as they interact with them in all the ways we have reviewed in this chapter. It is best answered, however, after several other types of primate social interaction have been investigated, as we do in the next three chapters. Nevertheless, it is important to emphasize one key point. As we have already noted in this chapter, much of primate social interaction is based on the ability to understand the animacy and directedness of the behavior of others: individual primates may learn from their observations of groupmates which behaviors serve as good predictors of which other behaviors for which individuals. Thus, over repeated occurrences in a certain context, when they see an individual engage in a particular behavior, they come to see the individual as "directed toward" a particular outcome—its usual consequence. This is obviously a key cognitive skill in virtually all of primate and other mammalian social interaction, and social knowledge of this type makes up a key part of the primate social field. The hypothesis that primates comprehend the animacy and directedness of behavior is potentially important theoretically because many theorists of primate cognition view all primate social behaviors as resulting from only two processes: mindless operant conditioning or an explicit understanding of the intentional or mental states of others ("theory of mind"). The ability to understand the animacy and directedness of behavior is an intermediate possibility in that it is clearly based on cognitive processes of inductive learning, but it concerns only the behavior, not the minds, of other individuals.

Complementing the understanding of the animacy and directedness of behavior is the ability of primates to use this knowledge to actively formulate various types of social strategies, such as those employed in the creation of coalitions and alliances. Again, however, this may in some instances be accomplished without any explicit knowledge of the psychological states of others. Thus, primate social strategies may be quite insightful, foresightful, and complex, based on informed decision making and mental representation, involving an understanding of the animacy and directedness of behavior and various third-party social relationships, without being aimed in any way at the intentions or mental states of others. Although we have documented in a general way some primate social strategies in this chapter, our focus has been mainly on individuals' understanding of their social fields and how this influences their social interactions within that field. We now turn more specifically to primate social strategies designed to influence the behavior of others, some parts of which may have become ritualized into intentional communicative signals.

8

Social Strategies and Communication

Primates' knowledge of their social fields enables them to anticipate and predict the specific behaviors of specific individuals in a variety of social circumstances. But primates may also use their social knowledge in more active, even proactive, ways as they formulate social and communicative strategies designed to actually manipulate or influence their groupmates. These social and communicative strategies, like the strategies employed in foraging, tool use, and other domains of physical cognition, potentially involve various types of insight and foresight into the social problems they are designed to solve.

Some examples of social strategies were reported in chapter 7 in the context of coalitions, alliances, and cooperative problem solving, and there are many other reported observations as well. For example, when they are engaged in some struggle over dominance, primate individuals may perform various behaviors to recruit an ally, perhaps even an especially powerful one, using that ally as a "social tool" (Kummer, 1967). More proactively, if an individual fears an opponent, it may stay close to a powerful individual from the outset, or it may groom the potential opponent ahead of time in an effort to prevent the aggression from occurring and thus demonstrate some kind of foresight into the social problem (Koyama & Dunbar, 1996). Grooming another individual in advance is also used on some occasions when valued resources such as food or sex are being sought from that individual, as was seen in section 7.3, "Reciprocity and Interchange." Also proactively, Kummer (1982) reported that when two male baboons were placed in the same small cage for transport, they turned their backs to one another and stayed very still, presumably as a way of preventing aggression in this potentially volatile context. In situations in which there is an active competitor for some resource, primate individuals may, for example, attract the opponent's offspring and then, when it abandons the resource to protect the offspring, quickly return and obtain the resource. These kinds of observations are reported with some frequency by observers of primates in their natural habitats and in relatively natural social settings in captivity (when they are one-time occurrences, they are often called "anecdotes"). The reports of Kummer (1982, 1995) and de Waal (1982, 1989a) are especially rich and well known.

There is a great deal of controversy over the use of these kinds of observations in the scientific investigation of animal cognition, since in many cases their interpretation is difficult and, moreover, they are often not repeatable. Recently some attempts have been made to collate many of these anecdotal observations in a systematic fashion and to see if any patterns allowing reliable scientific inferences emerge. Of special importance in this regard is the subset of social strategies that some researchers have labeled "tactical deception," which have been collated and analyzed by Whiten and Byrne (1988b) and Byrne and Whiten (1990). For some, the importance of these observations derives from their indication of some kind of "theory of mind" on the part of the deceiver: the deceiver intends that the dupe perceive some false information or have some false belief about the world. This is a difficult inference in all cases. But regardless of their status as true deception involving some kind of theory of mind (as presumably occurs in humans), it is relatively uncontroversial that many of these "deceptive" behaviors involve flexible social strategies in which one individual actively attempts to manipulate or influence the behavior of another, and thus they are of obvious interest and importance to an investigation of primate social cognition.

Another special subset of primate social strategies are intentionally produced communicative signals. In the current context these are special for at least two reasons. First, from a methodological point of view, communicative signals, almost by definition, are capable of being observed repeatedly and thus they may be studied with more rigorous scientific methods involving replications and experimental manipulations. Second and more importantly, from a theoretical point of view, communicative signals are especially interesting because they are not social behaviors designed to produce their effect directly, but rather they are ritualized social acts designed to induce others to act via their own self-generated powers; they thus embody some understanding of conspecifics as animate beings.

> [Communicative] acts achieve their ends indirectly. That is, the actor does not act physically to alter things to suit his needs, pushing or dragging other individuals about, beating them into submission, or the like. Instead, the actor's behavior provides other individuals with information, and the actions that *they* take on the basis of this information lead to any functions that are obtained. (Smith, 1977: p. 389; emphasis in original)

And so, for example, one individual might force another to leave its territory by attacking and driving it away (a social strategy perhaps), or it might simply bare its teeth and growl (a signal), providing a sign to the intruder of what will happen next if it does not leave. Communicative signals of this type may be under the voluntary control of the signaler to a greater or lesser degree, often depending on the extent to which the ritualization process occurred during phylogeny or during ontogeny. To the extent that communicative signals are produced automatically whenever an individual is in a certain mood state or perceives a certain stimulus (without flexibility, strategic choice, or voluntary control), they are not cognitive phenomena for the signaler and therefore not of direct concern to us here. To the extent that they are used as flexible communicative strategies, they are clearly relevant to an investigation of primate social cognition, and we will refer to the use of such signals as "intentional communication" (as opposed to communication more broadly defined). The process is basically the same for signals produced in the gestural and the vocal modalities.

Finally, some of the most intriguing and suggestive findings about primate social cognition have emerged from studies of nonhuman primates learning to communicate with human beings. For example, extended communicative interactions with human beings sometimes brings out communicative skills in primates that they do not typically employ with conspecifics, for instance, referential pointing, which might indicate some important skills in the ability to share attention with others. Moreover, in some cases human beings have actively taught some great ape individuals symbolic forms of communication based on human languages. The fact that great apes master some aspects of linguistic communication suggests to some researchers that nonhuman primates have skills of cognition and social cognition that simply have not been tapped in traditional experimental procedures, while other researchers reject this suggestion. The study of "ape language" has a long and contentious history, which we cannot review in detail here (see Savage-Rumbaugh, 1986, for one view of the early history of this work; see Ristau & Robbins, 1982, for a more critical look at this history). What we do here, therefore, is simply report the most complex and humanlike communicative achievements that have been documented in scientifically rigorous ways for the most skillful great ape individuals.

Our procedure in this chapter is thus to focus attention on those primate social and communicative strategies that are used in the most socially flexible ways, thus indicating some forms of social knowledge and strategic decision making. We first investigate primates' use of social strategies, with a primary focus on what some researchers have described as tactical deception. These sometimes involve ritualized communicative signals, but quite often they do not. We then consider the most flexibly used communicative signals of various primate species—what we call intentional communication—with an eye to how they are acquired, understood, and used. We do this first for gestural communication and then for vocal communication. Finally, we consider the ways that nonhuman primates communicate with humans, in some cases using human language-like forms of communication that humans have taught them.

8.1 Social Strategies: Deception

As primates compete with conspecifics for valued resources, in some cases they simply take what they want because they are dominant to the others present, and in other cases they are lucky enough to discover the resource in a noncompetitive situation. In other cases, however, their only hope is to outwit their conspecifics, as was first emphasized in the modern context by Dawkins and Krebs (1978). Recruiting a dominant or otherwise helpful ally is one strategy for achieving that end (as described in chapter 7). Another set of social strategies, however, is to influence the behavior of conspecifics directly. There are many ways to do this, but the primate social strategies that have been most hotly debated are those that have been classified under the general rubric of tactical deception (Whiten & Byrne, 1988b) (perhaps because they seem most clever to human eyes). The question of whether these strategies are intentionally deceptive from the behaver's point of view—whether the behaver actually intends to convey false information or create a false belief in the other—is a difficult and contentious one that Whiten and Byrne discuss at some length. The problem is that other interpretations for

Figure 8.1. Chimpanzee begging and screaming. (Photo courtesy of F. de Waal.)

most deceptive acts are also possible based on more straightforward processes of behavior, learning, and communication: one individual simply intends that another engage in or not engage in a particular act, with the issue of deception not arising.

The methodological problem in this area of research is that most of the evidence for tactical deception in primates comes from a unique source. Whiten and Byrne (1988b) and Byrne and Whiten (1990) were concerned that many primatologists had observed instances of tactical deception, but because they were relatively isolated examples, they were not published. A compendium of such observations might reveal some important patterns. Consequently, the two investigators contacted a large number of primatologists (mostly through membership roles in scientific organizations) and asked them for any observations they might have made over the years that implied tactical deception. They then sifted through the resulting reports (that met the basic functional definition of deception in that one animal did something and another perceived a situation wrongly as a result) and classified them in various ways with an eye to the cognitive mechanisms involved. They were especially careful to identify and eliminate from further consideration observations that had fairly simple interpretations not involving deception. In the most recent analysis, Byrne and Whiten (1992) had a corpus of 113 observations that qualified as tactical deception, from 32 informants, with just more than half coming from captive animals. They classified them as concealing, distracting, creating an image, social tool use, and counterdeception.

Many primatologists have noted the weaknesses of this methodology from a scientific point of view, including Byrne and Whiten. Among the most obvious problems are (1) observers often reported unrecorded observations made some time previously and thus had to rely on memory; (2) because they were not specifically looking for deception, observers did not record other uses of the supposedly deceptive behavior (or other instances of the same context in which there was not anything that looked like deception) and thus may have misinterpreted the intention behind it; and (3) given the fact that many reports were one or a few observations by researchers who had been observing primates for many years, it is possible that the reported observations were simply behavioral sequences that occurred by chance but had interesting human interpretations. Bernstein's (1988b) critique, in which he points out that he could poll a large sample of reputable scientists for instances in which a dream came true the following day and get many positive instances, is especially skeptical. Its major point may be summarized by his observation that "The plural of 'anecdote' is not 'data'" (p. 247).

In any case, in this important area of research, the only information we have is the compilations of Byrne and Whiten (and a few other observations reported by researchers in other forums) and a few sets of experimental observations. We therefore report first on the observations analyzed by Byrne and Whiten (1992) and a few others in three broad categories: active concealment, active misleading, and counterdeception. In each of these three categories, we report only selected examples and refer the reader to the original compilations for a fuller set. After describing the three sets of observations, we report on experimental evidence that might shed light on their proper interpretation. In all of this, two crucial issues must be kept constantly in mind: (1) is the reported observation an instance of a reliable social strategy, or is it simply an accidental confluence of events? and (2) does the reported observation involve one animal attempting to manipulate the psychological states of others? These two issues must be kept separate because it is likely that many of these observations are instances of reliable social strategies designed to influence the behavior of others (and thus they are of interest to us here), but they may not be instances of deception as it is typically understood from the human point of view (i.e., intentionally sending false information or creating a false belief in other individuals). We are mainly concerned in this chapter with the first of these issues; we address the second in chapter 10, "Theory of Mind."

8.1.1 Active Concealment

There are a number of observations by field primatologists of individuals not doing something they normally would do, seemingly so that others will not detect their presence (see figure 8.2). For example, Goodall (1986) reports that groups of chimpanzees on patrol for other chimpanzee groups remain more more silent than usual, and Boesch and Boesch (1989) report the same thing for chimpanzees hunting for monkeys. Milton (#25 in Byrne & Whiten, 1990) reports that a male spider monkey once traveled on the ground to follow a potential mate also traveling on the ground, when he normally would travel in the trees, possibly because travel in the trees is noisier than travel on the ground and thus risks detection by potential opponents. In all three of these cases, the individuals involved have presumably learned that noise-making in these contexts is likely to lead to adverse consequences.

Other reports of concealment involve more active behavioral processes. The best

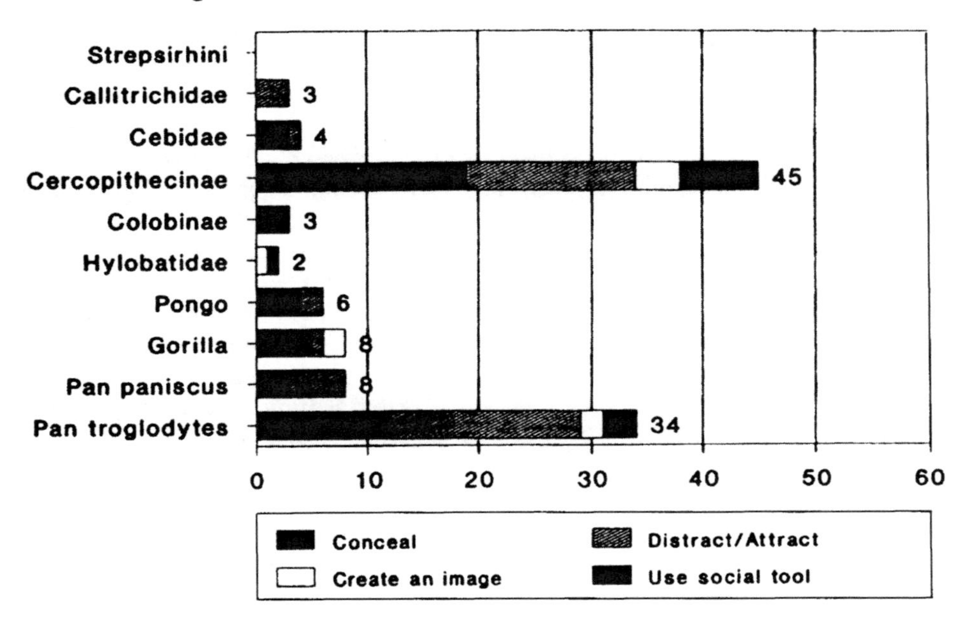

Figure 8.2. Deception in nonhuman primates. These data represent cases that were independently rated by both authors as cases of "tactical deception." The figure depicts all cases of intentional (level 2) and conditioned (level 1) tactical deception. (Figure courtesy of R.W. Byrne.)

known example is the observation of Kummer (#56 in Byrne & Whiten, 1990) in which a hamadryas baboon female very gradually over many minutes shifted her position so that she could groom a subadult male who was resting behind a rock, out of sight of the dominant male who would not tolerate such an interaction if he saw it. Another interpretation, however, is that the female simply monitored the dominant male's reaction as she slowly approached the subadult male, ready to flee if he responded; when he did not respond she continued to her goal. Two other observations of this type are (1) de Waal's (1982) observation of a male chimpanzee covering his erect penis with his hands so that a competitor could not see it, and, on another occasion, an individual covering its fear-grimace with its hands so as not to give away its emotional state; and (2) Tanner and Byrne's (1993) observation that a gorilla covered its play-face with its hands on a number of occasions, presumably to conceal its playful intentions. In these two cases, it is more difficult to envision a specific learning sequence in which an individual somehow noticed that when its hands were in the right place another animal refrained from doing something it might otherwise have done. Consequently, these observations seem to indicate that at least some apes know that others must be able to see their communicative displays before they can have their effect, and they know how to actively manipulate this. (Interestingly, Goodall, 1986: p. 583, reports specifically that she has never observed this kind of active hiding behavior in her 25 years of observing the Gombe chimpanzees.) There are no relevant observations for monkeys of this active hiding of their own displays, but Byrne and Whiten (1992) report a number of instances of monkeys altering their travel trajectories so as to keep a tree as an obstacle between themselves and a human observer, presumably in order to evade detection.

8.1.2 Active Misleading

Byrne (#67 in Byrne & Whiten, 1990) reports that an adolescent male savannah baboon was being chased aggressively by some adults. He suddenly jumped to his hindlegs and stared as if he had spied a predator. The pursuing adults stopped and stared also, and the chase ended. Byrne could see no predator or other reason for staring, although, of course, it is possible that the fleeing baboon had indeed seen or heard something. Similarly, Goodall (1986) reports that a young chimpanzee in a group feeding situation was not getting much food due to dominant individuals. The youngster then arose and walked off deliberately, a behavior that almost always induces others to follow no matter who performs it. The others followed and the youngster later circled back and got more food. He did this on repeated occasions. Goodall speculates that the first time may have been an accident—the youngster decided to go to another food source and noticed that the others followed—but that on the subsequent occasions the youngster was acting deliberately.

Rasmussen (#98 in Byrne & Whiten, 1990) reported an observation in which a male savannah baboon recruited three conspecifics, approached a rival male who was consorting with a female, and then screamed as if he had been attacked. The three allies converged on the rival and when they became engaged in a fight the challenger herded the female in the opposite direction. No such manipulation of a "social tool" has been reported for apes. De Waal (1986) reported that a chimpanzee who was angry at another held out its hand in an appeasement gesture and when the other approached, attacked it; importantly, this happened repeatedly. Altmann (#88 in Byrne & Whiten, 1990) reports that female savannah baboons approach mothers regularly with the apparent intention of grooming them, but once in proximity they seem more interested in touching the offspring. Cheney and Seyfarth (1990b) report that in the midst of a hostile encounter with another group, a vervet monkey gave an alarm call which served to disperse the individuals and end the potential fight; there did not seem to be a predator nearby, although there is no way the researchers could be sure.

The repeated use of stable social strategies such as those observed by Goodall, de Waal, and Altmann are not explainable in terms of a chance occurrence of events, as are the one-time occurrences. However, they are explainable as relatively stable social strategies that are designed to create a desired state of affairs (e.g., getting others to leave food so that it may be enjoyed alone, or to approach so that they can be attacked, or to tolerate an approach so that their infant may be touched) by manipulating the behavior of others. There is no need in any of these cases to conclude that these strategies are directed at the psychological states of others, that is, at creating false beliefs in others.

8.1.3 Counterdeception

In many of the cited observations the "dupes" are easily misled by the deceivers. But there are about four observations (three of which are from chimpanzees) that show some evidence of an individual taking active countermeasures. The most famous is reported by Menzel (1973c) in which experimenters hid food and informed only some chimpanzee individuals of its location (see section 7.1.1 for a full description of the experiments). When the chimpanzee Belle was the only individual in the group who knew the location of the hidden food, she soon quit uncovering it when Rock was nearby because he would steal it from her. She began to lead the others away from the food and sit patiently. When Rock followed her and began searching the area near her, she would then try to go to the food and uncover it. This sometimes worked and sometimes did not. To help along the process, Rock then began walking away from Belle and suddenly spinning around to see if she was moving to the food. Researchers such as Byrne (1994) believe that this is evidence of counterdeception and that "Rock understood Belle's (deceptive) intentions and can anticipate her thoughts" (p. 133). But the unfolding of these interactions between Rock and Belle took place over a period of several months (Menzel, 1973c), although this fact is not always reported (Menzel, 1974). It thus seems possible that over that amount of time, each individual could have learned something about how to predict the behavior of the other and how the other would respond to its own behavior—a prototypical case of the creation of social strategies through understanding the animacy and directedness of the behavior of others.

A second widely cited example was reported by Plooij (#253 in Byrne & Whiten, 1990). One chimpanzee was at a human feeding box when it opened with a click (controlled remotely by humans). Another chimpanzee entered the area. The first closed the box and sat idly by. The second chimpanzee left the area, but "hid" behind a tree and watched the first chimpanzee. When the first chimpanzee opened the box, the second came out from behind the tree and took it. The question in this case is simply why the second chimpanzee did not just open the box for himself, or at least try it before he left; the first chimpanzee had no special skills of box opening (although there may have been reason for the second to suspect that the box would not be opened for him). It is thus possible that he left the area for his own reasons and then decided to look back to see what the other was doing.

8.1.4 Experimental Approaches

There are currently many different evaluations of these kinds of anecdotes and observations. On the one hand, they are reported by reputable primatologists who know their subjects well and can spot a behavior that is unusual in a particular circumstance. On the other hand, if the observations are fairly rare, it is possible that they are instances of overinterpretation of a chance confluence of events. There are some of these observations that have been reported to be recurrent, however, and they would not seem to be subject to the interpretation in terms of chance occurrences. They would seem to be learned strategies, or perhaps even insightful strategies. But in all cases it is possible to see these social strategies not as attempts to deceive others—to create in their minds a false belief about the world or one's own intentions—but simply as strategies to get others to do something or not do something. Thus, if an individual gets attacked every time it mates with a particular female in the presence of a particular male, it can learn a strategy of only attempting to mate with her when the rival is not in sight, or even create the insightful and foresightful strategy of herding her out of sight before attempting to mate. Another example of an insightful strategy might be when an individual has noticed that others attack an opponent that it screams at, it may use that knowledge to ask for an attack in a novel circumstance (e.g., when there is no real fight going on but it wishes the other to be distracted). The point is *not* that these are not intelligent or insightful social strategies; they are. The point is simply that the deceptive interpretation may be added by the human observer.

This interpretation would seem to be supported by the few experimental manipulations that have been done. First, in Menzel's (1973c) studies, it is possible that Rock and Belle simply learned over several months how the other reacted when they did certain things, and they then learned strategies for making that happen when it suited their purposes. This interpretation is consistent with the findings of Coussi-Korbel (1994). Her white-collared mangabeys developed some of the same types of strategies for manipulating the behavior of others that Menzel's chimpanzees developed, but in this case a detailed day-by-day analysis was kept and reported. What she found was that one informed mangabey individual learned the strategy of leading her main competitor away from the food, as had one of Menzel's subjects, after only four trials. In the first trial she went to the food and her competitor followed and stole it from her. In the second trial she did not go to the food and the competitor stayed back with her while others managed to find the food. In the third trial she held back again, and when the competitor held back also she dashed for the food and got it first. And in the fourth trial she walked off in the wrong direction, the competitor followed, and at an opportune time she dashed for the food and got it again. This study documents in detail the very rapid and creative learning process by which one individual may develop a social strategy for manipulating the behavior of others.

Kummer, Anzenberger, and Hemelrijk (1996) reported a different kind of experiment in which longtail macaques interacted with humans. The individuals of this captive group had been trained so that when a human was facing and watching them, they could not drink from either of two juice nipples in their cage, or else they would be threatened, but when a human was present but facing away (back turned), drinking was allowed. The key test came in the transfer phase when an opaque screen was then introduced into the cage, obstructing the human's view of one of the juice nipples (in a variation of this procedure, a semitransparent screen was used so that the subject could see the human from the location of the nipple), thus providing the opportunity for deception at one of the nipples. Results were clear-cut. None of the seven subjects preferred to drink from the nipple that was behind either type of screen, as opposed to the visible nipple, during the transfer trials. Although the investigators considered several specific characteristics of the natural ecologies of these macaques that might have led to their failure in this test, they concluded overall that it is unlikely that these monkeys are capable of accounting for the visual perspective of others so as to deceive them.

The only other experimental investigation of deception in nonhuman primates is that of Woodruff and Premack (1979). In this study four juvenile chimpanzees learned to indicate to a naive human trainer which of two opaque containers contained food, at which point the human would remove the food and give it to them. They mostly oriented to the container with food, looked at it, became excited when the human approached it, and, after a while, some learned to "point" to the correct container by hanging their fingers through the cage in the direction of the food-baited container. After the chimpanzees learned to do this, two different types of human trainers were introduced. First, a cooperative trainer acted as before, attempting to locate the hidden food and, when he did, giving it to the subject. A competitive trainer also attempted to locate the food, but when he found it he kept it and ate it himself. The question was whether the subjects would learn to provide accurate information for the cooperative trainer but inaccurate information for the competitive trainer. In a later phase of the experiment, subjects were also exposed to a cooperative trainer who pointed to the location of the hidden food for them and a competitive trainer who pointed to the container not containing the food. The question in this case was whether they could comprehend someone else's attempts at deception (see figure 8.3).

In the first phase of the experiment (lasting 5 months), all subjects began by indicating the location of the food accurately for both the cooperative and competitive trainers, with no attempts to withhold information or actively deceive either. Three of four subjects continued in this manner throughout the first phase, but one subject began to withhold information from the competitive trainer in the second block of 24 trials. In the second phase of the experiment, which took place 6 months later and included some methodological changes (and lasted 14 months), all of the subjects began to withhold information from the competitive trainer. Two of the subjects induced the competitive trainer to behave with only chance-level accuracy by suppressing their tendency to orient to the baited container (sometimes turning their backs), but two others induced him to choose the wrong container by orienting or pointing to it (one beginning on trial 180 of phase 2, and one on trial 248 of phase 2). In the comprehension test in which subjects attempted to find the food based on the pointing of cooperative and competitive trainers (also administered in phase 2), all subjects began by choosing the container pointed to regardless of whether it was the cooperative or competitive trainer who was pointing. One of the subjects caught onto the deception of the competitive trainer and started choosing the container not pointed at on trial 37, one on trial 133, and one on trial 159. The fourth subject never caught on.

In phase 3 of the experiment, conducted 10 months later and lasting 1 month, each subject was given 48 trials of production (24 each with a cooperative and competitive trainer) and 48 trials of comprehension (24 each with a cooperative and competitive trainer), with the two types of trainers occurring in counterbalanced order within sessions (in the previous phases the trainer was the same for a given session and different across sessions). In production, all four of the subjects accurately designated the location of the food for the cooperative trainer and made some attempts to withhold information from the competitive trainer. Two of the subjects (the same two as previously) also actively misled the competitive trainer. In the comprehension test, three of the subjects quickly caught on to the competitive trainer's attempted deceit, whereas one did not (the same one as previously). The investigators interpret these findings as evidence for skills of intentional deception in chimpanzees. But it is important to note that the deceptive skills emerged only gradually over many trials in the three phases of the experiment over a period of 3 years, plenty of time for the subject to learn a social strategy to get the other to choose the bucket containing no food so that he could get the bucket with food. In this interpretation there is intentional communication, but the interpretation in terms of deception is superfluous.

A recent experiment by Mitchell and Anderson (in press) attempted to replicate these results with two capuchin monkeys (*Cebus apella*). After learning to "point" accurately for a human to one of two covered bowls, cooperative and competitive trainers were introduced. Although statistical analyses were not performed, both individuals seemed to learn to differentiate over several hundred trials between the two trainers.

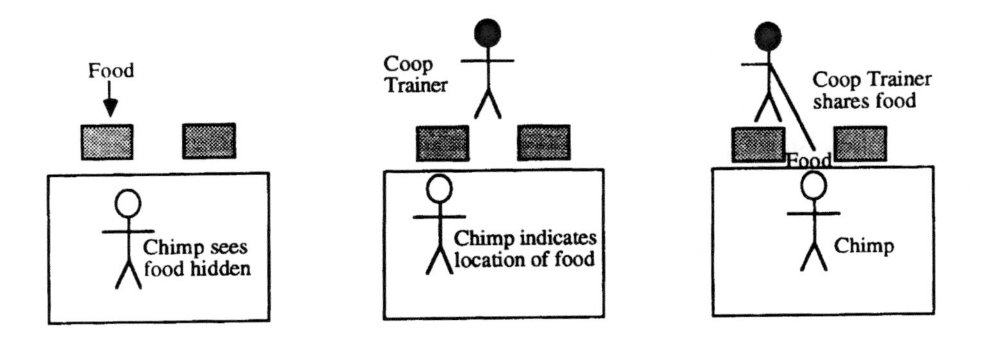

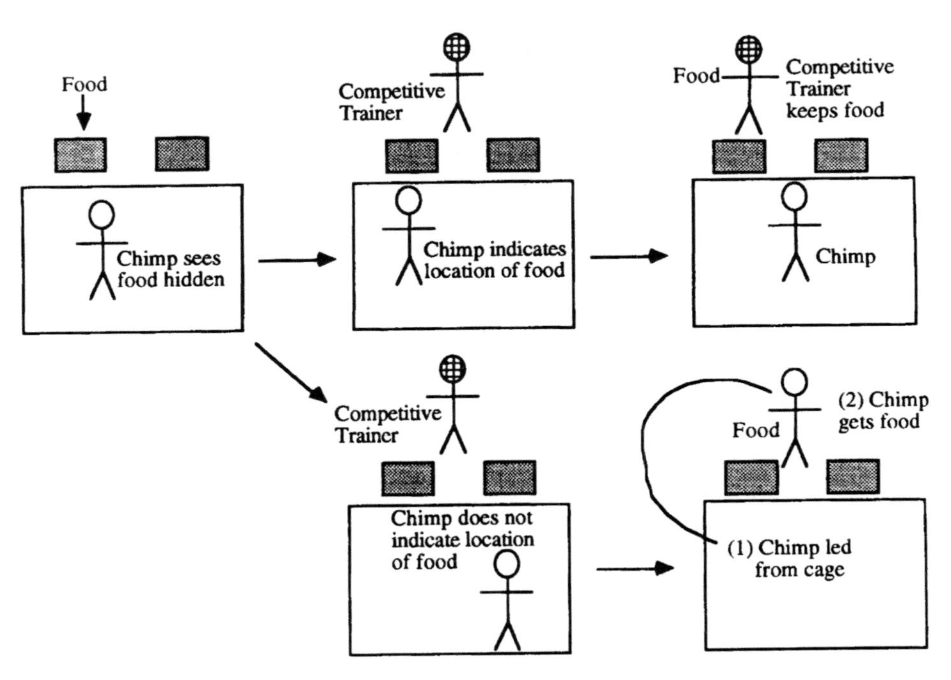

Figure 8.3. Woodruff and Premack's (1979) experimental setup.

One subject mostly refused to point for the competitive trainer, whereas the other actively misled him to the empty bowl. After about 700 trials, the competitive human trainer was performing with what appeared to be below-chance accuracy for both subjects.

8.1.5 Comparison of Species

Perhaps as much as any phenomenon of primate cognition, the interpretation of the scientific evidence for tactical deception is problematic and controversial. One problem is the prevalence of unsystematic observations ("anecdotes") in the evidential corpus. No one disagrees that these are suggestive observations worthy of further systematic observation and experimentation, but this simply has not been done. In the few cases of systematic observation, primates have been observed to learn a strategy over time that helps them defeat a competitor in ways that appear to humans to constitute deception. But it is not clear why we may not simply call these instances of social strategies in which one animal understands a social problem in a way that allows it to then search its repertoire of behavioral and communicative means for solving it, as they seemingly do in many social circumstances. Along these lines, Byrne (1994) reports an observation of a domestic cat: the cat wants in a chair but a human occupies it; the cat goes to the door and meows; the human rises to let the cat outside; the cat climbs into the unoccupied chair. This kind of strategy may be learned gradually or it may be a product of insight: the cat knows that it needs to get the human out of the chair and it recalls that in the past a meow (intended in that case to request being let out) has led to that result. In either case, and despite the unquestionable intelligence involved, there is no necessary reason to believe that the animal is intentionally deceiving in the sense that it intends to create a false belief in the other, and indeed Byrne does not believe that it is. The prevalence of strategies similar to those reported by Byrne and Whiten (1990) in many nonprimate species, many of them birds and many of them flexible strategies (see Cheney & Seyfarth, 1990b: p. 198–199), lends support to this interpretation in terms of a set of insightful social strategies that apes and monkeys invent and that humans interpret as deception.

A number of investigators have claimed that there is evidence that deception is somehow cognitively more complex in apes than in monkeys (there are no reported observations of any prosimian species). For example, de Waal (1986) believes that the examples are more varied and complex for chimpanzees than for any other species. Likewise, Byrne and Whiten (1992) conclude that while chimpanzees and baboons dominate the reports they have reviewed (possibly because of the specific scientists studying these species), chimpanzees have been observed to engage in more instances of deception involving the taking of the "mental perspective" of the other and counterdeception. But what constitutes complexity or the taking the mental perspective of another is never fully spelled out by any of these investigators, and there are only four instances of mentally complex counterdeception reported by Byrne and Whiten (1990): three for chimpanzees and one for baboons. And it should be recalled that baboons, but not chimpanzees, have been observed to use others as social tools in their acts that have been interpreted deceptively. In all, the evidence for monkey–ape differences in the production of seemingly deceptive acts is not strong. Something similar to tactical deception is first seen in human children sometime after their third birthdays (Chandler, Fritz & Hala, 1989).

8.2 Intentional Communication: Gestures

Animal communication is a large topic of study (Hauser, 1996), and much research has been devoted to primate communication in particular (e.g., Altmann, 1967; Snowdon, Brown & Petersen, 1982). Much of this research is concerned with communicative signals that have evolved over phylogenetic time to regulate evolutionarily important social interactions such as those involving sex and aggression. Many of these signals do not seem to be dependent on individual learning, however, and so they are not used flexibly, with individual decision making involving cognitive processes. These will thus not be our concern here. Our concern is with processes of intentional communication.

In intentional communication, the behavior of the sender must involve a goal and some flexibility in the means for attaining it. As in all intentional behavior, this suggests that alternative means may be used toward the same end, the same means may be used toward alternate ends, and some new exigencies may be flexibly accommodated (Bruner, 1972). Intentional communication specifically involves, for example, an individual using the same signal flexibly in different communicative contexts, using different signals within the "same" communicative context, and the capability to inhibit a signal when it is to its advantage to do so. In most cases this flexibility results from the fact that the signals have been ritualized not phylogenetically but ontogenetically, that is, they have been learned. In some cases the learning involves not the signal itself, but the appropriate communicative context for the signal; that is, flexibility in the use of communicative signals is assessed not with respect to their physical form but with respect to their use vis-a-vis a communicative goal. For instance, an individual may not have to learn a predator alarm call as a vocalization, but it may have to learn the situations in which it is appropriate, thus requiring individual decision making about when it should be used.

In identifying an act of intentional communication, it is also important to determine that the sender is directing the signal to a potential recipient, ideally doing some things to show its understanding of the effect of the signal on the recipient. Thus, in the prelinguistic communication of human infants, such understanding is often inferred from the infant's monitoring the response of the other to communicative signals; for example, an infant may point to an object and then turn to visually monitor its mother's response, becoming more insistent (or even using another signal such as whining) if she does not look where she is supposed to (Bates, 1976; Franco & Butterworth, 1996). Such behaviors are taken as evidence that the infant knows that achieving its goal in this situation has something to do with its ability to affect the mother's behavior or attention and thus that its pointing behavior is a truly communicative act whose function is to convey information. Other evidence for this same knowledge is manifest in situations in which an individual shows signs of expecting a particular response from a recipient, or uses a signal differently depending on the identity or attentional state of the recipient, so-called audience effects.

Studies of primate gestural communication concern facial, manual, and postural behaviors in a variety of social contexts. Although the anatomy of these behaviors has been extensively studied, their communicative functions have been much less well-studied (see Maestripieri & Call, 1996, for a partial review). Focusing on intentional communication specifically, the relevant studies are very unevenly distributed among species. There are almost no systematic studies of the intentional gestural communication of monkeys, but there are a number of studies of the intentional gestural communication of apes, especially chimpanzees, both in the wild and in captivity. In this section we therefore begin with chimpanzee gestural communication and then proceed to report what is known about other apes and monkeys. In all cases our focus is primarily on those gestural signals that are learned, flexibly controlled, and used in a way sensitive to the audience.

8.2.1 Chimpanzees

Goodall (1986; see also van Lawick-Goodall, 1968b) reported more than a dozen distinct gestures used by wild chimpanzees for various social ends in the Kasakela community in the Gombe National Park. It is not clear how many of these might show the kind of flexibility of current interest (although Goodall reports her general impression that they are flexibly used), but Plooij (1978) reports on a few especially creative uses of gestures among the juveniles of this same community. In one case he observed a juvenile who typically used an "arm-high" gesture to invite grooming under its arm use that same gesture in an appeasement context.

Tomasello, George, Kruger, Farrar, and Evans (1985) and Tomasello, Gust, and Frost (1989) observed many of these same types of gestures in a captive chimpanzee colony at the Yerkes Primate Center (see also Berdecio & Nash, 1981). Focusing on the juveniles of the group, they observed a number of gestural signals used to solicit food, play, grooming, nursing, and so forth. There were a number of lines of evidence that these signals were learned, most especially individual differences in which signals were used (with some signals used by only one individual; a phenomenon also reported by Goodall, 1986). Although no quantitative assessments were made, the investigators report that many of the gestures were used quite flexibly in different contexts. For many gestural signals, there was also evidence that the signaler expected a response from the recipient, as it typically waited expectantly and monitored the recipient's reaction after gesturing. For example, the initiation of play often takes place in chimpanzees by one juvenile raising its arm above its head and then descending on another, play-hitting in the process. This then becomes ritualized ontogenetically into an 'arm-raise' gesture in which the initiator simply raises its arm and, rather than actually following through with the hitting, stays back and waits for the other to initiate the play, monitoring its response all the while (see section 9.3 for more on the process of ontogenetic ritualization). If the desired response is not forthcoming, sometimes the gesture will be repeated, but quite often another gesture will be used. In other situations a juvenile was observed to actually alternate its gaze between the recipient of the gestural signal and one of its own body parts; for example, one individual learned to initiate play by presenting a limp leg to another individual as it passed by (an invitation to grab it and so initiate a game of chase), looking back and forth between the recipient and its leg in the process. Gaze alternation has been used by developmental psychologists to indicate that infants understand something of the recipient's role in the communication process (e.g., Bates, 1976, 1979). Tomasello et al. (1989) also reported the creation of new gestures as new materials were introduced to the group (e.g., using newly introduced wood chips to initiate play by throwing them at others), indicating much flexibility in chimpanzee communicative gesturing.

Tomasello, Call, Nagell, Olguin, and Carpenter (1994) observed a second genera-

Figure 8.4. Gorilla throwing items from water moat. (Photo by J. Call.)

tion of juveniles from this same group, with the goal of documenting more precisely the kinds of flexibility of use and audience effects they manifest. They found that many of these gestures were used very flexibly; in many cases the same gesture was used in different contexts, sometimes across widely divergent behavioral domains. For example, one juvenile whose mother was attempting to wean her, would throw wood chips at her (this was most typically a play gesture for this individual) to try to coerce her to comply with its whimpering requests to nurse. Juveniles had an average of 2.5 gestures each that were used for more than one communicative function. The converse pattern also held, that is, sometimes different gestures were used in the same context. All eight juveniles observed had a repertoire of gestures they could use seemingly interchangeably toward certain ends, which they sometimes did in rapid succession in the same context if the first gesture failed (e.g., initiating play first with a "throw chips" followed by an "arm raise"). The average number of signals that the juveniles had to initiate play was eight, to initiate nursing was three, to ask to ride on or walk with another was three, and to beg food was two. Thus, these gestural signals of chimpanzees seem to have a great deal of flexibility of use in that the same gesture may be used for different ends, and different gestures may be used for the same end. With regard to actively deceptive uses of gestural signals, de Waal (1986) observed a chimpanzee use a submissive gesture to indicate a desire to reconcile after a fight, but then it attacked the recipient as it approached.

Especially important for the issue of audience effects is a subclass of gestures called "attention getters," which serve to attract the attention of others to the signaler (from the wild, see the "leaf clipping" signal observed by Nishida, 1980, for the Mahale K group). Thus, "throw chips," "poke at," "ground slap," and "foot stomp," when produced in conjunction with the signaler's waiting expectantly for a response, all seemed to be aimed at having the recipient attend to the signaler. It was thus natural that they could be used quite flexibly in a wide variety of contexts where gaining the attention of the other would make available some information that would lead to the desired action on the part of the recipient. For example, many attention getters had as a primary function the initiation of play. In such cases the signaler already had a play-face and posture (presumably as an involuntary response to its "mood"), but knew that play could not ensue until the recipient had noticed the play-face and posture—thus the attention getter. Another example is that of Crawford (1937), who observed one juvenile chimpanzee attempt to communicate to another that it should come play its role in a cooperative task by simply poking it or banging the cage to attract its attention, relying on the fact that if the other looked in the appropriate direction it would do the right thing. Further evidence along these lines was provided in a separate set of observations by Tomasello et al. (1994). They found that juveniles actively chose among their attention-getting gestures based on the observational situation of the recipient. Thus, juveniles only gave a visual signal to solicit play (e.g., "arm raise") when the recipient was already oriented appropriately, but they used their most insistent attention getter, a physical "poke," most often when the recipient was socially engaged with others.

Evidence for how these gestural signals are learned will be presented in full in chapter 9 on social learning. For now, it is sufficient to note that the learning process most likely is one of ontogenetic ritualization, which typically takes place over many instances of the same social interactional sequence as the individuals involved come to anticipate what the other will do in certain circumstances and then actively do something to bring about those circumstances (cf., "conventionalization" Smith, 1977; see section 9.3 for evidence). Ontogenetic ritualization clearly relies on primates' understanding of others as animate and directed social beings, as the signal is intended as a spur to action for a being capable of self-generated action (see Tomasello, Call, & Gluckman, in press, for this argument).

8.2.2 Other Apes

Much less is known about the gestural communication of the other great ape species. Bard (1992) reported that a number of juvenile orangutans living in a rehabilitation camp begged for food from their mothers by holding their hands under her mouth, palm up and cupped. These gestures were accompanied by waiting for a response, and so seemed intentional. There is no information, however, on the flexibility of use of this gesture or on any potential audience effects. De Waal (1988) reported 15 distinct gestures in a group of captive bonobos, but again the flexibility of their use and their sensitivity to audience were not reported.

Savage-Rumbaugh, Wilkerson, and Bakeman (1977) reported a number of gestures used in a sexual context by bonobos. Of most interest here is the fact that the investigators sought to identify some gestures that were iconically related to the change of state in the partner's behavior that the signaler desired. For example, the male would often physically push the female's body into a position facilitating copulation. On other occasions he simply touched a side of the female's body and brushed his hand across the rest of the body, presumably a ritualized version of the physical manipulation strategy. Finally, on some occasions the male simply moved his hand across the female's body, a gesture the investigators' interpreted as an iconic indication of what he would like her to do. How these "iconic" gestures developed is not known, but it is likely that they derived as ritualizations of the physical manipulations themselves, not as gestures that the signaler thought would be interpreted iconically by the recipient. If this is so, then the "iconic" relationship of the gestures to the desired action may be from the human point of view only, as for the bonobo they may just signal the desired action in the same way as other ritualized signals, that is, based on the mutual shaping of behavior in previous interactive sequences (see section 9.3).

Tanner and Byrne (1996) have recently reported on a number of intentional gestures used by a captive gorilla with his groupmates. They report in detail on nine different gesture types, the majority in the context of play, emphasizing gestures that seem to be iconic in much the same way as those of the bonobos observed by Savage-Rumbaugh et al. (1977). In particular, an adult male gorilla often seemed to indicate to a female playmate iconically, using his arms or whole body, the direction in which he wanted her to move, location he wanted her to go to, or the action he wanted her to perform. Once again, the learning process is not known for these signals and so the role of iconicity in the gorilla's comprehension is not known. Tanner and Byrne (1993) also reported that a female of this same group used her hands to hide her play-face from a potential partner, indicating some flexible control of the otherwise involuntary grimace as well as a possible understanding of the role of visual attention in the process of gestural communication. Of the 22 instances of this behavior appropriately observed, 21 occurred when the potential partner would have been able to see the play-face if it were not actively concealed.

8.2.3 Monkeys

Almost nothing is known about monkeys' intentional gestural signaling, or indeed if they engage in this behavior at all. Although each monkey species has its own species-typical repertoire of facial expressions and gestural displays, no species has been observed to evaluate the flexibility of their gestures, their sensitivity to audience, or their use as acts of attention manipulation. The most interesting observations in the current context are those of Kummer (1968), in which hamadryas baboons engage in "notifying behavior" before they leave others in a troop (and sometimes before the approach and sit close by). This behavior consists in approaching the other and looking directly

Figure 8.5. Gorilla using hand clapping as auditory attention-getting gesture. (Photo by J. Call.)

in the face, sometimes sexually presenting as well. One interpretation of this behavior is that the sender wants to make sure that the other is looking before it engages in certain activities. This interpretation is given some support by the converse called "cut-off behavior," in which an individual looks to the ground when another is notifying (or otherwise signaling), presumably so that it is not then obligated to behave in accordance with its having been notified. Kummer and Kurt (1965) also observed a "ground-slap" behavior that seems to serve as an attention getter in much the same way that this behavior functions for chimpanzees, and a kind of teasing behavior in play in which one juvenile held an object in its mouth so that another, who might want it, could see it. Maestripieri (1996b) has observed that when pigtail macaque mothers want their infants to follow them and they do not, they sometimes return and stare in the infant's face (or even poke it) before leaving again (see also Maestripieri, 1996a,c).

There is little doubt that many monkey species have the social-cognitive skills to predict the behavior of others in certain situations; and when they observe others in the process of predicting their behavior on the basis of some initial step in the sequence, they are also able to take intentional control of that process by producing only the initial step. For example, Coussi-Korbel (1994) observed white-collared mangabeys in a food-finding game similar to Menzel's (1973c) studies with chimpanzees and found similar behavior, including both the ability of observers to predict the location of food and the ability of informed individuals to notice this prediction and attempt to influence it.

8.2.4 Comparison of Species

Great apes use their gestural signals with conspecifics in flexible and strategic ways that are in some ways sensitive to the audience. The same signal may be used for different ends, different signals may be used toward the same end, and many signals are given in ways that are appropriate to the observational situation of the recipient (i.e., the recipient must be bodily oriented for visual signals to be used). Monkeys have not been observed extensively enough to draw any firm conclusions about their skills of intentional gestural communication. It seems reasonable, however, that at least some of their gestural signals are learned, at least in the sense of learning the circumstances in which they may most usefully be deployed, and in some situations monkeys also seem to know that the recipient must be bodily oriented to a signal for it to have its intended communicative effect. Many animal species communicate using various behaviors in the visual modality (Hauser, 1996), although the flexibility of these behaviors, and their sensitivity to audience, is not always known. Human infants begin to gesture in flexible ways, sensitive to audience, late in infancy (Bates, 1979).

It is important to note at this point that there is an important commonality in the way communicative signals are ontogenetically ritualized and the way social strategies in general are created. For example, in Menzel's (1971a, 1973c, 1974, 1979) studies reviewed in section 7.1, the investigators make a special point of noting that in none of the experimental variations did informed individuals use overt communicative displays or signals to indicate to the others the location of food. But on some occasions they did engage in attempts to influence the behavior of others by leading them in "false" directions. They learned how to do this by observing the others reacting to their behavior in particular ways, and then on some future occasion, when they wished the others to react in that way, producing the behavior again. When this process happens in a particular way repeatedly (e.g., when an infant wishes to nurse on an hourly basis over many months), it may result in fairly ritualized behavioral signals that are typically intention movements of the strategist used to spur the other into action. In this case, the process is called ontogenetic ritualization (see section 9.3 for more details). In general, then, novel social strategies and learned ritualized signals are created in basically the same way, the difference being simply whether the eliciting situation is recurrent and frequent enough to lead to the ontogenetic ritualization of a specific communicative signal.

8.3 Intentional Communication: Vocalizations

Vocal communication has received extensive research attention from primatologists. Much of this interest seems to have derived from the analogy to human languages and the fact that languages evolved, at least in their latest stages, in the vocal modality (Seyfarth, 1987; Snowdon, 1993). Another influence may have been the well-developed methods for the analysis of vocalizations developed by ethologists studying bird song. In any case, there have been many studies of many different primate species that aimed simply to document the vocal repertoire of a species, without particular concern for communicative function. Our concern here, however, is the intentional use of vocal signals for particular communicative functions and what this tells us about primate social cognition. It should be noted at the outset that, in diametric opposition to gestural communication, in intentional vocal communication most of the studies have been done with monkeys, not apes.

8.3.1 Vervet Monkeys

Following up on some preliminary observations by Struhsaker (1967), Seyfarth, Cheney, and Marler (1980a,b) reported that in their natural habitats in east Africa, vervet monkeys use three different types of alarm calls to indicate the presence of three different types of predator: leopards, eagles, and snakes. A loud, barking call is given for leopards and other cat species, a short, coughlike call is given for two species of eagle, and a "chutter" call is given for a variety of dangerous snake species. Each call elicits a different escape response on the part of vervets who hear the call: to a leopard alarm they run for the trees; to an eagle alarm they look up in the air and sometimes run into the bushes; and to a snake alarm they look down at the ground, sometimes from a bipedal stance (see figure 8.6). These responses are just as distinct and frequent when researchers play back previously recorded alarm calls over a loudspeaker, indicating that the responses of the vervets are not dependent on actually seeing the predator. (Although they have not been studied nearly as intensively as vervets, a number of other monkey species also use different alarm calls for different predators; see Cheney & Wrangham, 1987, for a review.)

Vervet monkeys also have a number of different "grunts" that they use in various social situations (Cheney & Seyfarth, 1982a). Although the differences among the grunts are difficult for humans to detect, based on the different responses of vervets to calls that can be distinguished by humans using sound spectrographic techniques (including the playback of previously recorded calls), grunts are used in four communicative contexts: upon encountering a dominant groupmate, upon encountering a subordinate groupmate, to a groupmate moving to an open area, and to vervets who are not members of the group. (Note that a number of other monkey species use such "close calls" to modulate intergroup spacing and cohesion; see Seyfarth, 1987, for a review.)

The major issue addressed by Cheney and Seyfarth is that of reference: do the vervet alarm calls and grunts contain information about an external referent, or are they simply an indication of the caller's degree of arousal or probability of engaging in a certain behavior? A number of lines of evidence from naturalistic observation suggest that these vocalizations are truly referential in the sense that they are reliably associated with particular classes of environmental events and not used for inappropriate environmental events (although this fact does not rule out the possibility that monkey vocal signals also carry information about the caller's emotional state; Cheney & Seyfarth, 1990b; Marler, Evans, & Hauser, 1992). In addition, Cheney and Seyfarth (1988) provided experimental confirmation of this interpretation. In an experimental setting, vervet monkeys were presented with prerecorded calls all coming from the same individual. One group heard two different calls that concerned two different referents (e.g., an eagle alarm call and the "wrr" call, used in association with another vervet group). In another condition they heard two different calls that concerned the same referent (i.e., a wrr and a chutter, both associated with other vervet groups). In a habituation test, individuals treated the two calls with different referents differently, but they treated the two calls with similar referents similarly (even though they sounded different). Interesting confirmation of this finding was provided by Seyfarth and Cheney (1990), who found

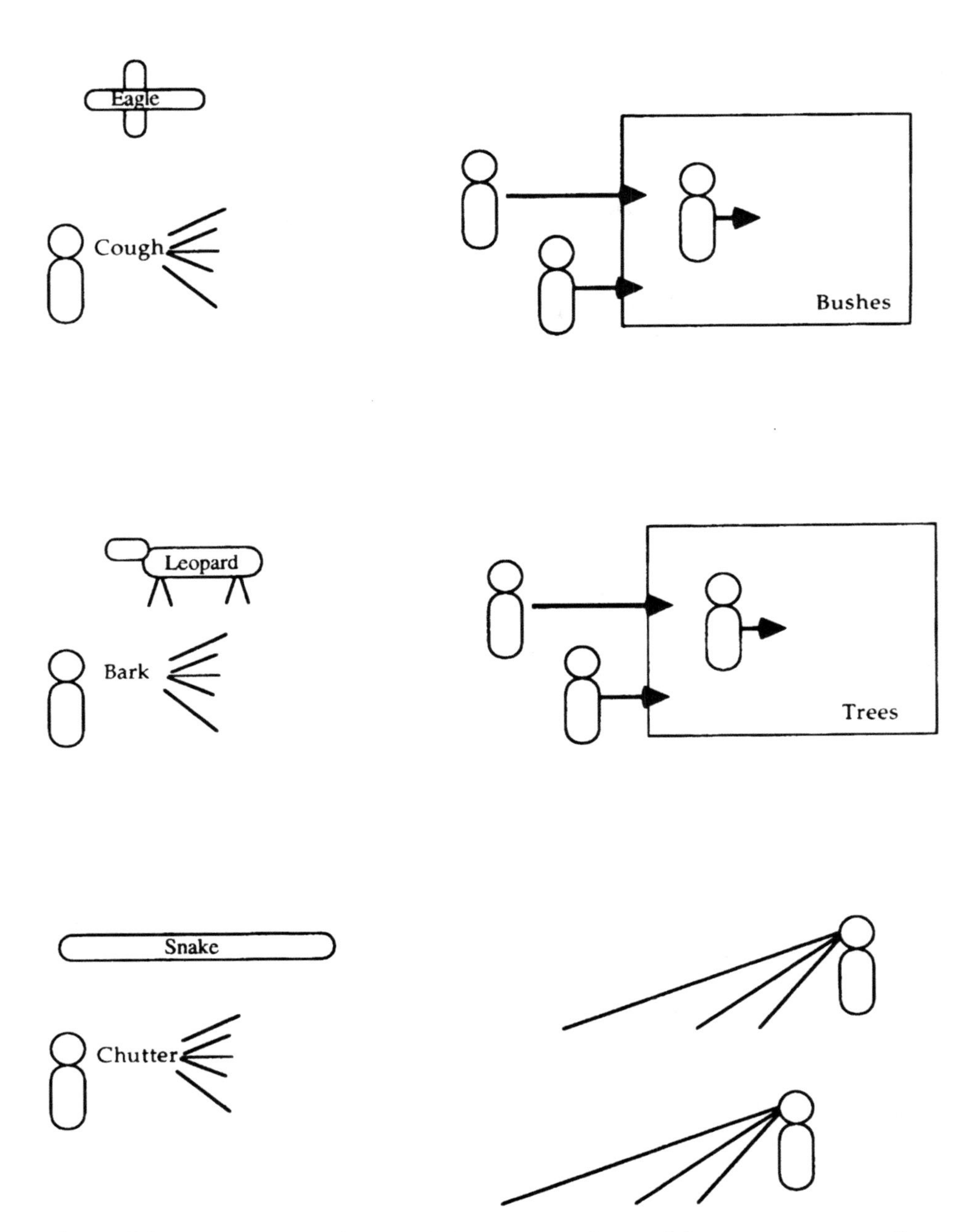

Figure 8.6. Different types of alarm calls by vervet monkeys and the most common reactions of listeners, as studied by Cheney and Seyfarth.

that vervets extended habituation of leopard alarm calls given by vervets to the very different-sounding leopard alarm calls of a sympatric bird species.

For purposes of investigating vervet social cognition, however, the central question is not reference per se. The kind of referential behavior seen in vervet monkeys could conceivably be accomplished by means of a mechanism in which animals are genetically equipped, or even learn, to make certain noises in the presence of particular visual stimuli or emotional states and to respond in certain ways to particular acoustic or visual stimuli (Hauser, 1996). This could be done with little or no understanding of conspecifics, and indeed it has been found that domestic chickens and ground squirrels give different alarm calls to different predators and respond in unique ways to playbacks of those calls (Evans, Evans & Marler, 1993; Leger & Owings, 1978), even though no one has hypothesized, as they have in the case of monkeys, that chickens and ground squirrels have a deep understanding of the behavior of conspecifics (although perhaps they do). The crucial question for issues of social cognition is what knowledge of other individuals and their behavior is implicit in vervet vocal communication. We must therefore examine evidence for the two key criteria of intentional communication: flexibility of use and use in socially sensitive ways (i.e., audience effects).

With regard to flexibility of use, it should be noted first that there is some evidence that vervet vocal signals are learned, at least in the sense of their comprehension and use in appropriate circumstances (this is reviewed more thoroughly in chapter 9, "Social Learning"). In describing the ontogeny of vervet alarm calls and grunts, Seyfarth and Cheney (1986) note that the vocal structure of the calls is similar to that of adults from a very early age. The referents for which the calls are used, however, shows development from more general classes containing some inappropriate referents to more adultlike usage during the first year of life. Their adultlike response to playbacks of adult calls also emerges gradually during the first year of life. Further, Hauser (1988a, 1989) found that vervet infants who had more experience with particular types of calls at earlier ages (one of another species and one of their own) produced those calls at earlier ages themselves. Also relevant to the issue of flexibility of use is one report that when a group was in the midst of a hostile encounter with another group, an individual vervet gave an alarm call which served to dissipate the violence (Cheney and Seyfarth, 1990b), although the interpretation of such anecdotes is in many ways problematic (as reviewed in section 8.1). We should also note with regard to flexibility that the results of the Cheney and Seyfarth (1988) habituation experiments may be interpreted as demonstrating a kind of flexibility of comprehension: vervets understand that different calls may have similar referents.

With regard to social sensitivity and audience effects, Cheney and Seyfarth (1990b) report that vervets in the wild do not give alarm calls to predators when they are alone. Further, Cheney and Seyfarth (1985) report that captive vervets give more alarm calls when they are with their offspring than when they are with unrelated juveniles, and males give more alarm calls when they are with females than when they are with male competitors. These audience effects would seem to indicate the voluntary quality of vervet alarm calls, as well as a comprehension on the caller's part that the function of the call is communication with others. One must be careful in interpreting the cognitive underpinnings of these phenomena, however, because several nonprimate species also give alarm calls differentially depending on the presence of others and differentiate kin from other conspecifics (e.g., ground squirrels and domestic chickens; Gyger, Karakashian, & Marler, 1986; Sherman, 1977), but it is not clear whether complex social cognition is involved. Because of the potential costs of alarm calls in drawing attention to the caller (and thus perhaps attracting other predators), it is possible that the "audience" is simply a part of some hard-wired system in the sense that the presence of certain discriminative stimuli are requisite for call production. This seems unlikely given the variety of audience effects observed: alone versus with others, offspring versus others, and potential mates versus potential competitors. However, it should also be noted that vervet monkeys often continue to give alarm calls well after their groupmates have already seen the predator and are already calling themselves and are safe (Cheney & Seyfarth, 1981), perhaps indicating a lack of knowledge that once conspecifics know of the predator's presence further calling is unnecessary. It is also noteworthy that vervet adults do not give alarm calls to baboons, which are dangerous to their offspring but not to adults (Seyfarth, personal communication). Exactly what these audience effects mean is therefore still an open question.

8.3.2 Other Monkeys

In looking at the vocal communication of other monkey species, two contextual situations have been most studied: recruitment screams and food calls. First, Gouzoules, Gouzoules, and Marler (1984, 1985) described the different vocalizations that juvenile rhesus macaques make when they are attacked by another individual. These monkeys use one of five different calls to recruit support from relatives as they are being attacked by another individual. Most clearly distinct are the "noisy screams" given to higher ranking individuals when there is physical contact and the "pulsed screams" given when they are being attacked by relatives. Using playback techniques, Gouzoules et al. determined that individuals respond to the acoustic properties of the calls, not to the arousal level of the caller or to anything in the caller's behavior, with a responder gauging its response on who the aggressor is and what kind of aggression is occurring. Gouzoules and Gouzoules (1989a) made similar observations of pigtail macaques. Juveniles of this species also use distinctive recruitment screams in different agonistic contexts, but the acoustic features of their calls and the aspects of the situation encoded are different. Like rhesus monkeys, their calls are used in association with both the relative rank of the aggressor and the severity of the attack (although with very different-sounding screams). There is no evidence that their calls encode the kinship status of the aggressor, as do rhesus calls in this context. The investigators note, in this regard, that aggression among closely related individuals is much more prevalent in rhesus monkeys.

The Gouzouleses report evidence similar to that for vervets regarding the developmental modification of call use by pigtail macaques during early ontogeny, that is, development toward more adultlike use during the first year of life. Gouzoules and Gouzoules (1989b) also found in this species that females called more accurately at earlier ages than males, and Gouzoules and Gouzoules (1990) found that members of different matrilines in a captive pigtail group had identifiably different vocal "signatures"—both implying the possibility of learning (see also Gouzoules and Gouzoules, 1995). In none of these studies did the investigators observe any clear signs of deceptive uses of recruitment screams, although they noted that this was not a domain in which "crying wolf" would be long tolerated by groupmates because allies put themselves in

Figure 8.7. Siamang vocalizing. (Photo courtesy of M. Seres.)

danger when they respond to the screams. With regard to audience effects, it is noteworthy that in using their recruitment screams, rhesus monkeys alternate their gaze between the attacker and the individuals being recruited (Gouzoules et al., 1984), and they do this only when there is a receptive audience present in their visual field. This would seem to be strong evidence that the signaler understands something about the function of the call in recruiting the aid of specific individuals, or at least the expectation of some effect of the call on others.

The other well-known context concerns food calls. Unlike alarm calls, which aid the survival of groupmates (including kin), and recruitment screams, which aid the survival of the caller (and responding to them often aids the survival of kin), food calls add a competitive component because they might serve to decrease the caller's intake of food. The first study was that of Dittus (1984), who observed toque macaques (*Macaca sinica*) in their natural habitat in Southeast Asia. Upon finding abundant and high-quality food sources, these animals gave calls that were clearly distinct from their other calls, which brought groupmates quickly to the feeding site. Individuals did not give calls more often when they were hungriest (i.e., had not eaten for some time), implying they were not a direct function of hunger states. However, the referential interpretation of the calls is made problematic by Dittus's (1988) observations that members of this group also called in this same way upon the arrival of others at the food site, called in response to the calls of others, and called even sometimes in non–food-related contexts. One interpretation, then, is that these calls reflect a general level of excitement. Similar observations have been made of a wild group of spider monkeys (*Ateles geoffroyi*; Chapman & Lefebvre, 1990) and laboratory colonies of golden tamarins (*Leontopithecus rosalia*; Benz, 1993) and cotton-top tamarins (*Saguinus oedipus*; Elowson, Tannenbaum & Snowdon, 1991). In these latter studies it was found that the rate of calling was a straightforward function of the desirability of the food. Marler, Dufty, and Pickert (1986a) report an identical finding for domestic chickens.

Hauser and Marler (1993a) observed the naturally occurring food calls of rhesus macaques in a free-ranging environment. Of special importance are three calls (warbles, arches, and chirps) used exclusively in food contexts, especially for encounters with rare and highly preferred foods. These food calls were used irrespective of the caller's hunger status, although call rate was greater when the caller was hungry. Calls almost invariably brought nearby groupmates to the scene of the food. Females called significantly more often than males. In field experiments with this same group, Hauser and Marler (1993b) presented individuals with preferred foods of different types and a few nonfood items in different contexts and confirmed these observations experimentally. There are no observations on the ontogeny of monkey food calls or on their flexible use in novel contexts. However, Hauser and Marler (1993b) observed that rhesus monkeys sometimes did not give food calls upon finding preferred foods (just over half the time), showing some voluntary control over production and suggesting the possibility of deception for purposes of hoarding. Interestingly, when a noncaller was detected with food, it often received aggression, resulting in its having less food to eat than if it had called (or had remained undetected). The investigators thus hypothesized that rhesus monkeys use their food calls to announce their possession of food to groupmates.

There are no observations of the kinds of audience effects that might lend support to this hypothesis (e.g., differential use as a function of who is nearby). Indeed, in the one experiment relevant to this question, Cheney and Seyfarth (1990a) found that rhesus and Japanese macaque mothers failed to inform their offspring of the presence of food they had seen hidden in the offspring's enclosure when the offspring had not seen the hiding and therefore were ignorant of the food's presence. Gouzoules, Gouzoules, and Ashley (1995) also point out that it is quite common in captive populations for macaques of various species to give food calls to a caretaker approaching with food even after all group members are present and the food is plainly visible to all. It should be noted in this context that Marler, Dufty, and Pickert (1986b) found that domestic chickens give food calls differentially depending on who is nearby.

Two other types of monkey vocalizations deserve mention. First, observations of social contact calls for Japanese macaques (Green, 1975; Masataka, 1989), stumptail macaques (Bauers & de Waal, 1991), and squirrel monkeys (Masataka & Biben, 1987) have shown that some of these calls reliably predict the subsequent behavior of the caller. There is no evidence, however, that any of these are referential, can be used flexibly, or are adjusted for different audience situations (see also the study of squirrel monkeys by Boinski & Mitchell, 1992). In fact, Talmadge-Riggs, Winter, Ploog and Mayer (1972) and Winter, Handley, Ploog and Schott (1973) found that squirrel monkeys raised is isolation still give species-typical vocalizations in most appropriate contexts, and Owren, Dieter, Seyfarth, and Cheney (1993) found few differences in the use of vocalizations by cross-fostered rhesus and Japanese macaques (see chapter 9, "Social Learning," for more details). In contrast, however, Boinski (in press) reported evidence that two New World monkey species (red-backed squirrel monkeys, *Saimiri oerstedi*, and white-faced capuchins, *Cebus capucinus*) do seem to visually monitor the effects of their travel coordination calls on groupmates.

Second, Snowdon and colleagues have found that (1) pygmy marmosets use one of three distinct variants of a trill vocalization depending on how near or far they are to groupmates in their natural habitats in the Peruvian Amazon (Snowdon & Hodun, 1981); (2) pygmy marmosets in captivity "take turns" in their trill vocalizations, indicating a sensitivity to groupmates (Snowdon & Cleveland, 1984); and (3) cotton-top tamarins respond with different vocalizations depending on whether a previous vocalization comes from a groupmate or a stranger (see Snowdon, 1993, for a review). The vocalizations of these New World monkeys thus show some signs of flexible use and sensitivity to social context, although there are few reports of deceptive uses or audience effects per se.

8.3.3 Lemurs

Evidence has been presented by Macedonia and colleagues that captive ringtailed lemurs, a prosimian species, use referential alarm calls in predator situations, and playback experiments have confirmed differential responsiveness to different calls (Macedonia, 1990; Macedonia & Yount, 1991). Upon initially spying a predator of any type, the lemurs give "gulps," and this is followed by different calls for aerial and ground predators, each eliciting a different set of evasive procedures. Pereira and Macedonia (1991) experimentally manipulated the presence and behavior of potential predators and found that the lemur calls were not affected by "response urgency"; that is, the normal calls were given for aerial and ground predators regardless of whether they were still some distance away from the group or whether they were close to and about the attack the group. There is no evidence that the alarm calls of ruffed lemurs (*Varecia variegata*) show this same referentiality. There are no studies of the ontogeny, flexible use, or social sensitivity of the alarm calls of either lemur species.

8.3.4 Apes

Compared to monkey vocalizations, ape vocalizations have been little studied with regard to how individuals understand and use them communicatively. There are no systematic studies of the type performed with monkeys on either the alarm calls or recruitment screams of apes (the one exception is a study of the long-distance alarm calls of Kloss's gibbon, *Hylobates klossi*, which focuses more on their function vis-a-vis kin selection; Tenaza & Tilson, 1977). Indeed, although vocalizations are made by a number of ape species in both of these contexts, in neither case has any scientist detected the kind of referential specificity that seems to characterize the monkey calls in these contexts (see Cheney & Wrangham, 1987, for a comparison of the predator-specific alarm calls of monkeys with the more generalized alarm calls of apes). Almost all of the systematic research that has been done on ape vocalizations concerns the food calls of chimpanzees.

Chimpanzees have three calls that occur with regularity in the context of food: the "pant-hoot," the "food grunt," and the "food-aaa." Only the first two have been studied systematically. In a study of feeding behavior in the Gombe National Park in eastern Africa, Wrangham (1977) first reported that chimpanzees often use the well-known pant-hoot vocalization when they discover food, and they give more of them when large amounts of food are discovered. However, pant-hoots are used in a variety of other circumstances as well, such as "social excitement" and "sociability feelings" (Goodall, 1986), and the use for food cannot be acoustically distinguished from the other uses (Clark & Wrangham, 1993). Moreover, Clark and Wrangham (1994) found that pant-hoots given by food discoverers did not increase the frequency with which groupmates arrived at the food site (although Ghiglieri, 1984, and Wrangham, 1977, did find an increase). More recent studies of pant-hoots have found that they are also often used by males during travel, often when their closest allies are nearby (Mitani & Nishida, 1993), suggesting the possibility that pant-hoots serve to announce location to important groupmates or even to recruit allies. But Clark and Wrangham (1994) found that pant-hoots given upon arrival at food were not given differentially depending on who was nearby. In all, it is safe to say that the function of chimpanzee pant-hoots is currently unknown, although they almost certainly serve to announce to groupmates the caller's location, and such announcements are made at times of elevating arousal such as when initiating travel (or a new travel direction) or upon arrival at a food site or upon the anticipation of encountering significant groupmates. It is unlikely that they announce the discovery of food in particular, and, if they do, they certainly are not sensitive to audience because in captivity pant-hoots are commonly given upon seeing the approach of food even when all groupmates are in full view and watching the approach of the food as well (personal observations of the authors).

The chimpanzee "food grunt" has also been systematically studied, in this case with captive groups. Hauser and Wrangham (1987) presented individual captive chimpanzees with either 5, 10, or 20 prunes. Discoverers called most often upon finding 20 prunes, called a bit less upon finding 10 prunes, and did not call at all upon finding 5 prunes. The investigators interpreted this result as indicating that chimpanzees are able to withhold calls if it is in their interest to do so, since in this case announcing the discovery of the smaller numbers would have meant little food for the discoverer. In an extension of this work, Hauser, Teixidor, Field, and Flaherty (1993) found that captive chimpanzees call differently depending on how easily the discovered food might be divided: when the same large quantity was presented as either one large chunk (of watermelon) or as 20 smaller pieces, discoverers called more often when there were pieces (see figure 8.8). While this result might suggest that the food call is being used to indicate food availability for others (with divisibility associated with easier availability for others), it may be that the 20 pieces simply appear to be more food and thus induce more excitement (a similar result has been reported for house sparrows; Elgar, 1986). Hauser et al. also found that chimpanzees used the pant-hoot, in addition to the food grunt, only when the amount of food was large, confirming observations from the wild that pant-hoots are given more for large amounts of food. In any case, many researchers now believe that food grunts are referentially specific to food (and under at least some voluntary control), and that

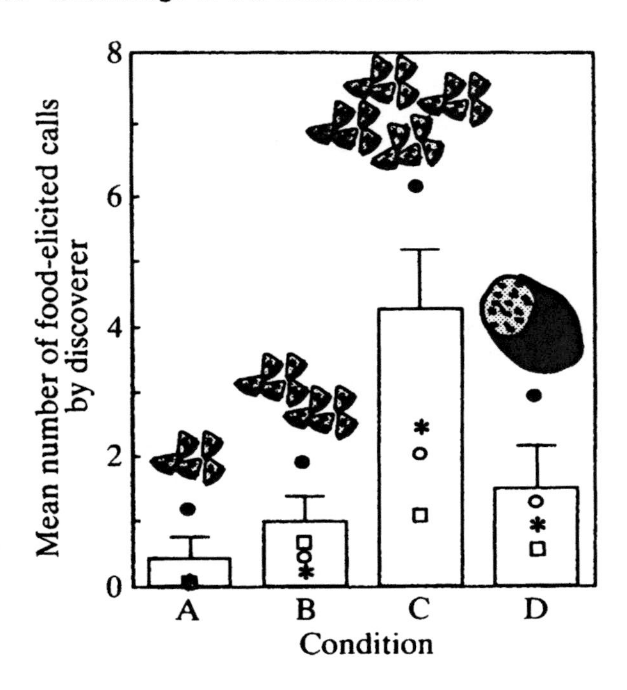

Figure 8.8. Hauser et al. (1993) results showing that chimpanzees call more frequently when same amount of food is in more pieces. (Reprinted from *Animal Behaviour* with permission from Academic Press.)

pant-hoots do something more general such as announcing location, status, or excitement (with less evidence that they are under voluntary control).

If we examine all of these chimpanzee vocalizations with respect to their flexibility and sensitivity to audience, there is little evidence that they involve complex cognitive or social-cognitive processes. In general, chimpanzee calls seem to be closely bound with their emotional states, so much so that Goodall states that "The production of a sound in the *absence* of the appropriate emotional state seems to be an almost impossible task for a chimpanzee" (1986: p. 125; emphasis in original). She also notes that Hayes and Hayes (1951) had an almost impossible time teaching their home-raised chimpanzee to make any sounds voluntarily and that in Menzel's (1964) study of chimpanzees raised in social isolation, the subjects used many species-typical vocalizations without previous exposure to them (although they seemed to have to learn in some instances the appropriate social contexts in which to use them). However, it should also be noted that recently a human-raised bonobo has demonstrated the ability to make distinct vocalizations voluntarily (Hopkins & Savage-Rumbaugh, 1991), and there is now some evidence for dialect differences in chimpanzee vocalizations in the wild (Mitani, Hasegawa, Gros-Louis, Marler & Byrne, 1992), although in neither case is flexibility with respect to communicative goal reported. Goodall (1986) also claimed that chimpanzees can *suppress* calls when it is advantageous to do so (e.g., when on patrol for neighboring groups), and a number of researchers have reported that female chimpanzees sometimes conceal copulation vocalizations, especially when copulating with younger males (de Waal, 1986; Hauser, 1990), presumably because they do not want to be punished by dominant males. There are no reported examples of apes of any species using vocalizations to actively mislead or deceive conspecifics, however.

With regard to audience effects in chimpanzee food calls, Clark and Wrangham

(1994) found no correlation between the pant-hoots of wild chimpanzees and who was nearby or far away. Mitani and Nishida (1993), however, report some relations between which conspecifics were nearby (defined by their affiliative or alliance relation to the caller) and the likelihood of calling. There have been no studies of the adaptability of food grunts for different audiences, and in fact Hauser (1990) reports that there were no discernible audience effects in when females did or did not produce their copulation vocalizations. Chimpanzee silence during patrols, on the other hand, would seem to reveal some sense of audience, at least in the sense of knowing that noise attracts unwanted others. Similarly, in a recent set of observations on captive bonobos, van Krunkelsven, Dupain, van Elsacker, and Verheyen (1996) report that individuals call more frequently when they discover food within the context of a social group than when they discover it alone.

A singular and interesting set of observations related to chimpanzee vocal communication was reported by Boesch (1991a), who attempted to discern how the chimpanzees of the Tai Forest coordinated their travel. He noticed that other chimpanzees reacted in a special manner when the dominant male of the group (Brutus) drummed loudly on trees, always accompanied by a pant-hoot which presumably allowed others to identify him. Although they did not react in special ways to single bursts of drumming, when Brutus hooted and drummed on two different trees in relatively close succession in the context of travel, the others, many of whom were out of sight in the dense forest, seemed to change their travel direction to match a trajectory from the first to the second drumming. Amazingly, if Brutus drummed twice on the same tree, the others at their various locations in the forest seemed to take this as an indication that travel would cease for the moment. Drumming after the rest period (even twice on the same tree) served to initiate travel once more. This behavior, which ceased after some adult males disappeared from the group, is the only observation of its kind and indicates further the role of pant-hooting and other forms of auditory communication in localizing individuals and coordinating travel.

Little is known about the communicative function of the vocalizations of great ape species other than chimpanzees. Mitani (1985) studied the long calling of orangutans, but no data pertinent to flexibility or audience effects were reported. Harcourt, Stewart, and Hauser (1993) studied the "close calls" of gorillas in socially intimate contexts, but again there were no data reported relevant to the cognitive processes involved (see also Seyfarth, Cheney, Harcourt and Stewart 1994). De Waal (1988) studied captive bonobos and identified a number of different vocalizations that seem to occur reliably in specific contexts, but none was studied systematically for its flexibility or use in different audience contexts.

8.3.5 Comparison of Species

Vocalizations are clearly an important way that primates communicate with one another. Many other animal species also communicate in fairly sophisticated ways vocally. Domestic chickens give referentially specific alarm calls (as do other mammalian species) and food calls, both of which are sensitive to the presence of other types of conspecifics (Evans, Evans, & Marler, 1993). Many monkey species also have alarm calls and food calls that are used flexibly and show these same characteristics. Apes have been much less studied, but if anything their vocal communication would seem to be less complex than that of monkeys. The problem is that for apes there have been no playback experiments to test for comprehension in the absence of eliciting stimuli (or differential responsiveness to different calls in the same context), no systematic studies of ontogeny or learning, no experimental studies of audience effects, and no experimental manipulations of comprehension such as the habituation studies of Cheney and Seyfarth (1988).

Nevertheless, with regard to the cognition, especially social cognition, underlying primate vocal behavior, the following may be said. The flexible use that monkeys show in their vocal communication, including the inhibition of some signals in some contexts and perhaps even deceptive use, along with their adjustability for different audiences, argues that they are using these vocal signals intentionally to affect the behavior of others. It is not so clear that any of their signals are intended to affect the psychological states of others, however. When some monkey mothers are given a chance to inform their offspring of the presence of food or of a predator in novel circumstances, when their offspring are unaware of the presence of the food or predator, they fail to do so—perhaps because they have already made their initial response to the new stimulus and it is no longer perceptually present. The conclusion of Seyfarth and Cheney (1993), therefore, is that, although monkeys produce specific vocalizations selectively in response to specific objects and events and conspecifics understand that they do this, the signaler does not seem to produce its vocalizations to modify the psychological states of its conspecifics. We have no evidence that apes are any different in this regard.

8.4 Communication with Humans

Because human beings think of language as the most cognitively complex form of communication in all of the animal kingdom, there is a long history of attempts to teach apes to comprehend and use human language. Kellogg and Kellogg (1933) and Hayes and Hayes (1951) attempted to teach some English words to their infant chimpanzees Gua and Viki, but mostly failed because the chimpanzee vocal apparatus is not capable of making many sounds important in human speech. Gardner and Gardner (1969) attempted to overcome apes' difficulties with speech by teaching a chimpanzee, Washoe, some parts of American Sign Language (see also Fouts and Budd, 1979). Their apparent success led to similar projects with a gorilla, Koko (Patterson, 1978), and an orangutan, Chantek (Miles, 1990), both of whom were at least as proficient as Washoe. Other attempts to overcome the speech barrier were made by Premack (1976), who used plastic tokens to stand for spoken words in communicating with the chimpanzee Sarah, and by Rumbaugh (1977), who created for the chimpanzee Lana a visual language based on graphic symbols (lexigrams) depicted on a computerized keyboard.

The early work on ape language was put to severe test by the critiques of Terrace and colleagues who raised their own chimpanzee, Nim, with American Sign language (e.g., Terrace, Petitto, Sanders, & Bever, 1979). On the basis of their experience with Nim, as well as a reanalysis of the data of other investigators (including videotapes of Washoe), Terrace et al. argued that the linguistic apes were not really linguistic at all. What the apes were doing was learning operant responses to get food and other human-controlled rewards. These responses just happened to correspond to human sign language, but they should not be considered linguistic symbols in any sense of the term. Almost all of Nim and Washoe's productions, at least, were requests for things, mostly food. Terrace et al. also found that the apes engaged in a lot of mimicking and repetition of these responses from the humans interacting with them that were then taken to be "sentences" and "conversations." Thus, in response to the question "Do you want to eat an apple?" an ape might produce "Apple eat eat apple eat apple hurry apple hurry hurry," which would have been reported as "Eat apple hurry" and which would appear to an outsider as a conversationally appropriate response in the form of a sentence. Terrace et al. also noted that almost all of these kinds of "sentences" were highly redundant within themselves, with longer productions conveying no more information than shorter productions. To prove all of these points in one study, Terrace et al. simply trained a pigeon to peck the keys of a linear array in particular sequences in the presence of particular visual stimuli (meant to be analogous to the visually based systems of Premack and Rumbaugh) and argued that the language of the apes was in no way discriminable from the productions of the pigeon (Terrace, Straub, Bever, & Seidenberg, 1977).

Part of what Terrace et al. argued was true, but only of their own chimpanzee, Nim, who had been taught its "language" by a fairly rigid set of behavioristic training regimes. Nim was indeed tied fairly tightly to producing signs for food rewards, and he did mimic and repeat a lot of human signs simply because he was rewarded for doing so. But this does not seem to be true of all of the linguistic apes; for example, some other linguistic apes do not mimic and repeat any more than human children, and they do so under similar pragmatic circumstances (Greenfield & Savage-Rumbaugh, 1991). In answering the Terrace et al. critique, one set of investigators in particular has documented that apes can use something resembling human linguistic symbols and can produce and understand combinations of those symbols in creative ways. Savage-Rumbaugh and colleagues took up the challenge of the Terrace et al. critique and set out to work with apes in different ways to see if they could learn more than operant responses. In reviewing the linguistic competence of apes, we therefore focus on the research of Savage-Rumbaugh and colleagues as representative of the most humanlike behavior yet demonstrated in the domain of language by any nonhuman animals. This is in terms of the nature of their comprehension and production of linguistic symbols and symbol combinations. To establish the communicative foundation for ape language research, however, we must begin with a brief look at the types of nonlinguistic gestural communication that apes develop with humans when they are raised in humanlike environments. This will establish that in using their humanlike symbols they are indeed attempting to communicate.

8.4.1 Gestural Communication with Humans

Primates raised or trained by humans retain their species-specific modes of intentional communication with conspecifics, but they may also ontogenetically ritualize some gestures with humans. Gómez (1990) reported a number of behavioral and attentional strategies developed by a young gorilla, Muni, with the humans who interacted extensively with her during her first few years of life. For example, when Muni wanted to exit through a locked door, she would sometimes gently lead the human to the door and alternate her gaze between the latch and the human's eyes. On some occasions she even placed the human's hand on the latch and looked to his eyes. Gomez argued that these were attempts to influence not the human's behavior directly (as, for example, when Muni early in her development tried to physically push the human to the door so she could climb up him to the latch), but rather to let the human know what she wanted by making contact with the human's eyes. In a similar vein, a human-raised bonobo was observed to give a nut to a person who was supposed to crack it open, and then slap the nut and place a stone on top of it (Savage-Rumbaugh et al., 1986). Given the types of gestural communication that apes employ with one another, there is no reason to believe that these are anything other than intentional attempts to communicate.

Of special interest is the gesture of pointing. In their natural habitats, apes have not been observed to point to relatively distant entities in the environment in order to draw the attention of conspecifics. Many apes who have had some contact with humans, however, do learn to point for humans, although the form of this pointing typically does not follow the human pattern of index finger extension but involves more of the arm and hand. This has been well documented for all four great ape species: chimpanzees (Savage-Rumbaugh, 1986), bonobos (Savage-Rumbaugh et al., 1986), gorillas (Patterson, 1978), and orangutans (Miles, 1990). In all cases the ape was first trained in some kind of communication with humans that involved close-range pointing, for example, human sign language or pointing to the keys of a keyboard used for communication. But then in all cases as well the ape began spontaneously to point for humans in seemingly flexible ways to more distant objects they wished to have or to locations that they wished to visit.

Some other apes who have had various types of human contact have not learned to point, however, raising the question of just what role humans play (Povinelli et al., 1992). In a systematic comparison, Call and Tomasello (1994a) found that an orangutan who had been raised in a very natural humanlike environment pointed more flexibly and comprehended human pointing better than an orangutan who had been trained by humans to point in a very specific context. Call and Tomasello (1994a) also found that the human-raised orangutan pointed differentially depending on whether the human was visually attending to him, whereas the orangutan who had been specifically trained to point was less sensitive to the human's observational readiness (see figure 8.10). It is also interesting that only apes raised with extensive human contact have been reported to engage in "declarative" pointing in which the animal's goal is not to obtain an object or initiate an activity but simply to share attention, for example, by "showing" an object or pointing to something toward which the individual has no particular desires (Call & Tomasello, 1996), although these episodes are infrequent and their interpretation has all the problems of interpreting anecdotes.

If we raise the question of whether the apes in these cases are intentionally communicating with humans, the answer seems to be yes. They clearly learn the gestural signals, use them in flexible and novel ways, and use them in ways that are sensitive to the interactant. And at least some of these gestures are clearly about external objects and events. It is important to establish this groundwork because all of the "linguistic" communication that takes place between apes and humans takes place within this context, at least in studies in which the use of "language" is not totally tied to food rewards. It is also important to establish that the nature of the social environment is important because it is likely that not all linguistic apes have similar linguistic skills.

It should also be noted that there have been several reports of captive monkeys pointing for the human interactants (e.g., Blaschke & Ettlinger, 1987). More recently,

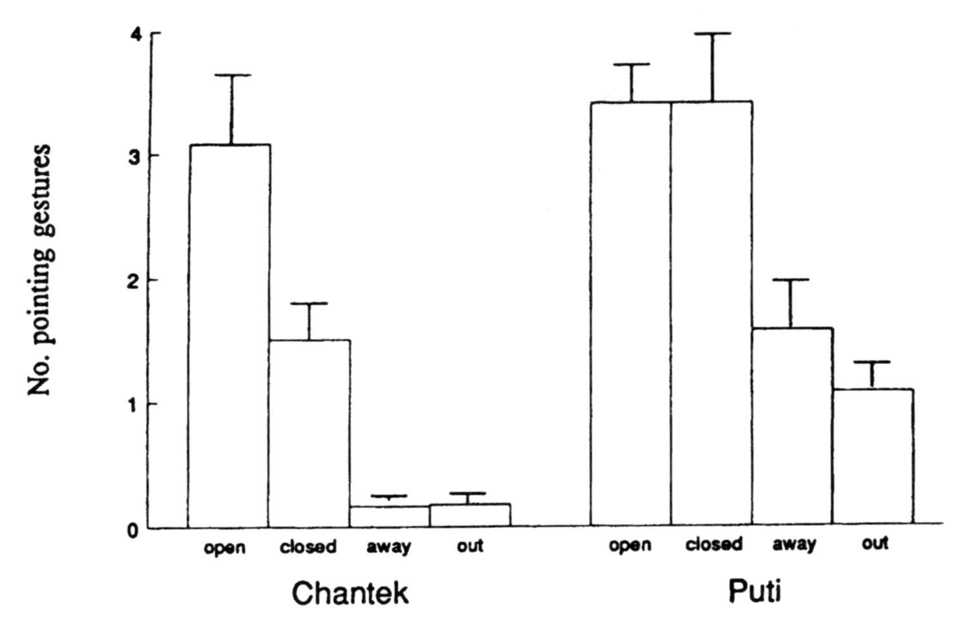

Figure 8.9. Call and Tomasello results for Chantek and Puti. Open, E (Experimenter) sitting in front of the subject with eyes open; closed, E sitting in front of the subject with eyes closed; away, E standing at the end of the room with the back turned to the subject; out, E is outside of the room. (Reprinted from *Journal of Comparative Psychology* with permission from the American Psychological Association.)

however, there have been conflicting reports on monkeys' comprehension of pointing by humans. Hess, Novak, and Povinelli (1993) found that a rhesus monkey who had learned to point for humans did not comprehend human pointing, but Anderson, Sallaberry, and Barbier (1995) found that three capuchin monkeys (*Cebus apella*) did learn to follow a human pointing gesture (as well as some other indicators) to a baited food well, with one subject doing so from the very first trials.

8.4.2 Linguistic Symbols

In response to the Terrace et al. (1977, 1979) critique, Savage-Rumbaugh and colleagues decided that the primary issue in ape language research was whether apes could learn to use something resembling human linguistic symbols. In working with the two chimpanzees, Sherman and Austin, they concentrated on individual lexigrams (created so as not to resemble their referents) and their symbolic status. Like Lana before them, these two chimpanzees were taught to use a specially constructed computer keyboard filled with lexigrams. Based on their analyses of what constitutes a human symbol, however, Savage-Rumbaugh and colleagues attempted to make sure that Sherman and Austin learned (1) to both comprehend and produce lexigrams; (2) to use their lexigrams in a wide array of situational contexts, even in the absence of their referents; and (3) to use their symbols spontaneously with each other. They had some success in this effort, but the nature of their efforts and successes, and their subsequent experiences with other apes, taught them much about what it means to use a symbol (Savage-Rumbaugh, 1986; Savage-Rumbaugh et al., 1980).

The most important of their subsequent experiences was their work with a wild-caught bonobo named Matata and her adopted son, the captive-born Kanzi. Savage-Rumbaugh et al. attempted unsuccessfully to teach Matata lexigrams. However, her son Kanzi, who had spent his first 2.5 years holding onto and observing his mother while she was interacting with humans around the keyboard (and also interacting with humans in other ways), learned many of the lexigrams his mother had not. Since he could only have acquired these observationally (he had not been directly trained in any way), the investigators decided to continue allowing Kanzi to acquire his symbols incidentally rather than through direct training with external reinforcement. The subsequent procedure used with Kanzi was based on child language research showing that human children acquire most of their early linguistic symbols in highly predictable routine interactions with adults, also without training (e.g., Bruner, 1983; Lock, 1978). Human adults do not teach their children language, but rather adults and children do things in the world and language is used to help them communicate about those things and to get them done; when these things take place in routine interactions, it is often the case that the child knows nonlinguistically, from previous experience, what the adult is doing and what she might be talking about (Tomasello, 1992b). Kanzi was thus exposed to the lexigram language in as natural a manner as possible as he and his caretakers went about their daily activities. His only reward was communication itself and the things it helped directly to accomplish.

The most spectacular result of the early research with Kanzi was that the lexigrams he produced seemed to be symbols in every sense from the beginning. While Sherman and Austin had showed many skills with their lexigrams, for the most part each skill had to be trained separately. For example, if they were taught to produce a lexigram in response to an object being held up to them, they could not then comprehend and respond appropriately to a human's use of that same lexigram by picking out the corresponding object. They could be trained to do so, but Kanzi did not need such training. Because he acquired his symbols spontaneously by observing and comprehending their use by his caretakers, from the beginning he understood them bidirectionally, that is, all of the symbols he produced he also comprehended without any specific training. His early symbols also were generalized and used in a decontextualized manner without specific training, including those for perceptually absent referents (Savage-Rumbaugh et al., 1986). His early vocabulary matched in many ways the early vocabularies of human children, including proper names for individuals; labels for common objects; words for actions, locations, and properties; and even a few function words such as "no" and "yes" (Nelson, 1987). Although no other linguistic ape has had its competencies assessed in all these ways, it would not be surprising to find that all the apes raised in more natural humanlike cultural environments, rather than in laboratory training environments, also use their symbols in this more humanlike manner.

There is no widespread agreement about what constitutes a symbol. Seidenberg and Petitto (1987) argued that Kanzi does not use linguistic symbols, but only instrumental procedures for getting what he wants. And indeed, the fact that even Kanzi uses the vast majority of his language to request rather than to comment (96% of all productions in the study of Greenfield & Savage-Rumbaugh, 1990), gives pause to inter-

Figure 8.10. The bonobo Mulika using a keyboard. (Photo courtesy of E.S. Savage-Rumbaugh.)

pretations of totally humanlike usage. However, the fact that Kanzi (and two other bonobos and a common chimpanzee since; Savage-Rumbaugh, Brakke & Hutchins, 1992) learn their symbols observationally and then use them bidirectionally seems to indicate that these apes are using humanlike symbols. Given the prevalence of requests in Kanzi's productions, one could certainly imagine that the reciprocity involved is on the behavioral level only: anything humans can use to direct my behavior I can also use to direct their behavior in an analogous situation. But still, the symbols Kanzi has learned seem clearly more "shared" than the naturally occurring gestures of apes which, like human infants' prelinguistic gestures, would seem to be effective procedures only, with no understanding that a single communicative device is the basis of two-way communication and that both participants use it in the same way.

8.4.3 Combinations of Symbols

The problem with investigations of "syntax" and "grammar" in children and apes is that these terms are used in different ways by different investigators. For our purposes, we need to distinguish two levels of grammatical competence. First, a subject may produce combinations of symbols in novel ways that create new meanings. The most well-known example is Washoe calling a duck "water bird," a combination that surprised even her caretakers. There are many other well-documented cases involving more mundane action–object combinations and the like, for example, as reported in Greenfield and Savage-Rumbaugh (1990, 1991). Second, a subject may produce combinations that have specific symbols to serve as grammatical markers. That is, although "water bird" is a creative combination, there are no symbolic indications in this utterance of how these two symbols are to be related to one another; the relating is left up to the listener. In adult human language, on the other hand, sentences have second-order grammatical symbols that indicate how the individual symbols are interrelated. For example, in the English sentence "John kissed Mary," who is doing the kissing and who is being kissed is indicated by the order in which the symbols are produced. The fact that word order is indeed being used as a communicatively significant symbol is demonstrated by the fact that the opposite ordering of elements would make Mary the kisser and John the kissed ("contrastive usage"). In other languages, word order is not used for this purpose, but rather special word endings (case markers) indicate the roles being played by John and Mary, and in still other languages this may be indicated by intonation contours (Bates & MacWhinney, 1979).

In the context of ape language research, it is especially important to recognize that using a consistent word order does not mean that word order is being used as a contrastive symbolic device indicating the relations among words (Tomasello, 1992c). Kanzi, as well as many other linguistic apes, has combined his symbols in some creative ways indicating some first-level grammatical competence (Greenfield & Savage-Rumbaugh, 1990, 1991). In these productions, which constitute about 10% of his total productions, he uses some consistent ordering patterns; for example, in his two-word combinations Kanzi uses the action words "tickle" and "chase" about twice as often as the first word than as the second word. In general Kanzi uses words for action as first words and words for the object of action as the second word, but this reflects rather directly the order in which he has heard those words used by caretakers. There is no evidence that a different ordering of words would produce a different meaning. Kanzi also combines his words with gestures in creative ways, mostly by using pointing to indicate who is to perform some action or the location toward which an action should be directed. He mostly uses the gesture after the lexigram, but the opposite ordering does not indicate a consistent change of meaning. Again, a consistent ordering pattern is not the same thing as the contrastive use of order as a grammatical symbol conveying a consistent second-order meaning.

One of the unique things about Kanzi is that he seems to comprehend large portions of spoken English, and this comprehension appears to go well beyond his productions. From the beginning, Kanzi's comprehension seemed to go well beyond just picking up some words, the way a dog might tune into the word "dinner." He seemed to be comprehending whole sentences in ways that suggested a creative synthesis of the words, not a rote learning of a recurrent sound pattern. After several years of working with Kanzi, Savage-Rumbaugh and colleagues systematically tested his comprehension of English commands, such as "Take the telephone outdoors" and "Give Rose a shoe" (Savage-Rumbaugh, Murphy, Sevcik, Brakke, Williams, & Rumbaugh, 1993). Over a period of several months and in separate administrations, Kanzi and a human 2 year old were presented with several hundred sentences. Almost all sentences were given as requests involving an action word and one or more object or person words, with the task of the subjects being to act out the sentence. For each trial for each subject multiple objects were present, affording many different types of actions. Several dozen different action verbs were used in the different sentences. Many sentences were constructed so that the actions named would not be deducible from knowledge of the objects involved

Table 8.1. Comprehension of verbal commands by Kanzi, a chimpanzee, and Alia, a 2-year-old human.

Sentence type and description[a]	Kanzi	Alia
1. "Put *X* in/on *Y*" (1a, 1b)	.66	.73
2. "Give *X* to *P*" (2a, 2b)	.66	.75
3. "____ P (with the X)" (2c, 2d, 3)	.80	.65
4. Information (variety of syntactic frames)	.67	.83
5a. "Take *X* to *L*"	.78	.71
5b. "Go to *L* and *get X*"	.82	.45
5c. "Go get *X* that's at *L*"	.77	.52
6. "Make *X*____*Y*"	.67	.56
7. Others (mostly complex sentences)	.78	.33

From Tomasello (1994), with permission.

[a] *X* and *Y* designate objects; *P* designates a person; *L* designates a location; and ______ designates an action verb such as "bite," "chase," "hammer," "hug," "hide," and "scare."

(e.g., "Put the money in the mushrooms"), and some were such that the role of the two objects involved could easily be reversed (e.g., "Put some milk in the water").

Results of the study are summarized in Table 8.1. In brief, no matter how the data were coded and scored, both Kanzi and the human 2 year old were very competent at carrying out requests of all types. Kanzi carried out 59% of all blindly presented sentences promptly and correctly, the child 54%. When generic prompting and encouragement by the experimenter were allowed, Kanzi was correct on 74% of the blind trials, the child on 65%. In attempting to determine what knowledge of language underlay Kanzi's performance in this study, Tomasello (1994a) reanalyzed the data to discriminate among a number of different hypotheses. He found that Kanzi's performance indicated that he was doing more than simply responding to the words in the sentences in some plausible way; he understood the use of word order as a contrastive symbolic device. For example, Kanzi responded correctly to the two requests: "Put the *hat* on your *ball*" and "Put the *ball* on the *hat*." Overall, he comprehended both members of 57% of pairs using the "put *X* on *Y*" formula (to the child's 39%), which is significantly different from the chance probability of acting out both sentences correctly (25%). Tomasello's conclusion was thus that Kanzi knows, for example, that the object being put (e.g., "hat") is named first and the place it is being put (e.g., "on the ball") is named only afterwards, even when the object and location could easily have been reversed. He knows that the person biting or chasing is mentioned before "bite" or "chase," and the one being bitten or chased is mentioned only after. He knows these things presumably because he has seen one order paired with one version of the key event (e.g., *X* chasing *Y*) and another order paired with another version of the same event (e.g., *Y* chasing *X*).

We may now ask a final question: Does Kanzi's knowledge of language transcend individual words, that is, does he know that there are classes of words such as noun and verb that allow the formation of ordering patterns (or other kinds of structural patterns) across different verbs? If he did, this would imply that he understood generalized syntagmatic categories such as agent, patient, and instrument (or even subject and direct object). Operationalized, the question is this: If Kanzi were taught a new verb without any object words involved (e.g., "Look! Daxing!" as a well-known person performed a novel action on a well-known object), would he automatically know the conventional order for mentioning the person and the object? Because 2-year-old human children do not do this, it would not be surprising if Kanzi did not (Olguin & Tomasello, 1993; Tomasello, 1992c). Children of this age apparently believe that each action word has its own syntactic devices associated with it, and it is likely that Kanzi is operating at this level as well in his comprehension of human speech. On the other hand, note that both Kanzi and 2-year-old humans seem to use their object words quite flexibly in all kinds of novel sentences, which might indicate something like a grammatical category of noun (Tomasello & Olguin, 1993).

The most plausible hypothesis is thus that, like 2-year-old children, Kanzi understands actions and events and the words that designate them. He understands that actions and events have participants and that the order in which these participants are mentioned for a particular event determines what role each plays in that particular event. He has something like a grammatical category of noun allowing him to comprehend (and perhaps produce) newly learned object labels in the same sentence frames in which previously learned object labels have been used. He has yet to form any class of verbs or sentence schemas, however, that would license grammatical generalizations across events. This means that his syntagmatic categories are of the type "hitter" and "thing hit with," "putter" and "thing put," not generalized categories such as "agent" and "patient" or "subject" and "direct object."

It is probably not the case that Kanzi understands language in every way like a human two-year-old child. In his production he uses mostly single words, he acquires his linguistic skills more slowly than human children do, he uses his lexigrams mostly to request rather than to report or make comments, he does not talk about some things that children are very fond of talking about (e.g., possession), his production of symbol combinations show no signs of grammatical marking, and his comprehension of symbol combinations shows word-specific structure only. In terms of conversation, even the analysis of Greenfield and Savage-Rumbaugh (1991) is not convincing that Kanzi or any other linguistic ape knows how to produce a full-fledged conversational turn in which the topic of conversation is acknowledged and specifically indicated, and simultaneously something is predicated of that topic. None of this is to say that these limitations relative to humans might not one day be overcome. In particular, the keyboard medium is somewhat clumsy for conversational purposes and the lexigrams that are available do not represent the full range of options that human children have at their disposal.

8.4.4 Comparison of Species

Apes who have been raised in various ways by humans develop a species-atypical set of gestural signals, including referential pointing, perhaps in some cases for declarative purposes. Precisely what in their ontogenies leads to these developments is not known at this time because the precise role of humans in encouraging or training pointing and other attention-directing behaviors of primates has yet to be studied systematically. Call and Tomasello (1996) hypothesize, however, that the experience of being treated intentionally by humans, in which humans encourage the sharing and manipulating of attention to outside objects and events, is a powerful experience for nonhuman primates and utterly unlike any kind of experience that they have in their natural social environments. The fact that only apes who have been raised by humans seem to comprehend as well as produce pointing is an important indication of the role of human social interaction in the development of certain communicative skills.

Apes raised by humans can, under certain conditions, also learn some important aspects of human language. When raised in a manner similar to human children in a social environment full of predictable interactive routines, apes learn to use humanlike linguistic symbols for purposes of intentional communication. They can learn to use them bidirectionally as well, that is, they automatically produce what they comprehend and comprehend what they produce. This seems to indicate an understanding of the other as a full communicative partner, although precisely how the other's role is understood is still an open question. The symbol combinations apes produce show creativity and flexibility in creating new meanings combinatorially, but they do not have productive grammatical devices. At least one ape individual has shown the ability to comprehend grammatical devices, but those devices are probably ties to individual words and not used generally across words in a productive fashion.

It is important to note that some marine mammals have learned to respond appropriately to sequences of human gestures in ways that are similar to Kanzi's responses; that is, they have learned to perform the appropriate responses to one of several commands using novel object labels, and they perform differently and appropriately depending on the ordering of the symbols (e.g., bottlenose dolphins as reported by Herman, 1986; Herman, Pack & Morrel-Samuels, 1993). The response of Savage-Rumbaugh et al. (1993) is that there is no convincing demonstration that these animals understand those gestures in a symbolic fashion—flexibly and in both comprehension and production. In any case, a closer comparison of the two species would be useful and might allow for some more definitive statements to be made about the symbolic status of the linguistic productions of Kanzi and other human-raised apes. No monkey species has ever been raised and trained in ways that would allow a fair assessment of its skills to acquire humanlike forms of communication.

8.5 What Primates Know about Others in Communication

Complementing primates' social knowledge, as reviewed in chapter 7, is their ability to use that knowledge in the formulation of active social and communicative strategies. In competitive situations, nonhuman primates employ social strategies to gain some advantage. Some of these strategies may be seen as deceptive. Although some of the observations that have been classified as instances of deception may be based on coincidence, others have been observed on repeated occasions and so clearly reflect active social strategies that individuals have either learned or created. In either case they involve cognition because the individual understands the situation and behaves flexibly in light of the social problem presented. Thus, an individual might want a group to cease fighting and so come up with the strategy of using an alarm call. Such a strategy might be based on past learning experiences in exactly this circumstance (e.g., one in which an individual gave an alarm call to a predator when members of another group were nearby and watched them scatter), or it might be based on the insight that this call has led others to scatter in the past in other situations and so it may also lead to that same result in the current circumstance. These are intelligent social strategies based on informed decision making, an understanding of the animacy and directedness of the

Figure 8.11. Bare teeth display by rhesus monkey. (Photo courtesy of F. de Waal.)

behavior of others, and sometimes on some form of insight or foresight into the social problem at hand. Whether or not they represent attempts to influence the psychological states of others deceptively is another question, and indeed we have our doubts. Whether or not primates differ from other mammals in the ways they create and use social and communicative strategies is at this point an open question as well, and again there is room for doubt.

Primates, like many other species, have phylogenetically ritualized communicative displays that convey to groupmates, and perhaps to other conspecifics, their location, their identity, and their mood (Byrne, 1982). In addition, they have some communicative signals that involve, to some degree, learning during ontogeny, which then leads to more flexible use in different contexts. This is true for both gestural and vocal communication, with most of the studies of gestures concerning great apes and most of the studies of vocalizations concerning monkeys. With regard to gestures, apes have many gestural signals that are ritualized ontogenetically and used flexibly in different communicative contexts. Of particular interest for issues of social cognition, chimpanzees (1) use gestures that function to attract the attention of others to self, (2) hide some of their own involuntary displays when others are appropriately oriented, and (3) use different gestures when the recipient is bodily oriented and when it is not. Much less is known about monkeys' use of such flexible gestural signals. The process by which these gestural signals are acquired is discussed in more detail in chapter 9, but the process of ontogenetic ritualization is essentially one in which the individual learns over repeated instances of a particular social interaction that the initial part of its behavior may induce the other into action. The evolutionary conditions under which gestures (as opposed to vocalizations) may be especially useful are discussed in an enlightening manner by Maestripieri (in press).

With regard to vocalizations, vervet monkey alarm calls and rhesus macaque recruitment screams have many fixed parameters in their actual physical manifestations, as do the fear calls of many species, but many of these calls also involve some degree of learning in terms of the appropriate ecological and social conditions for their use. They thus may be used somewhat flexibly and strategically. It is interesting in this context that Macedonia and Evans (1993) speculate that the evolution of call systems that differentiate external stimuli such as predators and aggressors referentially is likely to take place for species in which the response needed for the different predators are different. Thus, ground squirrels run into their burrow no matter what the predator, and the key parameter eliciting their vocalization is response urgency. Vervet monkeys, however, must behave differently in the presence of leopards and eagles or they will not survive. This may help explain why there are no reports of call systems of any primate species differentiating among different types of food (since the differences are not directly crucial for survival) and why apes have not evolved differentiated vocal alarm systems (since particular groups tend not to have multiple types of lethal predators). In any case, more studies of ape vocal communication that focus on flexibility of use and audience effects are sorely needed.

The apes that have become especially proficient in their gestural and symbolic communication with humans have been not just trained but "enculturated" by humans (Call & Tomasello, 1996). That is, they have been treated as intentional agents by humans who attempt to follow into, manipulate, or share their attention, which fundamentally changes the ritualization process. This does not happen to developing youngsters in the wild, nor does it happen in many training regimes set up by behavioristic researchers. It is possible that growing up in a humanlike cultural environment with humans as the major type of communicative partner creates a species-atypical set of communicative competencies that take advantage of the powerful cognitive skills of humans, for example, the alacrity with which they respond to the kind of orienting responses that might be ritualized into referential pointing or the enthusiasm with which they provide valued items based on particular hand configurations they have shaped and trained. Although some monkeys have been raised by humans, there are no systematic studies of the nature of their upbringing and the skills they develop in such environments. The linguistic skills of apes are in some ways similar to those of 2-year-old human children, although there are some important differences as well. It is also likely that this enculturation process leads apes down developmental pathways that are different in a whole host of other social-cognitive skills (see section 12.2 for more discussion).

In terms of species differences, in their natural habitats both monkeys and apes signal to conspecifics intentionally and flexibly. There is good evidence for monkey vocalizations and ape gestures. The evidence for deception and audience effects, though mostly anecdotal in both cases, is of the same nature for both groups. There are few, if any, observed differences between monkeys and apes evident in existing research on communication, with the possible exception that ape gestures may be used a bit more flexibly (across entirely different behavioral domains) than monkey vocalizations, the problem being that there is little research on the nonvocal communication of monkeys or on the vocal communication of apes that pays close attention to processes of social cognition. Nor do we know how apes would perform in either of the experimental studies performed with monkeys; that is, we do not know whether apes would warn their offspring or direct them to food as two monkey species did not, and we do not know whether in a habituation experiment apes would treat as equivalent different signals (either vocal or gestural) typically associated with the same referent. The clearest difference between monkeys and apes in the domain of communication might seem to be the susceptibility of apes to humanlike cultural environments, but monkeys simply have not been raised or studied in the appropriate ways to know if this is a reliable difference between the taxa.

As in the case of social knowledge reviewed in chapter 7, an important question is whether and in what ways nonhuman primates understand the psychological states of conspecifics as they employ social strategies and communicate with them. We defer our full consideration of this issue until chapter 10, "Theory of Mind." For current purposes the important point is that primate individuals clearly formulate various types of social and communicative strategies aimed at manipulating or influencing the behavior of others in particular ways, and these clearly rely on an understanding of conspecifics as animate and directed beings. Some of these social and communicative strategies are the result of trial-and-error learning or ontogenetic ritualization in particular contexts, but others are the result of an individual's insight into a social problem in which it foresees that some strategy that has been useful in another situation in the past may be adapted to the current problem, perhaps even representing a truly unique solution. As in all forms of insight and foresight, this requires perceiving how the to-be-manipulated items works—how it contributes to the structure of the problem. This might mean an understand of others in terms of their psychologies, but it might also mean simply an understanding of others in terms of their animate and directed behaviors. Looking at primate social strategies as relying on an understanding of others as animate and directed thus provides a middle cognitive ground between the stark alternatives of mindless operant conditioning and a humanlike theory of mind.

9

Social Learning and Culture

As manifest in the domains of activity reviewed in the previous two chapters, the social interaction of primates depends on individuals' abilities to learn about other members of their groups and how they are likely to behave in certain situations and to learn various social and communicative strategies for manipulating those groupmates for their own ends. In both of these cases the processes of learning involved may be broadly construed as social in the sense that the learning is about social interaction. However, most psychologists typically use a somewhat narrower definition of social learning to indicate situations in which one individual attempts to actually reproduce or match the behavior of another (e.g., Bandura, 1986). This narrower definition thus does not include interactions in which an individual learns a new behavior or behavioral strategy through reciprocal social interaction: an individual does *A* and a groupmate responds with *B*, as happens in the learning of a social strategy or the ritualization of a communicative gesture. Rather, in the narrower definition of social learning, the individual learns a new behavior or behavioral strategy by observing and attempting to reproduce a conspecific's behavior in situations in which they are both aiming toward a similar goal, either physical or social. The complement of social learning narrowly defined is the process by which an individual attempts to demonstrate for another, or to teach another, how some common goal might be accomplished. It is these specific kinds of social learning and teaching processes in which individuals attempt to match one another's behavioral strategies that researchers most often investigate under the aegis of social learning, and together they constitute the twin faces of what is sometimes referred to as cultural transmission.

From an evolutionary point of view, social learning and cultural transmission are topics of great interest to behavioral ecologists studying a wide range of animal species from birds to mammals (see the papers in Heyes & Galef, 1996, for a representative sampling), and primatologists have shared this interest. As organisms that live socially and depend heavily on learning, primates and other social mammals may increase their reproductive fitness by (1) taking advantage of things learned by other members of the group; (2) passing on skills and information to their offspring that would be dangerous or arduous for them to acquire on their own; and (3) exploiting the past experiences of elders in key circumstances such as drought that occur only rarely (King, 1991, 1994; Kummer, 1971; Nishida, 1987b). The potential is thus that processes of organic evolution may be supplemented by processes of cultural evolution in which organisms adapt to changing environments much more rapidly. If our interest is in the cognitive aspects of these phenomena, however, there are just two essential questions. First, we may ask about the ways primates augment their own individually acquired knowledge and skills by means of social learning, and second, we may ask about the knowledge of others that is implied by the types of social learning in which primates engage. Some recent reviews of primate social learning focusing on cognitive processes may be found in Visalberghi and Fragaszy (1990a) for monkeys, Tomasello (1996a) for apes, and Whiten and Ham (1992) and Tomasello and Call (1994) for all primate species.

One potential indicator that a species engages in some form of cultural transmission is the existence of population-specific behavioral traditions, that is, behaviors that are similar for the individuals within a population, including across generations, but that are not found in other populations of that same species. But population differences of behavior are not an infallible sign of social learning, despite the fact that some researchers describe all cases of population-specific behavioral traditions with the term "culture" (e.g., Bonner, 1980). The problem is that in many cases, the physical environments of the different populations present different problems and opportunities and it is these, in concert with the flexible learning skills of the species, that lead to the observed population differences of behavior. For example, the ways that chimpanzees construct their nests vary in consistent and reliable ways among different populations, but the nature of those differences is likely due to the individuals of each population adapting to the local ecologies in which they find themselves in terms of the trees and nest-building materials available (Baldwin, et al., 1981). In such cases there are population-specific behavioral traditions, but no social learning or cultural transmission is likely involved.

Matters are made even more complex by the fact that species such as primates, who are proficient and flexible learners in general, may acquire information from one another's behavior in a variety of different ways that fall under the general heading of social learning and cultural transmission. For example, in many animal species, including many birds and mammals (and even one species of octopus, an invertebrate), individuals are attracted to the locations of other animals or the objects with which they are interacting. This process is called local enhancement (Thorpe, 1956) or stimulus enhancement (Spence, 1937). The attraction does not involve any processes of learning, social or otherwise, but merely places individuals in a position to learn. Because all learners are the same species, what they then learn from the environment is often identical. On the other hand, there are also processes of mimicking and imitative learning in which individuals attempt to reproduce the actual behavior of conspecifics. The overall problem in the study of social learning is that it is not always easy to determine which social learning processes are involved in particular cases. The phenomena that some researchers have identified as "teaching" are perhaps even more elusive, as many behaviors of individuals help others to learn, but far fewer represent cases in which an individual intends that others learn from its behavior.

Figure 9.1. Chimpanzees in a position to learn from one another. (Photo by J. Call.)

Given our interest in cognitive processes, the focus of this chapter is on the different types of cultural transmission and social learning characteristic of different primate species. We begin our investigation with the natural phenomena: the population-specific behavioral traditions of primates in the wild. As much as possible we look at these with an eye to which types of learning or social learning might underlie them. We then turn to observational and experimental studies of captive primates that seek to identify more analytically the different types of social learning processes in which primates engage and which might account for population differences of behavior in the wild. We look first at the social learning of instrumental activities (mostly involved with feeding) and then at the social learning of communicative signals. We conclude with a brief look at primate behaviors that some researchers have referred to as "teaching," since these too have important implications both for how individuals learn from one another and for how they understand one another.

9.1 Behavioral Traditions in the Wild

In their natural habitats, the two main species of primate that have been observed to display population-specific behavioral traditions are Japanese macaques and chimpanzees. For a variety of reasons, genetic explanations of these differences are unlikely (e.g., populations living close to one another differing as much or more than populations living far apart), and so many researchers speak of primate "culture" (e.g., Boesch, 1992, in press; McGrew, 1992; Nishida, 1987b; see the papers in Wrangham, McGrew, de Waal, & Heltne, 1994). Other researchers have been more cautious with this term, preferring to call cultural only behavioral traditions in which particular types of social learning are involved (Galef, 1992; Tomasello, 1990, 1994b). Although words are not important in and of themselves, it is important to determine the nature of primate cultures and the proximate mechanisms on which they depend. We begin our investigation, therefore, with a look at the nature of primate behavioral traditions, focusing on the types of learning and social learning by which they may be maintained.

9.1.1 Japanese Macaque Food Processing

On a number of occasions, specific behaviors with food have been observed to "spread" through particular groups of monkeys, implying some form of social learning. The best known instance of this occurred in the Koshima troop of Japanese macaques and concerned their practice of "washing" human-provisioned sweet potatoes before eating them (Itani & Nishimura, 1973; Kawai, 1965; Kawamura, 1959). In 1953 an 18-month-old female, Imo, was observed to take pieces of sweet potato, given to her and the rest of the troop by researchers, and to wash the sand off of the pieces in some nearby water (at first a stream and then the ocean). About 3 months after she began to wash her potatoes, the practice was observed in Imo's mother and two of Imo's playmates (and then their mothers). During the next 2 years, seven other youngsters also began to wash potatoes, and within 3 years of Imo's first potato washing, about 40% of the troop was doing the same. The fact that it was Imo's close associates who learned the behavior first, and their associates directly after, was thought to be significant in suggesting that the means of propagation of this behavior was some form of imitation in which one individual copied the behavior of another. This was the first report of nonhuman primate culture.

This phenomenon is clearly of great interest to behavioral and cognitive scientists, but the interpretation in terms of culture and imitation has two main problems. The first problem is that potato washing is much less unusual a behavior for monkeys than was originally thought, which raises the possibility of individual learning. Brushing sand off food turns out to be something that many monkeys do naturally, and indeed this had been observed in the Koshima monkeys before the emergence of washing. It is thus not surprising that potato washing was also observed in four other troops of human-provisioned Japanese macaques soon after the Koshima observations (Kawai, 1965), implying that at least four individuals had learned on their own. Also, in captivity, individuals of other monkey species (longtail macaques and capuchin monkeys) learn quite rapidly to wash their food when provided with sandy fruits and bowls of water, with no signs of imitative learning (Visalberghi & Fragaszy, 1990b, who refer to the behavior as "dipping"). These further observations thus raise the possibility that individual learning was responsible for the acquisition of potato washing by the Koshima macaques, as each individual "reinvented the wheel" for itself. The timing of the learning coincided with first human provisioning, perhaps, because it was only then that the appropriate learning conditions (sandy food near water) were present. In addition, it should be noted that on his visit to Koshima, Green (1975) observed that the human provisioners seemed to favor throwing potatoes to those monkeys who washed them, often into the water. This human encouragement and reinforcement for washing might explain why the behavior was more widespread in the Koshima troop and persisted for longer than in the other troops of monkeys.

Figure 9.2. Japanese monkey "washing" potatoes. (Photo courtesy of E. Visalberghi.)

The second problem has to do with the pattern of the spread of potato-washing behavior within the group. The spread of the behavior was relatively slow, with a mean and median time of more than 2 years for acquisition by the members of the group who learned it (Galef, 1992). This would not seem consonant with a process of imitation, which is typically thought of as a rapid process. Moreover, and perhaps even more tellingly, the rate of spread did not increase as the number of users increased. If the mechanism of transmission were imitation, an increase in the rate of propagation would be expected as more demonstrators became available for observation over time. In contrast, if processes of individual learning were at work, a slower and steadier rate of spread would be expected—which was in fact observed (Galef, 1992; although see Lefebvre, 1995, for a different view). Imo's friends and relatives may have been the first to learn the behavior because friends and relatives stay close to one another, and thus Imo's friends likely went near the water more often during feeding than other group members, increasing their chances for individual discovery. Galef (1992) also notes that after a certain period, many youngsters grew up finding potatoes in the water, an excellent opportunity for individual learning and discovery of the type engaged in by the original inventor.

Two similar phenomena have been observed in Japanese macaques. In the same Koshima group of Japanese macaques, a behavior called wheat placer-mining was observed. Once again, Imo was first to engage in the behavior; this time when she was 4 years old. Imo took a handful of wheat and sand and placed it in the water so that the wheat floated to the surface, clean and edible. The behavior spread in the same pattern as potato washing: first to Imo's friends and relatives and then to their relatives, but without an accelerating rate over time. In this case, however, the spread took place at an even slower rate (39% of the troop after 6 years), and there were even fewer practitioners than potato washing (Kawai, 1965). Second, Huffman (1984, 1986; Huffman & Quiatt, 1986) observed the origin and spread of stone play in two different troops of Japanese macaques. In this behavior an individual gathers, carries, rolls, clacks, or rubs together several stones. Such play usually occurs immediately after human food provisioning, and an individual bout typically lasts a few minutes. The behavior spread only among the groups' youngsters at first, and because the inventor's frequent playmates seemed to learn first, Huffman (1986) claimed that cultural transmission, possibly involving imitation, was at work (see also Tanaka, 1995a, on possible social transmission in Japanese macaque grooming practices).

Both of these sets of observations may be analyzed in the same general terms as potato washing. There is an individual "inventor" who did not learn by any form of social learning. Other individuals were then exposed to the appropriate learning conditions at different times, with those more closely related to the inventor being exposed first. They then learned individually as well, likely facilitated by processes of stimulus enhancement in which their engagement with the wheat or rocks was stimulated by the engagement of the inventor. As in the other cases, individuals in closest proximity to the inventor and other practitioners would be in a better position to see and be attracted to the objects with which they were interacting.

9.1.2 Chimpanzee Tool Use

The other well-publicized example of a primate behavior that has been termed cultural is chimpanzee tool use. There are a number of population-specific tool-use traditions that have been documented for different chimpanzee communities, for example, termite fishing, ant fishing, ant dipping, nut cracking, and leaf sponging (see McGrew, 1992, for a review). Sometimes the "same" tradition even shows variability between groups (McGrew, Tutin & Baldwin, 1979). For instance, members of the Kasakela community at Gombe (as well as some groups elsewhere) fish for termites by probing termite mounds with small, thin sticks, whereas in other parts of Africa there are chimpanzees who perforate termite mounds with large sticks and scoop up the insects by the handful (Sabater Pi, 1974). Field researchers such as Boesch (1992, in press), Goodall (1986), McGrew (1992), Nishida (1987b), and Sabater Pi (1984) have claimed that specific tool-use practices such as these are "culturally transmitted" among the individuals of the various communities, and some use the term "imitation" as a key part of their explanation (although for the most part these fieldworkers have used the term imitation as a general term to include virtually all types of social learning).

As in the case of Japanese macaques, it is possible in all these cases that individual learning, perhaps reinforced by processes of local or stimulus enhancement, is responsible for the population-specific behaviors. Thus, it is possible that chimpanzees in some

Figure 9.3. Chimpanzee cracking nuts. (Photo courtesy of T. Matsuzawa.)

localities destroy termite mounds with large sticks because the mounds are soft from much rain, whereas in other localities other chimpanzees cannot use this strategy because the mounds they encounter are too hard. In this case, there would be population differences with no social learning involved. This same type of explanation is possible in other, perhaps all, known cases as well (Tomasello, 1990). With this possibility in mind, Boesch, Marchesi, Marchesi, Fruth, and Joulian (1994) investigated intensively the geographical distribution and local ecologies of the nut cracking of chimpanzees in western Africa (the only region of the continent in which the behavior is found). They found that nut cracking is confined to a particular geographic region on one side of a large river in western Africa. To test for ecological differences that might account for the nut cracking behavior, they chose two populations of chimpanzees, one whose members habitually cracked nuts and one whose members did not, living less than 50 km apart but on opposite sides of the river. They found few differences in what they considered to be relevant ecological variables. Both populations lived in the same kinds of forest and had available the same types and abundance of nuts and tools. Boesch (1995) supported this argument further with observations of several innovative tool-use behaviors of the Tai chimpanzees that seemed to spread through the group rapidly. Boesch (in press) notes that of the many different ways that chimpanzees could perform nut cracking (and sometimes do try to perform it in zoo settings), the Tai chimpanzees only use a few, arguing for the existence of "social norms" that constrain the techniques used.

The problem is that, as attempts to "prove the null hypothesis," ecological analyses by themselves can never be definitive. Thus, for chimpanzee populations that have the appropriate materials available but do not crack nuts; are the nuts in their region harder?

Are their trees located near predators? Are there other more easily obtainable sources of protein? Such possibilities may be multiplied basically ad infinitum, including the possibility that there were ecological differences between the groups in the past that are no longer present today. Moreover, it is also possible that the local ecologies are identical in relevant ways, but that one chimpanzee on one side of the river invented nut cracking (perhaps a genius or a lucky individual who discovered rocks and nuts in a felicitous arrangement left by humans of the area who are known to crack nuts), and her behavior left a stone hammer, some unopened nuts, and some opened nuts all in one place near a suitable substrate—very propitious learning conditions that might facilitate the individual learning of others. Moreover, the behavior of the inventor would also be quite noisy and attract the attention of groupmates. Thus, the combination of propitious learning conditions and processes of local enhancement might result in the acquisition of nut cracking by the inventor's groupmates, whereas in other groups, with no genius or lucky individual, the appropriate learning conditions simply never became available.

This interpretation is supported by Sumita, Kitahara-Frisch, and Norikoshi (1985), who looked closely at the spread of nut cracking in a captive group of chimpanzees. They found that learning for all individuals was gradual and that there were many idiosyncrasies in the way individuals performed nut cracking. They concluded that individual trial-and-error learning (along with local enhancement) was the learning process responsible for the spread of the behavior in the group: one creative individual made a discovery, which left propitious learning conditions for others, made salient through processes of local enhancement. It was not the case that individuals were copying the nut-cracking techniques of their groupmates. This same explanation may also apply to the other reported cases of the rapid spread of a tool use behavior by chimpanzees in relatively natural captive settings, for example, as reported by Menzel (1972, 1973b) and Hannah and McGrew (1987).

It must also be acknowledged, of course, that the nut cracking and other tool-use behaviors of chimpanzees in the wild might well be acquired via imitative learning in which individuals actually copy one another's behaviors or behavioral strategies. We simply cannot tell from naturalistic observations alone. What is needed to distinguish these possibilities are experimental manipulations in which investigators know precisely what individual primates have observed and not observed before they engage in a particular instrumental behavior. Such research is reported in section 9.2.

9.1.3 Chimpanzee Food Choice

A number of investigators have found that populations of chimpanzees in the wild have different diets or feed in different ways (e.g., McGrew, 1983; Nishida, 1980; Nishida, Wrangham, Goodall, & Uehara, 1983). Once again, however, the respective roles of ecology and social learning are difficult to determine through naturalistic observations alone. McGrew (1983) concludes from his survey that most of the differences in food choice are due to the different foods available in the different environments involved, but that some choices are likely attributable to social custom, because no good environmental explanations for some of the population differences are readily apparent (i.e., some populations eat a food that others avoid even though it is readily available). Again, however, an individual's choices are influenced by many factors, including alternative food choices (perhaps a food is ignored by a group because they have better alternatives), predators associated with specific foods at a given site, and so forth. Possible learning mechanisms for primate food choice is discussed in section 9.2.

9.1.4 Chimpanzee Communication

The gestural communication of chimpanzees also shows some population-specific behaviors that some researchers have called cultural. There are two reasonably well-documented cases from the wild (see Tomasello, 1990, for several less well-documented cases). First, Nishida (1980) observed "leaf clipping" in the Mahale K group of chimpanzees, and he reported that to his knowledge it was unique to this group. In this behavior an individual takes one to five very stiff, dried leaves and strips them in his mouth, making a distinctive and fairly loud noise—a very good attention-getter (see section 8.2 on chimpanzee gestural communication). Of the 30 individuals of the group (approximately half adults), 8 used it at least once in a sexual context and 5 used it in a play or frustration context. Soon after these observations, Sugiyama (1981) and Boesch (1995) observed leaf clipping in two other groups of chimpanzees across the continent in western Africa. The individuals at Bossou use it mainly in frustration or in play, whereas the males at Tai use it mainly as part of a drumming display. The second example is the "grooming hand-clasp" reported by McGrew and Tutin (1978). This is not really an instrumental gesture, but simply involves one individual raising the arm of another to access the underarm for grooming. Of the 19 noninfant individuals observed, 9 showed the behavior. The behavior was also thought at the time to be unique to the Mahale K group, and it does seem to be absent from the nearby Kasakela community in Gombe. But, as with leaf clipping, other researchers have since observed the behavior elsewhere, in this case in one wild and one captive community (de Waal, 1994; Ghiglieri, 1984).

In both of these cases the learning process involved is likely ontogenetic ritualization (see section 9.3). The basic problem in determining whether imitation or ritualization is responsible is that the data from naturalistic observations do not include precisely which individuals acquired the behavior at which times over an extended observation period. Given the evidence from captive chimpanzees (see below), our hypothesis is that leaf clipping is an attention-getting behavior that different individuals have discovered to be a useful way to get others to look at them. At all three study sites there was much individual variability in how this gesture was used, including even the goal being sought. In the case of grooming hand-clasp, it seems likely that individuals could easily learn this behavior by having conspecifics approach them with their arm raised in hopes of underarm grooming (because on previous occasions they have enjoyed being groomed under the arms) or by local enhancement as they watch others groom under the arms of groupmates and discover that this is where tasty parasites can be found. The fact that both leaf clipping and grooming hand-clasp have been observed in more than one group, which have not had the opportunity to observe one another, raises the possibility that in all groups the behavior is spontaneously invented by individuals through some kind of ritualization process. Based on all of these factors, Nishida (1994) recently expressed the view that ontogenetic ritualization (as opposed to imitative learning) may be the mechanism responsible for the acquisition of gestures in the wild.

It should also be mentioned that there is some evidence for population differences in the acoustic properties of chimpanzee vocalizations (Mitani, 1994). The causes of these differences are not clear, however, and there are no indications that beyond acoustic variation there are population differences in the communicative goals for which particular vocalizations are used.

9.1.5 Other Primate Species

It is interesting, and perhaps telling, that there are no well-documented cases of population-specific behavioral traditions in primate species other than Japanese macaques and chimpanzees, although many other primate species have been observed for extensive periods in the wild. There are a number of short-term observations of various species of primates doing things that might seem to be traditional or imitative, but in all cases individual learning (sometimes in conjunction with stimulus enhancement) is a viable alternative explanation. For example, Hauser (1988b) reported that during a period of drought, some members of a vervet monkey group dipped dry tortilis pods into the well of a tree containing a viscous exudate. Because this behavior had not been seen before in this group and because those that performed the behavior first were all from the same family, Hauser concluded that at least some individuals acquired the behavior by means of imitation. He also notes that the new and unusual circumstances may have created novel learning conditions for individuals as well.

Another set of observations is that of Byrne and Byrne (1993), who watched the complex manipulations that gorillas perform in gathering and processing plant foods (see figure 9.4). As documented in chapter 3, different plants require different techniques due to a number of factors, among them the size, location, and texture of the plants and how the edible parts are attached to the inedible parts. The researchers observed complex sequences of behaviors that are reliably used with different plants by all noninfant individuals of the observed group, raising the possibility that they might be learning from one another in some way (they hypothesize "program-level imitation": reproducing the general structure of the behavioral sequence, with the details being discovered individually). The problem with this case for inferring processes of social learning is that there are no variations of technique associated with particular families or groups of individuals or any comparison of gorillas who did and did not observe conspecifics to rule out the possibility that each animal is learning individually from its interactions with the plants.

It should also be mentioned that several researchers have reported that individuals from wild or seminatural captive groups have failed to learn a useful strategy used by an inventor, even though they were motivated and observed the inventor's behavior repeatedly. One recent set of observations is similar to those of the potato washing of Japanese macaques, but with an experimental dimension. In a captive colony of several dozen longtail macaques, Zuberbühler, Gygax, Harley, and Kummer (1996) observed an adult male spontaneously discover how to rake in fallen apples from under a tree outside the group's compound. He did this through the fencing with wooden sticks that could be found throughout the compound. The researchers then decided to control the behavior by allowing fruit to be available only once per week for approximately 30 minutes at a time; they observed all of these sessions over a 17-month period. What they observed was a marked effect of stimulus enhancement. At times when the tool user was raking

Figure 9.4. Mountain gorilla processing food, as studied by Byrne. (Photo courtesy of R.W. Byrne.)

in fruit with a stick, many other group members began manipulating sticks as well—more than they manipulated sticks at other times and more than they manipulated rocks during this same time. However, none of these other group members ever attempted to use its stick as a tool to rake in fruit in the first year of observations. After 1 year, one individual did learn the behavior; after an additional 9 months, a second individual learned it as well, followed by a third individual 10 months after that. The investigators concluded that these three individuals learned through a combination of stimulus enhancement (they were individuals who increased their manipulation of sticks significantly when the original tool user was using his tool during the first year) and individual learning.

Along these same lines, Boinski and Fragaszy (1989) noted that juvenile squirrel monkeys (*Saimirii oerstedii*) went through a painful process of trial and error in learning to rub noxious spines off caterpillars before eating them, even though they had seen adults rubbing them safely off with their tails on numerous occasions. Visalberghi and Fragaszy (1990a) reported that the rhesus macaques on Cayo Santiago have not learned to crack open coconuts, whose meat is highly prized, even though they have repeatedly observed two of their groupmates doing so by pounding them on rocks for a period of at least 5 years. Visalberghi and Fragaszy (1990a) also report on the unpublished research of Schonholzer, who observed two hamadryas baboons in a captive population, but none of their groupmates, soaking up water from a crevice with their tails and then licking it off. Nakamichi et al. (in press) report that the washing of grass roots by some individual Japanese macaques did not spread in the group. Finally, Milton (1993) reported that both the foraging and social behaviors of adult spider monkeys that had been released onto an island as youngsters were indistinguishable from that of adult spider monkeys that grew up in a natural group setting. This observation demonstrates the ability of individuals of this species to learn what they need to know by interacting directly with the physical environment and other individuals of their species, without invoking social learning from adults.

It should also be noted, following Visalberghi and Fragaszy (1990a), that overall there is no way of knowing all of the "negative" instances of a primate individual observing another perform some salient and useful activity in the wild, but then failing to exploit the situation and reproduce it for themselves. In their survey of the behavioral innovations of primates, two prominent researchers concluded that across many primate species including chimpanzees: "Of the many [innovative] behaviors observed, only a few will be passed on to other individuals, and seldom will they spread through the whole troop" (Kummer & Goodall, 1985: p. 213).

9.1.6 Comparison of Species

In some ways it is surprising that of all the primate species studied intensively in the wild, only a few species have shown clear evidence of population-specific behavioral traditions, and, with the exception of bird song, there are few examples from other animal species (see Terkel, 1996, for one very interesting example for wild rats). One factor may be that the best documented cases all involve very salient behaviors with objects, or else very distinctive gestures. It is thus possible that other species have some population-specific behavioral traditions but that these involve more subtle behaviors that are less easily observed by humans. It is also possible, however, that population-specific behavioral traditions are rare in primates because they require specific environmental conditions to support them. While it is clear that to exploit these conditions, flexible learning processes are required, it is not clear that social learning processes other than local or stimulus enhancement are required.

What is needed to help clarify the situation is experimental research in which it is known precisely what an individual has experienced with particular materials or situations. Although experimental research always may be criticized for being "artificial," if it is done in ways that seek to reproduce salient aspects of the situations of particular species in the wild, and there are several lines of evidence using different techniques in different behavioral domains, then such research may be used in combination with naturalistic observations to make inferences about learning processes in the wild.

9.2 Social Learning of Instrumental Activities

Experimental studies of observational learning have been conducted since before the turn of the century with a variety of animal species. But because the investigators did not see the need to make the discriminations among different types of observational learning that modern researchers make, they often did not employ all of the needed controls, and so their results are often inconclusive and have been superseded by recent investigations. Thorough reviews of this earlier imitation research can be found in Galef (1988), Hall (1963), and Whiten and Ham (1992), and a catalogue of the relevant studies with primates is presented in Table 9.1. We focus here on primate research that has taken place with some concern for the specific learning processes involved. Although there are some recent studies of chimpanzees that may help identify the social learning processes that underlie their behavioral traditions in the wild, unfortunately, there are no such studies of Japanese macaques. In this latter case, therefore, we have to rely on studies of related monkey species.

Before proceeding, we would like to deal briefly with two types of "prepared learning" in social contexts. The first is predator fear. It has been found that rhesus monkeys who observe conspecifics showing intense fear of snakes also develop this fear. The process is best thought of as "observational conditioning" in which individuals are automatically "infected" with the fear of others, and this becomes attached to whatever they are observing, with some objects of observation (e.g., snakes) being much more easily conditioned in this way (Mineka & Cook, 1993). The process would thus seem to be similar to process by means of which European blackbirds acquire some of their fears of particular predators (Curio, 1988). (Conversely, Menzel, Davenport, & Rogers, 1972, found that chimpanzees could overcome wariness of objects if housed with conspecifics who were attracted to them.) Processes of emotional relatedness to objects via such processes of contagion and observational conditioning are clearly important processes in the lives of many animals, but since they do not seem to involve complex cognitive processes, they are not our concern here (see Table 9.1).

Although detailed information is lacking, a second domain of prepared learning would seem to be food choice. That this is indeed a domain of prepared learning in some species is demonstrated by Galef's (1988, 1996) research with Norway rats involving the smell of particular foods on the breath of conspecifics or other forms of prepared learning. Laboratory studies with primates, however, have produced mixed results. For example, Fairbanks (1975) reported that pigtail macaques and spider monkeys did not learn to avoid distasteful foods on the basis of social cues. Similarly, Cambefort (1981) did not find any evidence of social learning of food choice in chacma baboons and vervet monkeys, whereas Jouventin, Pasteur, and Cambefort (1976) found some evidence supporting social learning of food choice by mandrills. Visalberghi and Fragaszy (1995) found more eating of novel foods by individual capuchin monkeys when feeding in a social situation, but there was no indication that processes of social learning were involved (their preferred explanation is a generalized increase in behavioral frequency in social situations, so-called social facilitation; see also Hauser, 1996). In the most experimentally rigorous study to date, however, Hikami, Hasegawa, and Matsuzawa (1990) found that Japanese macaque infants were directly influenced by the food choices of their mothers (and, to some degree, vice versa). Visalberghi (1994) provides a useful review of this literature, concluding that social facilitation and stimulus enhancement do work for many primate species in their attraction to certain foods, but that for most species the avoidance of specific foods is not learned socially but individually. Because the only social learning mechanism proposed in the domain of primate food choice is some kind of attraction to food that others are eating (stimulus enhancement), this is not a domain of central importance to our concern with social cognition (see Table 9.1).

Table 9.1 Social learning of attraction and avoidance to food and other objects.

Authors	Species	Task	Source	Explanation*
Harlow & Yudin, 1933	*Macaca mulatta*	Food intake	Experimental	SF
Yerkes, 1934b	*Pan troglodytes*	Food choice	Experimental	SE
Crawford & Spence, 1939	*Pan troglodytes*	Discrimination	Experimental	T&E
Darby & Riopelle, 1959	*Macaca mulatta*	Discrimination	Experimental	T&E
Miller et al., 1959	*Macaca mulatta*	Avoidance	Experimental	SE
Presley & Riopelle, 1959	*Macaca mulatta*	Avoidance learning	Experimental	T&E
Riopelle, 1960	*Macaca mulatta*	Discrimination	Experimental	T&E
Stamm, 1961	*Macaca mulatta*	Bar pressing	Experimental	SF
Hall, 1963	*Papio ursinus*	Food choice	Anecdotal	SE
Mahan & Rumbaugh, 1963	*Saimiri sciureus*	Discrimination	Experimental	T&E
Hall & Goswell, 1964	*Erythrocebus patas*	Food choice	Anecdotal	SE
Myers, 1970	*Macaca arctoides*	Discrimination		
	Macaca mulatta	Discrimination	Experimental	T&E
Feldman & Klopfer, 1972	*Lemur fulvus*	Discrimination	Experimental	T&E
Fairbanks, 1975	*Ateles geoffroyii*			
	Macaca nemestrina	Taste avoidance	Experimental	T&E
Jouventin et al., 1976	*Mandrillus sphinx*	Taste avoidance	Experimental	SE
Strayer, 1976	*Macaca nemestrina*	Discrimination	Experimental	SE
Cambefort, 1981	*Papio ursinus*			
	Cercopithecus aethiops	Taste avoidance	Experimental	T&E
Cook et al., 1985	*Macaca mulatta*	Avoidance	Experimental	SE
Lepoivre & Pallaud, 1986	*Papio papio*	Discrimination	Experimental	T&E
Mineka & Cook, 1988	*Macaca mulatta*	Predator avoidance	Experimental	SE
Hikami et al., 1990	*Macaca fuscata*	Taste aversion	Experimental	SE
Mineka & Cook, 1993	*Macaca mulatta*	Avoidance learning	Experimental	SE
Visalberghi & Fragaszy, 1995	*Cebus apella*	Food choice	Experimental	SE

* T&E = individual trial-and-error learning; SF = social facilitation; SE = stimulus enhancement.

9.2.1 Prosimians

As usual, we know little about the social learning of prosimians. There are three experimental studies. Feldman and Klopfer (1972) found that juvenile brown lemurs (*Lemur fulvus*) learned a stimulus discrimination task more rapidly if they observed their mothers performing it first. Welker (1976) found a similar gain in learning as small-eared bushbabies (*Otolemur garnettii*) watched conspecifics attempting to capture fish. Neither of these studies was designed to rule out stimulus enhancement as a possible explanation, and it is thus possible that the observers were simply more attracted to the relevant stimuli than those who had not observed.

Finally, Watson, Schiff, and Ward (1994) attempted to determine more precisely the nature of the learning processes involved in the observational learning of small-eared bushbabies by experimentally manipulating opportunities for their observation of the fishing behavior of conspecifics. Subjects were presented with a conspecific either fishing in a fishbowl or interacting with the fishbowl but not actually fishing in it. There was also a control condition in which subjects were exposed to the fishbowls but with no conspecific present. Results were that when later given the opportunity, the bushbabies fished more if they had seen a conspecific in the vicinity of the fish bowl than if they had not. Whether or not the conspecific actually fished was unimportant. The investigators interpreted their results in terms of stimulus enhancement effects (and the reduction of neophobia) in animals that watch a conspecific interacting with novel objects.

9.2.2 Monkeys

Researchers in the early part of the century reported both positive and negative results from studies of imitation in various monkey species (Dewsbury, 1984). Recent theoretical and empirical work suggests that the reason for this inconsistency is that a distinction was not made between stimulus enhancement and imitative learning. If this distinction is made, we may see that classic findings such as those of Warden and Jackson (1935), who found that a macaque seeing another pull a chain to reveal food would do the same, are most likely cases of stimulus enhancement, the observer had its attention drawn to the chain and learned to pull it on its own. To interpret such studies, what we need to know are what a monkey who saw another simply touch the chain or who saw the food being revealed with no demonstrator monkey present, or who saw a different technique for making the food accessible would do. On the basis of these kinds of questions, Whiten and Ham (1992) concluded that in all of the early studies of monkey imitation, stimulus enhancement cannot be ruled out as an explanation.

This explanation gains in plausibility when we examine more recent studies in which investigators have sought to distinguish different types of social learning. Beck (1972, 1973b) reported that hamadryas and Guinea baboons did not learn to use a tool to rake in food by watching a demonstrator (see also Chamove, 1974, for a similar negative finding for rhesus macaques). Beck (1976) did report one instance of an individual pigtail macaque performing a tool-use strategy soon after a cagemate, but with only one instance it is possible that they both invented the same strategy independently. Adams-Curtis (1987) reported that a number of capuchin monkeys (*Cebus apella*) who observed a cagemate solving a mechanical puzzle and were motivated to do so themselves nevertheless failed to copy the successful strategy. Westergaard (1988) reported the relatively rapid spread of a tool-use behavior in a captive group of liontail macaques, but again it is likely that simple stimulus enhancement was at work.

The most systematic program of research on the social learning of monkeys is that of Visalberghi and Fragaszy (reviewed in 1990a), who have observed capuchin monkeys (*Cebus apella*) in several tool-use tasks. Although there are a few isolated instances in which one animal has performed a behavior similar to that of another, the majority of Visalberghi and Fragaszy's observations support the view that capuchin monkeys acquire their tool use skills through trial-and-error learning and stimulus enhancement. For example, Visalberghi (1987) gave a task of nut cracking to two groups of capuchin monkeys (see figure 9.5). In each group one individual learned how to perform the task relatively rapidly and did so repeatedly (dozens of times) while others watched and eagerly sought the nuts after they were opened. Still, despite their motivation and the easy availability of numerous tools and nuts, no other individuals learned the task. Similarly, Fragaszy and Visalberghi (1989) had a skillful capuchin tool user poke a stick into a clear plastic tube to extract food while its captive groupmates watched, but found no

Figure 9.5. Capuchin monkey cracking nuts. (Photo courtesy of E. Visalberghi.)

evidence that the observers copied the techniques of the successful tool user or benefited from their observations in any way other than being attracted to the tool and the food. Adams-Curtis and Fragaszy (1995) also found no evidence of imitative learning as capuchins observed a conspecific performing a sequence of actions on a mechanical puzzle leading to a food reward.

One criticism of some of these studies is that they require monkeys to use tools or to perform other object manipulations that are difficult for them (although it should be noted that in the Visalberghi and Fragaszy studies, the tool-use behaviors demonstrated were all behaviors that at least some individuals had been observed to perform spontaneously). In an attempt to meet this criticism, Ham (1990; cited in Whiten and Ham, 1992) gave stumptail macaques a food acquisition task in which subjects observed one of two techniques for causing an apparatus to dispense a food reward. Although subjects who observed both techniques were similarly attracted to the apparatus via local enhancement, in their subsequent independent attempts they did not use the particular technique they had observed. It should also be noted that Hervé and Deputte (1993) found that a capuchin monkey (*Cebus apella*) raised mostly by humans from 3 months of age was subject to a stimulus enhancement effect when humans highlighted objects for him (i.e., he was attracted to them and subsequently manipulated them), but that he did not imitate any of their specific behaviors on the objects.

9.2.3 Apes

Apes have a reputation for being better imitators than monkeys. In some ways this reputation is deserved, but in other ways it is not. The issue is that all of the strong evidence for imitative skills comes from apes who have been raised and trained by humans in all kinds of instrumental and social skills. These studies likely do reveal ape capacities for imitative learning, but such capacities have not been found in apes who have not had extensive contact with humans, and monkeys have never been raised and trained in appropriate ways to see if they too could develop such skills in a humanlike cultural environment. Of the 13 anecdotal observations that Whiten and Ham (1992) report as possible instances of ape imitation, 11 are imitations of human behavior by human-raised apes (usually with human artifacts). And, indeed, the two studies that have compared apes with different amounts of human experience (which are reviewed below), have found dramatic differences (Hayes & Hayes, 1952; Tomasello, Savage-Rumbaugh & Kruger, 1993). We therefore review the evidence for ape imitation in two subsections, keeping separate the two types of ape.

There are five experimental studies of the social learning of instrumental activities by apes who have not grown up mainly in a human environment (i.e., excluding older studies in which different types of social learning were not distinguished, e.g., Haggerty, 1913; Wright, 1972; Yerkes, 1916). Tomasello, Davis-Dasilva, Camak and Bard (1987) trained an adult chimpanzee demonstrator to rake food items into her cage with a metal T-bar. When a food item was in the center of the serving platform, she learned to simply sweep or rake the item to within reach. When the food was located either along the side or against the back of the platform's raised edges, however, she had to employ more complex two-step procedures to be successful. Young chimpanzees were exposed to this adult demonstrator as she employed all three of her strategies (the experimental group). Several other chimpanzees in this same age range were exposed to the demonstrator in an unoccupied state throughout (the control group). Results showed that experimental subjects learned to use the tool (after only a few trials in most cases), while control subjects mostly did not. This finding is consistent with the hypothesis that, in the wild, chimpanzees acquire their tool use skills via some form of social learning.

Two additional pieces of information help to specify the type of social learning involved. First, chimpanzees in the control group manipulated the tool as often as chimpanzees in the experimental group. This would seem to rule out an explanation in terms of simple stimulus enhancement: both groups were attracted to the tool. Second, experimental subjects employed a wide variety of different raking-in procedures, and none of them learned either of the demonstrator's more complex, two-step procedures (even though they were trying and failing on these trials most of the time). This would seem to rule out full-fledged imitative learning in which the observer attempted to reproduce the demonstrator's behavior or behavioral strategies. Our hypothesis was that what the experimental subjects learned from the demonstrator was the affordances of the tool vis-a-vis the food, that is, some causal relations between movements of the tool and movements of the food, which they then exploited in devising their own behavioral strategies (see also Paquette, 1992a). We called this kind of social learning emulation learning because the observers were attempting to reproduce the change of state in the world that the demonstrator had produced, but were doing so with their own idiosyncratic methods. Emulation learning goes beyond stimulus enhancement in stipulating that the learner gains more from its observations than a simple attraction; it actually learns something about the changes of state of the world that are possible. It does not learn about the behavior or behavioral strategies of others, however.

In a second experimental study, Nagell, Olguin, and Tomasello (1993) attempted to investigate the emulation hypothesis more directly. Juvenile chimpanzees and 2-year-old

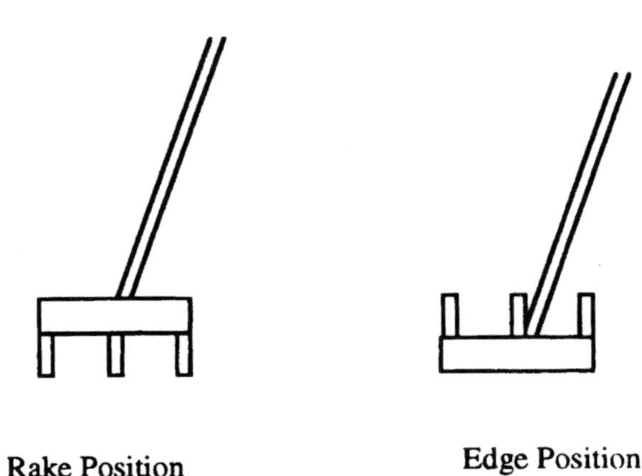

Figure 9.6. Two positions of the rake used in the Nagell et al. (1993) and Call and Tomasello (1994b) studies. (Reprinted from *Journal of Comparative Psychology* with permission from the American Psychological Association.)

human children were presented with a rakelike tool and a desirable but out-of-reach object. The tool was such that it could be used in either of two ways leading to the same end result (see figure 9.6). For each species, one group of subjects observed a human demonstrator employ one method of tool use (less efficient) and another group of subjects observed the other method of tool use (more efficient). The point of this design was that the stimulus enhancement of the tool was the same in both experimental conditions; only the precise methods of use were different. What we found was that, whereas human children in general copied the method of the demonstrator in each of the two observation conditions (imitative learning), chimpanzees used the same method or methods no matter which demonstration they observed. This was despite generally equal levels of tool use and tool success for the two species (again ruling out stimulus enhancement). It should be noted that the majority of children insisted on this reproduction of adult behavior even in the case of the less efficient method, leading to less successful performance than the chimpanzees (who ignored the demonstration) in this condition. From this pattern of results, we concluded once again, that chimpanzees in this task were paying attention to the general causal relations between tool and food, but they were not attending to the actual methods of tool use demonstrated; they were engaged in emulation learning.

Call and Tomasello (1994b) gave this same task to orangutans. Again, some subjects observed a human demonstrator use the tool the more efficient way, while other subjects observed the demonstrator use the tool in the less efficient way. It was found once again that subjects behaved identically in the two experimental conditions, showing no effect of the type of demonstration observed. Moreover, analysis of individual learning curves suggested that a large component of individual trial-and-error learning was at work, and this was also true of the data from the chimpanzees of Nagell et al. (1993) when they were reanalyzed in this way. One novelty of this study was that a follow-up was performed with an additional pair of subjects using an orangutan demonstrator familiar to the subject (i.e., its mother). The same pattern of results was found, leading again to the hypothesis that the orangutans acquired their tool use skills in this task via emulation learning.

One possible criticism of these three studies is that the tool-use task may have been too difficult for the apes to learn via observation. Although there is no evidence for this in these studies—and, in fact, the majority of subjects in all experimental conditions gained some success with the tool during the course of the experiment—Call and

Figure 9.7. Chimpanzee manipulating the "artificial fruit" used by Whiten et al. (1996) (Photo courtesy of A. Whiten.)

Tomasello (1995) nevertheless presented juvenile and adult orangutans with tasks involving simple behavioral strategies. The trick was that the causal relations of the apparatus—the way the tool contacted and made the food move—were hidden from the subject, thus making emulation learning impossible. What the subject saw was a human demonstrator manipulate a stick protruding from a box in one of four ways and then receive a reward. When subjects were then given their turn to manipulate the stick (in several different sets of randomized and blocked trials), there were no signs that they imitatively learned any of the actions demonstrated to them. They basically performed the actions randomly and had only a chance rate of success. Providing subjects with an orangutan demonstrator did not change this result. A majority of human 3- and 4-year-old children did learn to imitate the requisite actions and thus performed at above-chance levels. Again, then, this study provides further evidence for the emulation hypothesis since removing the possibility of emulation learning makes it difficult for apes to effectively learn from the behavior of others anything useful about a tool-use task.

Finally, Whiten, Custance, Gómez, Teixidor, and Bard (1996) presented chimpanzees with a transparent "foraging box" containing fruit (see figure 9.7). On any given trial, the box could be opened by one of two mechanisms, each of which could be operated in two ways. For each mechanism, a human experimenter demonstrated one way of opening the box to some subjects and the other way to other subjects (with the other mechanism being blocked). Subjects were then given the chance to open the box themselves. Results were that for one mechanism, there was no effect of the demonstration observed. For the other mechanism, there was some evidence that chimpanzees were more likely to use the manner of opening demonstrated by experimenters. However, in the analysis of Tomasello (1996a), they could easily have learned this via emulation learning: one group of chimpanzees saw that the stick afforded pushing through the clasp, while the other saw that it afforded twisting and pulling. Byrne and Tomasello (1995) also point out that the behaviors used in this task were probably not new for the subjects and thus there might also have been some process of "response facilitation" (a known behavior is made more likely to occur through observation) at work. It is also notable that human 3-year-old children given this same task reproduced the demonstrator's technique much more often and much more faithfully than the chimpanzees did.

9.2.4 Human-Raised Apes

As mentioned above, apes enjoy a reputation as good imitators. This is based mainly on one classic study involving the famous chimpanzee Viki. Hayes and Hayes (1952) intensively trained Viki to mimic various body movements and gestures for a period of more than 17 months. The training consisted of a human performing a behavior (e.g., blinking the eyes or clapping the hands; see below) and then using various shaping and molding techniques, with rewards, to get Viki to "do this." After she had become skillful at this game, the Hayes exposed Viki to six problem-solving tasks that she would be unlikely to solve for herself. She was given a 2-minute baseline period with each task on her own, and then a human demonstration (up to 3) was given. Viki performed one of the target behaviors during baseline, and had trouble with two others, but for three problems she solved them only after demonstrations. The difficulty is that there were no control conditions, that is, no tasks given after the baseline period without demonstration to see what Viki would do on her own with repeated opportunities. Drawing firm conclusions about the precise social learning processes involved is therefore risky. Nevertheless, it is important to note that another chimpanzee tested by Hayes and Hayes, Franz, who had not been raised or trained in the special ways that Viki had, showed no signs of any type of social learning in this set of problem-solving tasks.

Tomasello, Savage-Rumbaugh, and Kruger (1993) conducted a more systematic study of the skills of chimpanzees and bonobos to reproduce modeled actions. Subjects were three mother-reared captive apes (two bonobos and one chimpanzee), three enculturated apes (two bonobos and one chimpanzee raised like human children and exposed to a languagelike system of communication), and 2-year-old human children. Each subject was shown 24 different and novel actions on objects and encouraged to reproduce them: children were told to "do this," and the apes were pretrained to reproduce modeled actions (as well as having been informally encouraged to imitate earlier in their lives with some regularity). Each subject's behavior on each trial was scored as to whether it successfully reproduced (1) the end result of the demonstrated action, and/or (2) the behavioral means used by the demonstrator. The major result was that the mother-reared apes reproduced both the end and means of the novel actions (i.e., imitatively learned them)

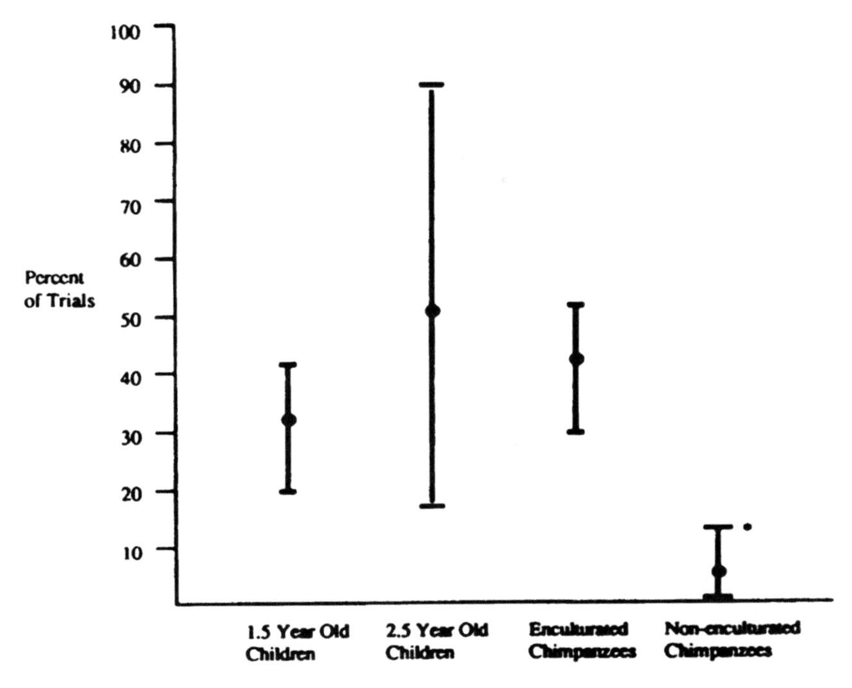

Figure 9.8. Tomasello et al. (1993) results for the imitation of complex actions on objects. (* = Lower than other three, $p < .05$.) (Reprinted from *Child Development*.)

hardly at all. In contrast, the enculturated apes and the human children imitatively learned the novel actions much more frequently, and they did not differ from one another in this learning (see figure 9.8). It is still important to note, however, that it is at least theoretically possible that the apes in this experiment were on some trials emulating the results experimenters produced and on other trials mimicking the body movements of the experimenter as they had learned to do in the past (see Heyes, 1993b).

A different type of evidence for imitative learning in human-raised apes was provided by Russon and Galdikas (1993, 1995). The subjects of their study were rehabilitant orangutans who had been raised in various ways by humans (mostly as pets) and then introduced to a human camp on the edge of a forest where they were free to come and go as they pleased. What Russon and Galdikas observed was a number of behaviors that were so unlike the typical behaviors of orangutans and so unlikely to have been learned by individuals on their own that imitative learning was inferred. These behaviors almost all involved the manipulation of human artifacts in humanlike ways—such things as "brushing the teeth," "applying insect repellent," "using a knife," and so forth. Although suggestive, these observations are difficult to interpret because there was no documentation of precisely what the orangutans observed of these behaviors by humans, there was no documentation of what orangutans could learn from exploring the objects by themselves, and there was no information available on the nature of their earlier experiences with humans (and this is important since there were only a few orangutans who produced these intriguing behaviors with regularity). These observations thus beg for replication under conditions that allow researchers to determine precisely what things these animals might learn from humans and how they might learn them.

Finally, Call and Tomasello (1995: study 3) gave the human-raised orangutan Chantek a number of arbitrary body movements to mimic (along with the sign to "do this"), which he did quite readily (thus replicating the findings of Miles, Mitchell & Harper, 1996). The interesting twist is that he was now given this same request ("do this") in the context of the problem-solving apparatus used by Call and Tomasello (1995), which in the original study he had basically failed to cope with successfully even after a human demonstration. Unexpectedly, the command to "do this" given before the demonstration of an action on the apparatus did not lead Chantek to above-chance performance. Apparently, Chantek had caught on to the mimicking game in the context of arbitrary body movements, but he did not apply that knowledge to a problem-solving context in which he was seriously trying to obtain a reward. One interpretation is that mimicking behaviors on the sensory–motor level is something that apes can be trained to do, but that understanding what another is doing in instrumental, problem-solving situations in a way that is relevant for one's own problem-solving attempts requires an understanding of the intentions of others, which apes may not be able to do without certain specific types of experience and training from humans, or at all.

9.2.5 Comparison of Species

Experimental studies of primate social learning support the hypothesis that population-specific behavioral traditions of primates in the wild that are concerned with instrumental activities are maintained by means of individual learning in common environments, local enhancement, and emulation learning, not by means of imitative learning. In the case of monkeys, it is clear that they can learn a number of adaptive behaviors individually and that their individual learning is facilitated by processes of local and stimulus enhancement. It is possible that on some occasions they may learn via emulation learning something of the structure of a task in which a conspecific is engaged. In the case of some apes, their especially intelligent and creative use of tools and their ability to understand changes in the environment brought about by the tool use of others may lead to especially powerful manifestations of emulation learning. Human-raised apes are much more skillful at imitative learning per se. This may be because they have learned more about objects and tools, it may be because they have been trained to mimic human actions, or it may be because they have learned more about the intentional behavior of others. Table 9.2 provides a catalogue of the studies reviewed in this section to document this point of view.

It is not likely that either monkeys or apes who have not had extensive human contact imitatively learn the instrumental behaviors of others. What are salient for monkeys and apes in such situations are the tool, the food, and their spatial-temporal-causal relations. When they see these changing in certain ways as a result of the manipulations of others, they understand the significance of that for their own actions. This is intelligent problem-solving behavior, but it is not attempting to reproduce the behavior or behavioral strategies of another animate being. It is also unlikely that other nonhuman species such as rats (Heyes & Dawson, 1990) engage in truly imitative learning in which they copy the behavior or behavioral strategies of conspecifics (see Byrne & Tomasello,

Table 9.2 Social learning of instrumental activities.

Author(s)	Species	Task	Source	Explanation*
Thorndike, 1898, 1901	*Cebus*	Puzzle box	Experimental	T&E
Hobhouse, 1901				SE
Kinnaman, 1902	*Macaca mulatta*	Puzzle box	Experimental	SE
Watson, 1908	*Cebus*	Tool use &		
	Macaca mulatta	puzzle box		
	Papio cynocephalus		Experimental	T&E
Haggerty, 1909	*Cebus*			
	Macaca mulatta			
	Macaca fascicularis	Puzzle box	Experimental	SE
Witmer, 1909	*Pan troglodytes*	Object manipulation	Experimental	IMIT?
Shepherd, 1910	*Macaca mulatta*	Tool use	Experimental	T&E
Witmer, 1910	*Macaca fascicularis*	Puzzle box	Experimental	
Haggerty, 1913	*Pongo pygmaeus*			
	Pan troglodytes	Tool use	Experimental	EMUL
Shepherd, 1915	*Pan troglodytes*	Object manipulation	Experimental	SE
Kempf, 1916	*Macaca mulatta*	Problem solving	Experimental	T&E
Yerkes, 1916	*Macaca fascicularis*	Object manipulation		
	Macaca mulatta	Object manipulation		
	Pongo pygmaeus	Tool use	Experimental	SE
Shepherd, 1923	*Pan troglodytes*			
	Pongo pygmaeus	Tool use	Experimental	T&E
Kohler, 1925	*Pan troglodytes*	Tool use	Experimental	T&E
Yerkes, 1927	*Gorilla gorilla*	Tool use	Experimental	T&E
Aronowitsch & Chotin, 1929	*Macaca*	Problem solving	Experimental	?
Kellogg & Kellogg, 1933	*Pan troglodytes*	Object manipulation	Observational	T&E
Warden & Jackson, 1935	*Macaca mulatta*	Puzzle box	Experimental	SE
Warden et al., 1940a	*Cebus apella*			
	Macaca mulatta	Puzzle box	Experimental	SE
Hayes & Hayes, 1952	*Pan troglodytes*	Problem solving		
		Object manipulation	Experimental	IMIT
Knobloch & Passamanick, 1959	*Gorilla gorilla*	Object manipulation	Experimental	T&E
Wechkin, 1970	*Macaca mulatta*	Object manipulation	Experimental	SE
Beck, 1972	*Papio hamadryas*	Tool use	Experimental	T&E
Menzel, 1972, 1973b	*Pan troglodytes*	Tool use	Anecdotal	SE/EMUL
Menzel et al., 1972	*Pan troglodytes*	Object manipulation	Experimental	SE
Wright, 1972	*Pongo pygmaeus*	Tool use	Experimental	SE/EMUL
van Lawick-Goodall, 1973	*Pan troglodytes*	Tool use	Observational	SE/EMUL
Beck, 1973b	*Papio papio*	Tool use	Experimental	T&E
Beck, 1973a	*Papio hamadryas*	Tool use	Experimental	SE
Beck, 1976	*Macaca nemestrina*	Tool use	Experimental	SE
Gardner & Gardner, 1978	*Pan troglodytes*	Object manipulation	Anecdotal	T&E
Galdikas, 1982	*Pongo pygmaeus*	Object manipulation & problem solving	Anecdotal	SE/EMUL
Mathieu, 1982	*Pan troglodytes*	Object manipulation	Observational	T&E
Anderson, 1985	*Macaca tonkeana*	Tool use	Experimental	SE/EMUL
Mignault, 1985	*Pan troglodytes*	Object manipulation	Experimental	T&E
Sumita et al., 1985	*Pan troglodytes*	Tool use	Experimental	EMUL
Antinucci & Visalberghi, 1986	*Cebus apella*	Tool use	Experimental	SE
Adams-Curtis, 1987	*Cebus apella*	Puzzle box	Experimental	SE
Tomasello et al., 1987	*Pan troglodytes*	Tool use	Experimental	EMUL

(*continued*)

Table 9.2 *(continued)*

Author(s)	Species	Task	Source	Explanation*
Visalberghi, 1987	*Cebus apella*	Tool use	Experimental	SE
Westergaard & Fragaszy, 1987a	*Cebus apella*	Tool use	Experimental	SE
Westergaard & Lindquist, 1987	*Macaca silenus*	Object manipulation	Experimental	SE
Hauser, 1988b	*Cercopithecus aethiops*	Food processing	Observational	T&E/SE
Perinat & Dalmau, 1988	*Gorilla gorilla*	Object manipulation	Anecdotal	T&E
Chevalier-Skolnikoff, 1989	*Ateles geoffroyi* *Ateles belzebuth* *Ateles fusciceps* *Cebus apella* *Cebus albifrons*	Object manipulation	Observational	T&E
Fragaszy & Visalberghi, 1989	*Cebus apella*	Tool use	Experimental	SE
Gómez, 1989	*Gorilla gorilla*	Object manipulation	Anecdotal	T&E
Visalberghi & Trinca, 1989	*Cebus apella*	Tool use	Experimental	SE
Visalberghi & Fragaszy, 1990b	*Cebus apella*	Tool use	Experimental	SE
Gibson, 1990	*Cebus*	Object manipulation & tool use	Anecdotal	SE
Paquette, 1992a	*Pan troglodytes*	Tool use	Observational	EMUL
Byrne & Byrne, 1993	*Gorilla gorilla*	Food processing	Observational	T&E
Hervé & Deputte, 1993	*Cebus apella*	Object manipulation	Experimental	SE
Nagell et al., 1993	*Pan troglodytes*	Tool use	Experimental	EMUL
Russon & Galdikas, 1993	*Pongo pygmaeus*	Object manipulation	Anecdotal	EMUL
Tomasello et al., 1993	*Pan troglodytes* *Pan paniscus*	Object manipulation	Experimental	IMIT
Call & Tomasello, 1994b	*Pongo pygmaeus*	Tool use	Experimental	EMUL
Watson et al., 1994	*Otolemur garnettii*	Problem solving	Experimental	SE
Call & Tomasello, 1995	*Pongo pygmaeus*	Puzzle box	Experimental	T&E
Russon & Galdikas, 1995	*Pongo pygmaeus*	Object manipulation	Anecdotal	EMUL
Whiten et al., 1996	*Pan troglodytes*	Puzzle box	Experimental	EMUL/IMIT?

* T&E = individual trial-and-error learning; SE = stimulus enhancement; EMUL = emulation learning; IMIT = imitation or imitative learning; ? = some difficulty of interpretation.

1995, which also includes a brief discussion of some well-known examples from birds, and Terkel, 1996, which documents the social learning of rats in one context in some detail). Why human-raised apes have more complex skills of socially learning is at this point a matter for speculation (see Call & Tomasello, 1996, and Tomasello, 1996a, for discussion). Human children begin to show strong skills of imitative learning soon after their first birthdays (Meltzoff, 1988).

9.3 Social Learning of Communicative Signals and Gestures

An important complement to these studies of learning in instrumental contexts are studies of the social learning of communicative signals and other gestures. Such studies are important because in communicative interactions there is less opportunity for an individual to learn from the physical environment in nonsocial ways (e.g., the affordances of a tool). The learning of communicative signals always takes place in social interaction with others, by means of either ontogenetic ritualization or some form of social learning. Only the latter involves one individual actually reproducing the behavior of another. And it is important in our examination of how primate communicative signals are learned that we remain aware of the two aspects of a communicative signal—its physical form (vocal or gestural) and its use in an appropriate communicative context—since in some cases it is only the use of a signal that may be learned or, possibly, socially learned. As noted in chapter 8, almost all of the relevant research for monkeys concerns vocalizations and for apes concerns gestures.

9.3.1 Monkeys

The evidence for monkey social learning in the domain of vocal communication is mixed. Some species show acoustically distinct "dialects" in the wild (Japanese macaques, Green, 1975; red-bellied tamarins, *Saguinus labiatus:* Maeda & Masataka, 1987), and even vocal "signatures" along matrilines within some monkey groups (pigtail macaques; Gouzoules and Gouzoules, 1990). There are clear developmental changes in how some pigtail macaque calls are produced across age (Gouzoules and Gouzoules, 1989a,b, 1995), but it is not clear whether the mechanism of this change is the maturation of the vocal apparatus or if indeed there is some social learning of vocal structure over age. Although infant vervet monkeys already produce some calls in adultlike ways at birth, other calls develop more slowly during ontogeny, and it is not until they are 6 months old that young vervets comprehend the various alarm calls of their species and respond in adultlike ways (Seyfarth & Cheney, 1986). Hauser (1988a, 1989) found evidence that vervet monkeys who had greater experience with two different vocalizations (the alarm calls of starlings and the "wrr" call of vervets, respectively) learned to comprehend and produce them, respectively, at earlier ages than those who had less experience. Elowson, Snowdon, and Sweet (1992) also found developmental patterns in the use of trill vocalizations by pygmy marmosets (*Cebuella pygmaea*).

The problem with these findings, as often in cases naturalistic observations, is that it is difficult to compare competing hypotheses precisely. Interestingly, a number of experimental studies have been conducted in which individuals are raised in different environments and their communicative skills subsequently assessed. The most studied species is the squirrel monkey (*Saimiri sciureus*). At least three different studies have found that individuals reared in isolation (or even deafened at birth) nevertheless show species-typical vocalizations throughout ontogeny (see Snowdon & Elowson, 1992, for a review). A similar study of rhesus monkeys (Newman & Symmes, 1974) found only minor disruptions in vocal behavior, and Larson, Sutton, Taylor and Lindeman (1973) used operant conditioning techniques to train rhesus macaques to increase the amplitude and duration of a variety of calls with some limited success.

More recently, a number of cross-fostering experiments have been conducted in which infants of one species were raised from birth by adults of another species. For example, Masataka and Fujita (1989) found some differences in the food calls of two species (one Japanese macaque and two rhesus macaque individuals) as a result of cross-fostering; the differences were mainly in the fundamental frequencies of the

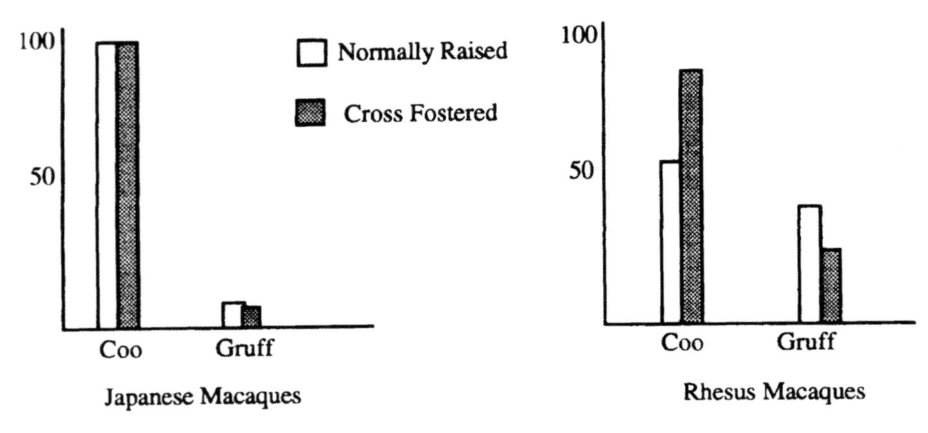

Figure 9.9. Owren's results showing limited modification in the calls of one macaque species and no modifications in those of another after cross-fostering. (Based on Owren et al., 1993)

calls. Using several methodological improvements, Owren and Dieter (1989) failed to replicate this result, however. Owren, Dieter, Seyfarth, and Cheney (1992a, b) found some modest modifications in the food calls of two cross-fostered Japanese macaques (but not two cross-fostered rhesus macaques), but they also found that the acoustic structure of the food calls of the adults of the two species were not discriminable. This suggests that the modifications observed in the Japanese macaques were not copied from adult rhesus but were either learned from rhesus peers or else conditioned by adults in the context of food to become more like those of other immature rhesus. Finally, Owren, Dieter, Seyfarth, and Cheney (1993) also found some limited modifications in the use of nonfood calls by these same cross-fostered subjects. In looking at the use of close calls such as "gruffs," often used by rhesus when they are playing, the investigators report that: "cross-fostered Japanese macaques almost never used gruffs even though they lived in an environment in which gruffs were used at a high rate . . . Similarly, cross-fostered rhesus macaques used gruffs during interaction with peers, when calling to a dominant male, and in play, despite not hearing this call from their Japanese macaque peers" (p. 402). The modifications that were found occurred in only a few limited contexts and involved only a subtle shift in some vocalization patterns and the increase in frequency of use of calls already in animals' repertoires (see figure 9.9). In general, this is the conclusion of Seyfarth and Cheney (in press), who make a special point of dissecting the process into the acquisition of skills of vocal production, usage, and response.

In the only study of this type that focused on monkey gestural communication, de Waal and Johanowicz (1993) investigated the reconciliation gestures of juvenile rhesus and stumptail macaques who were cohoused for a period of 5 months. They found that the frequency of postconflict reconciliation in rhesus monkeys increased when they were raised with stumptail macaques (who reconcile with one another more frequently than rhesus under natural circumstances), even when they were returned to their own group. The nature of the gestures and displays for reconciliation remained unchanged for both species, however. The investigators thus argued that some form of social learning may have been at work in encouraging the rhesus to reconcile more often, but there was not imitation of the means of reconciliation. This same type of explanation is also most plausible in explaining the common affiliative patterns of rhesus macaque mothers and infants (de Waal, 1996) and is basically in accordance with the classic findings of Mason (1963), who observed rhesus individuals raised in isolation from conspecifics:

> Rhesus monkeys display a number of stereotyped postures, gestures, and vocalizations . . . which serve a communicative function in day-to-day social interactions. Although the basic form of such responses and their general relations to arousal level or affective state are almost certainly unlearned . . . , their effectiveness in controlling and coordinating social activity is probably heavily dependent upon experience. (p. 171)

9.3.2 Apes

Ape vocalizations have not been studied as intensively as monkey vocalizations. Although there is some evidence for "dialects" among different groups of wild chimpanzees (Mitani et al., 1992), their nature and significance is currently under debate (Mitani, 1994). Goodall (1986) reported her impression that chimpanzee vocalizations are closely tied to mood and not under voluntary control (see also Hayes, 1951; section 8.1), and Menzel (1964) reported that isolation-reared chimpanzees produced many of the same types of vocalizations and gestures as their more naturally reared conspecifics. The most convincing piece of evidence for some modifiability in ape vocalizations comes from Hopkins and Savage-Rumbaugh (1991), who found that a human-raised bonobo used not only his species-typical vocalizations, but also some vocalizations that may have derived from listening to human speech from an early age; although without a thorough developmental profile of naturally occurring bonobo vocalizations, the connection with human vocalizations must remain speculative. One suggestive further finding was that another adult bonobo raised by humans only from the juvenile period on did not show these same novel vocalizations.

Much of the evidence for the social learning of communicative activities by apes comes from the studies of Tomasello and colleagues on the gestural communication of a colony of chimpanzees living in a seminatural group in a rich captive environment. Tomasello et al. (1985) observed the infants and juveniles of this group (1–4 years old), with special emphasis on how they used their signals (see figure 9.10). Looking only for intentional signals accompanied by gaze alternation or response waiting (both indicating the expectation of a response; see section 8.2), they found a number of striking developmental patterns. For example, they found that juveniles used many gestures not used by adults, that adults used some gestures not used by juveniles, and that some juvenile gestures for particular functions were replaced by more adultlike forms at later developmental periods. None of these patterns was consistent with the idea that infant and juvenile chimpanzees acquire their gestural signals by imitating adults.

In a longitudinal follow-up to these observations, Tomasello et al. (1989) observed the same juvenile chimpanzees in the same group setting 4 years later (at 5–9 years old), this time with a more direct focus on learning processes. They sought various forms of evidence that the youngsters acquired their gestures by means of imitative learning or ontogenetic ritualization. In ontogenetic ritualization, a communicatory signal is created by two organisms shaping each others' behavior in repeated instances of

Figure 9.10. Chimpanzee using an arm-raise gesture. (Photo by J. Call.)

a social interaction. For example, a juvenile may initiate play with a peer without a signal by simply jumping on another to wrestle, slapping him on the head as she jumps on. The recipient may notice that such initiations always begin with the initiator raising her arm in preparation for slapping and so anticipate by responding when only that "intention movement" is given (Tinbergen, 1951). The initiator notices the anticipation of the recipient and understands that the intention movement by itself is sufficient to initiate the play, and so at some future encounter raises its arm *in order to* initiate play. Another example is given by Plooij (1978), who described a sequence in which at around 6 weeks of age, chimpanzee infants begin to play bite everything that comes within the reach of their mouths, including their mothers' hands. In reaction, the mother begins to poke at and tickle the infant. Within a few weeks, the infant learns to grasp and pull at the mother's hand when she has stopped tickling, seemingly in order to induce her to continue. Another example is an infant initiating nursing by going directly for the mother's nipple by grabbing and moving her arm. In some future encounter the mother anticipates the infant's desire at the first touch of her arm, and so become receptive at that point. This then leads the infant to abbreviate its behavior to a simple touch on the arm in order to get the mother to be receptive.

In ontogenetic ritualization, two organisms essentially shape one another's behavior in repeated instances of a social interaction. There is no hint in any of this that one individual is seeking to reproduce the behavior of another (and thus it does not fit the narrow definition of social learning); there is only social interaction and shaping that results eventually in a communicative signal. The general form of this type of learning is:

1. Individual A performs behavior *X*.
2. Individual B reacts consistently with behavior *Y*.
3. Subsequently B anticipates A's performance of *X*, on the basis of its initial step, by performing *Y*.
4. Subsequently, A anticipates B's anticipation and produces the initial step in a ritualized form (waiting for a response) *in order to* elicit *Y*.

The main point is that a behavior that was not at first a communicative signal becomes one by virtue of the anticipations of the interactants over time. It is also worth noting that in the case of attention-getters the third step is not always necessary. For example, if A slaps the ground (for whatever reason) and B looks, A may exploit that involuntary reaction in the future by slapping *in order* to get B to look and so use it as an intentionally communicative signal. As noted in the previous chapter, this is the same basic process by which individuals create social strategies, although sometimes the original behavior may be insightfully related to its final function from the beginning, based on a generalization from other similar situations. In all cases, the learning processes involved rely on individuals' understanding of the behavior of a social partner as animate and directed.

Two observations of Tomasello et al. (1989) suggested that indeed ontogenetic ritualization, and not imitative learning, was responsible for most of the gestural signals. First, there were a number of idiosyncratic signals that were used by only one individual (Goodall, 1986, also reports this for the Kasakela community at Gombe). These signals could not have been learned by imitative processes and so must have been individually invented and ritualized. Second, many youngsters also produced signals that they had never had directed to them, for example, others never begged food, or solicited tickling or nursing from youngsters. The youngsters' signals for these functions thus could not have been a product of their imitation of signals directed to them (second-person imitation). In many cases it was also extremely unlikely that they were imitated from other infants gesturing to conspecifics (third-person imitation or "eavesdropping") because many were only produced in close quarters between mother and child with little opportunity for others to observe (e.g., for nursing).

In the third set of longitudinal observations, Tomasello et al. (1994) observed a completely new generation of youngsters in this same group of chimpanzees, that is, the 1–4 year olds who were not even born (with one exception) at the time of the previous observations. To investigate the question of potential learning processes, Tomasello et al. made systematic comparisons of the concordance rates of all individuals with all other individuals across the three longitudinal time points; that is, they constructed a matrix for each pair of individuals that revealed which gestures they used in common and which gestures were used by only one of the pair. The analysis of all the pairs revealed quite clearly, by both qualitative and quantitative comparisons, that there was much individuality in the use of gestures, especially for nonplay gestures, with much individual variability both within and across generations. As before, there were also a number of idiosyncratic gestures used by single individuals at each time point. Also notable is the fact that there was a fairly large gap between the two generations of this study (4 years), and many of the gestures learned by the youngsters of the younger generation were ones that youngsters of the older generation would have been using infrequently during the crucial learning period. It is also important that the gestures that were shared by many youngsters were also used frequently by captive youngsters raised in peer groups with no opportunity to observe older conspecifics (e.g., Berdecio & Nash, 1981).

Finally, in an experimental investigation, Tomasello et al. (in press) removed an adult female from this same group and taught her two different arbitrary signals by means of which she obtained desired food from a human (and repeated the procedure with one other adult female from another group). When she was then returned to the group and used these same gestures to obtain food from a human in full view of other group members, there was not one instance of another individual reproducing either of the new gestures (nor in the replication). The study has several limitations (e.g., this is a situation in which most chimpanzees had already used other gestures successfully in the past, it was a human recipient, only two individual demonstrators were used), but the fact that its findings are in general agreement with those of the naturalistic observations provides at least some further validity to the longitudinal studies.

It should also be mentioned that in addition to ontogenetic ritualization, in some cases emulation learning may be at work in the domain of primate communication. For example, if an individual already knows how to induce humans to give it food, observation of another individual receiving food might elicit attempts from that individual to reproduce that environmental effect for itself. But, as in all cases of emulation learning, the individual does not learn a new behavior or behavioral strategy in such situations but simply uses a behavioral strategy it already knows to reproduced an observed environmental effect.

9.3.3 Human-Raised Apes

There are two sets of studies relevant to the social learning skills of human-raised apes in the domain of communication. The first concerns the way they acquire humanlike symbols in the various ape "language" projects. The second is really not about communication per se but concerns some studies in which apes have been trained by humans to imitate arbitrary gestures and body movements on command.

The process by which apes learn humanlike communicative symbols has not been systematically investigated. For apes learning a manual sign language, the early reports were that by far the most effective technique was "molding" the hands (Fouts, 1972; Miles, 1990; Patterson, 1978), and that imitation was a poor way for the animals to learn signs (Fouts, 1972; Sanders, 1985). Later reports on the chimpanzee Washoe (after about 2–4 years of learning) were that she could learn new signs through imitative learning (Fouts & Godin, 1974, cited in Fouts, Fouts & van Cantfort, 1989). More recently, Savage-Rumbaugh et al. (1986) reported that the bonobo Kanzi acquired all of his earliest communicative symbols via imitative learning, in this case by manually contacting the lexigrams on a keyboard. Kanzi had not been trained directly on the keyboard because his mother was the object of training and study. Mostly from atop her back, Kanzi had observed the training of his mother (which was mostly unsuccessful), and he had been encouraged to interact with humans and to reproduce their behavior in other contexts. At about 2.5 years of age, Kanzi began to demonstrate that he had learned the appropriate use of many of the lexigrams his mother was being taught, presumably through observational learning of some sort. Although Kanzi's ability to learn lexigrams observationally has never been systematically tested, if indeed that is how he learned them, it goes beyond mere mimicking as he uses them in appropriate and novel communicative contexts.

With regard to mimicking movements, Hayes and Hayes (1952) trained their human-raised chimpanzee Viki to reproduce various body movements and gestures that they made, for example, blinking the eyes or clapping the hands. They trained her throughout her daily life in their home for a period of more than 17 months before systematic testing began. The training consisted of a human performing a behavior and then using various shaping and molding techniques, with rewards, to get Viki to "do this." After she had become skillful, some novel behaviors were systematically introduced. In general, she reproduced them faithfully and quickly; she had clearly "gotten the idea" of the mimicking game. Recently, Custance, Whiten, and Bard (1995) demonstrated in a more rigorous fashion similar abilities in two nursery-reared chimpanzees after they were trained for 3.5 months in a manner similar to Viki. Of the 48 novel actions demonstrated after the training period, one subject correctly reproduced 13 and the other correctly reproduced 20. Whether the subjects could see their own responses or whether the response was "invisible" (e.g., facial expressions) was not a significant factor. There are also some reports that the human-raised orangutan Chantek is able to mimic human actions on command in this same manner (Miles et al., 1996; see also Call & Tomasello, 1995), although the nature and extent of the training involved in this case is not known.

An intriguing recent finding was reported by Myowa (1996). A single chimpanzee, raised from birth by humans, was exposed weekly over a 10-week period to human demonstrations of facial gestures of the type used in studies of human neonatal imitation (Meltzoff & Moore, 1977, 1989). The investigator concluded that this chimpanzee demonstrated the ability to reproduce these facial gestures (tongue protrusion, mouth opening, lip protrusion) because of the pattern of its responses over several weeks early in the study (5–11 weeks of age, with no evidence of reproduction after that age). Analysis of the subject's performance at each week analyzed does not yield such strong results, however, even with the relatively crude scoring method used (a subject was given credit for imitation on a given trial if blind observers deemed that the appropriate model was either the most likely or the second most likely model to have occurred). A replication of the experiment with more subjects, so that each session may be analyzed more carefully with some statistical power, is needed.

9.3.4 Comparison of Species

Most of the communicative skills of primates emerge gradually during ontogeny, as they do in human infants beginning in their second year of life. In the case of monkeys, the evidence is that they do not learn the basic form of their vocal calls or gestural signals, although there may be some room for minor variations based on experience of some as-yet undetermined nature (perhaps of the same nature as the vocal learning of some bird species). Learning seems to play a more important role in monkeys' comprehension and use of vocal and gestural signals in their appropriate communicative contexts, although cross-fostering experiments mostly show the limited nature of this role. There is no evidence in any studies with monkeys that imitative learning is the basis for their learning to comprehend or use communicative signals of any type. The limited modifications that are observed are just as likely due to some form of individual learning in which individuals learn through the responses of others to associate particular signals with particular outcomes.

Apes in their natural habitats, on the other hand, clearly learn both the form and the communicative function of many of their gestural signals, especially those that are used intentionally (the evidence is less clear with respect to vocalizations). Chimpanzee youngsters acquire the majority, if not all, of their intentionally based gestures by individually ritualizing them with one another on the basis of recurrent social interactions. There is no evidence that they learn any of these communicative signals by means of imitation. Some apes who have been raised or trained extensively by humans, however, do seem to acquire many of their communicative signals by means of imitative learning. The amount and nature of human experience that is required before these skills emerge is unknown, nor are there explanations that go beyond speculation (see Tomasello, 1996a). Monkeys simply have not been raised or tested in the manner of human-raised apes, and so the possibility of their acquiring more humanlike skills of imitative learning in the communicative domain are unknown. Table 9.3 summarizes all of the relevant studies.

9.4 Teaching

In human cultures, child skills of social learning are complemented by adult skills of teaching. If teaching is defined broadly to include any behavior of one animal that serves to assist another animal's learning, teaching is not uncommon in the animal kingdom (Caro & Hauser, 1992; Maestripieri, 1995c). But if we restrict our attention to the flexible and insightful forms of instruction in which one individual intends that another acquire a skill or piece of knowledge and adjusts its behavior contingent on the learner's progress in skill or knowledge, few species actively teach conspecifics. In this intentional view, teaching would require some understanding of the intentionality or mental states of conspecifics, because the entire motivation for teaching rests crucially on the perception of a behavioral incompetence or lack of knowledge in another individual.

9.4.1 Monkeys

Maestripieri (1995b) reported that rhesus monkey mothers seem to encourage the independent locomotion of their offspring in a way sensitive to their competence. Early in their infant's life, many rhesus mothers walk away from them backwards, 'lip-smacking' as they go. If an infant gives a distress call, the mother quickly returns, but if not she continues to "encourage" his independent locomotion. It was found in this study that those infants who were more often treated in this way by their mothers early in life were earlier to break and then make contact with them independently (see Maestripieri,

Table 9.3 Social learning of communicative signals and gestures.

Authors	Species	Task[a]	Source	Explanation[b]
Hayes & Hayes, 1952	*Pan troglodytes*	"Do what I do"	Experimental	IMIT
Gardner & Gardner, 1969	*Pan troglodytes*	"Do what I do"	Anecdotal	IMIT
Fouts, 1972	*Pan troglodytes*	ASL	Experimental	T&E
Patterson, 1978	*Gorilla gorilla*	ASL	Anecdotal	T&E
Plooij, 1978	*Pan troglodytes*	Social behavior	Anecdotal	T&E
Gardner & Gardner, 1978	*Pan troglodytes*	ASL	Anecdotal	T&E
Fouts & Budd, 1979	*Pan troglodytes*	ASL	Experimental	IMIT
Miles, 1983	*Pongo pygmaeus*	ASL	Observational	IMIT
Sanders, 1985	*Pan troglodytes*	ASL	Experimental	T&E
Savage-Rumbaugh et al. 1986	*Pan paniscus*	Lexigrams	Anecdotal	?
Tomasello et al., 1989	*Pan troglodytes*	Gestural communication	Observational	EMUL
Miles, 1990	*Pongo pygmaeus*	ASL & "do what I do"	Anecdotal	IMIT
Mitani et al., 1992	*Pan troglodytes*	Vocal communication	Observational	T&E
Mitchell & Anderson, 1993	*Macaca fascicularis*	"Do what I do"	Experimental	T&E
Tomasello et al., 1994	*Pan troglodytes*	Gestural communication	Observational	EMUL
Custance et al., 1995	*Pan troglodytes*	"Do what I do"	Experimental	IMIT

[a] ASL = American Sign Language [b] T&T = individual trial-and-error learning; EMUL = emulation learning; IMIT = imitation or imitative learning; ? = some difficulty of interpretation.

1996b, for a similar finding with pigtail macaques). One alternative interpretation of the mother's behavior, however, is simply that she has conflicting motivations to travel and to stay close to her infant (although it is reported that on some occasions mothers seem to actively place their infants in these situations). Thus, it can be said that the mother's intentions are that the baby follow her, not that the baby learn something (although the infant may learn something anyway), and sometimes the mother's leaving behavior may become ritualized into an intentional social strategy or a communicative signal. Nishida (1987b) also reports observations in which macaque mothers pull their infants away from dangerous objects, and Fletemeyer (1978) reports that high-ranking Chacma baboons threaten juveniles away from potentially harmful food. While it seems clear in such cases that adults intend for the juveniles to avoid the object, it is not clear that they intend for the juvenile to learn something.

There have also been a number of observations in which monkeys fail to instruct their young about important environmental events. For example, adult vervet monkeys do not correct the incorrect alarm calls of their offspring (Cheney & Seyfarth, 1990b), and a number of species fail to prevent their offspring from eating noxious foods (Japanese macaques: Hikami, 1991; pigtail macaques and spider monkeys: Fairbanks, 1975; baboons: Jouventin et al., 1976; see Visalberghi, 1994, for a review). Also, Cheney and Seyfarth (1990a) reported that Japanese and rhesus macaque females who were shown either food or a predator, which were then hidden before their offspring arrived on the scene, did nothing to either direct their offspring to the food or warn them of the predator.

Figure 9.11. Japanese macaque mother and infant. (Photo by J. Call.)

9.4.2 Apes

Goodall (1986) reported that chimpanzee mothers also "encourage" their infants to walk in much the same way as the macaques reported above, but again it is possible that they are simply displaying a conflict between leaving and maintaining contact with their infants (see also Plooij, 1984; van de Rijt-Plooij & Plooij, 1987). Chimpanzee mothers also have been observed to take poisonous leaves away from their infants (Nishida et al., 1983), but again it is possible that what the mother intends is that the infant simply refrain from eating the leaves, not that it *learn* the behavior of refraining from eating the leaves. In some observations of ape and bonobo mothers with their offspring and human objects, Bard and Vauclair (1984) found very few interactions in which the mothers treated their infants intentionally or attempted to instruct them in any way.

The only systematic observations of teaching by any ape species are reported by Boesch (1991b). He reports a number of observations of adult chimpanzees in some way encouraging their youngsters to crack open nuts. In line with an intentional definition of teaching, he divided his observations into "facilitation" and "active teaching." Observations of facilitation are fairly common, and include mothers allowing infants to use their hammers or nuts (which they tend not to let other animals do). In his decade of observation, Boesch has seen only two instances of what he considers active teaching:

1. A mother was cracking some very hard nuts, with her son eating most of them. The son then tried to crack some nuts for himself, with limited success. At one point he placed a partly open nut on the anvil in a way not conducive to successful opening. Before he could

strike the nut, his mother picked it up, wiped off the anvil, and placed it on the anvil in the correct position. The son then successfully struck and opened the nut.

2. A daughter was attempting unsuccessfully to open nuts with an irregularly shaped hammer. Her mother then joined her. The daughter gave the mother the poor hammer. The mother then, in a very deliberate manner (for over one minute), rotated the hammer into its best position for pounding the nut. She then successfully opened a number of nuts, with both mother and daughter eating them. The mother then left and the daughter proceeded to make attempts on her own, with mixed success but always with the hammer in the orientation her mother had used. (paraphrased from Boesch, 1992: pp. 176–177)

Boesch claims that in the first example the mother anticipated her son's impending failure and intervened to ensure his success, although it is also possible that she simply noticed that the nut was not correctly positioned and positioned it for herself, which her son then proceeded to exploit. Boesch interprets the second example as the mother correcting her daughter's mistake and demonstrating the correct method. In this case it is possible that the mother was simply using the hammer to crack nuts for herself as she normally would. The only behavioral evidence for the instructional interpretation of this observation is that the mother slowly rotated the hammer into the most efficient position for her own cracking attempts. In either case, these two isolated observations would be much more significant and easier to interpret if they fell into a larger pattern of instructional activity in this and other contexts in the chimpanzees' lives. Further observations of this type are clearly needed before definitive conclusions may be drawn.

9.4.3 Human-Raised Apes

Fouts et al. (1989) reported one instance of an adult human-raised chimpanzee, Washoe, shaping the hands of another chimpanzee into an American Sign Language sign (see also Fouts, Hirsch & Fouts, 1982). Patterson and Linden (1981) also reported a case of a human-raised gorilla seeming to mold a sign for a naive human caretaker. Savage-Rumbaugh (personal communication) reports that, although she has seen a number of suggestive incidents, direct instruction of one another is not something that her chimpanzees and bonobos seem especially concerned with.

9.4.4 Comparison of Species

Although there is no question that many primate species do things that assist the learning of conspecifics (i.e., unintentionally), there is no convincing evidence that individuals of any nonhuman primate species engage in behaviors intended to assist the learning of others. In fact, on the basis of her thorough review, King (1994) concludes that individual primates must be skilled information gatherers on their own because they cannot expect others to "donate" information to them socially. Boesch's two observations and the two anecdotes reported by Fouts and Patterson suggest the possibility that apes may on rare occasions engage in something like intentional instruction of others. But until further systematic observations are made, it is prudent to remain either agnostic or skeptical (see Visalberghi & Fragaszy, 1996, for a similar view). Indeed, human children do not show signs of the ability to actively instruct other children until their fourth year of life (Ashley & Tomasello, in press).

9.5 What Primates Know about Others in Social Learning

Animals who live in social groups have the opportunity to learn things from others that they could not learn on their own. Evolutionary biologists are interested in such processes of social information transfer or cultural transmission because potentially they provide for a powerful form of the "inheritance" of information that may complement in important ways inheritance through the genome—a kind of transmission of individually acquired information. Boyd and Richerson (1985) argue and provide evidence that cultural transmission is an especially useful adaptation for species that inhabit ecological niches that change too rapidly for the relatively slow processes of organic evolution. Some evolutionary biologists choose to speak of cultural transmission whenever there is information transfer among members of a species that does not derive from the genome (e.g., Bonner, 1980), but our focus here is on social information transfer that occurs by processes of social cognition and learning, especially in domains that are not narrowly prepared biologically.

Primates and other animals may learn from the goal-directed behavior of others, and this may occur in a number of different ways. First of all, animals may learn some things as a direct result of the behavior of others, but what is learned concerns the physical environment only, not the behavior of others. Thus, many animal species learn things about the environment as a result of being attracted to a particular location or object and then learning something on their own (local or stimulus enhancement; see Zentall, 1996, for a review). We have also proposed a more cognitive version of this process called emulation learning in which an individual is not just attracted to a location but actually observes and understands a change of state in the world produced by the manipulations of another, which may be its only way of learning that such a change of state is possible. For example, individuals may learn by observing others such things as a nut can be opened and food found inside, a log can be rolled over and food found under it, sand comes off food when it is in water, or a stick hitting a piece of fruit can cause it fall to the ground. In the terminology of Gibson (1979), by observing the manipulations of others, individuals may learn all kinds of "affordances" of the environment that they would be unlikely to discover on their own. On some occasions the individual will then want to produce that same change of state for itself, using whatever behavioral strategies it has available, some of which may have been used in the past to produce that same effect. Regardless of whether the resulting behavior matches the behavior of the other, what the individual has learned is something about the environment, not about the behavior or behavioral strategies of other organisms.

In the domain of communication, primates may learn to produce signals by ritualizing them with others during ontogeny (although some of their signals seem to rely little on learning). As outlined above, this process, which has been strongly implicated in the gestural communication of chimpanzees but not much studied in other species, requires mainly that an individual be able to anticipate the behavior of others based on past experiences in similar situations (perceive it as "directed"), and then to gain intentional control over the sequence of events. Thus, an individual must be able to observe a consistent sequence $X \rightarrow Y$, predict that when X occurs Y is likely to occur, and then, on some future occasion, desire Y and generate X as a possible way of making Y happen. It is likely that many primate social strategies are created by this type of learning, although perhaps only when X is a behavior that the individual itself produces, including many instances that have been labeled as deception. Much of the creativity of primate communication results from the ability of individuals to apply the fruits of this learning to novel communicative contexts. Whereas ontogenetic ritualization clearly relies on individuals behaving intentionally, (indeed, this kind of flexible and creative behavior defines intentionality), it does not imply the understanding of others as intentional so that their social or communicative strategies may be copied for one's own use.

The only solid evidence of imitative learning in either instrumental or communicative contexts is for apes raised by humans in specific ways. It is unclear whether monkeys might acquire similar imitative learning skills if they were raised in similar ways. Apes raised by humans seem to be able to learn something from a variety of object-directed behaviors, in terms of both the means and ends of the demonstrator, as well as something resembling human linguistic symbols (although the data could be much better in this latter case). How they do this is not known. It might be speculated, however, that a humanlike environment in which apes are intentionally taught, encouraged to jointly engage in certain activites with humans, and rewarded for reproducing human behaviors, is in some way instrumental in the ontogeny of apes' humanlike skills of imitative learning, as it likely is for the ontogeny of these same skills in humans (Tomasello, Kruger & Ratner, 1993). It is interesting that skills of teaching, in which individuals seek intentionally to instruct others or provide them with needed information, are not a prominent aspect of primate behavior in any setting, whether natural or artificial.

Because they are based on different types of social learning and cognitive skills, the cultures that some nonhuman primates create in their natural habitats are different in important ways from human cultures. Tomasello et al. (1993) point out in particular that nonhuman primate cultures do not seem to contain artifacts that accumulate modifications over time in the manner of many human artifacts (the "ratchet effect"). This is because what is "transmitted" across generations in nonhuman primates consists mainly of (1) affordances of the environment that individuals discover vicariously by observing changes of state in the environment that result from the behavior of others, as well as other behaviors that are arrived at through local enhancement in concert with processes of individual learning; and (2) communicative signals that have been ontogenetically ritualized through repeated interactions in which interactants shape one another's behavior. These processes of transmission simply do not involve the fidelity of transmission characteristic of imitative learning and teaching, so they cannot serve to "ratchet up" specific instrumental or communicative activities across generations. Whether and in what ways these same differences with human culture would hold for enculturated primates if they were put in a position to create cultures on their own, without further human intervention, is an interesting question for future research.

As in the case of the previous two chapters, we conclude our consideration of primate social learning and culture by pointing out that a crucial factor in all of this is how primate individuals understand the behavior, or perhaps the psychological states, of others. As argued by Tomasello et al. (1993), the way an organism learns from another depends crucially on how it understands the other's behavior. Imitative learning, as we have defined it, requires that the learner perceive and understand not just the bodily movements that another individual has performed (mimicking) and not just the changes in the environment in which the behavior has resulted (emulation learning), but the learner must also understand something of the "intentional" relations between these (i.e., how the behavior is designed to bring about the goal). This then determines precisely which aspects of the other's behavior the learner seeks to reproduce and how flexibly the learner may differentiate means from goal in its attempts at reproduction. In all, we have found basically no evidence of such understanding in nonhuman primates in social learning situations, with the possible exception of some great ape individuals raised and enculturated by humans. The two social learning processes for which there is a great deal of evidence, and which once again strike a middle ground between operant conditioning and theory of mind as social cognitive processes, are emulation learning and ontogenetic ritualization. Both of these processes rely on an understanding of the behavior of other beings as animate and directed, but neither requires an understanding of their psychological states. How nonhuman primates understand the psychological states of others is the topic to which we now turn.

10

Theory of Mind

Many of the social cognitive skills reviewed in the previous three chapters shed light on what different species of primates understand about the behavior of conspecifics. In chapter 7 it was established that most primates recognize familiar individuals and their relationships to one another and use this information, along with general skills in "reading" behavioral cues, to *predict* the behavior of conspecifics. In chapter 8 it was established that most primates formulate social strategies and communicate with one another intentionally not just in order to predict, but also in order to influence and *manipulate*, the behavior of conspecifics. In chapter 9 it was established that primates are capable of *learning from* the behavior of conspecifics many things about both their physical and social environments. The question posed in this chapter is whether in addition to this knowledge of the behavior of conspecifics, primates also have knowledge of their psychological states—their minds. On most accounts, an understanding of the psychological states of others would allow for more accurate and flexible social predictions and manipulations, especially in novel contexts in which there has been no opportunity to learn particular behavioral sequences. There is no question that primates behave intentionally; the question is whether they also understand the behavior of others as intentional.

One problem with most theorizing about this question is that the theoretical tools used are decidedly blunt. Most often the question is posed in a dichotomous fashion such as "Does species X attribute mental states to others?" or "Does species Y have a theory of mind?" But mental life is potentially complex, with many different types of psychological states that may be perceived or understood in others (Povinelli, 1994b; Whiten, 1994). We believe it is possible, therefore, to discern a somewhat more differentiated catalogue of the psychological states that primates may potentially understand in others, including at least three levels: (1) behavior and visual perception (gaze), (2) intentions and attention, and (3) knowledge and beliefs (see Tomasello, 1995a; Tomasello et al., 1993). The evidence is different at each of these different levels of analysis, and so to proceed beyond simple dichotomies, separate accounts of each level are needed.

Also relevant to current concerns are how primates understand themselves as psychological beings. Unfortunately, most of the research on primates' knowledge of self comes from studies of their behavior in front of mirrors. It is arguable whether this is properly classified as social cognition at all, as opposed to simply the use of a human artifact to gain knowledge of nonvisible parts of one's own body. But because of the theoretical arguments of Gallup (1982) to the effect that awareness of self implies awareness of others, mirror self-recognition has been thought of in the context of social cognition by many researchers, and so we review that research here. We attempt in this review to point out the ways in which mirror self-recognition is and is not related to social knowledge in general and also point out some ways that a sense of self as social agent, not just physical object, may be assessed.

Our procedure in this chapter is to review evidence for primate knowledge of the psychological dimensions of other animate beings separately at the three levels of analysis described above. Systematic data at each of these three levels are limited, but in each case there is some information based on inferences from primate social knowledge, social strategies, and social learning, and other information based on more or less direct experimentation. Almost all of what we know experimentally is about the skills of great apes, in most cases apes with an extensive history of human training and interaction. In almost all cases the species whose mental states the apes are attempting to understand is *Homo sapiens*. We conclude with a brief review of research on mirror self-recognition and other forms of primate self-knowledge.

10.1 Understanding Behavior and Perception

The most straightforward way in which one organism may understand another is simply in terms of its animacy, that is, its capacity to behave spontaneously, sometimes in the presence of particular stimuli. This differentiates animate from inanimate objects and represents a different principle of operational causality in the two cases. In many places in the preceding chapters we have pointed out the ways in which primates understand the behavior of others as animate. We also have shown how primates can predict the behavior of conspecifics in some situations based either on past experience in similar situations or on generalized cues such as emotional displays and current direction of travel in particular contexts—what we have called an understanding of the directedness of others. In addition, in some cases individuals create social strategies and communicative signals of various types, through either learning or insight, to influence the self-animated behavior of others. And primates learn various things from conspecifics that rely on an understanding of them as animate and directed beings as well. Evidence for these skills was the central concern of chapters 7, 8, and 9, and so we may conclude without reservation that primates know much about the directed behavior of other animate beings.

Given this evidence for primate understanding of behavior, what we are mainly concerned with in this section is evidence that primates understand visual perception (more accurately, direction of gaze) as an important aspect of the animacy and directedness of others. Most primates are highly visually oriented organisms compared to other animals, and so it would seem natural that they might come to some understand-

Figure 10.1. Gorilla.
(Photo by J. Call.)

ing of the visual perception of conspecifics and how it might work to guide their behavior. There are two main contexts in which we might expect to see this understanding manifest. The first derives from communicative situations and concerns primates' understanding of one of the preconditions of communication, namely, that the potential recipient actual perceives the communicative signal. The second derives from a variety of social interactions and concerns primates' skills in following the visual gaze of others to interesting or important external phenomena. We discuss these each in turn.

10.1.1 Detecting the Gaze of Others on Self

As reported in chapter 8, apes perform a number of behaviors that evidence their understanding that others need to be visually oriented to them for communicative signals to achieve their intended effects. For example, Tomasello et al. (1994) found that in a captive colony of chimpanzees, visually based gestures (e.g., arm raise) were performed only when potential recipients were oriented toward the performer. When recipients were not so oriented, tactually based gestures (e.g., poke) or auditorily based gestures (e.g., ground slap) were used. In a similar vein, Tanner and Byrne (1993) report repeated cases of captive gorillas hiding their play-faces so that others could not see them (see also de Waal, 1982, for a similar report on a chimpanzee hiding a fear grimace). Fernández and Gómez (1983) reported that human-raised gorilla infants seem to make sure that humans are properly oriented before they perform gestures toward them (see also Gómez, 1990). On the other hand, Premack (1988) reported that when captive juvenile chimpanzees needed a human to help them open a container across the enclosure but the human's eyes were covered by a blindfold, only one of the four subjects removed the blindfold from the human, suggesting the possibility that it was not the eyes per se that the apes were tuned into.

The problem is that in all these studies, the direction of visual gaze of the other was confounded with the general bodily orientation of the other. Two experimental studies have tried to unconfound these to some degree. First, Call and Tomasello (1994a) studied systematically the effect of the orientation of the recipient on the pointing gestures of two captive orangutans. One of the orangutans was nursery-raised and trained as an adult to point to food for human experimenters, while the other orangutan was human-raised and learned to point without specific training (see Miles, 1990). The task was for the subject to point for a human to which of two drinks it wanted (one always contained substantially more). Immediately after placing the two drinks on opposite ends of a platform in front of the subject's cage, the observer did one of four things: (1) left the room; (2) walked to the opposite side of the room and turned his back; (3) sat down behind the platform facing the drinks and the subject, but with eyes closed; (4) sat down behind the platform facing the drinks and the subject, with eyes open. A video camera recorded what the subject did in each of these conditions, and an observer blind to experimental condition later watched the videotapes and counted how often they pointed to a drink. It was found that both orangutans mostly refrained from pointing when the observer left the room or crossed the room and turned his back. Both pointed quite frequently when the observer sat opposite them with eyes open. In the condition in which the observer was bodily oriented to the subjects but with eyes closed, the home-raised orangutan pointed infrequently, whereas the other captive orangutan continued to point at a relatively high rate. Thus, the captive orangutan showed little understanding of the role of the eyes in the process of visually based communication, whereas the human-raised orangutan showed somewhat more understanding. Since the being whose eyes were important was a human, it is perhaps not surprising to find that the human-raised ape performed in more sensitive ways.

Second, in a series of 15 experiments, Povinelli and Eddy (1996b) tested 4- and 5-year-old chimpanzees' understanding of how humans must be bodily oriented for successful communication to take place. They trained seven captive subjects to approach a Plexiglas barrier and extend their hands toward one of two experimenters (each with a hole in the Plexiglas in front of them) to request that a human give them food (see Figure 10.2). In the critical test trials, one human stood behind each hole and food was available on a table between them. The subjects had to choose which human to beg from, as the first begging gesture was the only one responded to. In the first experiment Povinelli and Eddy presented subjects with one human facing forward and one facing backward (among other conditions). The chimpanzees consistently gestured toward the human who was facing forward, thus confirming the findings of Call and Tomasello (1994a) and Tomasello et al. (1994). In a number of other experimental conditions, however, chimpanzees did not seem to distinguish between more subtle differences between the humans. For example, they did not gesture differentially for a human who wore a

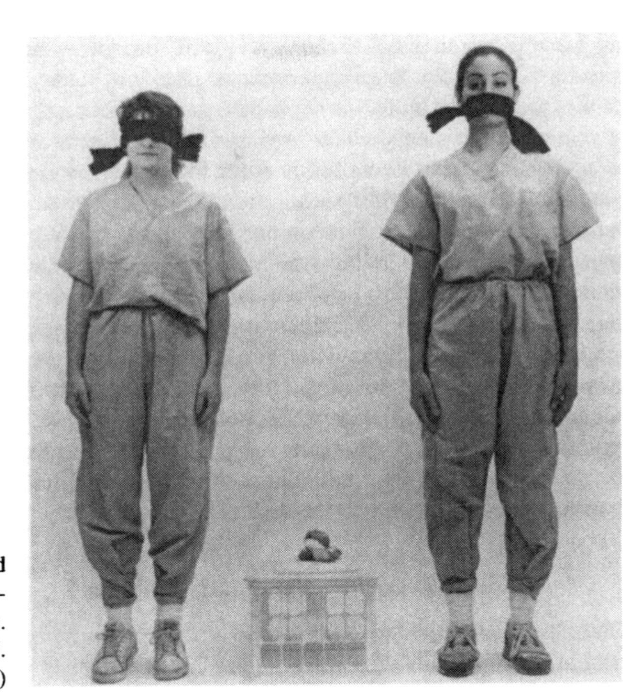

Figure 10.2. Povinelli and Eddy's experimental arrangement used in several studies. (Photo courtesy of D.J. Povinelli.)

blindfold over his eyes (as opposed to one who wore a blindfold over his mouth), or for one who wore a bucket over his head (as opposed to one who held a bucket on his shoulder), or for one who held his hands over his eyes (as opposed to one who held his hands over his ears), or for one who had his eyes closed (as opposed to one who had his eyes open), or for one who was looking away (as opposed to looking at the subject), or for one whose back was turned but who looked over his shoulder to the subject (as opposed to one whose back was turned and was looking away). A number of control experiments ruled out possible artifactual explanations of these results. The investigators concluded that although these young chimpanzees have learned one or more cues that signal conditions conducive to communication with humans (e.g., bodily orientation facing toward versus away from the subject), overall they do not seem to understand precisely how visual perception works. Povinelli and Eddy (1996b), provide evidence that direct eye contact and certain orientations and movements of the head are both important cues for chimpanzees in this communicative situation.

There are only a few observations of monkeys' understanding of the visual perception of others in communicative situations. Perrett and Mistlin (1990) reported that stumptail macaques react to the presentation of pictures of faces (both conspecifics and other primates, including humans) in a way that suggests their ability to differentiate between when they are being looked at and when they are not (see Thomsen, 1974, for similar evidence for other monkey species). The basis of this differentiation does not appear to be head direction, but eye-gaze direction independent of head direction. Kummer (1968) reported that baboons will sometimes approach groupmates and stare into their face before setting out from camp, and Maestripieri (personal communication) reported that when a pigtail macaque mother "pucker faces" to its infants as she is leaving but the infant is not looking, she will sometimes move in front of the infant and "pucker face" again. In an experimental situation, Keddy-Hector, Seyfarth, and Raleigh (1989) found that male vervet monkeys were less aggressive toward an infant if its mother could be seen visually monitoring their interactions through a Plexiglas barrier, as opposed to control conditions when the mother was not in sight. In an intermediate condition, the male could see the mother through a one-way mirror but she could not see the male and infant (and so her behavior was not contingent in any way on their interactions). In this condition the male was aggressive an intermediate number of times relative to the conditions in which the mother was visibly present or absent. These findings might be interpreted as indicating the male's appreciation that the mother's perception of his interactions with the infant would have later consequences. An experiment by Cheney and Seyfarth (1990a: study 2) with rhesus and Japanese macaques ruled out the overt behavior of the mother as the key variable affecting the males' reticence, as it replicated this result with the one-way mirror reversed so that the animal in the test area with the infant could see the mother in the indoor area but not vice versa (so she made no threatening behaviors).

10.1.2 Following the Gaze of Others

There has been little systematic study of the ability of nonhuman primates to follow the gaze of others to outside entities. There are a few anecdotal observations. For example, in the observation cited in chapter 8 for deception, a savannah baboon being chased by another stopped and looked in the distance, at which point the chaser stopped and looked also (Byrne & Whiten, 1990, see also de Waal, van Hooff & Netto, 1976, Kummer, 1967). In this same vein, Byrne and Whiten (1990) reported a number of strategic behaviors in which monkeys traveled in such a way as to keep an occluder, such as a tree, between themselves and a human observer, seeming to understand something about the observer's line of sight. Plooij (1978) observed chimpanzees in the wild following both his own gaze and that of conspecifics. In Menzel's (1973c, 1974) studies, it was hypothesized that chimpanzees were using the gaze direction of knowledgeable groupmates to determine the location of food, although this was not specifically tested (as against, for example, direction of travel). In a slightly more systematic study, Hayes (1951) reported that focusing her attention on a location reliably led the home-raised chimpanzee Viki to approach and inspect that place. Carpenter, Tomasello, and Savage-Rumbaugh (1995) found numerous instances of both human-raised and captive chimpanzees and bonobos following the gaze of humans to objects (see also Bard & Vauclair, 1984).

It should be noted that in all these observations of gaze-following, both anecdotal and more systematic, bodily orientation and visual gaze direction were always confounded. It is thus unclear precisely what cues were being used, assuming that indeed subjects were cueing on the other and not on some external stimulus that they both could see. Moreover, because looking in the same direction that a conspecific is looking may be learned fairly easily as a response to a discriminative stimulus signaling the presence of important objects or events, it is only in experimental situations that the mechanism underlying gaze-following behavior can be systematically assessed. Only a few such studies have been published, all with human experimenters as the animate beings whose gaze is followed.

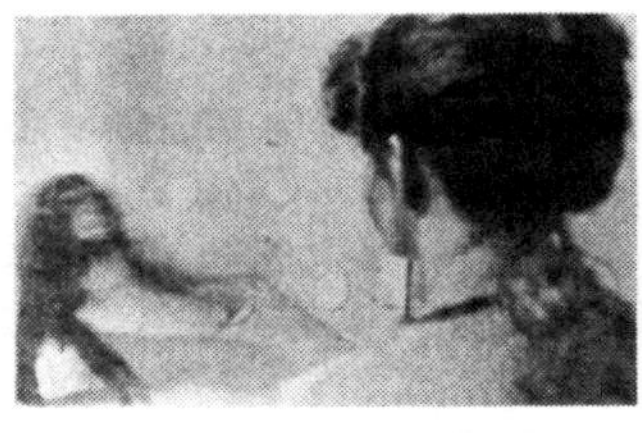

Figure 10.3. Gaze following in a chimpanzee. (Photo courtesy of D.J. Povinelli.)

First, using the same seven chimpanzee juveniles and the same experimental arrangement as in their other studies (reported in the previous subsection; see figure 10.2), Povinelli and Eddy (1996a: study 1) experimentally manipulated gaze direction for 5- to 6-year-old chimpanzees (see figure 10.3). Each subject saw a human look to a location above and behind it. In one condition the human moved both his eyes and head, and in another condition he moved his eyes only. In both of these conditions the chimpanzees looked where the human was looking more than in a baseline condition in which the human looked at the subject. Moreover, subjects looked where the human was looking equally often whether or not the head was moved, demonstrating the efficacy of eye direction alone. Povinelli and Eddy (in press) extended these findings by showing that young chimpanzees can even follow human gaze to the space behind them, something human infants do not do until 18 months of age.

Second, in a series of five studies using capuchin monkeys as subjects (*Cebus apella*), Anderson, Sallaberry, and Barbier (1995) secretly baited one of two food wells and then presented the pair to subjects. In a baseline condition, the experimenter simply stared between the food wells. In experimental conditions the experimenter provided a cue of some sort. Over a series of studies using several different cues, direction of eye gaze (including head direction) never proved to be an effective cue by itself (whereas touching the baited food well was). Itakura and Anderson (1996) were able to train a single capuchin monkey to follow human eye gaze to the food hidden in one of two opaque containers, but it took them more than 120 trials to do so, suggesting the possibility that gaze direction was learned as a straightforward discriminative cue. Itakura (1996) observed individuals of 11 different primate species (2 lemurs, 1 squirrel monkey, 2 capachins, 4 macaques, 2 great apes) as they responded to a human looking to one or the other side of their cage. Although many species did follow a pointing gesture, only a single orangutan subject followed the human's gaze appropriately in the absence of pointing.

10.1.3 Comparison of Species

It is difficult to come to any firm conclusions about species differences in this domain because the data are so sketchy. A variety of primate species seem to know that for their communicative signal to work the recipient must be bodily oriented to them, although the evidence is much more extensive for apes than monkeys. It is unclear the extent to which any of these species understand the role of the eyes in this process. There are at least some indications that, for orangutans, experience with humans makes the role of the eyes more salient. Also, some chimpanzees seem to follow the gaze direction of humans, and this behavior seems to be based on the eyes specifically, although the directional specificity of this skill may be limited. There is one report of capuchin monkeys not following a human's gaze direction. For most primate species there are no systematic data one way or the other, and there are basically no systematic observations of any primate species following the gaze of conspecifics rather than humans. We are aware of no experimental data on any nonprimate species following the gaze of conspecifics. Human infants become skillful at following gaze during their second year of life (Corkum & Moore, 1995).

In combination with the data from the previous three chapters, it may be concluded that primates know a good deal about conspecifics as animate and directed beings. In particular, they understand a good deal about the behavior of others and many of the conditions that precipitate particular actions. They also have some understanding that the bodily orientation of others (and perhaps their eye gaze direction) may be used as a cue both for their subsequent behavior and for discovering events and objects at other locations in the environment.

10.2 Understanding Intentions and Attention

Beyond the understanding of others as animate beings who are directed to particular entities in the world—based on readily observable actions—it is possible that primates understand others as intentional agents. Intentional agents are organisms whose behavior and perception are organized in terms of their goals, which are not readily observable. What may be observed of intentional agents is that their individual behaviors are not invariantly followed by particular outcomes. Rather, their behavior is directed toward bringing about a specific change of state in the environment regardless of what particular behavior is needed to accomplish this (so-called equifinality). This type of behavioral organization is apparent when an organism persists in its directed activities until a certain change of state is brought about, using a number of different behavioral means if necessary, and it even behaves in predictable ways if obstacles requiring certain kinds of novel adjustments present themselves (Bruner, 1972; Gergely, Nádasdy, Csibra and Biró, 1995; Powers, 1973). Although there is no question that the behavior of primates is organized in this way, it is still an open question whether they understand this fact. Primates' understanding of the intentions of others has been assessed experimentally in a few studies, and, in addition, there are a number of observations that various investigators have taken as evidence of an understanding of intentions.

One peculiarity of our account relative to many others is that we consider attention to be an intentional phenomenon. Following Gibson and Rader (1979), we note that two organisms, or the same organism at different times, may focus their attention on different aspects of the same object or event without changing their gaze direction or bodily orientation at all. This reflects decision making of a kind analogous to that made behaviorally: in light of its current goal, or any other considerations, an organism may choose to focus on one or another aspect of the current perceptual situation. Thus our main question is whether there is any evidence that nonhuman primates understand that others can shift their attention intentionally, even while maintaining the same gaze direction.

10.2.1 Understanding Intentions

Premack and Woodruff (1978) showed the laboratory-trained, language-trained chimpanzee Sarah videotapes of human actors coming upon obstacles in problem solving situations (see figure 10.4). For example, Sarah saw a human looking up to an out of reach banana hanging from the ceiling, or a human wanting to exit through a locked cage door, or a human trying to operate a hose that was unattached to the faucet. The videotape was then stopped and Sarah was presented with a pair of photographs, one of which represented, from the human point of view, a solution to the problem: such things as the person mounting a box under the banana, a key, and the hose attached to the faucet. Sarah saw each tape four times and was given feedback about correctness after each trial. In general, Sarah performed in a seemingly insightful fashion on these tasks (often from the beginning, before the feedback could have been effective). In a follow-up study, it was also found that when a trainer that Sarah disliked was used, she chose the inappropriate actions instead of the appropriate ones. This control was intended to rule out the possibility that in the first phase Sarah was simply associating objects that normally go together or identifying a familiar sequence of actions, since most of the videotapes presented objects and action sequences with which Sarah was familiar. Premack and Woodruff (1978) argued that Sarah's success constituted evidence that she "recognized the videotape as representing a problem, understood the actor's purpose, and chose alternatives compatible with that purpose."

There are, however, alternative to this interpretation of the study. Savage-Rumbaugh, Rumbaugh, and Boysen (1978b) pointed out some serious shortcomings in the methodologies of the experiments. They noted that in the first phase of the study, Sarah may have been choosing "solutions" solely on the basis of the associations among objects developed from her experiences with caretakers and their behavior with keys, hoses, and the like, as Premack and Woodruff themselves feared. They examined each item Sarah was presented with and found that, in general, the items for which such associative procedures were most straightforward were the ones on which Sarah performed best (e.g., key with lock, hose with faucet); those that were more obscure associations were items for which Sarah performed more poorly. To confirm this possibility, Savage-Rumbaugh et al. presented two language-trained chimpanzees, Austin and Sherman, with a match-to-sample procedure in which they were shown, for example, a picture of a foot and asked to choose between pictures of a shoe and a key. The animals had no training in this task (although they, like Sarah, had experienced previous training in match-to-sample procedures in general). Both subjects chose the closely associated objects on 25 of 28 trials. Because Premack and Woodruff's control condition with the "good" versus "bad" experimenter was supposed to show that Sarah did not always choose the correct alternative based on associative strategies, Savage-Rumbaugh et al. also looked closely at this phase of the experiment. The problem in this case was that the items presented were the same as in the previous phase and the good trainer was the same actor as before. For the good trainer the correct solution thus remained as before (e.g., the key with the lock). The bad trainer was new and the bad alternative was also new and always was some sort of "mishap" (e.g., lying on the floor with a box on top of him). There is thus the possibility that Sarah paired the new actor with the new alternative. Moreover, and perhaps even more importantly, Premack and Woodruff point out that Sarah could have behaved in this phase simply on the basis of the photographic alternatives at test without reference to the videotaped "problem" at all: she might simply have chosen to see the bad actor in a bad situation (i.e., she simply chose the situation she wanted to see).

This experiment thus has some very plausible interpretations in terms of Sarah's experimental strategies that do not involve her seeing the videotaped sequences as problems, defined intentionally from the human actor's point of view. It thus begs for replication or confirmation using a different methodology. In the one attempt to do this, Premack (1986) reported (very briefly) of an attempt to train Sarah to discriminate between videotaped sequences that depicted intentional actions versus those that depicted nonintentional actions. Although no details of the procedure are given, Premack reports that Sarah never learned the discrimination (1986: p. 85). Another study that is sometimes cited in this context is an unpublished study with a single subject. Povinelli (1991) presented the chimpanzee Sheba with two alternatives. In one, a human experimenter, on her way to deliver juice to a subject, accidentally spilled it. In the other, the experimenter intentionally poured the juice on the floor and then threatened Sheba to further demonstrate her bad intentions. When Sheba was later asked to choose between experimenters from whom she might receive juice, she chose the "clumsy" over the "mean" experimenter on all three trials. This could mean that she understood the intentions of the two actors and chose to receive juice from the well-intentioned one. It also might mean that she simply disliked the actor who threatened her and so avoided him. To control for this possibility, Povinelli and Perilloux (in press) attempted to replicate the study with six captive juvenile chimpanzees but using as the "mean" actor one who simply poured out the juice intentionally without threatening the subject (and another "clumsy" actor who again spilled the juice accidentally). In this study, none of the six subjects showed a preference for the "clumsy" over the "mean" human experimenter.

Supporting this view in an indirect way, Tomasello et al. (1993) argued that fully imitative learning of actions on objects in problem-solving situations, in which an individual chooses from among several novel strategies the one that it sees another demonstrating, requires an ability to perceive the actions of others as intentional. This reasoning is similar to that of Premack and Woodruff in their problem-solving situation, that is, to see what the demonstrator is "trying to do" (what strategy she is attempting), an observer must be able to understand the goal, and perhaps subgoals, being pursued. There is no evidence for such behavior in primates raised in species-typical environments (see section 9.2). However, Tomasello et al. did find that three enculturated apes learned to reproduce some actions on objects that humans demonstrated for them, raising the possibility that they perceived the intentions of the human demonstrator. The apes in this study had been trained to reproduce human actions, however, and they were not in a problem-solving situation (they were simply trained to "do as I do" with an object). The possibility thus arises that the apes were either emulating the changes of

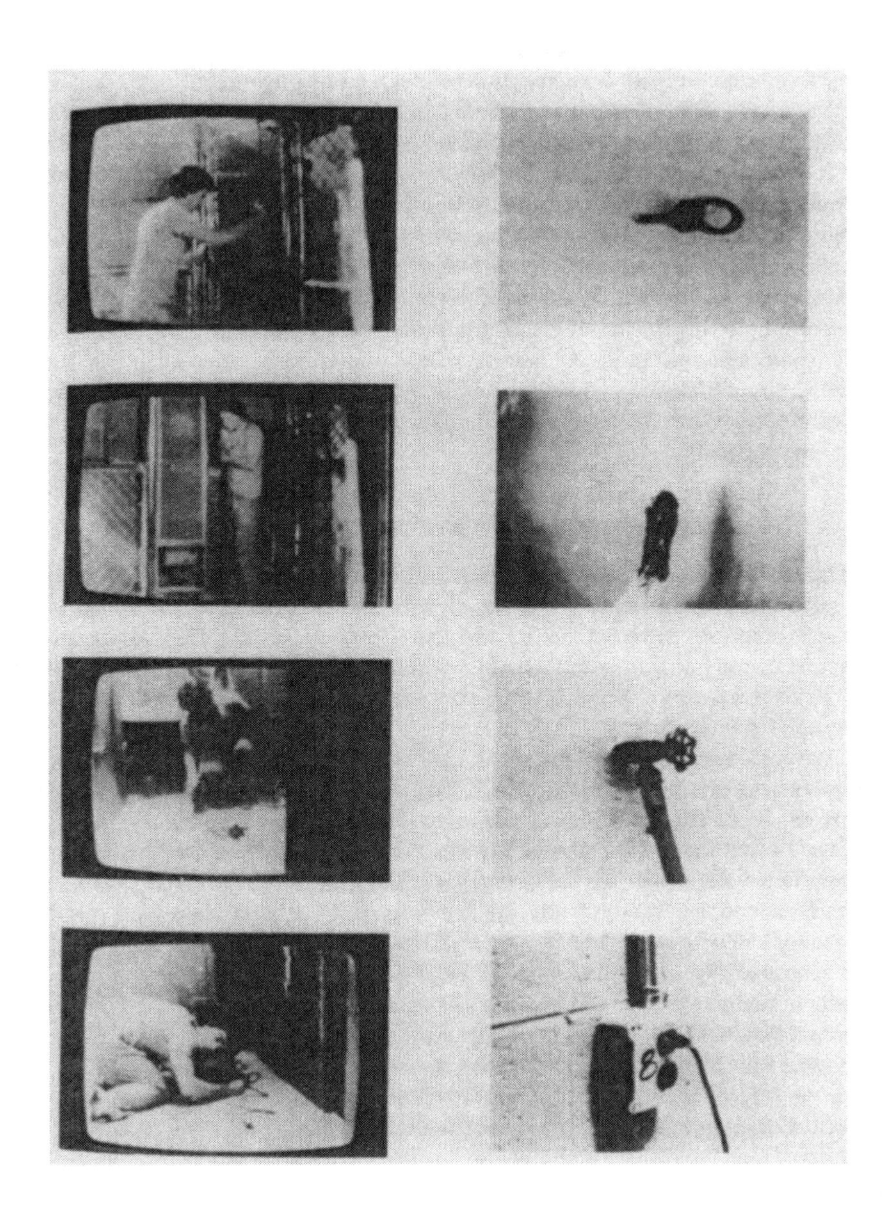

Figure 10.4. Sample of problems and solutions presented to the chimpanzee Sarah in the Premack & Woodruff (1978) study. From top to bottom: Locked door/key; cold room/burner; dirty room/hose; unplugged fan/plug. (Reprinted from *Behavioral and Brain Sciences* with permission from Cambridge University Press.)

state they observed in the object or mimicking human body movements (implying attention to behavior only). If either of these possibilities is true, it would undermine the interpretation in terms of the perception of behavior as intentional.

There are several other anecdotal observations suggesting that primates may understand the intentions of others. Savage-Rumbaugh (1984) reported two observations in which a female bonobo refrained from attacking or threatening a human that had hurt her infant, which she normally would do, seemingly because she perceived that the human's offending behavior was accidental. Also, Whiten (1993) has argued that the detection of the deceptive behavior of others can be taken as evidence of perception of intention; that is, the definition of deceit is that an individual really has one intention but is attempting to display another, and therefore the detection of deception by a groupmate would seem to require an understanding of intentions. Anecdotal observations of the detection of deception have been reported for a number of primate species, including both monkeys and apes, although other interpretations of these behaviors are possible (see chapter 8).

10.2.2 Understanding Attention

As alluded to above, some theorists see attention as a kind of intentional perception, that is, a perceiving organism may intentionally modulate its attention even while looking in the same direction or even at the same entity (Gibson & Rader, 1979). There is only one systematic study of this phenomenon in nonhuman primates, although there are some related observations concerning the way some human-raised apes understand and use communicative symbols.

Povinelli and Eddy (1996a: study 2), using the same subjects and experimental apparatus as previously (see figure 10.2), had a human experimenter face a single chimpanzee as it entered its side of the test room to beg for food through the Plexiglas wall (as it had been trained to do in this situation). The experimenter then immediately stared in a particular direction. The trick was that in the human's line of sight was a solid partition affixed to the Plexiglas wall. If the partition had not been there, the experimenter's focus would have been to the far corner of the room, behind the subject. The investigators reasoned that if the chimpanzees were simply using gaze direction as a cue that something interesting was somewhere along the line of sight of the experimenter, they should follow that line of sight all the way to the rear of the room (since there were no interesting objects in the line of sight). However, if they stopped and inspected the partition, the inference was that they must know that the experimenter was looking at it (under the assumption that it would not be an interesting object in and of itself). Of special importance would be chimpanzees' attempts to look at the experimenter's side of the partition, since this would demonstrate that the subjects were not just treating the partition as an interesting object to which his gaze direction had led them, but rather that they were actively trying to see what the experimenter saw. Results showed that the subjects did look to both sides of the partition more than they looked to the back wall (and more than in some control conditions). Unfortunately, there was no statistical comparison of their looks to the experimenter's side of the partition as opposed to their own side of the partition. The interpretation that the subjects simply considered the partition an interesting object in and of itself (having been led to it by following the experimenter's line of sight) is thus still a viable possibility.

Another line of evidence of apes' understanding of attention is the linguistic productions of apes learning a humanlike system of communication. It is well known that the productions of these apes are almost invariably requests (see section 8.4; one estimate is requests account for more than 95% of all productions) and that the intention behind those productions that are not requests is difficult to discern. Given that the early communication, including prelinguistic communication, of human infants includes many declaratives in which the sole motive is to share interest and attention with others (Tomasello & Camaioni, in press), the paucity of true declaratives in nonhuman primates may evidence of a lack of understanding that others intentionally perceive and attend to the world in ways that may be influenced and shared.

10.2.3 Comparison of Species

The ontogenetically first way in which human infants understand the psychological states of others is in terms of their intentions and attention, typically during the second year of life (Tomasello, 1995a). Many researchers believe that nonhuman primates understand the behavior of others in this way as well. The original Premack and Woodruff (1978) experiment with one language-trained chimpanzee originally gave rise to this belief, but the finding has not proven to be robust. The ability of chimpanzees to acquire some skills of mimicking as demonstrated by Hayes and Hayes (1952) was also influential in fostering a belief in ape understanding of other persons, but subsequent experiments on the social learning of apes have not provided evidence that they understand the behavior of others as organized around intentions. The experiment by Povinelli and Eddy (1996a) raises the intriguing possibility that chimpanzees know something about the content of the visual attention of others, but further analyses would be required to make that result reliable. Bringing into the picture other behavioral phenomena such as deception, which some investigators have interpreted as requiring the understanding of intentions, does not clear up the picture, nor does an analysis of ape linguistic productions.

There are few observations of nonhuman primates in their natural habitats that are relevant to their understanding of the intentions of others and almost no observations of any type for species other than chimpanzees. The understanding of intentions should have consequences for the ability of individuals to predict the behavior of others and to create social strategies so that, for example, if individuals understood the behavior of conspecifics as intentional they should be able to predict how a conspecific who was headed for food would deal with a novel obstacle that needed to be surmounted to obtain the food (perhaps even before that conspecific started moving). An understanding of others as intentional agents should also lead individuals to create social strategies that attempt to manipulate the intentions of others, for example, showing them food (or placing it in sight) when they want to divert another's attention. It would also be desirable that studies of the understanding of intentions be conducted with individuals and their conspecifics rather than with humans, perhaps especially for monkeys about whom we know so little. We are aware of no studies with nonprimate animals that attempt to assess their understanding of others as intentional agents.

10.3 Understanding Knowledge and Beliefs

Another level of complexity and abstraction is added to the picture if we consider primates' understanding not just of the behavior/perception or intention/attention of others but their understanding of the knowledge and beliefs of others. In the human developmental literature, in fact, it is only the understanding of the thought and beliefs of others that is considered to be indicative of a full-blown "theory of mind" containing metarepresentations (mental representations of the mental representations of others). In particular, researchers have come to believe that the best test for theory of mind is the understanding of the false beliefs of others, which children usually do soon after their fourth birthdays (Wimmer & Perner, 1983). For example, in one widely used task, a treat is hidden in a location in full view of two witnesses: the subject and another child. The other child then leaves the room, at which time the subject watches as the treat is moved to a second hiding place. Then, the subject is asked where the other child will search for the treat when she returns. If the subject answers that the other child will search in the original location, where she saw it hidden, and not in the new hiding place, this is interpreted as an indication that she understands a psychological process of belief that is distinct from reality. The importance of false belief is essentially methodological. In any test of the understanding of the beliefs of another, if those beliefs correspond to those of the subject of the study, there is always the possibility that the subject is simply responding in terms of her own beliefs. The crucial test, therefore, must come in situations in which the subject and another individual have different knowledge or beliefs. As in the case of intentions, there are only a few experimental studies to report (all but one with great apes attempting to understand the knowledge and beliefs of humans), along with a smattering of other, less systematic observations that may be relevant.

10.3.1 Experimental Evidence

Premack (1988) reported briefly on an experiment in which four juvenile chimpanzees watched one of two containers being baited with food by a human and then chose the container they wanted. Experimenters then occluded the baiting process so that all the subjects could do was choose randomly. Two experimenters then stood on either side of the room, with only one of them able to see the baiting process (the vision of the other being occluded by a panel). The chimpanzees were then given the choice of which of those two experimenters should indicate the container they should choose. Two of four individuals reliably chose the experimenter who had witnessed the baiting process basically from trial one, possibly indicating their understanding that that individual was in possession of some knowledge that neither they nor the other experimenter possessed (i.e., which of the containers had food). One problem with this interpretation is that when these subjects were later presented with a similar scenario and choice between experimenters, except that they themselves saw the baiting process and thus knew which container held the food, they nevertheless chose the trainer who had witnessed the baiting. This is in spite of the fact that they did not need his advice; they could have simply chosen the container in which they saw the food hidden. This control condition raises the possibility that their earlier choices were based on some general preference for the witnessing trainer. (Also, in his brief description Premack does not report many methodological details of the experiment that would be necessary for a full evaluation.)

Povinelli, Nelson, and Boysen (1990) performed a similar experiment and reported its results more fully (see figure 10.5). They had four chimpanzees, all with much human experience, witness an experimenter bait one of four cups behind an occluder (so that the chimpanzees could not tell which cup contained the food). Another human remained outside the room and consequently did not know which cup was being baited. After this naive human entered the room, subjects could choose either the naive or the knowledgeable human to inform them of the whereabouts of the food. If they chose the knowledgeable experimenter, they invariably got the food; if they chose the naive experimenter, they invariably did not. Results indicated that three of the four chimpanzees correctly chose the knowledgeable experimenter significantly more often than the naive experimenter, two after 100–150 trials and the third after 250–300 trials (with accuracy rates hovering at around 70% correct during these late trials). The interesting twist in this experiment was that there was a transfer phase consisting of 30 trials in which a neutral experimenter again baited the cups. The difference was that this time, instead of remaining outside the room during the baiting process, the naive experimenter stayed inside the room but with a bag over his head; a knowledgeable experimenter again watched the baiting process. All three subjects maintained their successful performance during this transfer phase. Thus, although the subjects may not have come to this experiment with the necessary social-cognitive skills (they took many trials to become successful), what they learned during the first phase seemed to be more general than a simple discrimination based on the physical presence of an experimenter. If they learned a discrimination, it had to do with something more general such as "witnessing" the baiting process visually.

There are two problems with this interpretation. First, in response to a critique by Heyes (1993a), Povinelli (1994a) reanalyzed the transfer data and reported that in the first five trials of the transfer phase, subjects behaved randomly; only after that did they consistently choose the experimenter who had observed the hiding. These results raise the possibility that in the transfer phase, as in the original phase, subjects were simply learning a discriminative cue for obtaining the food. The fact that they learned the requisite discrimination more quickly in the transfer phase than in the original phase may reflect a process of learning to learn (see section 4.1). The second problem is that when this study was replicated with 4-year-old nursery-raised chimpanzees, six subjects failed to differentiate between the knowledgeable and ignorant experimenters (Povinelli, Rulf & Bierschwale, 1994). Although the failure to replicate the previous findings may be attributed to the age of the subjects, these data, in concert with the reanalysis of the transfer data of the original experiment, call into question the ability of chimpanzees to understand that humans can be either knowledgeable or ignorant about an environmental event.

There are two other relevant sets of data for apes. First, Premack (1988) reported (again briefly and sketchily), on a task in which good items (e.g., preferred foods) were invariably placed in one side of a cabinet (painted white) and bad items (e.g., feces) were invariably placed in the other side (painted black). The cabinet was outside the subject Sarah's cage so that only humans could open it, but the lock was controlled by Sarah electronically from inside her cage. The routine was that a human would enter at "teatime" every day, Sarah would unlock the cabinet, and the human would give Sarah something good. After 18 trials of this type, in which Sarah's latency to open the cabinet was about 7 seconds, a "villain" (wearing a mask) entered a few minutes before teatime,

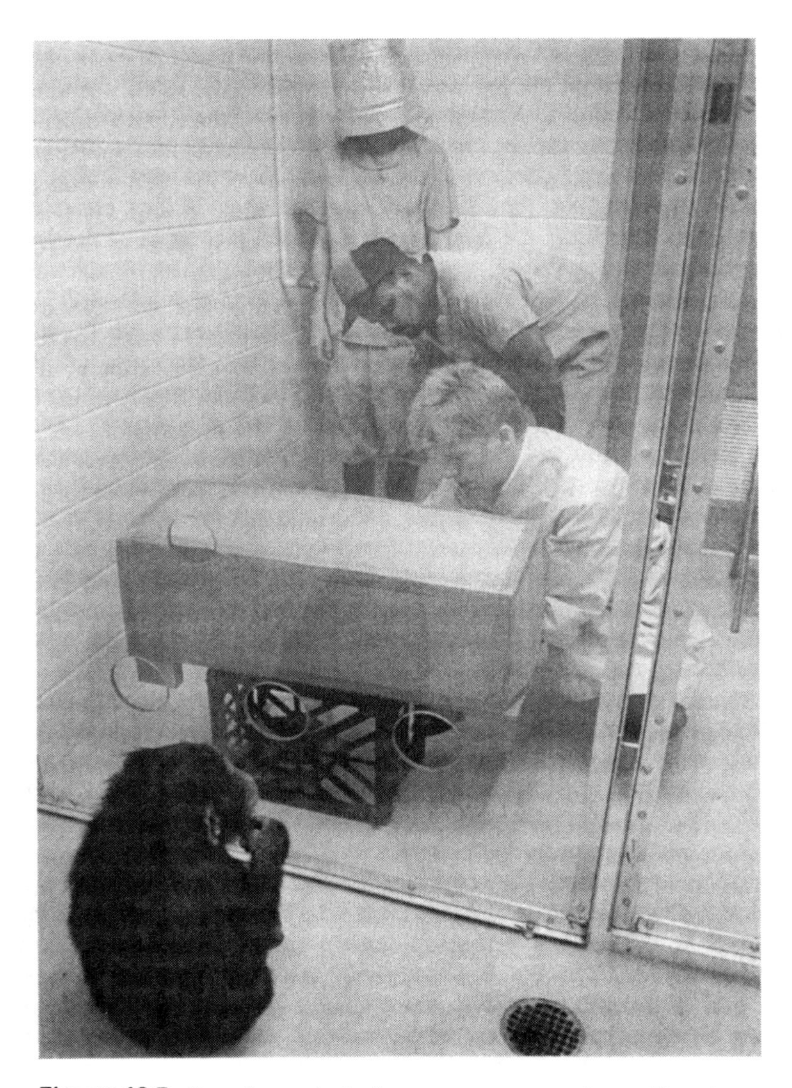

Figure 10.5. Transfer test in the knower-guesser experiment. Experimenter baits one container in the presence of both human informants although only one of them is able to see where the food is placed.

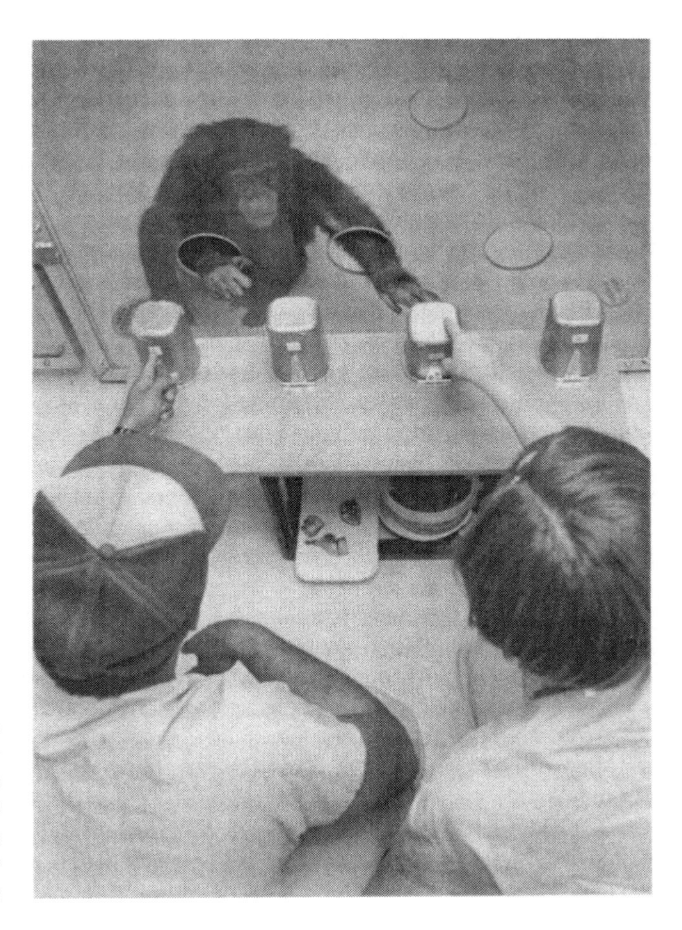

Figure 10.5. (*continued*) Both informants (the knower and the guesser) point and the chimpanzee chooses one container. (Photo courtesy of D.J. Povinelli.)

pried open the cabinet with a crowbar, and rearranged the items within so that the good and bad things were now in the other's usual location. When the normal experimenter entered at teatime as usual, Sarah did not modify her behavior in any way to signal that something was awry; her general demeanor and latency to respond were just as they always were. Premack (1988) suggested that this study is actually a better test of apes' understanding of false belief than previous tests because in this situation successful performance would require that the subject understand the difference between the belief states of themselves and another being. He is thus more cautious than in his previous publications about what chimpanzees may know about the knowledge of others.

The other relevant study was reported by Call and Tomasello (1994a) (see figure 10.6). In this study each of two captive orangutans (one human-raised) was confronted with three opaque containers within a wire cage. Working through a door in the cage, a human experimenter placed food in one of the containers, in full view of the subject, and then locked the door and left. Another experimenter, blind to the location of the food, then entered. Based on previous training, the subject pointed to the container with food. In response to this pointing, the experimenter picked up a nearby rake, raked the designated container across the cage to himself, extracted the food, and gave it to the subject. The key variation was then introduced. In these new trials, the first experimenter followed the same general procedure, except that before he left, he hid the rake under one of three cloths hanging from the facing wall. The second experimenter then entered the room and observed the subject pointing to the location of the food as usual. However, in this case he was unable to retrieve the food; he both struggled unsuccessfully with the locked door and looked for the rake on the floor. After only a few trials in this situation, the human-raised orangutan spontaneously began to point to the rake; the other orangutan learned to point to the rake only after some additional training. In interpreting an unpublished study with a similar design and result, Gómez and Teixidor (1992) concluded that subjects were displaying an understanding that the experimenter held a false belief about the whereabouts of the rake that needed correcting. Call and Tomasello (1994a), however, interpreted their results more cautiously in terms of the subject simply using its pointing in a flexible manner to direct the human to the food via an intermediary tool. That is, in their interpretation the subject knew that to get the food the human first had to get the rake, and that pointing at things directed humans to them. This implies an understanding that humans are animate and directed and that they make things happen, but it does not imply an explicit assessment of the human's belief states.

Finally, there are three experimental studies of knowledge attribution with monkeys. First, Povinelli, Parks, and Novak (1991) attempted to replicate the Povinelli et al. (1990) experiment with chimpanzees using rhesus macaques. As in the original study, subjects watched one experimenter hide food while another was out of the room or had a bag over his head; the subjects then could choose to be informed of the food's location by either of the experimenters. Despite several hundred trials of training, none of the four monkeys tested learned reliably to choose as their informant the experimenter who had watched the baiting process. Second, Kummer et al. (1996), as reported in detail in section 7.1, found that longtail macaques did not learn to use an opaque partition in their cages to shield their forbidden drinking activities from human experimenters, suggesting that they did not know that the barrier would prevent the humans from seeing, and thus from knowing, about their illicit behavior.

Third, in the only study in which primates were asked to assess the knowledge states of conspecifics, Cheney and Seyfarth (1990a) tested Japanese and rhesus macaques (mostly mothers) to see if they would inform other individuals (mostly their offspring) of the presence of food or a predator. The experimental arrangement consisted of two conditions. In one condition both mother and infant sat in one cage and watched as an experimenter hid food in an adjoining cage or a predator (a human in a mask with a capture net) approached the adjoining cage and hid behind a barrier close to it. In the other experimental condition the infant was absent when these events took place and thus only the mother witnessed them. In both conditions the infant was then released into the adjoining cage, and the question was whether the mother would behave differently in the two conditions: would she make special attempts to use food calls or predator calls (or any other means of communication) to inform her ignorant offspring about the hidden food or predator? Cheney and Seyfarth found that the mothers behaved identically in the two conditions, and they did this even though their offspring behaved differently in the two conditions; that is, they searched for food when informed but not when ignorant, and they stayed close to their mothers when informed

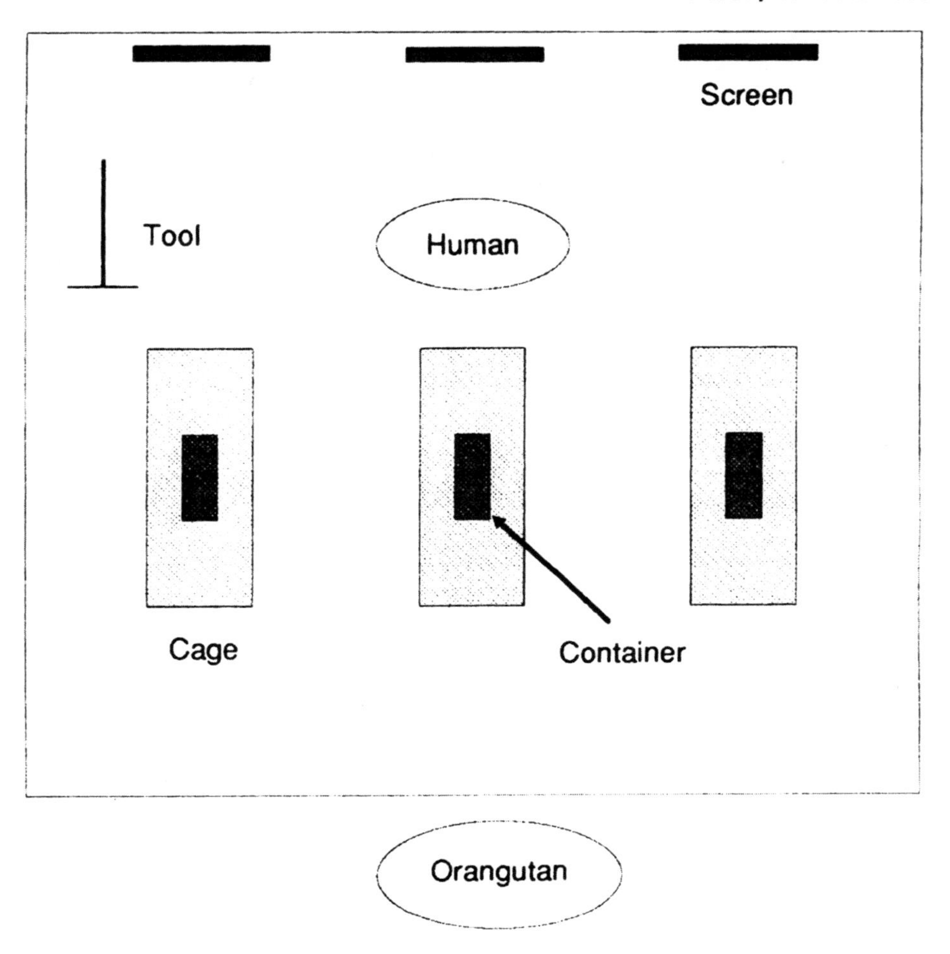

Figure 10.6. Call and Tomasello's (1994a) experimental arrangement. (Reprinted from *Journal of Comparative Psychology* with permission from the American Psychological Association.)

of the predator but not when they were ignorant. The investigators concluded that monkeys do not understand and take into account the knowledge states of conspecifics.

10.3.2 Other Evidence

The main source of observational evidence that has been used in support of primate understanding of the knowledge states of others is deceptive social strategies. As outlined in section 8.1, primate social strategies that have been labeled as deceptive may be interpreted in many ways. In the most generous interpretation, observations of deception indicate that one animal is attempting to create a knowledge state in another that does not correspond with what the first individual knows to be the case. The problem is that even in human developmental psychology there is still much debate over whether deception needs to be interpreted in this generous manner (e.g., Chandler, Fritz,

& Hala, 1989). Byrne and Whiten (1990, 1992) think that their several cases of "counterdeception" (in which one individual outwits another before it can carry out its act of deception) are especially revealing of primate skills of knowledge attribution. It is also possible, as outlined in section 8.1, that what primates are doing in these anecdotal observations is using intelligent social strategies to affect the behavior of others, not their mental states. Moreover, Whiten (1993) points out that especially good evidence that a particular act of deception was intended to create a false belief in another would be an unsuccessful attempt to deceive followed by a different type of attempt at deception aimed at creating that same false belief. This criterion of different means to the same end is one used widely to infer the intentions of organisms, but to our knowledge it has never led to the inference that any nonhuman primate was intending to create false beliefs in others.

A second source of evidence concerns teaching. Many investigators believe that intentional teaching presupposes an ability to compare the mental states of self and other: "I have some knowledge that another does not have" (e.g., Cheney & Seyfarth, 1990b). Although it is possible to interpret some teaching as focusing on the level of sensory–motor skills only (e.g., teaching someone to walk would not seem to involve mental state attribution), there is little evidence of intentional teaching in any species of nonhuman primate (see section 9.4), and what evidence there is does not involve the psychological level (nor is there good evidence for other animal species). Consequently, intentional teaching would not seem to be a behavioral domain that provides further evidence for nonhuman primates' understanding of the knowledge states of others.

10.3.3 Comparison of Species

Other than the anecdotal observations of deception, which have other plausible interpretations, there is basically no evidence that monkeys understand the knowledge states of others, and, indeed, in two experimental studies there is some evidence that they do not (Povinelli et al., 1991; Seyfarth & Cheney, 1990b). The evidence for apes is equivocal. Premack found some positive evidence in two of four chimpanzees, but subsequent control analyses raised the possibility of some kind of experimental strategy. Povinelli et al. (1990) also found some positive evidence for three of four chimpanzees, but they took a long time to learn to discriminate knowledgeable from ignorant humans, the transfer to a new but logically identical situation was also relatively slow, and the findings did not replicate with juvenile chimpanzees (Povinelli et al., 1994). Although we should not prejudge the issue given the small amount of systematic evidence available, at this point there is little convincing evidence that any nonhuman primate species understands the knowledge states of either humans or their conspecifics in the way that human children do from four years of age.

10.4 Understanding Self

There has recently been a resurgence of interest in topics such as self-perception, self-recognition, and self-concept, both in humans and in nonhuman animals (e.g., Neisser, 1993a; Parker, Mitchell & Boccia, 1994; Rochat, 1995c). A widely used methodology involves the reaction of individuals to their reflections in mirrors, as pioneered by Gallup (1968, 1970). In studying human self-understanding, a variety of other types of observations are also made involving such things as children's self-consciousness (embarrassment) in the presence of others during late infancy, their self-references in language during early childhood, and their verbal self-evaluations in later childhood (Harter, 1983). Because they are nonverbal, nonhuman primates have been studied almost exclusively within the mirror self-recognition paradigm. Because Gallup (1982) speculated that mirror self-recognition implies awareness of one's own mental states, and that this in turn implies awareness of the mental states of others, mirror self-recognition has been viewed as a dimension of social cognition. We thus examine first the evidence from this paradigm and follow this with a few observations on other lines of evidence for self-perception and self-understanding in nonhuman primates, mostly involving a more clearly social sense of self. Tables 10.1 and 10.2 summarizes studies of mirror self-recognition by nonhuman primates.

10.4.1 Mirror Self-Recognition

Gallup (1970) studied the reactions of four chimpanzees and six monkeys (four stumptail macaques and two rhesus macaques) to their mirror reflections. Each animal was individually exposed to a mirror for 8 hours per day for 10–14 days. Gallup noticed a marked difference in the behavior of the chimpanzees during this period. On about the third day of exposure, he saw a marked increase in their self-directed behaviors that relied on use of the mirror, for example, grooming parts of the body that would otherwise be visually inaccessible, picking bits of food from between the teeth while viewing the mirror image, and so forth. These replaced largely social behaviors in which the chimpanzees acted as if the mirror reflection was a conspecific. In the case of monkeys, however, Gallup saw few self-directed behaviors at any time during their days of exposure; their behaviors were almost all social. After the exposure period, all subjects of all three species were anesthetized and bright red marks were placed on one eyebrow ridge and on the opposite ear—both visually inaccessible to the subject. The marker left no olfactory or tactile cues of its presence. After recovery from the anesthesia, subjects were placed in front of a mirror once again and their behavior observed. In this "mark test," chimpanzees behaved in a number of ways that evidenced their understanding that the reflection in the mirror was their own body. For example, they touched the mark much more often than they touched that part of their body in a baseline period in which their bodies were not marked. Chimpanzees often inspected their fingers visually (and sometimes smelled them) after touching the marks. As a further control, chimpanzees who had never before experienced mirrors were also given the mark test and showed no particular behavior to the marks. The monkeys who had been exposed to the mirrors for as long as the chimpanzees also failed to behave in special ways to the marks. Gallup et al. (1971) replicated these findings for socially raised chimpanzees but failed to replicate them with chimpanzees raised in social isolation.

Subsequent research has corroborated and extended Gallup's initial findings. Orangutans behave in the same manner as chimpanzees in front of the mirror and in the mark test (Lethmate & Dücker, 1973; Miles, 1994; Suarez & Gallup, 1981). There is some evidence that bonobos also behave similarly in front of mirrors (Hyatt & Hopkins, 1994), although they have never been given the mark test. The interesting case for apes is the gorilla. Suarez and Gallup (1981) found that gorillas did not increase mark-directed behaviors in the mark test. To control for the possibility that the gorillas simply were not motivated to touch marks on their bodies, they also marked the gorillas'

Table 10.1 Self-recognition studies in apes.

Authors	Species	Age range (years)	N	Self-exploration (N)	Mark-directed (N)
Suarez & Gallup, 1981	*Gorilla gorilla*	13–19	4	no	no
Ledbetter & Basen, 1982	*Gorilla gorilla*	10–11	2	no	no
Patterson, 1990	*Gorilla gorilla*	19	1	yes	yes*
Evans, in Swartz & Evans, 1994	*Gorilla gorilla*	22	1		yes*
Patterson & Cohn, 1994	*Gorilla gorilla*	19	1	yes	yes*
		18	1	yes	no*
Nicholson & Gould, 1995	*Gorilla gorilla*	26	1		yes*
Lethmate & Dücker, 1973	*Hylobates lar*	adult	1	no	no
	Hylobates agilis	adult	1	no	no
Hyatt & Hopkins, 1994 and Westergaard & Hyatt, 1994	*Pan paniscus*	2–3	2	no	NA
		7–9	3	yes(2)	NA
		18–34	5	yes (2)	NA
Walraven et al., 1995	*Pan paniscus*	2	1	no	NA
		8–9	2	yes(2)	NA
		15–23	4	yes(3)	NA
Gallup, 1970	*Pan troglodytes*	juvenile	4	yes	yes
Hill et al., 1970	*Pan troglodytes*	1	1[a]	no	no
		1	2	yes	yes
Gallup et al., 1971	*Pan troglodytes*	3–6	3[a]	no	no
		3–6	3	yes	yes
Lethmate & Dücker, 1973	*Pan troglodytes*	8–9	2	yes	yes
Suarez & Gallup, 1981	*Pan troglodytes*	9	1	no	no
		16–21	3	yes	yes (3)
Robert, 1986	*Pan troglodytes*	1	1	no	no
Calhoun & Thompson, 1988	*Pan troglodytes*	3–4	2	yes	yes (2)
Swartz & Evans, 1991	*Pan troglodytes*	4–9	5	yes (1)	no
		10–19	6	yes (1)	yes (1)
Lin et al., 1993	*Pan troglodytes*	1–2	6	yes (4)	yes* (2)
		3–5	6	yes (6)	yes* (6)
Povinelli et al., 1993	*Pan troglodytes*	1–5	48	yes (1)	NA
		6–7	10	yes (2)	NA
		8–15	12	yes (9)	NA
		16–?	35	yes (9)	NA
		?	30		yes (10)
Boysen et al., 1994	*Pan troglodytes*	3–6	2	yes (2)	yes (2)
		4	2	yes (1)	NA
Hyatt & Hopkins, 1994	*Pan troglodytes*	11–15	2	yes (2)	NA
		20–37	5	yes (3)	NA
Povinelli et al., 1994	*Pan troglodytes*	3–4	6		yes (1)
Thompson & Boatright-Horowitz, 1994	*Pan troglodytes*	1	1		no
		3–5	5		yes (1)
		?	4		yes (4)
Eddy et al., 1996	*Pan troglodytes*	2–3	12	no	NA
		7–10	11	yes (4)	NA
Lethmate & Dücker, 1973	*Pongo pygmaeus*	subadult	2	yes (2)	yes (2)
Suarez & Gallup, 1981	*Pongo pygmaeus*	12	1		yes
		22	1		no
Robert, 1986	*Pongo pygmaeus*	2	1	no	no*
Miles, 1994	*Pongo pygmaeus*	juvenile	1	yes	yes*

Bracketed numbers indicate the number of subjects that showed the skill; NA = not tested; * = no anesthesia used; [a] = subjects reared in isolation.

Table 10.2 Self-recognition studies in monkeys.

Author (s)	Species	Age range (years)	N	Self-exploration	Mark-directed (N)	Reaching or Observing (N)
Lethmate & Dücker, 1973	*Ateles geoffroyi*	adult	1	no	no	NA
Eglash & Snowdon, 1983	*Cebuella pygmaea*	adult	12	no	NA	yes
Lethmate & Dücker, 1973	*Cebus apella*	adult	1	no	no	NA
Anderson & Roeder, 1989	*Cebus apella*	1–2	2	no	NA	NA
		5–19	5	no	NA	NA
Anderson, 1983	*Macaca arctoides*	infant	10	no	no*	NA
Anderson & Chamove, 1986	*Macaca fascicularis*	infant	8	no	NA	NA
Gallup, 1977	*Macaca fascicularis*	juvenile	1	no	no	NA
Anderson, 1986	*Macaca fascicularis*	juvenile	1		NA	no
		subadult	1		NA	yes
Mitchell & Anderson, 1993	*Macaca fascicularis*	12	1	no	no	NA
Itakura, 1987a,b	*Macaca fuscata*	5–6	2	no	no	yes(2)
Gallup et al., 1980	*Macaca mulatta*	infant	3	no	no	NA
		adult	1	no	no	NA
Menzel et al., 1985	*Macaca mulatta*	?	?		NA	no
Suarez & Gallup, 1986	*Macaca mulatta*	adult	2	no	NA	NA
Novak, 1996	*Macaca mulatta*	adult	1		yes	NA
Anderson, 1986	*Macaca nemestrina*	adult	3		NA	no
Boccia, 1994	*Macaca nemestrina*	subadult	6		no	yes (5/5)
		adult	8		yes (1)	yes (8)
Thompson & Boatright-Horowitz, 1994	*Macaca nemestrina*	9–18	3		yes (1)	NA
Lethmate & Dücker, 1973	*Macaca silenus*	adult	1	no	no	NA
Bayart & Anderson, 1985	*Macaca tonkeana*	adult	1	no	no	NA
Anderson, 1986	*Macaca tonkeana*	juvenile	1		NA	yes
		subadult	2		NA	no
		adult	1		NA	no
Lethmate & Dücker, 1973	*Mandrillus sphinx*	adult	1	no	no	NA
Benhar et al., 1975	*Papio anubis*	2	2	no	no	NA
Lethmate & Dücker, 1973	*Papio hamadryas*	adult	1	no	no	NA
Hauser et al., 1996	*Saguinus oedipus*	juvenile	2	yes	yes (2)	NA
		adult	4	yes	yes (3)	NA

Bracketed numbers indicate the number of subjects that showed the skill; NA = not tested; * = no anesthesia used

wrists and found that they did touch and rub those marks quite frequently. Other investigators have also failed to find evidence of mirror self-recognition in gorillas (e.g., Ledbetter & Basen, 1982). Patterson (1990) and Patterson & Cohn (1994), however, reported that the human-raised gorilla Koko did pass the mark test and therefore recognizes herself in the mirror. This has now become accepted by most investigators, see Parker et al., 1994. Recently, Evans (cited in Swartz & Evans, 1994; see also Parker, 1994) also indicated that another gorilla had passed the mark test. The reason for this difference between gorillas that passed the mark test and other gorillas is not clear (but see Povinelli, 1993, for some speculations). Gibbons have not displayed mirror self-recognition (Lethmate & Dücker, 1973).

a

b

c

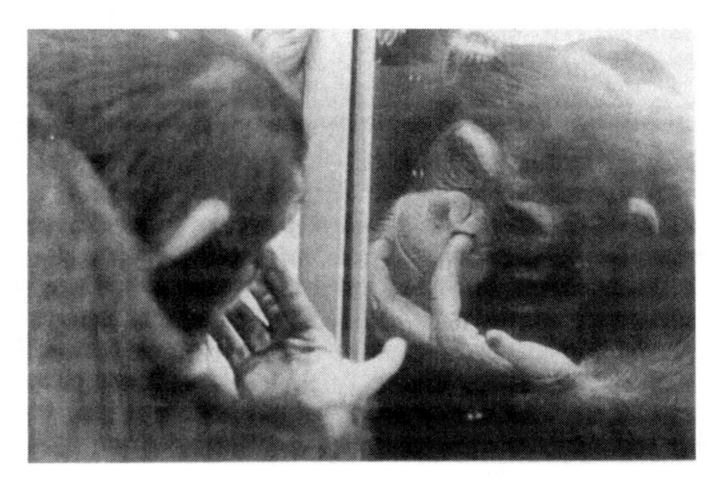

Figure 10.7. Young chimpanzees presented with a mirror engage in contingent facial (a) and body (b) behaviors and facial (c, d, e) and body (f) self-exploration. (Photos courtesy of D. J. Povinelli.)

d

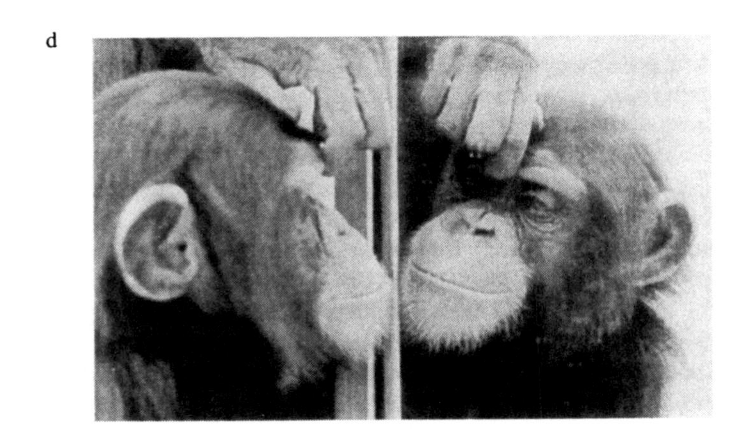

e

f

Figure 10.7. *(continued)*

Subsequent research has found that most monkey species under most conditions do not show mirror self-recognition. The species tested to date include spider monkeys, capuchin monkeys, squirrel monkeys, stumptail macaques, rhesus macaques, pigtail macaques, longtail macaques, liontail macaques, mandrills, olive baboons, and hamadryas baboons (see Anderson, 1984; Gallup, 1982, 1987; table 10.2). However, recently several investigators have reported some possible instances of mirror self-recognition in monkeys (e.g., Boccia, 1994; Thompson & Boatright-Horowitz, 1994). Of special interest is the possibility that monkeys have not displayed mirror self-recognition because they do not find marks in the traditional tests salient enough, or else they regard direct looks to the face of conspecifics as an aggressive display. Thus, Hauser, Kralik, Botto-Mahan, Garrett, and Oser (1995) found that cotton-top tamarins did pass the mark test if the marking was done in a particularly salient way on the white hair on top of the subjects' head—a species-typical and highly distinctive trait. Similarly, Novak (1996) trained rhesus monkeys to look at another individual face-to-face before presenting the mirror, and consequently found evidence of mirror self-recognition. Other investigators have wondered if perhaps monkeys do not pass the mark test because they do not know how to use mirrors to guide their reaching behavior. Tests of their ability to use mirrors to reach for visually inaccessible objects have produced mixed results (e.g., positive for pigtail, Japanese, longtail, and Tonkean macaques: Boccia, 1994; Itakura, 1987a,b; Anderson, 1986, respectively; and negative for "longtail and rhesus macaques": Anderson, 1986; Menzel et al., 1985 respectively). The question of whether there are some monkey species that recognize themselves in the mirror is thus undergoing renewed investigation.

Recent research has also shown some individual and developmental differences in the mirror behaviors of species who do seem to self-recognize. Swartz and Evans (1991) reported that only 1 of 11 chimpanzees, mostly wildborn but captive from an early age, passed the mark test. In the most thorough developmental analysis to date, Povinelli, Rulf, Landau and Bierschwale (1993) found that even after many hours of exposure, only 26% of their 35 fully adult chimpanzees (older than 16 years) showed clear signs of self-recognition in front of mirrors, operationalized as the performance of at least 5 self-exploratory behaviors directed at body parts not visually accessible without the mirror. This is as compared with about 75% of their 12 subadults (8–15 years), about 20% of their ten 6–7 year olds, and only 1 of 48 subjects in the 1–5 years old. With the exception of one individual, no chimpanzee passed the mark test before 4.5 years of age, and of all the subjects who showed positive signs of self-recognition in front of the mirror, about half passed the mark test. The developmental picture presented by Povinelli et al. is somewhat at variance with the findings of Lin, Bard, and Anderson (1992), who claimed to find signs of mirror self-recognition in 2.5-year-old chimpanzees. Povinelli et al. point out, however, that Lin et al. used a more generous definition of self-directed behaviors in their observations of spontaneous behavior and failed to provide control observations during the mark test to establish each individual's propensity for touching the places on its body that would eventually be marked (especially important because they also applied the marks when the animals were fully awake). Eddy, Gallup, and Povinelli (1996) have also replicated their developmental findings, as opposed to those of Lin et al., in another paradigm (addressing the criticisms of Heyes, 1994b, in the process). In a control procedure using a video display of conspecifics, as opposed to a mirror, Eddy et al. found that chimpanzees' seemingly mirror-contingent face and body movements were not reliable indicators of self-recognition, but that self-exploratory behaviors in which the direction of gaze to the body is more specific (and, of course, the mark test) are more reliable.

It is noteworthy that all the species tested with the mark test who have also been given the control procedure in which they are marked on their wrists as well invariably stare at, lick, and otherwise inspect those marks. These behaviors suggest that the subjects know that the wrist is a part of their body and they know this visually. The point is that primates know a good bit about their own bodies. Some parts they know visually, and probably all parts they know tactually and/or proprioceptively. In the most conservative accounts of mirror self-recognition (e.g., see Loveland, 1993; Neisser, 1993b; Rochat, 1995a), the organism is simply using the mirror as a tool to gain visual access to a part of its body (e.g., the face or rump) that it previously could perceive only tactually or proprioceptively. Mirror self-recognition is thus about perception of the body, which all primates are likely skillful at with no special training or experiences. Why vision should be privileged over other forms of self-perception and why the face should be privileged over other parts of the body, so that some researchers have begun talking about mirror self-recognition as "self-awareness" in a humanlike sense (e.g., Parker et al., 1994), is a question that researchers should address more directly. Other forms of self-knowledge involving aspects of the self other than the body are much more relevant to these loftier aims (Neisser, 1988).

10.4.2 Self as Social Agent

The aspects of self-knowledge that humans identify with most strongly are the social aspects. Thus, most investigators of human children find it significant that at around 2 years of age infants not only begin to pass the mark test in front of mirrors, but also begin to show signs of self-consciousness and embarrassment (Lewis, Sullivan, Stanger, & Weiss 1989). They do this both during the mark test in front of the mirror (e.g., monitoring the attitudes of others toward themselves, as if aware of the way the mark changes their appearance for others) and in other ways (e.g., coyness and shyness in front of strangers in ways that indicate their knowledge that others are perceiving and evaluating them). When communicating with others, children at around this same age begin to refer to themselves by various means in their language. Ironically, this aspect of self has long been known as the "looking-glass self" (Cooley, 1902), but the mirror in this case is not physical but social.

There is little evidence that nonhuman primates have a sense of self as a social agent. Although human-raised primates often show a kind of "shame" after disobeying their caretakers, this behavior is a simple submissiveness in anticipation of punishment, as it is in many pets such as dogs. There have been no reports of nonhuman primates showing anything resembling embarrassment in connection with the mirror or in any other contexts (even in entertainment contexts in which humans have made every effort to make them appear ridiculous to other humans). Linguistic apes have learned gestures or signs to indicate themselves, but the meaning of these signs is unclear. For example, the orangutan Chantek learned manual signs that humans glossed as 'Chantek' and 'me' before he showed any signs of recognizing himself in a mirror (Miles, 1994). It is thus possible that these signs are used by human-raised apes to indicate their bodies, or parts of their bodies, not their selves as social agents. Human children's early use of self-referential expressions could also be interpreted in this way, except that there are so many other signs of an emerging sense of a social self. None of the linguistic apes have gone on, to our knowledge, to talk about themselves and their own peculiar characteristics the way that human children do sometime after their second birthdays.

10.4.3 Comparison of Species

Whatever the meaning of mirror self-recognition, it is now being studied in a number of species. Epstein, Lanza, and Skinner (1981) taught a pigeon to peck a mark on itself in front of a mirror, but Gallup (1982) and others have pointed out not only the artificiality of the behavior given the training, but also the complete absence of any other signs of self-recognition in front of the mirror. Recent evidence suggests the possibility that dolphins may recognize their own bodies in front of mirrors, although the problems in interpreting the mark test for a species without extended forelimbs are great (Marino, Reiss & Gallup, 1994; Marten & Psarakos, 1994). Human children are well known to show all kinds of evidence of self-recognition in front of mirrors in the months immediately preceding their second birthdays (e.g., Amsterdam, 1972; Lewis et al., 1989).

Mirror self-recognition is one of the areas of research where most has been made of differences between monkeys and apes. By all accounts, the monkeys who recently have displayed some mark-directed behaviors in the mark test still differ significantly from chimpanzees and orangutans in the amount of interest in the mark as well as in other behaviors indicative of self-recognition (Gallup, 1994). However, factors such as the tendency to not look directly into the eyes of conspecifics displayed by some monkey species and difficulties in using mirrors to guide reaching behavior in other monkey species are problems that should be further investigated. Theoretically, it is important that theorists who link mirror self-recognition to social cognition do so in a more explicit way and propose ways to test for this connection.

10.5 What Primates Know about Others' Mental States

Theory of mind has become important in current theorizing about primate cognitive development for at least three reasons. First, a number of influential theories of primate evolution have proposed that the evolution of primate cognition took place mainly to solve social, not physical, problems, and having a theory of mind would be a powerful tool in solving social problems (e.g., Whiten & Byrne, 1988a). Second, cognitive scientists interested in the issue of cognitive representation have become interested in the question of how human beings form mental representations of the mental representations of others, so-called metarepresentations (e.g., Leslie, 1987). This process is thought to be especially important because of its potentially recursive nature, and recursion is vitally important in much of human cognitive development during the school-age years and beyond (Perner, 1993). Third, many theorists believe that how individuals of a species understand the behavior, and perhaps psychological states, of conspecifics will determine how they interact with them in such fundamental domains as communication and social learning. Something like a theory of mind might thus lead a species to the kind of social structure epitomized by human culture in which individually acquired skills and information are passed along culturally in addition to genetically (Tomasello et al., 1993).

Figure 10.8. Possible opportunity for animals to see their own reflection in nature. (Photo courtesy of F. de Waal.)

Throughout chapters 7, 8, and 9 we documented some of the ways that nonhuman primates understand the behavior of conspecifics as animate and directed. It is less clear, however, that nonhuman primates understand the psychological dimensions of the behavior and experience of others, in terms of either their intentions or beliefs. Clearly, apes understand something about visual perception, using bodily orientation to decide when a communicative signal should and should not be given, and even on occasion actively conceal from others one of their own involuntary displays (with one home-raised orangutan showing evidence of understanding the role of the eyes specifically in this process). Several laboratory chimpanzees have also followed the gaze of humans to interesting sights, again with some evidence that they understand the role of the eyes in this process. Monkeys show some evidence of understanding some of these same things, for example, stumptail and other macaques know when someone is looking at them directly or not, and some other species (e.g., hamadryas baboons and pigtail macaques) seem to engage in some communicative efforts to make sure that others perceive their communicative signals. These are all skills that human children show in one form or another soon after the end of infancy.

However, with regard to understanding the less observable aspects of behavioral functioning, there is no solid evidence that nonhuman primates understand the intentionality or mental states of others. Premack and Woodruff's (1978) study supposedly showing that one chimpanzee understood the behavior of humans as intentional has other fairly plausible interpretations; there have been several other failures to replicate these findings; and the studies of apes' understanding of knowledge have all found that they took many trials with feedback to learn to discriminate knowledgeable from ignorant individuals (with at least one failure to replicate these findings as well). Because the number of relevant studies in this domain is small, it is also possible that apes and other nonhuman primates do understand the psychological states of others, but we simply have not tested them in the appropriate ways. In general, our knowledge in this area is based on so few studies, with difficult and sometimes conflicting interpretations, that definitive conclusions are impossible. It should also be noted that there is basically no research with nonprimate animal species that could be used in this area for comparative purposes. Human children typically master these kinds of theory of mind skills in early childhood.

Mirror self-recognition is a well-established phenomenon in some great apes (although the question of the gorilla remains) but until recently has not been established in other primate species. Recent investigations of monkeys, in which the task has been adapted for their particular behavioral propensities, seem to be showing some skills in some monkey species as well. However, the importance of this behavior for questions of theory of mind is not well established, and, indeed, in the one study in which an attempt was made to correlate mirror self-recognition with an understanding of the psychological states of others, results were equivocal (Povinelli et al., 1994). One plausible view is that mirror self-recognition reflects the ability of an individual to use a mirror to inspect visually inaccessible places on their bodies, but this behavior has basically no interesting implications about social cognition or theory of mind. Mirror self-recognition emerges in human children at the end of infancy, and there are few data from nonprimate animals.

Because these issues are so important for social cognition, it is surprising that there is so little rigorous empirical research on nonhuman primate theories of mind. There is a special need for research focusing on animals interacting with members of their own species. It is possible that when nonhuman primates interact with humans they do not display the abilities that they would show with a conspecific, and thus we are currently underestimating their abilities. It is also possible that the human interaction and training that is required for experimentation creates species-atypical skills in nonhuman primates, and so we are currently overestimating their skills. It is also possible that by focusing on humanlike tasks, we are looking for the wrong thing altogether. Of course it is not easy to think of interactive situations that do not have both "rich" interpretations in terms of an understanding of the intentions and beliefs of others and at the same time a "lean" interpretation based on an understanding of the behavior of others. However, as in all areas of scientific endeavor, the appropriate procedure is for many different investigators each to try their own approach, including both naturalistic and experimental observations, with converging evidence from several different sources needed before conclusions may be reached with confidence. As of now, with respect to nonhuman primate theories of mind, this convergence of evidence does not exist, and so a healthy agnosticism, or even skepticism, would seem to be the wisest position.

11

Theories of Primate Social Cognition

Most primate species live in complex and dynamic social groups that present them with difficult cognitive problems on a continuing basis. All theories of primate social cognition assume this fact, although whether and in what ways primate social life is more complex than that of other mammalian species is a difficult question. Cheney et al. (1987) noted that although differences between primates and other mammals are not especially obvious, there are three characteristics of primate social life that together may make for special complexities. First, in a purely quantitative sense, most primates live in complex social groups that result in "a complex network of interactions, with many alternative strategies for survival and reproduction." Second, primates form various types of long-term social relationships with others involving kinship, dominance, and short-term coalitions, and these may change over the life cycle. Third, primates have an extremely varied repertoire of means for expressing themselves socially in terms of species-typical displays that make manifest their emotional states and behavioral intentions. Although it would be difficult to document primate uniqueness in any one of these characteristics, together they make for a highly complex form of sociality that may encourage the development of all kinds of social knowledge, social strategies, and social learning (see Cords, in press, for a review).

Theories of primate social cognition, like theories of primate physical cognition, concern both the proximate mechanisms by means of which primates learn to deal with social complexity on a day-to-day basis and the kinds of ecological conditions that might have led to the evolution of primate social cognition in the first place (ultimate causation). Unfortunately, in neither of these cases are the theories as well developed and articulated as they are in the physical domain. Nevertheless, as in chapter 6, our goals in this chapter are to (1) review the evidence presented in the previous four chapters for primate social adaptations that rely on individuals making flexible and informed choices, prototypically involving some form of mental representation; (2) present the major theoretical perspectives that have been proposed to account for this evidence; and

(3) make suggestions for future empirical research and theoretical development in the field of primate social cognition.

11.1 Summary of Primate Social Cognition

Table 11.1 provides a compendium of the primate species that have been investigated in each of the domains of social cognition reviewed in the previous four chapters. By way of summarizing the data, we briefly relate the major characteristics of the cognitive mechanisms employed by the various species in each of the three major groups of primates in each of these cognitive domains (i.e., in our summary description we combine New World and Old World monkeys).

11.1.1 Prosimians

As might be expected, we know little about prosimians in the domain of social cognition. Of the 16 domains listed in Table 11.1, we have information on prosimians in only 2 (as compared with 13 for monkeys and 16 for apes). Two species of lemurs use communicative vocalizations that seem to be associated with particular referents (types of predators). However, none of the investigators who have actively watched prosimians for many years reported to Byrne and Whiten (1990) any instances of deceptive communication or behavior. One species of lemur and one species of bushbaby have been observed to engage in social learning of instrumental behaviors, probably involving stimulus enhancement only. Although several theorists have hypothesized that prosimians do not possess the kinds of complex social-cognitive skills of monkeys and apes (e.g., Whiten & Byrne, 1991), nothing systematic is known about their understanding of social relationships, their skills of coalition formation or cooperation, the flexibility of their communication, the extent of their skills of social learning, or anything about their "theories of mind." In some of these cases, many nonprimate mammalian species have primatelike skills, suggesting that some common mammalian ancestor possessed them and so perhaps prosimians may have capacities as yet undiscovered in these domains.

11.1.2 Monkeys

A number of cercopithecine monkey species (little is known about New World monkeys and colobine Old World monkeys) form coalitions and alliances in their quest for resources and dominance status, sometimes using social strategies to recruit specific allies. There is some evidence for reciprocity in these coalitions and alliances across time, as well as the interchange of grooming for coalition support, with evidence that mental record keeping is required in at least some cases. Some these same monkey species have been shown to recognize individual conspecifics and to understand both direct and third-party social relationships. There are a few observations of cooperation among individuals of some monkey species, but rhesus monkeys who learned to cooperate could not later reverse roles without further training. There are many reported observations of "deception" by monkeys, although the extent to which these behaviors involve something other than basic social strategies is an open question. Several monkey species use their vocalizations for specific environmental referents, in some cases in

Table 11.1. Catalogue of species studied in each social-cognitive domain (see appropriate section for details of findings).

	Prosimians	New World monkeys	Old World monkeys	Apes
I. Social knowledge and interaction				
The Social field		*Ateles* *Cebuella pygmaeus* *Saimiri sciureus*	*Cercocebus albigena, torquatus* *Cercopithecus aethiops* *Macaca arctoides, fascicularis, fuscata, mulatta, nemestrina* *Papio anubis, hamadryas, ursinus* *Theropithecus gelada*	*Gorilla gorilla* *Pan paniscus, troglodytes* *Pongo pygmaeus*
Coalitions & alliances			*Cercopithecus aethiops* *Macaca arctoides, fascicularis, fuscata, mulatta, sinica* *Papio anubis, hamadryas, ursinus* *Theropithecus gelada*	*Gorilla gorilla* *Pan paniscus, troglodytes* *Pongo pygmaeus*
Reciprocity & interchange			*Cercopithecus aethiops* *Erythrocebus patas* *Macaca arctoides, fascicularis, fuscata, mulatta, sinica* *Papio anubis, hamadryas, ursinus*	*Pan paniscus, troglodytes*
Cooperative problem solving		*Ateles geoffroyi* *Cebus capucinus*	*Colobus badius* *Macaca fuscata, mulatta, nemestrina, tonkeana* *Papio anubis, hamadryas, papio*	*Pan troglodytes*

II. Social strategies and communication				
Deception		*Ateles geoffroyi* *Cebus apella*	*Cercocebus torquatus* *Cercopithecus aethiops* *Macaca fascicularis* *Papio anubis, cynocephalus, hamadryas*	*Gorilla gorilla* *Pan troglodytes*
Gestural communication			*Macaca arctoides, nemestrina*	*Gorilla gorilla* *Pan paniscus, troglodytes* *Pongo pygmaeus*
Vocal communication	*Lemur catta* *Varecia variegata*	*Ateles geoffroyi* *Cebuella pygmaea* *Cebus capucinus* *Leontopithecus rosalia* *Saguinus oedipus* *Saimiri orestedii, sciureus*	*Cercopithecus aethiops* *Macaca arctoides, fuscata, mulatta, nemestrina, sinica*	*Gorilla gorilla* *Hylobates klosii* *Pan paniscus, troglodytes* *Pongo pygmaeus*
Communication with humans		*Cebus apella*	*Macaca mulatta*	*Gorilla gorilla* *Pan paniscus, troglodytes* *Pongo pygmaeus*

(*continued*)

Table 11.1. *(continued)*

	Prosimians	New World monkeys	Old World monkeys	Apes
III. Social Learning and Culture				
Traditions		*Ateles geoffroyi* *Saimiri oerstedii*	*Cercopithecus aethiops* *Macaca fuscata, mulatta* *Papio*	*Gorilla gorilla* *Pan troglodytes*
Social learning, instrumental	*Lemur fulvus* *Otolemur garnettii*	*Ateles geoffroyi, belzebuth, fusciceps* *Cebus apella, albifrons* *Saimiri sciureus*	*Cercopithecus aethiops* *Erythrocebus patas* *Macaca arctoides, fascicularis, fuscata, mulatta, nemestrina, silenus* *Mandrillus sphinx* *Papio cynocephalus, hamadryas, papio, ursinus*	*Gorilla gorilla* *Pan paniscus, troglodytes* *Pongo pygmaeus*
Social learning, communication		*Cebuella pygmaea* *Saguinus labiatus* *Saimiri sciureus*	*Cercopithecus aethiops* *Macaca arctoides, fascicularis, fuscata, mulatta, nemestrina*	*Gorilla gorilla* *Pan paniscus, troglodytes* *Pongo pygmaeus*
Teaching		*Ateles geoffroyi*	*Cercopithecus aethiops* *Macaca fuscata, mulatta nemestrina*	*Gorilla gorilla* *Pan paniscus* *Pan troglodytes*

IV. Theory of mind				
Behavior & perception	*Lemur fulvus, macaco*	*Cebus apella, capucinus* *Saimiri sciureus*	*Cercopithecus aethiops* *Macaca arctoides, fascicularis, fuscata, mulatta, nemestrina, tonkeana* *Papio hamadryas, papio*	*Gorilla gorilla* *Pan paniscus, troglodytes* *Pongo pygmaeus*
Intentions & attention				*Pan paniscus* *Pan troglodytes*
Knowledge & beliefs			*Macaca fascicularis, fuscata, mulatta*	*Pan troglodytes* *Pongo pygmaeus*
Mirror self-recognition		*Ateles geoffroyi* *Cebuella pygmaea* *Cebus apella* *Saguinas oedipus*	*Macaca arctoides, fascicularis, fuscata, mulatta, nemestrina, silenus, tonkeana* *Mandrillus sphinx* *Papio anubis, hamadryas*	*Gorilla gorilla* *Hylobates lar, agilis* *Pan paniscus, troglodytes* *Pongo pygmaeus*

flexible ways attuned to the audience and contextual situation (and some nonprimate species do this to some extent as well). We have basically no systematic observations of monkeys' skills of intentional gestural communication or their skills of communication with humans after training with a human-based symbol system.

Japanese macaques have been observed to create and maintain some population-specific behavioral traditions over time (culture), and a number of different monkey species have shown some skills of social learning in their vocal communication as well as in instrumental contexts. It is unlikely, however, that any of those skills involve imitative learning strictly defined. There are observations of several macaque species that some investigators have interpreted as teaching, although it is unclear if teaching was their intention. Monkeys do not seem to have a theory of mind concerning their conspecifics, if what is meant by this term is understanding them in terms of their intentions and beliefs. The majority of monkeys tested do not seem to recognize their own faces in mirrors, although some recent methodological innovations and resulting observations threaten to overturn this conclusion.

11.1.3 Apes

The data for apes are very spotty in some domains as well and almost nonexistent for the lesser apes (there are only some limited data on the vocal communication of one species of gibbon). However, we have more data for the great apes than any other group of primates, with data for chimpanzees available in all 16 of the domains of Table 11.1 (and 11 domains for gorillas and 10 domains for orangutans and bonobos). Like cercopithecine monkeys, all apes seem to recognize individual conspecifics, to understand both direct and third-party social relationships, to form coalitions and alliances, and to engage in various altruistic interactions reciprocally. Chimpanzees have been observed to cooperate with one another in some complex ways in the wild (mostly in hunting), and in one study of captive chimpanzees there is some equivocal evidence that they may be able to understand the role of a human who is cooperating with them. Ape vocal communication has not been studied as much as monkey vocal communication, and there is much less evidence that it is a flexible, intentional system sensitive to audience and communicative context. On the other hand, the intentional gestural communication of apes (especially chimpanzees) has been more extensively studied and clearly involves the flexible use of a variety of different communicative strategies. Apes engage in what some investigators have called deception as frequently as monkeys, and great apes of all four species have learned in various training contexts some aspects of human-based systems of communication.

Chimpanzees have a number of population-specific behavioral traditions, some involving the use of tools and some, perhaps, involving gestural communication. In captivity apes have shown complex skills of social learning in which they learn about the changes of state that may be brought about in instrumental situations (emulation learning) and in which they create new communicative signals with conspecifics (ontogenetic ritualization). After human training of various sorts, three species of ape have shown the ability to mimic human gestures and movements, but it is unclear whether they can reproduce the intentional behavioral strategies of others. As in monkeys, there are a few reported observations of active teaching in apes, and all these have other explanations. In a variety of ways, apes have demonstrated that they understand

Figure 11.1. A pair of adult stumptail macaques engaged in "hold-bottom" behavior. (Photo by J. Call.)

something about visual perception: they understand that another has to be oriented to them for a communicative signal to achieve its goal, and they can in some circumstances follow the gaze of another to an outside entity. Although there are a few experiments suggesting that apes may understand the intentions and beliefs of others (i.e., humans), there are other studies showing that they do not. At least one individual from three great ape species has shown evidence of recognizing its face in a mirror; many other great ape individuals do not show such evidence, however, and this is especially true of gorillas.

11.1.4 Comparison of Species

It may be useful at this point to make some overall comparative statements about the social cognition of the different primate species. Because there are so few data available on the prosimians and lesser apes, it is best to compare the cognitive skills of monkeys and great apes (although the monkey data in the social domain are almost all from Old World monkeys of the cercopithecine branch), which is in any case the comparison of most interest theoretically to many researchers.

De Waal & Aureli (1996) have stressed that in dominance and agonistic contexts, great apes (chimpanzees) may engage in more complex forms of reconciliation involving third-party mediation than monkeys (two species of macaques). They also note, however, that there may be noncognitive reasons for this difference in terms of different social constraints concerning aggression. In their studies of trained cooperation with humans, Povinelli et al. (1992) have stressed that apes (chimpanzees) may understand the role of the other more deeply than monkeys (rhesus monkeys). The one study of chimpanzees on which this conclusion is based, however, concerned individuals who may have come to the experiment with skills that made them appear to reverse roles when in fact they did not. The vocal communication of monkeys is more complex and flexible than that of great apes, whereas for gestural communication the reverse may be true (although there are almost no studies of the intentional gestural communication of monkeys). Byrne and Whiten (1992) have claimed that great apes may engage in some kinds of deception that monkeys do not, although this inference is based almost totally on anecdotal observations and their classification into types by a single set of investigators. Human-raised great apes have learned some humanlike skills of symbolic communication, whereas monkeys have not been raised and tested in comparable ways.

A number of investigators have claimed that great apes are better imitators than monkeys, but only a two species of primates have been observed to have population-specific behavioral traditions in the wild—one monkey (Japanese macaques) and one ape (chimpanzees). In addition, in experimental contexts, he only individuals to show humanlike skills of imitative learning are human-raised great apes; monkeys simply have not been raised and tested in comparable ways. Based on comparable experiments with chimpanzees and rhesus monkeys, Povinelli and colleagues have claimed that apes may understand other minds in a way that monkeys do not, but subsequent criticisms (Heyes, 1993a), reanalyses (Povinelli, 1994a), and failed replications (Povinelli et al., 1994) have led to a more cautious assessment of possible monkey–ape differences (Povinelli, 1996). Three of the four great ape species have shown more robust skills of recognizing their faces in mirrors than most monkey species; however, recent evidence is that some monkey species have some self-recognition skills as well.

As in the case of physical cognition, it is widely believed that great apes have more complex social cognitive skills than monkeys. However, it is not widely appreciated that the database on which this conclusion depends is very small (Tomasello & Call, 1994). Moreover, it is possible that this belief reflects not the weight of the scientific evidence, but rather a systematic bias in favor of apes. Great apes have a physical size and appearance that more closely resembles that of humans, which may also lead to more humanlike attributions by researchers (Eddy, Gallup & Povinelli, 1993). Moreover, in many cases the great apes observed in tasks of social cognition have received much human training and enculturation of a type monkeys have never received (why monkeys have never been chosen for such training is an interesting question in itself). This training may lead to social cognitive skills that are qualitatively different from species-typical skills (Call & Tomasello, 1996). It should also be pointed out that the researchers making observations of monkeys and great apes in the field may come from different research traditions, with ape researchers more concerned with human evolution and thus more likely to see human traits. In any event, we see no compelling evidence that great apes as a group possess any social-cognitive skills that are not also possessed by monkeys as a group.

11.2 Theories of Proximate Mechanism

As mentioned previously, theories of primate social cognition are not well articulated at present. As in the case of physical cognition, this is due at least in part to the focus on "intelligence" as a unidimensional phenomenon that organisms have more or less of, rather than a focus on the particular cognitive mechanisms that various species employ in different situations. As in the case of physical cognition, there is also an excessive concern with determining whether species "have" mental representations—in this case whether they "have" metarepresentations (mental representations of the mental representations of others). What is needed are more articulated views of primate social cognition that specify different types of social knowledge and skills, without an excessive concern for their computational aspects. After a brief review of current theoretical approaches to primate social cognition, from both proximate and ultimate perspectives, we make some modest proposals in this direction (and these will continue into the succeeding chapter, which is concerned with primate physical and social cognition as a whole).

11.2.1 The Social Function of Intellect

The first systematic statement of the social function of primate "intelligence" was made by Jolly (1966). Based on her extensive observations of several prosimian species, both in the wild and in the laboratory, Jolly noted that complex intellectual processes are required for many social interactions, and, further, that some prosimian species show complex intelligence in their social interactions without showing comparable intellectual complexity in their interactions with inanimate objects. As befits the time at which this paper was written, Jolly basically equated intelligence with learning and the solving of discrimination learning problems. She did not posit any processes of social intelligence qualitatively distinct from physical intelligence, but simply observed that for all primates the social domain presented certain problems whose solution required sophisticated learning capacities, whereas this was not true for all species in the domain of objects and the "gadgetry" of experimental psychologists. Basically, Jolly's attempt was to expand our definition of intelligence to include the capacity to deal with social as well as technical problems and discriminations.

Humphrey (1976) came to some similar conclusions, but by a different route. His main empirical observation was that many primates display many complex skills of learning, problem solving, and creative intelligence in the laboratory that they do not seem, on cursory inspection, to display in the wild. There seems to be nothing in the lives of primates in their natural habitats to correspond to the many perverse and artificial problems they solve with regularity for humans. His solution to this conundrum was to posit that primates had evolved their skills of creative intelligence in social situations in the wild and that they were simply transferring those skills to the physical domain to solve various problems involving technical gadgetry in the laboratory. Humphrey also pointed out that the social world does not remain inert as the physical world does, and, thus, for any social strategy one individual may develop, another individual may attempt to countermand by developing its own strategy. His analogy was to the human game of chess involving plot and counterplot. Humphrey did not discuss any unique content that social intelligence or social cognition might have in terms of theory of mind or the like, but he implied that there may be some unique manifestations of intelligence in the social domain.

We should also note that more recently a number of investigators have focused on more cooperative interactions in which individuals exchange favors over time (see section 7.3 "Reciprocity and Interchange" as manifestations of some general (i.e., not

Figure 11.2. Japanese macaque helping sibling. (Photo courtesy of H. Gouzoules.)

specifically social) intellectual skills. There are two aspects of general intelligence that reciprocal social interactions are hypothesized to require: long term memory and quantitative skills. Whereas no theory of intelligence has been specifically built out of reciprocal interactions alone, they should at least be added to the list of general cognitive skills that complex social interactions require.

11.2.2 Metarepresentation and Social Cognition

Following Premack and Woodruff's (1978) question "Do chimpanzees have a theory of mind?" many researchers have gone beyond Jolly and Humphrey's early statements to inquire whether primate social cognition might involve some structures or processes distinct from those of physical cognition. The "theories" of social cognition that have derived from this perspective are still crude.

Following the research in theories of mind of human children, primatologists such as Whiten and Byrne (1991) have investigated whether the complex of social cognitive skills displayed by various primate species allows us to say that they employ metarepresentations as part of their cognitive apparatus. Metarepresentation, sometimes known also as mental state attribution or theory of mind, is the ability to form mental representations of the mental representations of others. Whiten and Byrne (1991) argue that apes, but not monkeys, have evolved the ability to employ metarepresentations in their social cognition. Their evidence is mainly that reviewed above for possible differences in monkey and ape abilities of deception, imitation, and theory of mind tasks involving the attribution of intentions, beliefs, and other psychological states. Whiten (1993, 1994) goes on to elaborate this view of mental state attribution as an aid in predicting the behavior of others. Byrne (1994) focuses on the role of social interactions in developing both skills of metarepresentation and skills of planning and calculation in general, which, when put together, lead to such things as the mental simulation of intentional actions of others.

A similar argument has also been made by Gallup (1982) and revised and expanded by Povinelli (1993), that apes understand the mental states of others in a way that monkeys do not. Gallup and Povinelli's evidence is mainly monkey–ape differences in mirror self-recognition, along with similar differences in such things as deception, imitation, pretense, and mental state attribution. Their argument revolves around the hypothesis that mirror self-recognition is an indication that the organism understands itself as a mental being, and so likely understands others in this way as well, perhaps by attributing its own self-perceived mental states to others. Other theorists have contested the view that mirror self-recognition involves anything resembling mental state attribution (Call & Tomasello, 1996; Heyes, 1994b), arguing instead that what is involved is something like mirror-guided body inspection.

Cheney and Seyfarth (1990b) also conceptualize the issues in terms of mental state attribution. They are less positive than either Whiten and Byrne or Gallup and Povinelli that any nonhuman primates understand the mental states of others. Their major focus is on monkeys, and their conclusion is that "Monkeys cannot communicate with the intent to modify the mental state of others because, lacking attribution, they do not recognize that such states exist" (p. 310). They are somewhat agnostic with respect to apes, but in general they are not convinced that apes are qualitatively different from monkeys in this respect. Cheney and Seyfarth go on to posit that nonhuman primates possess a knowledge of the world that is highly domain-specific, a kind of "laser beam intelligence" in each domain, with little transfer across domains. They base this theory on the fact that vervet monkeys can show great intelligence in one domain (e.g., vocal communication) but fail to apply that intelligence in other domains that do not seem to require any more complex computational procedures (e.g., inferring the location of predators from tracks and other telltale signs). Human cognition, in this view, is unique precisely because of its relative domain generality.

Some other investigators have rejected the metarepresentation view as the best way to look at social cognition and instead have focused on more basic aspects of social understanding. For example, Tomasello, Kruger, and Ratner (1993) posited three distinct levels in the understanding of others. First, an organism may simply perceive the behavior of another animate being and, based on experience and general social knowledge (including, perhaps, use of gaze direction), learn to predict its future behavior. Second, an organism may perceive and understand the behavior of others as intentional, conceiving the other as an intentional agent whose behavior and attention are intentionally controlled. Third, beyond intentions, which are in a sense manifest in behavior, an individual may look beneath behavior to the beliefs involved, conceiving of the other as a mental agent whose thoughts and beliefs may differ from reality (roughly equivalent to theory of mind). Tomasello and Call (1994) interpret much of the data on primate social cognition more conservatively and skeptically than many other investigators, and conclude that no nonhuman primates understand the behavior of conspecifics as intentional or mental, with the possible exception that some human-raised apes may understand something of the intentions of their human interactants. Another part of their argument is that if nonhuman primates had a humanlike theory of mind, they should engage in more humanlike forms of communication, cooperation, social learning, and teaching than they seem to do in their natural habitats. In this book we have added to this account the possibility that the understanding of third-party social relationships may be uniquely primate, which, if true, would add a whole new layer of complexities to their social fields and might thereby account for the special social complexities that seem to characterize primates as an order (at least simian primates).

It is debatable whether any of the foregoing accounts of primate social cognition should be given the status of a full-fledged theory. In addition to a paucity of data, which lends itself to many plausible interpretations, theorists in this domain of primate cognition also suffer from underdeveloped theories of the different aspects of social cognition. In our view, as theorists of social cognition in general articulate more clearly some of the different ways that organisms may understand the behavior and mental life of others, primatologists will begin to find ways to test them in the species they study. Our own more articulated view of primate cognition, including both the physical and social domains, is presented in chapter 12.

11.3 Theories of Ultimate Causation

Unfortunately, theories of the evolutionary foundations of primate social cognition are as sketchy as the theories of its proximate mechanisms. Again, a major limitation is the view of "intelligence" as a single trait. Another limitation is the view that there should be a "single cause" of the evolution of intelligence in recent evolutionary history. This formulation of the problem, for example, structures the debate among theorists who attempt to correlate primate sociality with some measure of brain size as a measure of intelligence. Thus, Sawaguchi and Kudo (1990) found correlations between sociality and relative brain size in several species of primates, and, using different measures of sociality and brain size, Dunbar (1992) found essentially the same result. Byrne (1993, in press) found that the number of instances of tactical deception reported for different primate species correlated highly with this same measure of brain size (section 12.3 presents more details of these studies, as they compare ecological with social causes). All of these theorists use their correlations to argue for the special importance of sociality in the evolution of primate intelligence as a unidimensional entity.

A more fruitful approach, in our view, results from more fine-grained views of primate social cognition and more historically sensitive accounts of the evolutionary pressures that have been at work for different primate species at different periods of their phylogenies. The theories that move us at least partially in this direction may be roughly divided into (1) ecologically oriented theories that emphasize the influence of predation pressure and feeding ecology on primate social interaction and structure in general (i.e., why primates became social in the ways they did), and (2) more cognitively oriented theories that emphasize the direct competitive and cooperative interactions that take place within primate groups and how these formed the evolutionary context within which different aspects of primate social cognition evolved.

11.3.1 Ecological Contributions

Many ethologically oriented theorists have asked why primate groups are organized in the way that they are, which forms the background for the evolution of primate social cognition. Van Schaik (1983; see also van Schaik & van Hooff, 1983) argued that the evolution of sociality in primates, as in many mammals, has mostly to do with defending against predation. For example, living and foraging in groups, as opposed to individually, makes for the earlier detection of approaching predators, as well as for more efficient strategies for escape and defense in which different classes of individuals might play different roles (e.g., based on gender and rank). In addition to the general question of group living, however, we may also go on to ask the question of why primates have evolved the particular types of social groupings, interaction, and cognition that they have.

Wrangham (1980, 1982) argued that the evolution of particular types of sociality in primates derived from the advantages of foraging in groups. In particular, because many primates forage for patchy resources (Milton, 1981, 1993), and patchy resources may be easily dominated by a small group of individuals to the exclusion of other conspecifics, primate groups came to dominate over individuals in feeding competition. Wrangham (1980, 1982) also emphasized that such competitive exclusion strategies may favor in particular the evolution of the kind of matrilineal kin groups that are present in many cercopithecine species, since group foraging by matrilineal groups also adds a dimension of inclusive fitness. Van Schaik (1989) proposes a more general theory in which both defense against predators and feeding competition play important roles. In particular, it is only in species where relatively high predator pressure forces relatively cohesive groups that intragroup competition is a major factor in social organization. The most distinctive feature of this account is the importance placed on two different types of feeding competition that may obtain in highly cohesive groups: scramble competition, in which individuals simply scramble to obtain food before others, and contest competition, in which individuals directly confront one another aggressively for access to food. Van Schaik suggests, and provides some evidence, that in the former case the species often evolve more individually based ranking systems, whereas in the latter they more often evolve something like matrilineal-ranked groupings.

The overall point is that both predation pressure and feeding competition may be at work in determining the social structure of a given species, and they may even do this in a complex mix depending on the particular ecological conditions that may obtain at various periods of phylogeny, and this, in turn, shapes their cognition. For example, it is possible that at one time orangutans were group living, and then when their major predators disappeared they turned to solitary life to reduce feeding competition (van Hooff, 1988). What is important for current purposes is that in all of these theoretical accounts, group living in the context of patchily distributed food resources creates the conditions for a complex mix of intragroup competition and cooperation. Although we clearly do not possess the data needed to make such specific determinations, these theories at least open the possibility that the ecology of a particular species, especially in terms of how they acquire food and avoid predators, could be specifically related to its particular type of social organization, which could in turn be specifically related to its social-cognitive skills.

11.3.2 Competition

Many cognitively oriented theorists have emphasized the role of intragroup competition for resources and status in creating social problems for primates, thus providing a more specific evolutionary basis for primate cognition and social cognition. These theorists in general have not attempted to relate specifics of types of social competition to specific types of social cognition. Most importantly, Humphrey (1976) emphasized that social strategies involving plot and counterplot with conspecifics require all kinds of cognitive skills of problem solving and memory if food and mates are to be obtained, and perhaps some specifically social-cognitive skills of intention reading as well. He also emphasized the "arms race" that results from social strategies in that every advance in predictive and manipulative skills on the part of one individual encourages skills to countermand that advantage by other individuals. De Waal's (1982) notion of primate "politics" is of this same general type. It is presumably the case that something like the ecological considerations outlined above have led to especially fierce competition among primates.

The most direct descendants of Humphrey's view are Byrne and Whiten (1988), who have dubbed this approach Machiavellian intelligence, after the Italian Renaissance politician who catalogued in cold-blooded fashion the steps one needed to take to accumulate social power. Following Dawkins and Krebs (1978) and other theorists of the selfish gene, Whiten and Byrne have emphasized not just competition, but outright deception and trickery as key aspects of the evolution of primate cognition and social cognition. These kinds of interactions in particular should lead to a cognitive arms race as individuals who can be easily duped by others will lose out on food and mates and so not pass along their genes, whereas those who can see through attempts at deception and trickery and devise similar strategies of their own others cannot see through, will have a huge competitive advantage.

11.3.3 Cooperation

The hypothesis of group selection in evolutionary theory is currently highly controversial (Sober and Wilson, 1994). But it is at least conceivable that primates have become social for cooperative ends such as group defense against predators, group foraging so as to gain an advantage over solitary foraging conspecifics, and "social memory" as embodied in behavioral traditions adapted to particular ecological niches. Although no primatologist has to our knowledge articulated a well-developed theory about the evolution of cognition in group selectionist terms, this may be an important part of the picture.

More prominent have been theories that emphasize the role of cooperation within a general framework of intragroup competition. For example, both Jolly (1966) and Humphrey (1976) pointed out the competitive advantages that would accrue to individuals who could socially learn from groupmates, and there are presumably some inclusive fitness advantages to individuals who teach their offspring useful skills and information as well. Chance and Mead (1953) and Kummer (1967) both emphasized the important role of cooperative coalitions in competition for sex, food, and dominance, with special emphasis on the fact that such interactions involved an individual simultaneously monitoring its relationships to two or more individuals. De Waal (1982, 1989b) emphasized the role of these cooperation-for-competition interactions as well, but he has posited in addition some advantages that accrue to individuals who know how to cooperate and make peace with groupmates (and so avoid many dangerous agonistic encounters), as well as to those who engage

Figure 11.3. Savannah baboon eating meat. (Photo courtesy of A. Whiten.)

in reciprocally exchanged favors over time as an additional form of cooperation for competition.

11.4 Directions for Future Research

In the introduction to part II, we argued that conspecifics are not only physical objects to be located, identified, and quantified in the manner of physical objects, but also social objects with whom one may interact, communicate, and exchange information. Even as physical objects only, animate beings may present some special cognitive problems because they move on their own and so may be difficult to locate—requiring, perhaps, stage 6 skills of object permanence to comprehend their movements outside the perceptual field. But it is as potential social partners that conspecifics make special demands on primate cognitive skills. How to identify these problems in terms of discrete adaptive functions is a difficult issue, but for current purposes we maintain the organization we have assumed from the beginning of this part of our investigation, specifically, social knowledge and interaction, social strategies and communication, and social learning.

In identifying these as adaptive problems for primates in general, we do not wish to preclude the possibility that in some cases adaptive functions for particular species should be conceptualized more narrowly; for example, it is possible that some of the predator alarm calls produced by some species of prosimians and monkeys should be thought of as relatively narrowly adapted cognitive skills. On the other hand, we should also remain open to the possibility that some cognitive skills may be an aspect of the solution to a number of different adaptive problems. In the domain of social cognition in particular, how an individual primate understands the way conspecifics work—for example, as animate, directed organisms or as intentional organisms—should affect the nature of its social interactions in all types of social situations.

With these considerations in mind, we would like to close our account of primate social cognition with some suggestions for future research. These will be placed within the context of what we know about each of the three basic domains of primate social cognition as we have reviewed them. Once again, we know next to nothing about prosimians, New World monkeys, or colobine Old World monkeys.

11.4.1 Understanding and Predicting the Behavior of Others

We have based our account of primate social cognition on their understanding of conspecifics as animate, directed beings whose behavior may be predicted on the basis of their communicative displays and some other recurrent behavioral sequences. Predictive skills of this type are clearly useful in all kinds of competitive and cooperative interactions. In terms of particulars, however, we know very little. First, we know very little about the specific behavioral cues to which different species are sensitive and how they go about predicting behavior from them. For example, there are no studies that investigate experimentally whether particular primate species use the gaze direction of conspecifics as a cue to their impending behavior, and there is only indirect evidence that they use conspecifics' bodily orientation as a cue to their direction of travel. Further, and along these same lines, we know little about what kinds of social interactive experiences are necessary for the ontogeny of skills of prediction and anticipation. For example, if primates use gaze direction and bodily orientation to predict the behavior of conspecifics, do they learn this individually, and, if so, how much and what kinds of experience are required?

We have hypothesized that the understanding of third-party social relationships is a uniquely primate social cognitive skill; however, this hypothesis is on shaky ground for at least two reasons. First, we have virtually no information on any primate species other than Old World cercopithicine monkeys and great apes, and second, the information we have on nonprimate mammalian species is not systematic with respect to this question. What is needed is more research, especially experimental research, on a wider range of primates and other species. Experiments on primates' choice of coalition partners in competitive situations would be especially enlightening since this is a specific type of social interaction in which the understanding of third-party relationships would be of most immediate benefit. Also important would be more naturalistic experiments of primates attempting to cooperate with conspecifics to solve various types of instrumental and social problems.

We also have almost no information on how primates categorize different social relationships in their groups so that different individuals' relationships may be seen as of a similar type, even without all of the relevant past experiences (e.g., predicting that two individuals who have been seen grooming will also form a coalition if a fight breaks out in the future) because the relationship of "friendship" and all that it implies may be inferred on the basis of observing only one of the defining interactions. We also know little about the range of the social relationships that primates may perceive and understand beyond the basic three: kinship, friendship, and dominance rank.

Figure 11.4. Grooming by stumptail macaques. (Photo by J. Call.)

At this point we can only imagine what kinds of evolutionary situations would encourage the development of the understanding of third-party social relationships and other forms of complex social knowledge and interactions. The main hypotheses concern the special feeding ecologies of some primate species that set the context for all kinds of complex cooperative–competitive interactions (Wrangham, 1980, 1982). There are, of course, some fossils relevant to this question, but a more important first step would be the systematic investigation of the range of feeding ecologies of various primate species and how these relate to their social organizations. This organization could then be related to various indications of the type of social knowledge and interactions engaged in by those species. Such a systematic survey would be a good first step in documenting the various ecological, and thus possibly the evolutionary, factors involved in the phylogeny of primate social knowledge.

11.4.2 Manipulating and Influencing the Behavior of Others

Animate beings behave spontaneously and, on occasion, despite one's best predictive efforts, somewhat unpredictably. It is thus an obvious advantage from a competitive point of view to be able to influence or manipulate the behavior of others. Social strategies created for this purpose may be based on fairly direct learning experiences in which an individual observes others reacting to a certain behavior in a certain situation and so simply repeats that behavior when it wants them to repeat their reactions (what some have characterized as operant conditioning in social situations). But such learning may also be generalized in various ways to new situations. These generalizations may even be so abstract that we refer to them as involving insight or foresight, analogous to the kinds of insight and foresight many primates show in solving physical problems. Social strategies of this type have not been specifically investigated in primates in the same manner and to the same degree as their problem solving in the physical domain. Thus, with the exception of Coussi-Korbel's (1994) and some of Kummer's studies (summarized in 1995), there are almost no experimental studies of primates' acquisition of social strategies in relatively natural social contexts that parallel those of the acquisition of strategies for solving problems in the physical domain. That is, there are very few experiments in which a social problem involving conspecifics is presented to an individual, the information and learning opportunities available to it are experimentally controlled, and its problem-solving strategies are observed as a result. We thus have little knowledge of the skills of different primate species learning or creating social strategies in difficult social situations.

We should note that the process of ontogenetic ritualization is simply one special instance of the development of a social strategy. That is, when two individuals, both capable of flexible learning, interact with one another, on some occasions a behavior they use to manipulate the other's behavior may become more or less ritualized. The process is a matter of degree, with some social behaviors becoming ritualized fairly tightly into communicative signals, and others simply becoming preferred strategies in particular situations. In all, we know little about the process of ontogenetic ritualization in terms of which species develop communicative signals in this way, which kinds of interactions and behaviors are most conducive to the process, and, in general, the role of ontogenetic ritualization in the social interaction and communication of primates. We are especially lacking in information about any monkey species, who clearly do learn some of their gestural signals and also learn when and where to use some of their vocal signals.

As in the case of predicting the behavior of others, it is presumably the case that if primates have evolved especially flexible and powerful skills of social cognition for the creation of social strategies and communicative signals, the most likely candidates to account for these skills involve the complex processes of cooperation and competition that make up the social lives of many primates. And again, in looking for the ultimate instigator of the evolution of this social complexity, the only well-developed hypotheses have to do with feeding ecologies and their effects on social organization, which suggests again that the investigation of the relationships among feeding ecology, social organization, and the development of social strategies could be profitable if studied across a wide range of primate species.

11.4.3 Learning from Others

In the case of the human primate, the ability to learn from others, and the propensity to teach others and provide them with artifacts for solving problems, is a major aspect of the cognition of individuals. It has thus been a major concern of primatologists to investigate the extent to which the cognition of nonhuman primates also has this cultural dimension. There is no question that many primate species benefit cognitively from living in the midst of conspecifics, but the processes by which they do so have yet to be determined.

Unlike the case of social strategies and communication, in which most of our knowledge comes from naturalistic observations, in the case of social learning we have a more even mix of observational and experimental investigations. The problem is that these two sets of observations do not seem to be telling a wholly consistent story. Fieldworkers have made many observations that they believe indicate complex skills of imitative learning by several primate species, and perhaps even indicate some skills of intentional teaching. Experimentalists, on the other hand, find that in laboratory tasks these same species only show evidence of possessing modest skills of social learning, including local enhancement, emulation learning, and ontogenetic ritualization, and few if any skills of intentional teaching. An important task for future research is thus to try to bridge this gap. The most straightforward ways of doing this are to conduct rigorous experiments in the field with animals living in their natural habitats (a difficult, but not impossible, task) and to conduct rigorous experiments with captive primates in more naturalistic environments, including some of the modern zoo environments that simulate in many important ways the natural habitats of primates.

The way that apes raised by humans respond to cultural environments containing human artifacts and intentional human teaching is of obvious interest and importance to these issues as well. In this case, however, we know little about the actual processes by which these individuals acquire their skills. Although there have been a number of experimental studies documenting the skills of these individuals in a systematic fashion, few experimental studies have attempted to document the processes by which they acquire the skills in the first place. Thus, we know little about how apes who have learned some humanlike linguistic skills actually learn new symbols or produce new symbol combinations. And, as we have mentioned repeatedly, no monkeys or prosimians have ever been raised in these same kinds of rich cultural environments to see what kinds of skills they would acquire.

In terms of the evolution of social learning and culture, the major questions all await a clear formulation of which species have which skills. If it turns out that all primate species have similar skills, then one type of evolutionary scenario will have to be constructed, whereas if humans have a special propensity for culture, then another type of story will have to be told. We expand our own view on this matter in chapter 13, "Human Cognition."

11.4.4 Summary

Underlying primate social skills in all of these domains is the way primates understand the behavior of conspecifics. There are animal species that use few cognitive skills in interacting with conspecifics (e.g., some invertebrates); nature has preprogrammed them with a relatively fixed and rigid set of social behaviors. But primates, at least the simian primates for whom we have a reasonable amount of information, seem to have evolved some ability to understand the behavior of others and to base their attempts at social prediction, manipulation, and learning on this understanding. Their ability to do this based on an understanding the behavior of others as animate and directed is not in question; the issue is whether they go beyond this to consider the psychological states involved.

Understanding behavior as intentional allows an individual to understand that different means may be directed toward the same end and that the same means may be used for different ends. A particularly clear instance of intentionality is in the behavior of organisms as they seek to remove obstacles on the way to a goal, showing that what is in one sense an end (e.g., moving an object to a new location) is on another level a means to an end (e.g., removing an obstacle to a goal). Studying organisms' understanding of intentionality is not an easy task, but there have been a number of recent studies of human infants that may be of relevance. Gergely et al. (1995) employed a habituation paradigm and found that only after 9–12 months of age are human infants surprised when a simulated animate being continues to behave in the same way after disappearance of the obstacle that necessitated the behavior in the first place. Meltzoff (1995) found that human 18-month-old infants imitated not just what an adult was doing, but what it was trying to do, thus showing their understanding of the adult's intentions. And Tomasello and Barton (1994) found that 2-year-old children could distinguish intentional from accidental behavior in their social learning. All of these are paradigms that could be used, with appropriate modification, to investigate the extent to which nonhuman primates understand the behavior of others as intentional. In a more naturalistic setting, individuals could be tested for their ability to predict what another individual would do as a number of different kinds of novel obstacles to its goal attainment were introduced. Paradigms for studying primate understanding of the attention of others are only now being developed, but they could certainly be studied easily in individuals who have learned some means of humanlike communication so that the precise objects of their attention at any one moment might be determined with some specificity.

At the level of knowledge and beliefs, the methodological issues may be even more troublesome. In the study of human children, the norm is to present subjects with so-called false-belief tasks in which they must follow a story narrative and imagine how other persons would behave if certain events took place. These tasks obviously rely on a large measure of verbal instruction and response, in addition to skills of imagination and prediction for events in which they are not participating. It is also noteworthy that for children to be considered competent in these tasks, they must respond in an adult-like fashion from the outset. In constructing nonverbal theory of mind tasks for nonhuman primates, it is likely that no single task will ever be considered totally equivalent to the human task. A variety of nonverbal tasks is needed, so that even though each will have its own particular limitations together they should tell a relatively complete story. The tasks need to be constructed so that subjects who understand the beliefs of others can respond correctly from trial one (perhaps by administering extensive pretraining). It is likely that nonhuman primates will behave most competently if the tasks allow them to participate directly in the interaction.

If it is also the case, as we contend, that different ways of understanding others will make themselves felt in all areas of social interaction, communication, and learning, then in attempting to determine how a primate species understands its conspecifics, their behavior in all domains of social behavior should be considered. This will provide for additional converging operations on the same question. For example, if a species understands its conspecifics as intentional, it should cooperate and communicate in ways that take into account the role or perspective of the other, and in its social learning it should attempt to reproduce not just the motor movements or environmental outcomes of another but rather the intentional strategies of another defined in terms of different means toward the same end. It is less clear what differences should be manifest in the behavior of individuals who understand others in terms of their beliefs, although certain forms of teaching and cooperation are obvious candidates. But in any case, progress in understanding nonhuman primates' theories of minds will only be made if the problem is approached from a variety of methodological perspectives in a variety of behavioral domains simultaneously.

11.5 Conclusion

There is no question that the most exciting recent developments in the study of primate cognition, indeed in the study of animal cognition in general, have concerned the social domain. Many theorists now agree that whatever features of primate cognition are unique to the order (and these of course have yet to be fully determined), they have likely derived from primate adaptations to their social environments. These may have been shaped by the demands of their feeding ecologies and the predators they need to avoid. We add our own version of such a theory in chapter 12.

The domain of social cognition differs from the domain of physical cognition in some important respects. In both cases attempts to characterize "intelligence" as a unidimensional entity are not overly helpful, but in the domain of social cognition they are particularly problematic. The reason is that there is such a wide range of social problems to be solved and they differ so widely among different species with different types of social organization that any such generalized attempt will ultimately resort to positing the kinds of all-purpose learning mechanisms that have proven to be so restrictive in the study of cognition. Currently the major problem is that we know so little about different primate species. Thus, the majority of what we know about primate social cognition concerns roughly a dozen or so species of cercopithecine monkeys and great apes. If the whole range of primate species were considered, it is likely that we would find a much wider range of strategies than we now know for such things as defense against predators, cooperative foraging, competition for mates, social learning of various activities, and coalitions and alliances for dominance. In keeping with our ecological approach, therefore, we believe that the most profitable first step for future research is to identify the different social problems that particular species encounter in their natural habitats. We may then proceed to investigate the kinds of flexibility, complexity, and representational skills employed in their behavioral solutions to those problems, using systematic experimentation wherever possible.

That having been said, in the domain of social cognition there is a key question that underlies all the others, and that is the question of the way individuals understand the psychological functioning of their conspecifics. As we have stressed repeatedly, major differences in this most basic social cognitive skill will have ramifications for how a species deals with social problems of all types. We do not know what specific alternatives are available at this point, but our view is that they must certainly go beyond the two alternatives of operant conditioning and theory of mind. Our own version of a middle view is that an understanding of others as animate and directed leads directly to all kinds of complex social knowledge, social strategies, and social learning processes, without an explicit understanding of the psychological states of others. Given the fact that some human-raised apes may have some special ways of understanding the psychological functioning of others, a comparison of primate individuals raised in social environments that differ from one another in systematic ways might be important for determining some of the ontogenetic factors responsible for developing certain types of social understanding (Call & Tomasello, 1996).

We now proceed to attempt to bring together all the facts we have catalogued about both the physical and social cognition of primates into some kind of coherent theoretical account of primate, including human, cognition.

PART III

A THEORY OF PRIMATE COGNITION

It is probably premature to propose a theory of primate cognition. Ideally, the database for such a theory would be extensive observations of the problem-solving behavior of all 180+ primates species in both natural and experimental settings for both physical and social problems. The fact is that we have extensive observations of perhaps one-tenth of the extant species; these species are not representative of the full range of primates; and even for the most-studied species there are large gaps in our knowledge. Just to give the most egregious examples, we know next to nothing about the cognition of any prosimian or lesser ape or colobine monkey species (in either the physical or social domain in either natural or experimental settings) and next to nothing about the social cognition of any New World monkey in nature or in the laboratory (with the possible exception of some capuchin monkeys). To make matters even more complex, the field of primate cognition, perhaps especially social cognition, is a rapidly moving and changing one in which new empirical results are being reported at a rapid rate.

Nevertheless, we feel obligated to round off our empirical review with some theoretical speculations that attempt to synthesize the various sets of empirical findings into one coherent account of primate cognition. We do this mainly in chapter 12, in which we seek to identify those aspects of nonhuman primate cognition that distinguish it from the cognition of mammals in general, asking in the process whether there are any significant cognition differences within nonhuman primates themselves. We examine these uniquely primate cognitive processes in some detail, seeking to establish their various interrelations. We conclude with a few speculations on the evolutionary origins of primate cognition, deriving mainly from comparisons of present-day species.

Because of the often anthropocentric accounts of cognition that dominate the study of primate cognition (including perhaps our own), we also feel an obligation to seek to identify those aspects of cognition that may be unique to the primate species *Homo sapiens*.

We do this mainly in chapter 13, in which we focus on the early ontogeny of human cognition, seeking to identify any unique processes within the general framework outlined in chapter 12 for primates in general. We also investigate where these processes might come from ontogenetically and provide some speculations on the evolution of human cognition as well, with special reference to the evolution of culture.

We conclude our investigation of primate cognition in chapter 14 by making some proposals for this field of study. These are mostly of a noncontroversial nature, for example, arguing for a closer collaboration between field and laboratory workers. But they also include some more controversial proposals in favor of theoretical orientations that exclude noncognitivist approaches and focus on specific areas of research that seem especially critical within the theoretical framework we have proposed.

12

Nonhuman Primate Cognition

Given all of the studies and observations we have reviewed in the preceding chapters, it is clear that primates as an order have evolved a number of cognitive adaptations of central importance to their survival and procreation. In the physical domain, virtually all species studied have demonstrated a basic knowledge of permanent objects and some of the ways they may be related to one another spatially, quantitatively, and in terms of their perceptual similarities. In the social domain, virtually all species studied have demonstrated a recognition of individual groupmates and a knowledge of some of the important ways they may be related to one another socially and behaviorally. In both domains individuals have demonstrated the ability to use this knowledge to formulate various types of strategies, ranging from efficient foraging and tool use to coalition strategies and social learning, that help them attain their goals with respect to such basic adaptive functions as feeding and mating.

Within this context, many scientists have presumed that primates possess some unique cognitive skills relative to other animals—mainly because they just seem to be "more intelligent." But just what those unique skills might be is seldom specified precisely. The fact is that many of the cognitive skills that primates display so impressively are also displayed by other mammalian species, as we have noted informally in various places in the preceding chapters (see also Cords, in press). Moreover, in many cases it is likely that the supposition of primate uniqueness derives from the fact that the cognition of primates has been studied more extensively than that of other animals and has been studied more often by more anthropologically inclined researchers who are more comfortable in attributing complex cognitive abilities to nonhuman animals than are the behavioral ecologists and ethologists who study other mammalian species. Consequently, the question of whether primates possess any cognitive capacities unique to their order is, in our view, still open.

A related question concerns the possibility of significant cognitive differences among different species within the primate order. More specifically, a central question for many current researchers and theorists is whether there is a large-scale cognitive difference between two very large primate taxa, monkeys and great apes. Although a number of investigators have assumed that there is some difference, and there have been a number of proposals about what that difference might be, the data supporting these proposals are in most cases drawn from only a small number of relevant species, and in many cases the data from apes are drawn from individuals who have been raised and trained by humans. Consequently, in our view, the question of monkey–ape differences is also still open.

Several other issues of fundamental importance to the study of primate cognition are also currently unresolved. Foremost among them are three questions: (1) Are the cognitive skills of primates mostly specific to particular behavioral domains, or are they more widely applicable to a number of different behavioral domains? (2) Do primates regulate their interactions with their social and physical worlds by means of an understanding of social events as intentionally determined and physical events as causally determined, or is their understanding based on a more straightforward understanding of antecedent-consequent sequences? and (3) Does raising nonhuman primates in humanlike cultural environments merely reveal their latent cognitive abilities, or does it actually create species-atypical, humanlike cognitive skills? In our opinion, all three of these issues will have to be addressed if we are to create a complete and coherent theory of primate cognition that explains how and why it evolved the way it did.

In this chapter, our first and main goal is to investigate the possibility that there are cognitive mechanisms unique to primates. Our second goal is to determine whether and in what ways these and other cognitive mechanisms might vary among the primates, especially between monkeys and apes. We pursue both of these goals in the first section. In a second section we address three fundamental and currently contentious issues concerning the proximate mechanisms that underlie primate cognition: the degree to which they are domain specific or domain general, the extent to which they are girded by an understanding of causality and intentionality, and the ways in which they are changed by human intervention. We conclude with some speculations on the evolutionary origins of primate cognition and with an outline for a general theory of primate cognition.

12.1 Uniquely Primate Cognition

The cognitive revolution has come to the study of animal behavior in piecemeal fashion, influenced in part by the different theoretical predilections and explanatory concerns of different groups of researchers. One consequence is that the cognitive capacities of different mammalian species have been studied to differing degrees. In general, the majority of mammalian species have not been studied with an eye to their cognitive skills at all, in part because most of the scientists who study them are field biologists mainly concerned with evolutionary issues of adaptive fitness and the like. Primates, on the other hand, have more often been studied by physical anthropologists and psychologists often concerned with human evolution and proximate mechanisms, and this concern typically emphasizes issues of the evolution of cognition, culture, language, and other important human traits and activities. Consequently, in many cases in which it appears that primates have a cognitive skill that other mammals do not, the issue is simply that the nonprimate mammals have not been studied in the same manner as primates (see Rowell & Rowell, 1993, for an enlightening discussion). Nevertheless, despite the incomparability of the information available, it is still possible to defend the hypothesis that primates—at least simian primates—have some cognitive capacities and skills that are unique among mammals. We do this in the two subsections that immediately follow.

Comparisons among monkeys and apes have some difficulties as well. Most especially, many of the claims for significant differences among these very large taxa are based on comparisons of only a few species, and in many cases researchers who study apes have more anthropological interests than those who study monkeys. Nevertheless, in this case we believe that a systematic review of all the evidence leads to the hypothesis that there are no significant cognitive differences among large-scale taxa within the primate order itself, with the possible exception of prosimians, about whom we know so little cognitively, and with the clear exception of humans, whom we treat in detail in chapter 13. We present evidence for this hypothesis in the final subsection of this chapter.

12.1.1 Mammalian Cognition

Although we cannot provide a thorough review of mammalian cognition in this brief account, it is clear from a variety of observations and experiments that mammals possess more complex cognitive skills than has previously been recognized. For example, many mammalian species have complex skills of spatial navigation and spatial memory, perhaps in the form of cognitive maps containing a number of remembered locations and the specific foods located at each (e.g., rats: Galef & Wigmore, 1983; Olton & Samuelson, 1976). Many mammalian species search for objects when they disappear from sight in the space around them, looking in the place they saw them disappear, and some even track the invisible displacements of objects (e.g., dogs: Doré & Dumas, 1987). Some mammalian species engage in various forms of object manipulation and tool use and other forms of physical problem solving (e.g., raccoons and sea otters: Beck, 1980). All mammals studied have been found to have various skills of discrimination, learning, categorization, quantification, and memory (Fobes & King, 1982; Roitblat, 1987). Overall, individuals of many mammalian species are capable of learning any number of specifics about space and physical objects as they interact with them directly, and these specifies may translate into the formation of various types of problem-solving strategies.

In the social domain, many mammalian species engage in complex social interactions with conspecifics. It is likely that most social mammals recognize individuals of their group and form relatively stable social relationships with them based on their direct interactions and in many circumstances can predict their behavior on the basis of evolved communicative displays and various other behavioral and contextual cues (e.g., elephants: Moss, 1988). Many mammalian species form coalitions and alliances in disputes over resources or dominance (e.g., hyenas and dolphins: see papers in Harcourt & de Waal, 1992), and at least one has been observed to show reciprocal altruism in food exchanges (e.g., vampire bats: Wilkinson, 1984). Some mammalian species hunt cooperatively (e.g., lions: Packer, 1994), warn their groupmates when specific species of predators are nearby (e.g., prairie dogs: Owings & Hennessy, 1984), and engage in a variety of social and communicative strategies. Many mammals are capable of learning about the environment by observing the behavior of conspecifics, leading in some cases to population-specific behavioral traditions (e.g., rats: Galef, 1996). Overall, individuals of many mammalian species are capable of learning about and from conspecifics as they interact with them directly, and this information may translate into the formation of various types of social and communicative strategies.

12.1.2 Primate Understanding of Tertiary Relations

In general, then, we may characterize the physical knowledge of primates as a result of their mammalian heritage in terms of an understanding of objects and locations arrayed in a traversable spatial field and existing when they are not perceived. We may characterize the social knowledge that primates possess as a result of their mammalian heritage in terms of an understanding of individual conspecifics arrayed in a traversable social field that includes their social relationships with them. Individuals may learn specific things about both objects and conspecifics in direct interaction with them, and these may lead to the formation of various types of problem-solving strategies directed at both objects and social partners. In addition, however, we have reviewed evidence in several of the preceding chapters to show that primates also understand the interactions and relationships that other objects and individuals have *with one another*. Following Piaget (1952), we refer to this as the understanding of "tertiary" relations because they involve interactions and relations among third parties in which the observer is not directly involved.

Several major theorists have noted the importance of tertiary relations for primate cognition. In the physical domain, Parker and Gibson (1979) focus on the Piagetian tertiary sensory–motor intelligence of some primates (see chapter 6). In an attempt to account for the evolution of human cognition, they focus on a few primate species' active manipulation of objects with tools as intermediaries as the major manifestation of tertiary understanding, leading to a primary focus on great apes and capuchin monkeys as especially "intelligent." Also in the physical domain, Thomas's (1980, 1986) learning theory account of primate cognition focuses on the acquisition of relational categories, also a type of tertiary relation. The data in this case come from the performance of primates and other species in discrimination learning experiments in which organisms form categories to human specifications to earn food rewards (see chapter 6). Primates seem especially skilled in these tasks. In the social domain, Cheney and Seyfarth (1990a,b) have focused on the importance of an understanding of third-party relationships for primate social interaction and communication. They point out the added problems that arise when an individual has to monitor not only its own relationships to others that are present in an interactive situation, but also the relationships of these others among themselves.

None of these three theorists has hypothesized that primate understanding of tertiary relations is a general cognitive skill that cuts across the physical and social domains. Parker and Gibson point out a rough analogy between tool use and intentional communication as "social tool use," but they do not deal with the understanding of tertiary relations that do not involve tools (thus ignoring relational categories). Nor do they deal with third-party social relationships. Thomas does not deal with the social domain either, whereas Cheney and Seyfarth stress the domain specificity of primate social knowledge, making the strong claim that such knowledge does not apply to the physical domain (we return to this issue in section 12.2.1). Of these theorists, only Thomas has hypothesized that the understanding of tertiary relations may be unique to primates as an order, that is, in the form of the ability to form abstract relational categories (Thomas, 1994). Cheney and Seyfarth posit that the understanding of tertiary relations is not exclusively primate, citing an observation of one bird species that supposedly demonstrates comprehension of a third-party social relationship.

Our hypothesis, which attempts to extend and build on these three proposals, is that all primates (at least all simian primates, since we know so little about prosimians) understand and form categories of tertiary relationships in both the social and physical domains, and it is this ability that is responsible for much of the cognitive complexity in both their everyday lives and in their experimental performances. Moreover, our hypothesis is that primates are the only order of mammal (and perhaps animal) that possess this type of understanding. In the social domain, evidence for the hypothesis comes mainly from naturalistic observations, with support from a few experiments. In the physical domain the evidence is mainly from laboratory experiments, mostly concerned with discrimination learning. Together, these two sources of evidence make for a very strong case. A central issue in all of this is the formation of *categories* of tertiary relations, because it may be that some nonprimate mammals perceive and understand something of some tertiary relations, but do not do so in a manner that leads to a categorical understanding of them. In evaluating our overall hypothesis, we deal first with the social domain and then with the physical domain.

In their natural social behavior, primates engage in a number of complex interactions that demonstrate an understanding of third-party social relationships. For example, as reviewed in chapter 7, in a number of primate species, individuals (1) redirect their aggression preferentially toward the kin of their past enemy; (2) simultaneously threaten a dominant individual while appeasing another individual dominant to the first; (3) preferentially groom partners that outrank the others present; (4) are surprised when individuals who should be dominant to other individuals seem to vocalize in ways indicating submission; (5) understand different recruitment screams that encode the dominance relations between the fighting individuals; (6) respect one individual's "ownership" of other individuals and objects; (7) seek to prevent certain relationships between third parties from forming; and (8) encourage reconciliation between third parties who have just been fighting. There is also some evidence that primates engage in redirected aggression of a type suggestive of relation matching; that is, if an individual's relative is attacked it may retaliate by attacking the relative of that attacker, suggesting that the individual matches its own relationship with the victim to that the relationship of another individual with the attacker (although other explanations of this behavior are possible).

None of these observations of primates in their natural habitats, however, necessarily indicates that they understand *categories* of third-party relationships. That is to say, in each of the phenomena observed, it may be that an individual knows each of the third parties involved as individuals and has perceived and remembered something of their past interactions with each other as individuals. This would be an important skill, but categorizing third-party relationships would lead to a much more powerful understanding of the social field. Evidence that primates do indeed perceive categories of social relationships comes from the experimental studies of Dasser (1988a,b) in which longtail macaques responded similarly to the mother–offspring relationship involving different individuals (and then to the sibling relationship involving different individuals). These findings demonstrate that these monkeys perceive some similarity in differ-

Figure 12.1. Coalition of rhesus macaques. (Photo courtesy of F. de Waal.)

ent exemplars of mother–offspring relationships and in different exemplars of sibling relationships involving members of their group. (It should be kept in mind, however, as outlined in section 7.1, that we still do not know how generative these relational categories are for primates. That is, we do not know whether organisms are capable of making inferences about the behavior of a novel pair of individuals on the basis of information about the category of their relationship alone, so that knowing that this pair of individuals is an instance of the mother–daughter category, without having observed any of their behavioral interactions, would generate novel predictions about a whole range of their behaviors.)

Although it is possible that scientists studying the social behavior of nonprimate mammals have simply not looked for these kinds of behaviors or have not looked at them in the right ways, it is nevertheless the case that despite extensive and intensive observations, none of these types of interaction has been reported for such well-studied mammals as elephants (Moss, 1988), dolphins (Connor, Smolker, & Richards, 1992) or lions (Packer, 1994) (see section 7.1 for more extensive documentation). Moreover, to our knowledge, no nonprimate species has ever been studied in an experimental paradigm similar to Dasser's in which their understanding of categories of social relationships was investigated. Consequently, while it would be prudent to await further investigations of the cognition of nonprimate mammals (absence of evidence is not evidence of absence), our current hypothesis is that the understanding of third-party social relationships is a uniquely primate cognitive skill.

The physical domain presents a similar picture. In discrimination learning tasks, many mammals can learn to make a relational discrimination, for example, always choosing the larger of two objects, and many can respond to complex stimuli composed of multiple objects spatially arrayed in specific ways (Reese, 1968). But nonprimate animals do not seem to be able to create or understand external relations or relational categories as distinct phenomena. As reviewed in section 4.3, it has been found that pigeons do not readily form a generalized concept of identity, whereas capuchin monkeys do (D'Amato et al., 1985, 1986). This same monkey–pigeon difference has also been found in the ability to make transitive inferences in which relations have to be combined to make another relational judgment, for example, the judgment that "A is rewarded more than B" must be combined with the judgment that "B is rewarded more than C" to yield the judgment that "A is rewarded more than C" (Terrace & McGonigle, 1994). Some investigators have found that primates immediately generalize the relational concept of oddity to novel stimuli (e.g., rhesus, capuchins, squirrel monkeys, orangutan, gorilla, chimpanzee), whereas many nonprimate mammals do not (e.g., rats, raccoons, cats) (see Thomas & Noble, 1988, for a discussion). And finally, both squirrel monkeys and chimpanzees are able to match relations to one another when they are presented simultaneously (e.g., match AA with BB, not CD; Burdyn & Thomas, 1984; Oden et al., 1988, 1990), whereas no nonprimate animal has been tested for this relations-matching ability. (The one possible exception to this hypothesis is one African Grey parrot who has been extensively trained in "language" by Pepperberg, 1987, although this nonmammalian study differs in some important ways from the assessment of relations-matching in primates; see section 4.3.)

The methodological issues in the experimental literature are difficult. There are investigators who still maintain that pigeons and other nonprimate mammals do form generalized concepts of identity and oddity and that rats form rule-based categories of various sorts (see section 4.3 for details). But in a number of publications, Thomas (e.g., 1986, 1992a, 1994) has made a convincing argument that no nonprimate animal has demonstrated the kind of trial-one generalizations that are needed to infer categorical understanding in these kinds of experimental paradigms. That is, having solved several oddity problems with different triads of objects, when presented with a new triad of objects, many primates pick out the odd object on the first trial, implying the ability to see the oddity relation in the new problem immediately; no nonprimate animal reliably does this (their improvement over problems is due to the acquisition of a learning set). Moreover, primates can perceive a relation between two objects and then find another pair of different objects that represents the same relation, a procedure that some theorists believe is the best test of an understanding of relational categories because it involves the simultaneous matching of relations (e.g., Premack, 1983); it would thus be useful if some nonprimate mammals were given such a task. In any case, the hypothesis that results from these data is that the understanding of tertiary relations among objects (or at least the ability to categorize tertiary relations among objects) is a uniquely primate cognitive skill.

Our overall hypothesis is thus that the understanding and categorization of relationships of outside objects and conspecifics to one another (tertiary relations) is the most important cognitive skill that differentiates primate from nonprimate mammals. This does not mean that nonprimate animals cannot behave in ways that depend on a comparison of objects or individuals, but only that these behaviors do not entail an understanding of tertiary relations as distinct entities. And they almost certainly do not entail an understanding of relational categories. Thus, nonprimate mammals may respond in unique ways to social situations involving three parties based on fairly straightforward forms of contingency learning (if I touch infant X, adult Y will attack), but there is no evidence that they perceive in these kinds of interactions the relationship between the third parties—much less that they have a categorical understanding of such third-party relations. One interesting possibility is that some nonprimate animals perceive and understand some tertiary relations, but not enough of them, or not enough of the right kind of them, to provide sufficient raw material for the formation of relational categories. If this were the case, it would mean that a quantitative difference, the number of tertiary relations of a certain type understood, might be the basis for a fairly important qualitative difference—the ability to understand the world in terms of relational categories.

The understanding of tertiary relations makes the lives of primates very complex cognitively because they have to deal not only with how they interact with and relate to the world, but also with how the other objects and individuals interact with and relate to one another. In the social domain this added complexity is obvious. Individuals who understand third-party relationships have to take into account such things as attacking another individual will likely lead its kin to retaliate (and not other group members); consorting with a groupmate in the presence of a particular individual will likely lead to that individual forming a coalition with another particular individual; and two individuals who have just fought are likely to cause more trouble if they come close together too soon after the fight and before reconciliation. In the physical domain the application to the real world is less obvious because the relevant phenomena have mostly been studied in laboratory situations. Nevertheless, it would seem that the understanding of tertiary relations, and categories of those relations, should have consequences in many situations including locating food and avoiding predators. In addition, the understanding of tertiary relations seems to be of obvious relevance in making quantitative judgments in which the amount of food at two locations is compared, as well as in the organism's active manipulation of objects to obtain or process food, as, for example, in tool use in which one object is manipulated relative to another. The categorization of these types of tertiary relations would allow for generalization across situations, including to novel situations, on the basis of the relational structure among objects alone.

It should be noted, finally, that there is no evidence of strong differences in any of these tertiary skills across large primate taxa such as monkeys and apes. In the social domain, there is no evidence for monkey–ape differences in any of the naturally occurring phenomena concerning redirected aggression, separating interventions, and the like (the one possible exception concerns third-party mediation, discussed in section 7.1). Indeed, the only experimental studies demonstrating relational categories in the social domain are with longtail macaques (Dasser, 1988a,b); if we were so inclined, we could claim this as a skill that has only been demonstrated in monkeys. In the physical domain, there is experimental evidence of equivalent relational category formation skills in oddity and identity, as well as in relation-matching skills as demonstrated by both squirrel monkeys (Burdyn & Thomas, 1984) and chimpanzees (Oden et al., 1988, 1990). Nor are there naturalistic data suggesting any significant monkey–ape differences in comprehending tertiary relations in the physical domain. In all, the understanding of tertiary relations seems to be a cognitive skill characteristic of all monkeys and apes about whom we have systematic evidence. Given the importance of issues of cognitive differences among primates for theories of primate cognitive evolution, we now turn to that issue explicitly.

12.1.3 Possible Differences Between Monkeys and Apes

Related to the hypothesis that primates are cognitively unique among mammals is the hypothesis that there are differences among species within the primates. Given that we know so little about prosimians, a number of recent investigators have proposed one major dichotomy: the cognition of great apes is more complex than that of monkeys in some fundamentally important ways. These proposals concern everything from processes of learning to theory of mind. Each of these proposals has been introduced and briefly reviewed at some point in previous chapters, but it is important for current purposes to revisit them here, all in one place, to see if some general pattern may be discerned. We look first at specific domains in the realm of physical cognition and then at specific domains in the realm of social cognition.

Working in a generally Piagetian framework, some investigators have claimed that apes solve problems using representational strategies, whereas monkeys solve these same problems using more perceptually based, practical, sensory–motor strategies. For example, on the basis of their studies of object permanence, Antinucci and colleagues (1989) claimed that apes are able to follow the invisible displacements of objects representationally, whereas monkeys are not. Their conclusion is based on the performance on one ape individual and several macaques and capuchins. But in a subsequent study, Schino et al. (1990) found that one capuchin monkey showed the same level of skill in this task as the one gorilla on whose performance Natale et al.'s (1986) conclusion was based. Also, the fact that dogs solve invisible displacement problems, (but show no other signs of a generally "representational intelligence") makes problematic the more general claim that apes' stage 6 object permanence skills indicate a representational capacity characteristic of their cognition as a whole (see chapter 2).

In a similar theoretical framework, Visalberghi (e.g., 1993) claimed that great apes understand causality in a more deeply representational manner than do capuchin monkeys. "Representational" in this context means approaching a tool-use task with an understanding of how the tool works, sometimes manifesting itself as foresight in anticipating potential problems before they arise by using a cognitive representation of the task. As noted in chapter 3, however, the comparisons on which these claims are based involve one New World monkey species (*Cebus apella*) and various great apes who have been raised with much human contact and training, only some of whom performed skillfully (with none of Reaux's, 1995, subjects performing skillfully). No Old World monkeys have been tested in any of these tasks. Also, it must be remembered that monkeys who fail in some causality tasks are quite skillful in others; for example, capuchins do quite well in the support problem, which requires an understanding of the spatial support relation (Spinozzi & Potì, 1989). This performance illustrates that the understanding of causality is not an all-or-none affair, but in different situations this understanding may involve different kinds of tertiary relations. Until many species have been tested in a wide variety of causal tasks, we believe it is premature to posit general differences in the nature of monkey and ape cognitive representation with respect to tool use and causality.

There have also been claims of a difference between monkeys and apes in their skills of classification, which are important in the current context because they require the subject to make implicit tertiary comparisons among the features of objects to determine their group membership. This claim has *not* been made with respect to the natural categorization of phenomena, at which all mammals seem proficient. The claim has been made with respect to tasks in which subjects have been trained to physically sort objects into groups (Ettlinger, 1983) and situations in which subjects group objects in their spontaneous play (e.g., Spinozzi, in press). But, as outlined in chapter 4, in the first case both chimpanzees and monkeys learned to sort objects equally well; it is just that the chimpanzees learned to do so more quickly. In the second case, the monkeys and apes performed in all ways comparably except that one of the five chimpanzees created two groups of similar objects in a way that no capuchin monkey or longtail macaque did. It is noteworthy that one human-raised chimpanzee developed complex classification skills involving multiple, different groupings of the same objects (Hayes & Nissen, 1971); however, no monkey has ever been raised and trained in a comparable manner.

Finally, in the physical domain, working in a learning theory framework, Rumbaugh and colleagues have presented data supposedly showing that great apes make discriminations on the basis of mediated cognitive representations, whereas monkeys make discriminations on the basis of simple associations (e.g., Rumbaugh, 1995; Rumbaugh & Pate, 1984a,b). The strongest findings concern reversal learning at which some great apes seem more proficient than some monkeys. However, in Rumbaugh's own data, rhesus monkeys performed as well as the great apes, and the overall differences between monkeys and apes as taxa involved relatively minor quantitative differences. Our own analyses (based on those of Fobes & King, 1982) also showed no differences in learning set formation in general (see section 4.1.1), and the study of Essock-Vitale (1978) concluded that both monkeys and apes used "hypothesis-based responding." It should also be mentioned that in the domain of spatial cognition, Tinklepaugh (1932) found a larger spatial memory in chimpanzees as compared with macaques (see chapter 2). But in regard to spatial memory, some nonprimate mammals may be more skillful than primates (at least in terms of number of locations remembered), and thus small quantitative differences between monkeys and apes would not seem to be of great cognitive significance.

Overall, then, although there may be some differences between monkeys and apes in physical cognition, in general these are fairly minor quantitatively. Characterizations of apes as somewhat quicker learners, or more flexible learners, than monkeys find some support from their quicker adaptation to experimental situations in several different domains of physical cognition. But we do not believe that any of the empirical findings support the hypothesis that there is a fundamental difference in the knowledge or strategies with which monkeys and apes cognize or interact with the physical world. In some cases there may be species differences, but these may not generalize to larger taxa—a special problem in the physical domain when the ape species involved is chimpanzees, the only primate of any taxa to use tools universally and regularly in the wild. There is certainly no evidence to suggest that great apes as a whole are the only primates that interact with the world on the basis of cognitive representations and representational strategies.

In the social domain, there have also been several claims that the social cognition of great apes is more complex than that of monkeys, although, again, the data on which these claims are based are scanty. One such claim concerns the presence of third-party mediation in chimpanzees but not monkeys (e.g., de Waal & Aureli, 1996), which is especially important in the current context because it concerns third-party social rela-

Figure 12.2. Social interaction among mangabeys. (Photo by J. Call.)

tionships. But at issue here is really just one behavior in which chimpanzees engage but two species of macaques do not (third-party mediation in reconciliation situations). There are various plausible noncognitive reasons for this difference, however, including the relative rigidity of macaques' dominance hierarchies and their relative inconsolability after a fight (see section 7.2). The fact that some monkey species engage in redirected aggression, grooming competition, separating interventions, and the like shows that in other contexts many monkey species understand third-party social relationships quite well.

Also in the social domain, some researchers have claimed that apes understand the intentional and mental states of others in a way that monkeys do not, based mainly on Povinelli's comparative studies and on anecdotal observations and classification of instances of tactical deception. But, as noted in chapter 10, with regard to both apes' understanding of intentions and their understanding of beliefs, the evidence is not strong. Basically, all the relevant experimental findings may be explained as subjects learning to use various cues over trials in the experimental context, and, moreover, there have been several failures to replicate key studies. With regard to deception, we can only say that when the anecdotal observations compiled by Byrne and Whiten (1990) are examined carefully, substantial differences between monkeys and apes are not readily apparent, especially if these observations are interpreted not as deception but as creative social strategies. For example, of the four instances of counterdeception, three were from chimpanzees and one was from a baboon. We should also mention in this regard that the well-known difference between monkeys and great apes in skills of mirror self-recognition may reflect differences in the understanding of how mirrors, not minds, work, and that recent data with some monkey species using new methodologies suggest the possibility of self-recognition in some monkey species.

In recent years, many investigators have also claimed that apes are better imitators than monkeys (see chapter 9). But the evidence cited in support of this claim comes mostly from the study of Hayes and Hayes (1952) and subsequent replications and extensions (e.g., Custance et al., 1995; Tomasello et al., 1993), in which a home-raised chimpanzee learned to imitate arbitrary human gestures during a 17-month period of human training (in a way that a non–human-raised chimpanzee did not). Research on apes who have not been so trained, however, does not reveal imitative learning skills of this or any other type (Tomasello, 1996a). This raises the general point that in assessing primate cognition, we must be careful to note under what conditions a particular skill is demonstrated or not demonstrated. If the animals involved were raised under conditions of impoverishment, we should not expect them to behave in their species-typical manner (e.g., see Boesch's, 1993, critique of Tomasello's social learning experiments with chimpanzees). If the animals involved grew up in the wild and simply do not cooperate with humans in an experiment, this is certainly not evidence that they do not have a capacity or skill. We must also be sensitive to the opposite possibility, however, that the individuals involved are learning some species-atypical skills only because of certain kinds of species-atypical experiences with humans, perhaps from an early stage of ontogeny. With specific reference to the issue of monkey–ape differences of imitative learning, we must note again that no monkey of any species has ever been raised and trained in ways comparable to the human-raised, language-trained apes. Of particular interest would be to raise some baboons in a cultural environment, because of their complex technical skills, or possibly Japanese macaques, because of their tendency to form population-specific behavioral traditions in the wild.

Overall, in the social domain we find little evidence of a fundamental difference between monkeys and apes in social cognition, either experimentally or in terms of naturalistic observations (see also Tomasello & Call, 1994). All of the claims for apes having greater complexity in the social domain are based on observations or experiments that have other very plausible interpretations, or else they find only some small, quantitative difference. In some cases the comparisons are based on only one or two species representing apes and monkeys as taxa. It is also important that in some cases the evidence for more complex ape skills comes from apes who have been raised and trained by humans in ways that monkeys have not.

Across the physical and social domains, then, we see little evidence of qualitative differences between monkeys and apes as taxa in any important dimension of cognition. It might plausibly be argued, however, that even though no one piece of evidence is perfect, the finding in so many domains that apes are cognitively more sophisticated than monkeys (with little evidence in the other direction) is itself a convincing argument. But it is just as likely, in our view, that what this consistency indicates is not a fact about the world but rather a deep-seated human bias toward viewing apes as more humanlike. More specifically, our view is that most of the differences between monkey and ape cognition that have been claimed derive from one of three sources: (1) a tendency of humans to see apes as "more intelligent" because they look more like humans, they express emotions in ways that humans relate to more closely, they behave at a tempo more like humans, and they are more closely related to humans phylogenetically (Eddy, Gallup & Povinelli, 1993); (2) a tendency to treat all monkeys as a group, without regard to the many important differences in the cognitive skills of, for example, capuchin monkeys and Old World monkeys, along with a tendency to use chimpanzees as representative of great apes in general; and (3) a tendency to ignore differences in the experiential background of subjects, especially the fact that apes who have been raised and trained by humans may not be good exemplars of their conspecifics in the wild, and monkeys have not been raised in these same ways to see if they would respond to human culture and training in the same way.

There may still be some differences between monkeys and apes in such quantitative aspects as speed of learning, number of locations that can be remembered, range of tertiary relations that can be understood, and so on (see Ettlinger, 1983, for the view that there are a few such small differences). However, in terms of the basic kinds of things that are understood, how they are understood and cognitively represented, and how they are deployed in overt cognitive strategies, we see no clearly qualitative differences between monkeys and apes. And we must stress that this conclusion is not dependent on a theoretical framework with an impoverished theory of cognition that does not allow for the possibility of qualitative differences, as, for example, in the behavioristic learning theory of MacPhail (1987) in which all animals learn in similar ways because there is only one way that learning can be described theoretically. We have described primate cognition in any number of domains in ways that are open to the possibility of qualitative differences but have found no consistent evidence for any such differences between monkeys and apes.

12.1.4 Summary

Our overall view is thus that the cognition of nonhuman primates (or at least simian primates) is distinguished from the cognition of other mammalian species in its reliance on the understanding and categorization of tertiary relations. Primates understand the social relationships among third parties, and form categories of these, in ways that nonprimate mammals may not. Primates also understand the relations among physical objects and form categories of these relations in ways that nonprimate mammals may not. This kind of understanding leads to much added complexity in the problems that primates must deal with in their physical and social worlds. However, there is little evidence for substantial qualitative differences in the cognition of monkeys and apes with respect to this or any other cognitive skill or strategy. Although apes may be somewhat more flexible and quicker learners in some situations, in all cases the fundamental nature of the knowledge, strategies, and cognitive representations involved seem to be the same across all primate species.

12.2 Issues of Proximate Mechanism

We have characterized primate cognition in terms of mammalian cognition in general, with the additional cognitive ability to discern and categorize tertiary relations. In providing a full characterization of the proximate mechanisms of primate cognition, however, three additional issues need to be addressed. First is the issue of domain specificity. Much debate in the cognitive sciences today revolves around this issue, and we have in the preceding chapters, and indeed in the whole organization of the book, treated physical and social cognition as relatively distinct from one another. But it is also possible that there are cognitive processes that span the two domains, for example, in tertiary relations as just reviewed. Second is the issue of what primates understand of causality and intentionality—tertiary relations in their dynamic instantiations, with the addition of "intervening variables." Many researchers have attributed these important kinds of understanding to primates on the basis of less than fully convincing evidence, in our view, and so we would like to revisit these phenomena anew. Third is the issue of the role played by different kinds of experience with humans, including their artifacts and instruction, in the cognitive development of primates. It is possible that certain kinds of interactions with humans may lead some primates down species-atypical ontogenetic pathways, perhaps even with respect to the understanding of causality and intentionality, and that their ontogenies may tell us something important about primate cognition as well. We discuss each of these three issues in turn.

12.2.1 Domain Specificity

Issues of domain specificity and modularity are currently of central importance in the study of human cognition (e.g., Karmiloff-Smith, 1992). A central form of argumentation is that many behavioral adaptations are quite narrowly specialized in all animal species, and so cognitive adaptations should be no exception (e.g., Tooby & Cosmides, 1991). But cognitive adaptations, as we have defined them in any case, always involve some degree of flexibility and generality to novel circumstances, the only issue being the nature and the degree of this flexibility and generality. The problem is that there are no clear definitional criteria for what constitutes a domain or module, and so resolving the issue (perhaps it is a pseudo-issue) will be difficult.

The most explicit account of domain specificity in primate cognition is that of Cheney and Seyfarth (1990b). They provide an in-depth study of the cognition of vervet monkeys, and they generalize from that study to other monkeys and apes. They propose two central hypotheses. The first is that most primate cognitive adaptations evolved in the social rather than in the physical domain. The second is that most primate cognitive adaptations in current-day primates are still specific to the social domain, with many of them not generalizing to the physical domain at all, or only poorly. Together these two hypotheses lead to a characterization of monkey cognition as a kind of "laser beam" social intelligence, focused on solving mostly social problems. Cheney and Seyfarth also allow that there may be some small number of domain-specific processes of physical cognition. The key in all cases is that knowledge in one domain is "inaccessible" to other domains (Rozin, 1976), with the major domains being the physical and the social.

It is likely true that much of primate cognition evolved for dealing with social problems, as Cheney and Seyfarth suggest, but the hypothesis of current-day domain specificity is much less plausible. Thus, although there is little direct evidence one way or the other, it is likely that the understanding of tertiary relations—which, in our hypothesis, comprise most of the unique cognitive skills of primates—evolved for dealing with third-party social relationships. Support for this view is the fact that the understanding of third-party relationships is of such direct importance to the everyday lives of primates, intruding on practically every behavioral decision they make, whereas the understanding of relational categories in the physical domain seem to be of much less immediate significance. This does not mean that primates cannot use their tertiary understanding in the physical domain, however; we have reviewed much evidence showing that primates can indeed form relational categories with physical objects (section 4.3). Our hypothesis is thus that the understanding of tertiary relations evolved first in the social domain, but under some circumstances it may be applied to physical objects (perhaps involving some different kinds of tertiary relations). As one clear example of this, a tertiary relation singled out by Cheney and Seyfarth is the understanding of quantitative transitivity. They argue that this is really an adaptive specialization for recognizing the place of individuals in dominance hierarchies. We agree, but at the same time point out that the same cognitive skill can be brought out by experimentalists with inanimate objects in special circumstances as well, for example, in studies of associative transitivity (as Cheney and Seyfarth recognize).

Cheney and Seyfarth (1990b) also give the example that there is a much greater need for the categorization of individuals in primate social interaction than for the categorization of objects in physical interactions: often an individual's effectiveness in fighting and mating depends crucially on its knowledge of those around them. Skills of categorization therefore might have evolved for social, not physical, ends. This may be an accurate evolutionary hypothesis, but Cords (1992) points out that primates have just as great a need for classifying different types of food and predators in their daily lives as they do for classifying social partners, and the experimental studies we reviewed in chapter 4 provide ample evidence that they can categorize a variety of types of objects. So again, the questions of evolutionary history and current-day proximate mechanisms are dissociable, with the hypothesis of social priority in primate evolution being quite plausible and the hypothesis of current-day domain specificity being less plausible.

Finally, Cheney and Seyfarth also claim that in the social domain vervet monkeys make all kinds of inferences about the behavior of conspecifics involving both visual cues and auditory cues, whereas in the physical domain they seem to make many mistakes and omissions. For example, they observed that although vervet monkeys respond to auditory cues of impending danger such as the sound of cow bells signaling the approach of human herders, they seem to take no notice of visual cues of impending danger such as the cloud of dust that cows raise as they approach. Vervets also do not seem to take visual notice of fresh python tracks and the carcasses left in trees by leopards, even though it would seem evolutionarily advantageous to do so. However, Gouzoules (1992) pointed out that there are other situations in which vervet monkeys do respond to visual cues of impending danger and that other species of monkeys clearly show sensitivity to a variety of visual cues of danger; Tomasello (1992a) pointed out that it is unlikely that individual vervets have had much experience with carcasses or python tracks followed immediately by the presence of a predator; and Cords (1992) noted further that vervets sometimes do not use important visual cues in their interactions with conspecifics (e.g., the tracks that conspecifics leave). The point in all these cases is that there are likely some domain general principles of cue salience and the like that account for the asymmetries in vervets' use of cues in the social and physical domains that Cheney and Seyfarth have observed. This is as opposed to their hypothesis of domain-specific cognition, with domain-specific cues, leading to the cognitive inaccessibility of primates' physical and social worlds.

Finally, even though Cheney and Seyfarth (1990b) stress the ways in which the social cognition of monkeys is more sophisticated than their physical cognition, following in the thesis of domain specificity, they also speculate that there are some behaviors monkeys perform in the physical domain that they do not perform in the social domain. For example, they claim that in foraging for food, primates have very good cognitive maps, but that they may not have such good maps for interacting with conspecifics. But much research with a variety of primate species that shows their sophistication in keeping track of the locations of their conspecifics through vocalizations and the like (section 2.1). Although the relevant study has not been done, most primatologists would probably predict that if a monkey or ape were exposed to conspecifics in particular locations in captivity, they would learn these locations just as quickly as if they were shown food in these same places. In this case it is likely that the skill of cognitive mapping evolved in the physical domain for foraging but now is applied to all objects including social objects. Again, we conclude that there is less domain specificity than Cheney and Seyfarth suggest.

Clearly, there is much domain specificity in the behavior of animals, as is argued by many researchers who study behavioral ecology, and in many cases this may include to some degree the underlying cognitive mechanisms involved. But if cognition implies flexibility, then there will always be some generality involved—confinement to a single, narrow context typically implies a behavioral mechanism relying on tight genetic programming rather than flexible cognitive mechanisms. Whether this flexibility involves moving across cognitive domains is difficult to determine because there are no clear criteria for defining domains or identifying instances of the cross-domain applicability of a single cognitive skill. Nevertheless, animate beings are also physical objects, possessing all the properties of shape, size, permanence, and so forth, that physical objects possess, and so everything that an organism knows about physical objects in general (their permanence, where they are located, the category to which they belong) is applicable to social objects as well, with no processes of "transfer" across domains needed at all. The reverse need not be true, however, since physical objects are not necessarily social objects. It still may be the case, however, that even though a number of primate cognitive skills arose evolutionarily in the context of social interaction and are still used mostly in that domain by most primate species, some species may use these skills in the physical domain as well, especially if they are engaged in human-designed experimental paradigms that provide special motives and opportunities. It is possible that in many of these cases what experimenters are observing is the organism's evolutionarily prepared ability to understand social entities and relationships, whose generalization to physical stimuli humans have encouraged through laborious training procedures. But this still argues against strict domain specificity (lack of accessibility) in the social cognitive skills of primates.

It is thus our general hypothesis, following Cheney and Seyfarth (1990b), that many of the most important cognitive adaptations of simian primates evolved for dealing with social problems, especially their unique skills involving the understanding of tertiary relations. We differ, however, in hypothesizing that some current-day primate species in some circumstances are now able to use these skills in the physical domain as well. We differ as well as in hypothesizing that basically all physical cognition is automatically applicable to the social domain since animate beings are also physical objects. The extent to which the social and physical knowledge of primates are accessible to one another is at this point still an open question, but our view is that there is much less domain specificity in primates' cognitive adaptations than in their other behavioral adaptations.

Figure 12.3. Bonobo food processing. (Photo courtesy of E.S. Savage-Rumbaugh.)

12.2.2 Understanding Intentionality and Causality

We have argued that what distinguishes primate cognition from that of other mammals is not the way they understand and categorize objects, events, and social partners in the environment directly, but rather the way they understand and categorize how outside objects, events, and social partners relate to one another. We have called this type of cognition the understanding of tertiary relations.

The question now arises of whether nonhuman primates also understand the dynamically expressed versions of tertiary relations known as intentionality and causality. There are several deep similarities in these two concepts. First, the understanding of intentionality and causality both concern temporally ordered events, that is, antecedent–consequent relations. Second, in both cases these events are indeed tertiary, that is, the antecedent and the consequent are both events that are external to the observer. And third, in both cases there is some inferred intermediary cause or goal that organizes and "explains" the event sequence such that different antecedents may lead to the same consequent (different causes lead to the same effect, different behaviors lead to the same goal), and the same antecedent may lead to different consequents in different circumstances (the "same" cause leads to different effects in different contexts, the "same" behavior accomplishes different goals in different contexts). And these intermediaries are explanatory in the sense that the cause or goal in some sense "forces" the outcome. All of which is to say that intentionality and causality both have to do with the "why" of happenings in the world: Aristotle's efficient and final causes, in fact.

With regard to intentionality, there is currently much debate about the degree to which nonhuman primates understand others as psychological agents whose behavior is mediated by underlying intentions, perceptions, beliefs, and other such mental entities (see chapter 10). On one hand, theorists such as Byrne and Whiten (e.g., Byrne, 1995; Byrne & Whiten, 1992; Whiten, 1993) propose that nonhuman primates operate in a social world in which individuals understand that others have intentions and beliefs, and they are constantly attempting to manipulate those mental entities. These theorists are especially generous in attributing complex social-cognitive skills to great apes. On the other hand, theorists such as Tomasello (1990, 1994b, 1996a; see also Tomasello & Call, 1994) do not believe that nonhuman primates understand others as intentional or mental agents. Rather, they propose that nonhuman primate social interaction, including that of apes, depends on more basic processes of social cognition involving the interpretation of behavioral cues, the prediction of the impending behavior of others, and the like (see also Povinelli, 1993). This more cautious view may be characterized, as it was throughout chapters 7–11, by saying that nonhuman primates perceive and understand others as "animate" based on their ability to move and do things spontaneously and as "directed" toward external objects and events in the sense that they have learned certain antecedent–consequent sequences of behavior; that is, they have learned certain behavioral and contextual cues that may be used to predict the impending actions of conspecifics.

The evidence relevant to this debate is of two kinds: "theory of mind" experiments with nonhuman primates in the laboratory, and observations of primate social complexity in the wild. The laboratory experiments that have been used to support ape understanding of intentions and beliefs have been reported in detail in sections 10.2 and 10.3, along with some critiques of their findings. Most importantly: (1) the original Premack and Woodruff (1978) experiment in which Sarah chose "solutions" for humans having "problems" has other, more mundane, explanations (Savage-Rumbaugh et al., 1978b), and there have been failures to replicate the findings using different methodologies (Premack, 1986; Povinelli & Perilloux, in press); (2) the Povinelli et al. (1990) experiment in which chimpanzees chose between a guesser and a knower as sources of information has not withstood further analysis and attempts at replication (Povinelli, 1994a; Povinelli et al., 1994); and (3) the experiment of Woodruff and Premack (1979) on "deception" is a demonstration that over many trials over many months chimpanzees may learn strategies for manipulating the behavior, not the mental states, of human experimenters (and the anecdotes of Byrne & Whiten, 1990, do not make a mentalistic interpretation more plausible).

In terms of the complexities of social behavior in the wild, there are several lines of evidence that argue against nonhuman primate understanding of intentions and beliefs. First, if the theory of mind experiments just cited reflected the true state of affairs in nature, then great apes understand the intentions and beliefs of others in a way that monkeys do not, since, for example, Povinelli et al. (1991) found that rhesus monkeys did not discriminate between a guesser and a knower and Byrne and Whiten (1992) claim that the deceptive skills of great apes are more sophisticated than those of monkeys. Consequently, we should expect to see more complexity in the social lives of apes than in the social lives of monkeys as they attempt to influence the mental states of conspecifics. But we do not. Even Byrne (1995: p. 144) admits that apes supposed possession of a theory of mind "does not result in a plethora of socially complex behaviors in apes, compared with other simian primates." We should also add that if apes possessed an understanding of the intentional or mental states of others, it is logical that they should engage in an array of other social behaviors that they do not engage in, for example, declarative attempts to draw the attention of others to outside entities, imitative learning of novel behavioral strategies, teaching on a regular basis, cooperation in which they take account of the perspective of the other, and perhaps even symbolic communication. Although it may be that human-raised apes have learned through extensive human tutelage at early stages of ontogeny to engage in something similar to these behaviors (see next subsection), these are clearly not everyday ape behaviors in their natural habitats.

We thus do not find strong evidence, either experimental or naturalistic, to support the position that apes or any other nonhuman primates interact with social partners based on an understanding of their intentions or beliefs. However, monkeys and apes do engage in all kinds of complex social interactions, and this raises the question of the cognitive basis of these interactions. We believe that the complexity in nonhuman primate social life comes from two interrelated sources. First and most important is primates' understanding of the behavior of others as animate and directed. Understanding the behavior of others as animate and directed, in combination with their knowledge of certain recurrent contextual situations, is a type of tertiary understanding that allows an individual to predict where others are going and what they are likely to do next in particular contexts, for example. Predictions of this sort are based on such behavioral cues as the phylogenetically ritualized communicative displays of others, the ontogenetically ritualized communicative signals of others, and a host of other behavioral and contextual cues whose significance is learned from experience (e.g., inferring the destination of another given its current travel direction and a knowledge of the location of food). Whatever the specifics, the ability of individuals to read cues of this type in the behavior of others in particular contexts and to make accurate predictions of their impending behavior and directedness as a result may be considered the glue of primate social life.

The second source of complexity in primate social life is based on the first, and that is the understanding of social relationships and categories of those relationships, especially those involving third parties. That is, understanding the behavior of others as directed is tertiary understanding in that one individual understands that another is related to or interacting with some third party. But, in addition, the evidence is that many primates understand that other individuals are related to one another in more abstract terms such as "kinship," "friendship," and dominance rank. But these types of understanding depend, in all cases, on observing the behavioral interactions that other individuals have with one another over time and forming some coherent conceptual package of those interactions. Thus, perhaps juvenile X is always seen in close proximity to female XX. It is also observed that X nurses from XX, that X runs to XX whenever they separate by much distance, that XX defends X if another adult aggresses against it, and they sleep together each night. Knowledge of these interactive sequences would allow for all kinds of predictions, for example, who XX is going to aggress against given who just attacked X, the location to which X is traveling given the location of XX in a tree, and the nest in which X will sleep tonight given knowledge of who made which nests. In addition, it may happen that this same cluster of behavioral interactions is observed between juvenile Z and female ZZ, and between other juveniles and females, so that a category of this type of relationship may be formed. This might even allow predictions to be made when a new pair of individuals is observed in one type of behavioral interaction (e.g., nursing) and another never-observed situation arises for that pair (e.g., the juvenile is attacked).

Together, then, the understanding of behavior as animate and directed and the resulting comprehension of various third-party relationships means that primates may predict the behavior of others in many complex ways, making for what de Waal (1982) has called primate politics, what Byrne and Whiten (1988) have called Machiavellian intelligence, and what we, following Kummer (1982), have called a complex social field. It does not, however, necessarily make for a theory of mind, or even a theory of intentions. To take a hypothetical example: Suppose that a subordinate individual has been excluded from a watering hole by more dominant individuals. Now suppose, by chance, a rock rolls down a cliff leading to the water's edge and scares away the drinkers. The subordinate, far enough away not to be frightened by the rock, then proceeds to drink in peace. Perhaps the same scene repeats itself a few times over the course of several weeks. From these experiences the observing subordinate may come to understand that the falling rock reliably predicts the scattering of others, but it is questionable whether it understands that the rock leads to fear in others, which then leads to their scattering. Most observers of nonhuman primates would be astounded, we believe, if one day the subordinate individual climbed the cliff and deliberately rolled the rock down toward the others to get them to scatter, since the rolling rock was not something it originally produced but only observed. (If in the original event the subordinate's own behavior was responsible for the rock's falling, then, of course, traditional types of learning could occur.) It would be even more astounding if the subordinate individual found some novel way to make its groupmates fearful and so scatter, for example, by making some other noise from behind the cliff. A creative strategy such as this would imply that it had understood that the fear of the others was the "mediating variable" (to use Whiten's, 1993, term) and that manipulating that variable in some way (not necessarily involving the rock that had led to the result previously) would also be effective.

The same point may be made by asking why nonhuman primates do not imitatively learn their social strategies and communicative signals. All the evidence presented in section 9.3 suggests that in learning new social strategies and communicative signals, apes and monkeys first behave, then notice the way others respond, then come to anticipate the response of others, and then use the original behavior (or some shortened version of it) to produce that response. They may even do this in novel and insightful ways in novel circumstances, as attested to by the various observations of "deception," which we have interpreted as creative social strategies. But the development of such social strategies does not rely in any way on an understanding of the intentions of others. Thus, if an individual behaves in a certain way and others react in a certain way, that individual may learn to use this behavior habitually as a social strategy to produce that same reaction in others in the future. If this entire sequence is realized enough times, and in the right ways, the individual may even learn to produce an abbreviated and stylized version of the original behavior, thus creating a communicative signal by means of ontogenetic ritualization. But what nonhuman primates do not seem to do is acquire social strategies and communicative signals by observing and imitating other individ-

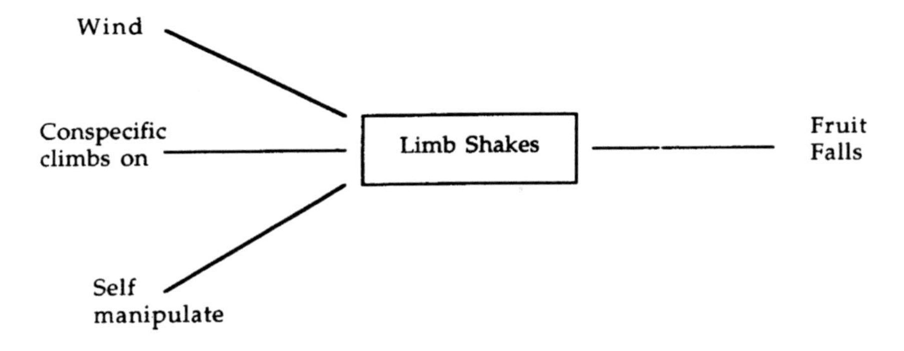

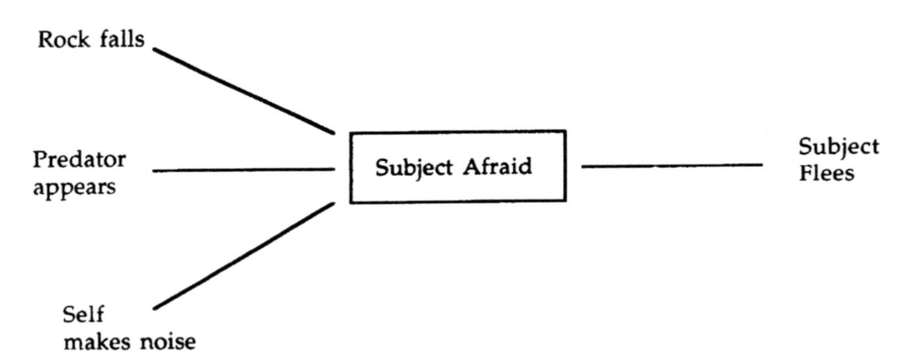

Figure 12.4. Intentionality and causality as manifest in the examples from the text.

uals as these individuals employ some strategy or signal in pursuit of a social or communicative goal; they have to behave themselves and experience the consequences. Nonhuman primates in their natural habitats do not imitatively copy the instrumental behavior of others because they do not perceive or understand the original behavior as goal-directed in the sense that it consists of a goal and various possible behavioral means to that goal that may be duplicated; they do not understand the other as an intentional agent who is similar to themselves as an intentional agent.

To summarize, nonhuman primates can understand the animacy and directedness of others, but they do not understand the intentionality of others. If they understood others' intentionality, they should be able to develop novel strategies that take into account the intentions or beliefs of others, learn novel strategies by observing others' behavior in communicative and other problem-solving situations, and engage in communicative behaviors such as showing an object to another, offering an object to another, pointing to an object or conspecific for another, or teaching another some skill or strategy it might find useful. Nonhuman primates are thus able to understand antecedent–consequent relations in the behavior of others, but there is no understanding of a psychological component in terms of the intentional and mental states of others that mediate their interactions with their environments (see figure 12.4).

The other dynamic tertiary relation that must be considered is causality. What is entailed in the understanding of causality is notoriously difficult to specify. The philosopher Hume (1739–40) could see in causal sequences nothing more than a consistent contiguity between two events, even though he felt something more must be involved. The philosopher Collingwood (1972) argues that the something more that Hume felt but could not see is a kind of "force," and he further hypothesized that the notion of force in causal judgments actually derives from the force humans feel when they coerce others socially to do things or are so forced themselves ("Causal propositions . . . are descriptions of relations between natural events in anthropomorphic terms", p. 322). If Collingwood's hypothesis (or something similar) is true, we should not expect nonhuman primates, who do not have a humanlike understanding of intentionality, to have a humanlike understanding of causality either.

There has been much less research and much less theoretical debate about causality than about intentionality, at least in the context of primate cognition. Most of what we know about nonhuman primates' understanding of causality comes from the work of Visalberghi and colleagues (reviewed in section 3.3). Based on her experimental studies, Visalberghi (1993) concludes that capuchin monkeys do not demonstrate an understanding of causality when they manipulate objects and tools, but that some great apes might. Operationally, the criterion is whether her subjects show evidence of foresight into the solution of a series of tool-use problems, that is, whether they show evidence of understanding, without overt trial-and-error learning, of the tertiary relations among objects and events in a tool-use situation. Her results show that some human-raised, human-trained apes are slightly better at this than capuchin monkeys, and so she claims that some great apes have a deeper representational understanding of causality. One issue is that the apes involved were human-raised and language-trained and thus they may not be representative of other members of their species (and the same may be said of the subject whose "causal" understanding was studied by Premack, 1976).

The central issue, however, is whether tool-use tasks are sufficient to infer casual reasoning. Under the influence of Piaget's (1954) analysis, most theorists have assumed that they are. The reasoning is that manipulating a tool foresightfully requires an understanding of the tertiary relations between tool and goal. But if the prototype of causal reasoning involves two events independent of the observer, tool use is a relatively poor exemplar. That is, when a subject manipulates a tool, it becomes an extension of the subject's own appendages; the subject controls the causal event in much the same way as if it had used its own hands. To be sure, in Visalberghi's studies there are several aspects of the situation that must be understood independently of the subject's activities, for example, the spatial relation between the tool and the food and various impediments between the tool and the food. But in all cases the relation to be understood is a spatial one of the same type that needs to be understood for effective reaching with an appendage. It is thus not clear that apes in these studies are doing more than learning about the spatial relations involved and how those affect their own direct manipulation of the tool. Nonhuman primates' ability to understand the efficacy of their own behavior in producing environmental events is not in question (see Dickinson & Shanks, 1995, for an interesting discussion); the issue is whether they understand that one external event is effective in producing another, that is, *causes* it, without the involvement of their own manipulations of the world. (Premack's, 1976, and Premack and Premack's, 1994, tasks presented to the human-raised chimpanzee, Sarah, involved events totally external to her, consisting of static pictures depicting antecedent and consequent situations, and she was asked to choose a picture of the "cause," but these tasks could have been solved with various types of noncausal strategies, see section 3.3.)

As in the case of intentionality, therefore, we are not convinced that apes need to be using a concept of causality in the experimental tasks purporting to illustrate its use, at least not in the humanlike sense of one independent event forcing another to occur. More convincing would be a situation in which an individual observes a contiguity of two events, infers a cause as intermediary, and then finds a novel way to manipulate that cause. For example, suppose that an individual ape, who has never before observed such an event, for the first time observes the wind blowing a tree such that the fruit falls to the ground. If it understands the causal relations involved, that the movement of the limb is what caused the fruit to fall, it should be able to devise other ways to make the limb move and so make the fruit fall. As in the case of the similar hypothetical example we used to illustrate intentionality, we believe that most primatologists would be astounded to see the ape, *just on the basis of having observed the wind make fruit fall*, proceed to shake a limb, or pull an attached vine, to create the same movement of the limb. Again, the problem is that the wind is completely independent of the observing individual and so causal analysis would have to proceed without reference to the organism's own behavior and the feedback it might receive from that (thus, it might be able to learn to shake the limb if its own movements had previously led to a limb shaking and the fruit falling as a result). Moreover, performing some novel behavior to make the fruit fall would involve an even deeper causal analysis of the web of possible ways that the cause could be repeated so as to reinstate the desired effect.

Again, in this domain, the same point may be made by looking at social learning. Thus, based on the evidence presented in section 9.2, if an individual monkey or ape were to observe a conspecific shaking a limb and fruit falling, it could not on the basis of this observation alone learn and acquire a possible strategy to make the fruit fall itself. If the observing individual already knew about limbs shaking and/or fruit falling from its own previous behavioral interactions with branches and/or fruit, however, then such an observation would expose it to an effect that it already knows how to produce—and indeed its observation of the effect as produced by the other is highly similar to its observation of the effect produced by its own behavior previously. The choice of a strategy for producing the effect thus comes from the organism's previous learning experience, not from its observation of the strategy used by a conspecific. This is one version of what we have called emulation learning. Consistent with this account, Visalberghi (personal communication) reports that the rhesus macaques on Cayo Santiago, who love "cooked" coconut and so wait anxiously for coconuts that humans have placed in the fire to burst open and scatter, have never been observed to place coconuts in the fire themselves, presumably because they have not ever learned how to produce this interesting and desirable effect on their own, but have only observed humans doing so.

Our hypothesis is thus that nonhuman primates understand causal relations only in the sense of one external event typically leads to another (the events are seen as ordered and therefore predictable), but they do not understand why one event leads to another in the human sense in which there is some force acting as intermediary that may be manipulated in various ways to cause the effect. Some nonhuman primates (e.g., the apes studied by Visalberghi and colleagues) may approach this understanding when they manipulate some object or tool, because this allows them to invoke an extended understanding of the efficacy of their own behavior to produce changes of state in the world, but not when the events are totally external to themselves. This level of understanding is analogous to the social domain in which individuals may ritualize communicative signals and use creative social strategies when the raw material is their own previous behavior, but not when they are outside observers of events only and they must act on the intentional states of others in flexible ways to achieve their desired social ends. It is interesting to note that when human children's understanding of causality is evaluated by using devices in which all of the events happen independently of the children's own activity, they have much more difficulty than in tool use tasks in which they control the antecedent event behaviorally (e.g., Frye et al., 1995).

Overall, then, we believe that monkeys and apes understand much about the ordering of both social and physical events in the worlds in which they live, and so they are very good at predicting outcomes and, in cases where their own behavior is involved, at manipulating antecedent events to produce consequent events. They are not so proficient at the human form of these concepts in which there are completely external events connected by an intermediary such as an intention or a cause. For this kind of understanding, there must be an appreciation of different causes for the same effect (or different means to the same goal) and that the same cause may lead to different effects in different circumstances (or the same means may be used toward a different goal in different circumstances). The key issue in the human forms of intentional and causal understanding is that there is a web of possibilities that the intermediary constrains or determines. Understanding this web of possibilities and the relevant intermediaries in a particular case enables individuals to devise novel ways of producing the intermediary and thus the end result. That is, in a particular circumstance in which the usual antecedent event is not present or not possible, an organism should in some cases be able to create a different antecedent event leading to the same intermediary and thus to the same result as usual (e.g., creating fear in conspecifics so that they will scatter or creating the movement of a limb so that fruit will fall). Nonhuman primates do not seem particularly skilled at such strategies when the relevant events are independent of their own instrumental behavior, but rather they seem to rely only on one-to-one mappings between antecedent and consequent events.

12.2.3 Human Enculturation and Training

One of the most interesting facts of primate behavior is that some species engage in complex cognitive activities in captivity that they do not engage in in the wild. The best known example is tool use. A large number of primate species (orangutans, bonobos, gorillas, capuchin monkeys, some species of baboons and macaques) engage in tool use only rarely, if at all, in the wild, but can become quite proficient in captivity. It is important that none these species require extensive training by humans. In most cases, simply exposing them to the tools for some period of time in situations in which they need to use the tool for some desired end (usually food) is sufficient for them to develop skills on their own (Call & Tomasello, 1996). More exposure to a wider variety to human tools and artifacts leads nonhuman primates to more complex and varied uses of tools and other artifacts, as does exposure to human manipulation of artifacts and instruction. It thus seems likely that what is emerging in such cases derives from a general form of

Figure 12.5. Enculturated apes. (Photo courtesy of E.S. Savage-Rumbaugh.)

sensory–motor cognition that many primates possess in the wild, but it manifests itself overtly only when certain contingencies obtain, and this often happens in captivity.

There are other skills that nonhuman primates develop in captivity, however, that seem to be more dependent on human training or other types of social interaction. For example, the skills of various primate species at placing quantities into ordered series must be meticulously trained by humans in the laboratory, in most cases over many thousands of experimental trials. A similar but different example is the well-known skills of some apes with languagelike systems of communication, which require several years of exposure to particular types of human interaction. What is different in this case is that these individuals sometimes develop from their social interactions with humans some more general cognitive skills that are not directly trained. For example, several ape individuals raised by human caretakers have demonstrated, in addition to their skills of "language," a number of skills of imitative learning (reproducing novel actions on objects; Tomasello, Savage-Rumbaugh & Kruger, 1993) and communication (referential pointing; Savage-Rumbaugh et al., 1986) that presumably result from some more generalized effects of human interaction. This generalized effect might justify speaking of these human-raised apes as "enculturated."

The question is what to make of these very humanlike skills that primates, mostly apes, have acquired in interaction with humans. One extreme is to say that they are simply "circus tricks" that the animals have been taught to perform without any real understanding of what they are doing. The other extreme is to say that these are skills that the individuals bring to their interactions with humans—they are skills that primates also have in the wild—but that in order for the animals to express them in ways that humans can recognize, some training in a human experimental arrangement or other form of communication is required. A middle view is that nonhuman primates do not have these skills in the wild, and so they do not bring them to the experiment or training, but they do bring a capability to learn certain skills if exposed to them in the appropriate circumstances. A variation on this middle view is that nonhuman primates may have the ability to learn some of these skills only if they are brought up from an early age, during some "critical period," in a certain kind of social-cultural, pedagogical environment.

In a thorough review of all of the available research literature concerning apes raised or trained by humans, Call and Tomasello (1996) found evidence for all these views, depending on the animals and skills involved. There are basically four types of effects. First, there are certainly some primates (e.g., those in circuses) who have been taught to perform activities whose significance they do not understand at all; these are uninteresting performances from the cognitive point of view, and it is possible that some primate performances in experimental arrangements are of this same general nature. Second, as alluded to above, some primate cognitive skills seemingly lie latent in many primate species, and all that is needed for their expression is motive and opportunity. For example, tool use does not need extensive human training to appear in many primate species who do not use tools in the wild, and no critical developmental period for exposure seems to be necessary. Third, nonhuman primates may learn some specific skills through explicit training in which humans, typically in laboratory settings, assist them in identifying what to pay attention to and what to do in solving particular cognitive tasks. Examples might include some of the mathematical tasks that individuals learn to solve over many trials of training. This training cannot create skills out of nothing, however, and so even in these cases it is clear that the organism brought some relevant skills to the laboratory in order to learn the task.

Fourth, some individual nonhuman primates may have their ontogenies channeled down species-atypical pathways by means of something like the process of human enculturation (Tomasello, Kruger & Ratner, 1993). A number of studies have shown that apes raised from an early age in a humanlike cultural environment develop some humanlike, social-cognitive skills that are not characteristic of their species-typical ontogenies, and indeed they do not develop these skills if exposed to human culture at later ages (Rumbaugh and Savage-Rumbaugh, 1992). For example, some individual apes have learned to use human symbols in very humanlike ways, point to objects for humans, and imitatively learn novel object-directed behavioral strategies. The effective factors that lead these human-raised apes into these more humanlike developmental pathways are not known at this time. Premack (1983) hypothesized that the acquisition of languagelike skills provides apes with an abstract conceptual code that fundamentally changes the nature of their cognition. This hypothesis may have some validity, but it does not account for the other changes in the cognitive development of enculturated apes such as skills of imitation and referential gestural communication.

In our view, the key is that apes raised in the context of human culture and instruction may begin to view human beings as intentional agents who can be imitated or communicated with referentially (although there are still differences with humans; see below). This understanding develops because apes bring to the process some skills of social cognition, and in their interactions with humans they are (1) encouraged to attend to what a human is attending to, with the human trying to make this happen (e.g., by waving an object to capture attention) and rewarding the subject with praise when she complies; (2) exposed to human artifacts and shown how they work, again with social praise and other rewards when the subject reproduces the human's conventional actions; (3) exposed to forms of gestural and symbolic communication that presuppose a reciprocity of understanding. These kinds of interactions are integral to the process that Vygotsky (1978) called "the socialization of attention", and may be instrumental in illustrating to social beings, who are capable of benefitting from such illustrations, that others have interactions with objects that they may in some sense reproduce. Given our analysis of the relation between the understanding of intentionality and causality, the partial understanding of intentionality by human-raised apes may also signal at least a partial understanding of causality as well, for example, in the studies of Visalberghi and colleagues concerning human-raised apes' comprehension of the workings of tools that they are manipulating (Limongelli et al., 1995; Visalberghi et al., 1995).

The most plausible hypothesis at present, therefore, is that human-raised apes understand the intentions of others in ways that their wild conspecifics do not. Even in the most positive perspective, however, it is clear that apes are not turned into humans by this process, as there are still some important differences. Although the limitations of human-raised apes' cognitive skills have never been systematically investigated, some differences with human children are the following: (1) although human-raised apes request objects and activities from humans via various joint attentional strategies, they still do not regularly (if at all) simply show something to a human or ape companion declaratively, or point to something just for the sake of sharing attention to it; (2) human-raised apes do not spontaneously cooperate with conspecifics in ways that show an understanding of the perspective of the other; and (3) there is little, if any, behavior of human-raised apes that one would call intentional teaching. The common factor in all of these phenomena is a sense of simply sharing experience with another psychological being, often for no reward outside the sharing itself. Thus one hypothesis is that, although apes can master the "referential triangle" in their interactions with humans for instrumental purposes when they are raised in humanlike cultural environments, they still do not attain a humanlike social motivation for sharing experience with other intentional beings. There would thus seem to be some aspects of the intentionality of others that human-raised apes understand, but other aspects that not even the most enculturated apes seem capable of or perhaps motivated to tune into (see table 12.1).

In the current context, the main point is that we do not know the full extent of the cognitive skills of apes raised, and possibly enculturated, by humans, and we also do not know their limitations. Current evidence seems to point to the possibility that exposure to human artifacts and intentional instruction leads nonhuman primates, at least the several species of great ape who have been raised in these ways, down ontogenetic pathways that are different from those of their conspecifics who grow up in more species-typical circumstances. They seem to learn something more about the intentionality of others, perhaps as a result of being treated intentionally themselves, although there are still some types of human intentional interaction in which they seem unable to or disinclined to participate.

Table 12.1. Summary of the effects of humans on the development of apes in different cognitive domains.

Domain	Effect of interaction with humans
Object permanence	Interaction with humans not necessary.
Object manipulation, tool use, mirror use, symbolic play	*Exposure* to human artifacts and *emulation* of their use leads to quantitative increases in knowledge of objects and their properties and dynamic affordances. May occur in many types of captive environments.
Categorization/classification, quantitative skills	Interaction with humans not necessary for natural categorization, but some *training* may be necessary for explicit classification and quantification based on abstract properties (e.g., training in attention management skills). Typically occurs in laboratory environments.
Social attention, social referencing, understanding visual gaze	Interaction with humans not necessary.
Intentional communication, imitative learning, understanding intentions	Being *enculturated* by humans may lead to an understanding of others as intentional and thus to qualitatively more humanlike skills of social learning and intentional communication. Typically occurs only in home-raised environments.
Cooperation, teaching, understanding beliefs	Human interaction has no significant effect because humanlike skills in these domains may not be attainable by apes of any kind.

12.2.4 Summary

In this section we have briefly discussed three key issues in the study of primate cognition. With regard to domain specificity, primate cognitive skills that have evolved for social or physical ends do not seem to be domain specific but seem to be applicable to the other domain under some circumstances. In particular, skills for dealing with physical objects (e.g., cognitive mapping) are automatically applied to social objects because social objects are also physical; skills for dealing with social objects may also be applied to physical objects in some as yet not fully understood circumstances (perhaps sometimes only in human-created experiments). With regard to the understanding of intentionality and causality, it seems that nonhuman primates understand the relationships among external events in simpler and more straightforward ways than do humans, that is, mainly in terms of the contingency and directedness of events on a one-to-one basis, with no comprehension of a wider web of possible events controlled by intermediaries such as intentions and causes. Finally, with regard to human training and enculturation, it seems that being raised in a humanlike cultural environment has some effects on primate cognitive development, leading to more humanlike skills in those behaviors such as social learning and communication, which depend on an understanding of others as intentional agents, and thus perhaps to more humanlike skills in the causal understanding of certain types of inanimate events as well (i.e., those such as tool use in which the subject's own behavior is a part of the causal sequence). There are still some important differences between these individuals and human beings, however, mainly with regard to the tendency to share experience with others for its own sake.

12.3 Issues of Ultimate Causation

Speculations about ultimate causation are always perilous. It is thus uncertain what kinds of ecological conditions might have led to the evolution of the unique set of cognitive skills possessed by primates. Recently, some researchers have attempted to answer this question by identifying levels of "intelligence" in different animal species and by seeking evidence for ecological conditions from the evolutionary past that may have led to these differences. As outlined in sections 6.3 and 11.3, researchers typically have used as their measure of intelligence some quantification of brain size and then have looked for variables from both the physical and social ecologies of different species that might be associated with brain size. Some studies have even attempted to contrast directly hypotheses about the physical versus the social origins of primate intelligence. We report on these studies here, but we also express our view that treating intelligence as a unidimensional entity and treating the social and ecological theories as mutually exclusive alternatives are both simplifications that may not be helpful in our search for the ultimate causes of primate cognition.

12.3.1 Ecological and Social Determinants

As reported in section 6.3, Clutton-Brock and Harvey (1980) found that frugivorous primate species have, on average, larger brains relative to body size than do folivorous primate species. The hypothesized reason for this difference is that ripe fruits are more patchily distributed in space and time than are leaves and thus finding them takes greater foraging efforts and skills. These investigators also found a general correlation between the relative brain size of primate species and their average day-range size. Again, the hypothesized reason is that species that have to search more widely for food have to develop a greater repertoire of cognitive skills, including especially skills of spatial memory and anticipation. Milton (1981) provided further support for this view in a detailed study of the foraging practices of two sympatric New World monkey species: the spider monkey, who eats mostly widely dispersed fruits and has a relatively large brain, and the howler monkey, who eats mostly easily obtainable leaves and has a relatively small brain (see section 6.3 for more details of these studies).

On the other hand, as reported in section 11.3, Sawaguchi and Kudo (1990) found another set of correlations, in this case between sociality and relative brain size. They used as a measure of brain size the ratio of neocortex to body size because the neocortex appears to be devoted less to general maintenance functions and more to information processing per se. They also thought it important to control for phyletic relatedness by performing correlations only within fairly narrow phyletic groups. What Sawaguchi and Kudo found was that social prosimians have larger relative neocortex sizes than solitary prosimians, and polygamous anthropoid primate species have larger relative neocortex sizes than monogamous anthropoid primate species. The reasoning in this case is that species that have to remember and deal with many different conspecifics should have larger brains, especially with regard to the neocortex, to allow them to do this (see section 11.3 for more details of these studies).

In an attempt to sort out these conflicting claims, Dunbar (1992) noted first that measures of foraging such as home range size and amount of fruit in the diet are highly correlated with measures of sociality such as group size. Because both of these variables have been found to be correlated with relative brain size, it is impossible to determine the true underlying relationships without a multivariate statistical approach in which foraging and sociality, are directly compared regarding their ability to predict relative brain size. Dunbar (1992) attempted to do this by correlating with one another the relative sociality (group size), foraging demands (home range size), and size of neocortex (relative to the size of the rest of the brain) for several dozen primate species. Consistent with previous results, what he found was that group size correlated with neocortex ratio at 0.76, and home range size correlated with neocortex ratio at 0.79, (see Figure 12.6, a and b). The difference was that when body size was controlled for statistically, the correlation of neocortex size with group size remained at the same high level, but the correlation of neocortex size with home range size disappeared. Dunbar argued that controlling for body size is crucial here because home range size must be made relative to the animals motory capabilities (foraging over 1 km^2 means something very different to a bushbaby and a chimpanzee). (Dunbar,1995, refined and extended these findings in several other ways.) Consistent with this general point of view, Byrne (1993, in press) found that the number of instances of tactical deception reported for different primate species, a more direct measure of social-cognitive skills than group size (although a very controversial one; see section 8.1), also correlated quite highly (0.77) with neocortex ratio.

Correlational findings such as these are always open to interpretation. For example, Dunbar did not take into account, as did Sawaguchi and Kudo (1990), the phyletic relationships of the different species. Thus, it appears that much of the correlation he found between group size and neocortex ratio (see Figure 12.6a) is due to the fact that the prosimian species sampled live in smaller groups and have smaller neocortices than the simian species sampled. There do not appear to be high correlations within the prosimians or within the simian species by themselves. Nevertheless, because Sawaguchi and Kudo (1990) found a relationship between group size and relative neocortex size while controlling for phyletic relatedness, and Byrne (1993, in press) found a similar relationship using another measure of social cognition, as of now the best hypothesis for the neocortical expansion in primates relative to other mammals is the social hypothesis. This accords with our view, following Cheney and Seyfarth, that the understanding of tertiary relations probably emerged first in the social domain. We nevertheless have some problems with this approach to the ultimate causes of primate cognition.

12.3.2 The Importance of Species History

There are two interrelated difficulties with this correlational approach to the ultimate causation of primate cognition. The first is one that has arisen repeatedly: treating cognition ("intelligence") as a single, unidimensional entity. The second is similar: treating the evolutionary cause of primate cognition as a single, unidimensional entity acting at one point, or similarly at all points, in evolutionary history. Few scientists explicitly espouse either of these simplistic views, but they are implicit in much research and theorizing.

If we consider, for example, the correlational research just reported, it is possible

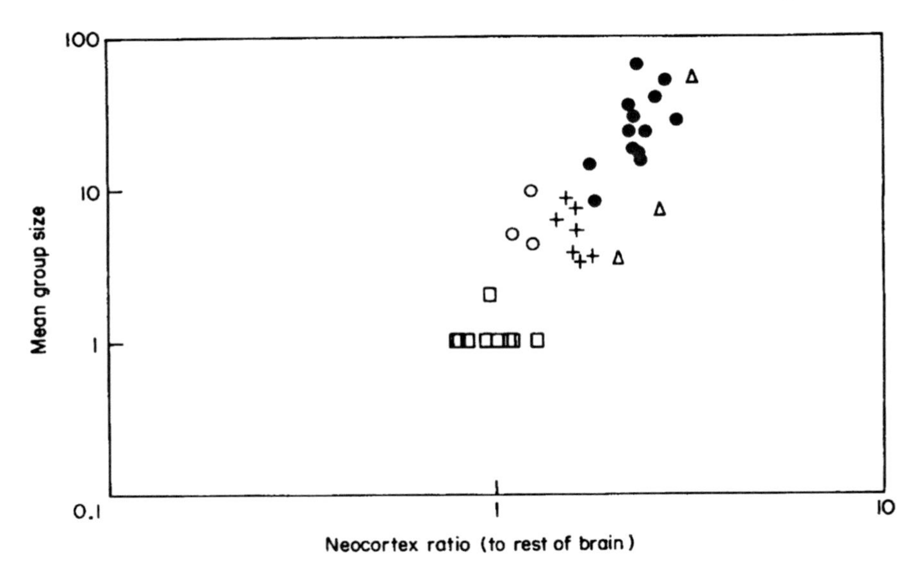

Figure 12.6. Relationship between mean group size in primates and neocortex ratio according to Dunbar (1992). (Reprinted from *Journal of Human Evolution* with permission from Academic Press.)

that both the ecological theory and the social theory are correct—it is just that they concern different cognitive skills (presumably manifest in different parts of the brain) that evolved at different periods in primate evolution. As just one possibility, perhaps it is the case that primates began their cognitive trajectory as they moved from nocturnal to diurnal lifestyles in the distant evolutionary past, and so began needing acute visual perception and hand–eye coordination to capture their insect prey in the light. There then followed the two crucial steps. First, some of these primates moved to a more herbivorous niche requiring skills of spatial memory and categorization as they foraged for patchily but predictably distributed fruits and new leaves. There would then be an expansion in the appropriate parts of the brain. Second, it may then have happened that this new way of foraging led some of these primates to become involved in complex social situations involving third-party relations and other social complexities. These would involve the intragroup competition that the patchy-resources niche encourages, as Wrangham (1980, 1982) has suggested, including various strategies for cooperating with allies in order to compete with adversaries. We should then see the evolution of skills of social cognition, perhaps those involving understanding third-party relations, and again the consequent expansion of the appropriate parts of the brain. (Interestingly, all the tests of the social hypothesis have measured the size of the neocortex, an evolutionarily newer part of the brain more likely to be involved in social cognition, whereas the tests of the ecological hypothesis have mostly measured relative brain size more globally, which would include such areas as the hippocampus, which many scientists suspect is involved in spatial memory.) The point is that both hypotheses may be correct, just for different cognitive skills and for different evolutionary periods.

Regardless of the truth of this particular hypothetical scenario, the point is that "single gun" theories of the evolution of complex cognitive mechanisms may be helpful in creating suggestive hypotheses, but the real work of evolutionary analysis is to construct a history for each species that details the specific evolutionary pressures it confronted at different evolutionary periods and the cognitive mechanisms that evolved as a result. The evidence we have at our disposal is limited and so it is not a task that may be easily carried out. But an ecological approach to the evolution of cognitive processes in both the physical and social domains will have to detail both what evolved and at how it evolved at different periods of evolutionary history for each species of interest. Presumably this task will be accomplished via the general cladistic methodology of phylogenetic reconstruction in which hypothetical common ancestors for various taxa are hypothesized based on commonalities in living species. The basic idea is that if a number of closely related species all display a certain characteristic (e.g., almost all cercopithicine monkeys display a certain kind of facial expression), whereas other species do not, then we may assume that the common ancestor to those species had that characteristic (see Byrne, 1995, and Povinelli, 1993, for application of this method to primates). These types of investigations should, of course, be used in combination with whatever fossil evidence we have on the common ancestors.

In the specific case of the cognitive adaptations of primates, creating such reconstructions would mean much more systematic investigations into the cognitive skills of different primate species, especially the relatively neglected prosimians, New World monkeys, and colobine Old World monkeys. More investigations of the lesser apes would also be important, especially to the large-scale hypotheses about a fundamental divergence in the cognitive skills of monkeys and apes that have been proposed by some researchers.

12.3.3 Summary

Measures of the brain size of various primate species, corrected for body size, correlate with the foraging demands those species face in terms of the amount of space they must cover to forage successfully. The size of the neocortex of these same species as compared with the rest of their brains, however, does not correlate with these foraging demands if the space they forage is scaled relative to their respective body sizes. Various measures of social behavior and cognition (e.g., group size, frequency of tactical deception, monogamy–polygamy) do correlate with the relative size of the neocortex (as compared with the rest of the brain) for many of these same species. There is thus some support in these studies for the proposal that much of primate neocortical expansion is due to complexities of their social, not physical environments. This might be seen as support for Cheney and Seyfarth's proposal that much of primate cognition, especially in its unique aspects, evolved in the social domain.

The challenge in this approach to problems of the ultimate causation of primate cognition, however, is to go beyond this rather simplistic approach in which one evolutionary factor is responsible for all the characteristics we see in current-day species. Obviously, for a full account, theorists need to specify for each species of interest both the particular cognitive skills of interest (with their associated brain regions, perhaps) and the associated evolutionary pressures involved in each case. They must do this in each case in the context of particular evolutionary periods in the history of that species, because each current-day species is a complex product of many evolutionary pressures and responses that have taken place over the entire course of their phylogenies.

12.4 The Structure of Primate Cognition

Our theory of primate cognition borrows from many theories of mammalian and primate cognition proposed by other researchers, but it attempts to organize the facts and phenomena in a new way. The theory may be summarized in six propositions.

1. *Primates are mammals and thus many of the cognitive mechanisms that underlie their behavioral adaptations are general mammalian cognitive mechanisms.* These mechanisms include the basic understanding of space and objects; the ability to discriminate, categorize, and quantify objects; the ability to recognize individual conspecifics and remember past interactions with them; the ability to communicate with and learn from conspecifics; and the ability to create flexible strategies to deal with problems in both the physical and social domains based on both learning and insight. These basic skills of physical cognition likely evolved in the context of foraging (in the case of many simian primates for relatively patchily distributed vegetation of one sort or another) whereas these basic skills of social cognition likely evolved in the context of general mammalian-type social interactions.

2. *What most clearly distinguishes primate cognition from the cognition of other mammalian species is the understanding and categorization of third-party social relationships and tertiary relational categories among physical objects.* Understanding these kinds of tertiary relations likely evolved first in the social domain to deal with the social complexities that arise when group-living individuals must compete for patchily distributed resources which allow domination by subgroups. A major competitive strategy in this context is cooperating with allies against opponents, and so understanding the various affiliative and agonistic relationships of groupmates to one another becomes of crucial importance. The result is a highly complex social field involving third-party relationships within which strategic social interactions take place. This understanding of third-party social relationships then was used by at least some primate species for understanding various tertiary relationships and categories among physical objects, making for a more complex spatial field of objects as well.

3. *Although some of the cognitive adaptations of primates may be narrowly specialized and domain specific, many others are more broadly based and domain general.* There is an asymmetry between skills developed in the physical and social domains in that animate beings are also physical objects, and so cognitive processes evolved for interacting with physical objects automatically apply to animate beings as well (e.g., skills of cognitive mapping for locating both physical and animate objects). On the other hand, physical objects are not automatically social objects, and so cognitive processes evolved for interactions with animate beings may or may not apply to the inanimate world. In general, the question of how narrowly or widely adapted is a particular cognitive skill is an empirical question, and there are some cognitive skills such as categorization and, perhaps, the understanding of tertiary relations that would seem to be widely applicable across domains.

4. *Apes and monkeys do not differ significantly in how they cognize their physical and social worlds.* The experimental foundation for claims that apes are "more intelligent" than monkeys is not a solid one, and there are few if any naturalistic observations

that would substantiate such broad-based, species-general claims. Apes may appear to be more intelligent than monkeys in general because of their greater similarity to humans in appearance and emotional expression and because many ape individuals, but no monkeys, have been enculturated by human beings. There is no evidence that apes and monkeys perceive or cognitively represent their physical or social worlds in qualitatively different ways.

5. *Although all primates understand the behavior of conspecifics as "animate" and "directed" and the behavior of physical objects as contingent on antecedent events, no nonhuman primate understands the behavior of conspecifics intentionally or the behavior of objects causally in humanlike ways.* This is because the understanding of intentionality and causality requires an understanding of the dynamic relationships among two or more outside entities by means of an intermediary (intention or cause) that organizes a web of possible antecedent–consequent sequences into a meaningful set. Nonhuman primates only understand one-to-one antecedent–consequent sequences, although because of the complex spatial and social fields in which they operate, predictions of the behavior of animate beings and objects based on these understandings may be quite complex.

6. *Ecological and social theories of the evolution of primate cognition are both correct, but about different cognitive skills and different evolutionary periods.* It is likely that foraging for patchy but predictable resources led to the evolution of complex cognitive maps and learning skills early in the evolution of simian primates. At some later evolutionary period, intergroup competition for those patchy resources led to social organizations and patterns of interaction for some species that encouraged the evolution of complex social strategies relying on an understanding of third-party social relationships.

13

Human Cognition

Approximately 6 to 8 million years ago, a population of African apes began evolving into the species known today as *Homo sapiens*. The process of speciation resulted in the emergence of a number of unique characteristics along the way, including bipedal locomotion, the loss of much body hair, and the making of stone tools. The most dramatic developments, however, occurred during the latter stages of this period as human beings began to create a bewildering array of cognitive products not seen in other primate species: such things as natural languages, symbolic art, multifunctional technologies, higher mathematics, and cultural institutions such as governments and religions. One of the deepest mysteries in the evolution of primate cognition is how it happened that this particular species came to specialize so dramatically in the complex cognitive processes that are embodied in these products. The basic puzzle is that, although 6–8 million years is a very long time historically, it is a very short time evolutionarily. Thus, *H. sapiens* is as genetically similar to its nearest primate relatives, *Pan troglodytes* and *P. paniscus*, as horses are to zebras, lions are to tigers, and rats are to mice, sharing something on the order of 98% of their genetic profiles (King & Wilson, 1975). Given this evolutionary proximity, any cognitive differences that we observe must be based on a very delimited set of biological adaptations. In attempting to characterize human cognition as a special case of primate cognition, therefore, the challenge is to find a small difference that made a big difference—a small change, or set of changes, that transformed the process in fundamental ways.

In our view, the only possible candidate for such an adaptation, or set of adaptations, is one that changed cognition from a basically individual enterprise to a basically social-collective enterprise. Human cognition is what it is because human beings are adapted for cultural life and all that entails. Most important, human children grow up in and participate in a social world that is full of all kinds of collective cognitive products and processes that others have created previously and that serve to amplify and transform their individual cognitive skills (e.g., Bruner, 1990; Cole, in press; Rogoff,

1990; Vygotsky, 1978). These include all kinds of cultural artifacts, including both material artifacts such as spoons and hoes and symbolic artifacts such as language and Arabic numerals. The point is that each of these cultural creations embodies the collective wisdom of the many different individuals who contributed to its invention and modification over historical time, including language, which embodies the way a culture construes and categorizes the world. As children learn to use these artifacts to mediate their interactions with the world, they participate in this wisdom. Also important are the various types of structured social practices in which children participate and in which they are exposed to many artifacts, such as games, scripts, rituals, and, perhaps most important, instructional formats. These communal activities serve to motivate children toward the conventional goals for which artifacts were designed, to instruct them in the pursuit of those goals, and, in general, to potentiate new cognitive skills. Together, the artifacts and structured social practices that human children inherit from their forebears, along with their ability to "tune in" to those artifacts and practices, of course, means that human cognition is socially constituted in ways that are unique in the animal kingdom (Cole, in press; Tomasello, Kruger & Ratner, 1993; Vygotsky, 1978).

Our general point of view is that human beings share with nonhuman primates all of their cognitive adaptations for space, objects, tools, categorization, quantification, understanding social relationships, intentional communication, social learning, and social cognition. But building on these adaptations, they also have a unique adaptation, or set of adaptations, for social-cultural life by means of which they "tune in" to other persons in their group, for example, by imitating them, learning to use their material tools, learning to use their symbols to communicate with them, teaching and being taught by them, and collaborating with them in complex ways. The outcome of this type of social life is not just that individuals benefit from the information gathered by others, but they also begin to employ novel forms of cognitive representation and organization, for example, forms based on intersubjectively understood symbols and the taking of multiple perspectives on cognized entities (Donald, 1991; Tomasello, Kruger & Ratner, 1993). The hypothesis is thus that if a human child were raised from birth outside of human contact and culture, without exposure to human artifacts or communal activities of any kind, that child would develop few of the cognitive skills that make human cognition so distinctive (see Candland, 1993, for an interesting discussion, although the scientific data on "feral children" is problematic in many respects).

Obviously, human cognition is a field of study in its own right and we cannot attempt to cover it thoroughly here (see Barsalou, 1992, for an especially useful review). But because human beings are primates too, and their cognitive skills in many ways constitute the background against which the cognitive skills of other primates are analyzed by scientists, it is important that we make some attempt to integrate human cognition into the general theoretical framework we have established for nonhuman primates. Consequently, in this chapter we attempt to identify some of the aspects of human cognition that make it unique, especially with respect to its cultural dimensions. We do this mainly from an ontogenetic point of view because we believe it is important to see what human cognition is like both before and after certain key ways of understanding emerge. We thus begin with infancy and the way human infants interact with and cognize both objects and other persons in the first year of life. We then proceed to the second year of life when, on the basis of a new understanding of others as intentional agents, infants begin to learn from others in new ways, including the use of conventional symbols such as those that make up language; and language then gives them a powerful new means of conceptual representation and organization. We look finally at cognitive development in the third and fourth years of life when children begin constructing "theories of mind" in their social cognition and all kinds of abstract categories and relations of inanimate objects in their physical cognition, which in both cases are predicated on their ability to take multiple perspectives on external events and to reflect on their own behavior and cognition. The final two sections are devoted to a number of important issues in the ontogeny and phylogeny of human cognition.

13.1 Human Cognitive Development

Our thumbnail sketch of the ontogeny of human cognition revolves around a set of cognitive skills that emerge in nascent form at the end of the first year of life and that are responsible for human children's ability to become cultural beings by participating in various collective cognitive activities. It is our hypothesis that all these skills derive from infants' new understanding of how other animate beings work, that is, an understanding that persons, as opposed to inanimate objects, are intentional beings who have goals, choose among different behavioral means for achieving those goals, and choose to pay attention to only some things as they monitor their progress toward those goals. A number of patterns of behavior and interaction reflect this new understanding as infants develop during the second year of life, including new forms of social learning that enable the acquisition of language and the special forms of cognition that it potentiates, including both event categories and narratives. These developments then set the stage for children's unique ways of cognizing their physical and social worlds during their third and fourth years of life, as they begin to engage in more abstract thinking about categories, relations, and numbers, as well as about the thoughts and beliefs of other human beings.

13.1.1 Cognizing Objects and Persons in the First Year of Life

Some of the most exciting research into human cognition in the past two decades has concerned the cognitive development of young infants. In general, researchers have found that human infants are much more cognitively competent than previously believed. This is true in both the physical and social domains.

In his classic works on human infancy, Piaget (1952, 1954) provided an account of the way human infants interact with the physical world that is the starting point for all subsequent theories. Piaget noted that at around 4 months of age, infants begin reaching for and grasping objects; at around 8 months of age, they begin looking for objects that have disappeared, even removing various obstacles in their attempts to grasp them; and at around 12–18 months they begin to follow the spatial displacements of objects, both visible and invisible, to new locations and to understand something of the tertiary relations among objects (both spatial and causal). Piaget hypothesized that all of these developmental changes in sensory–motor cognition were a result of infants' active manipulations and explorations of the physical objects as they constructed reality through converging lines of information (e.g., visual, auditory, haptic, proprioceptive) about the physical world.

Figure 13.1. A three-year-old human primate. (Photo courtesy of Victor Balaban.)

A major challenge to the Piagetian view has come from researchers who have found that human infants have some understanding of an independently existing physical world at an age that coincides with their earliest manipulations of objects—before they could have had time to use those manipulations to "construct" that world. For example, Baillargeon et al. (1985; see also Baillargeon, 1995) found that if infants are not required by researchers to manipulate objects, but only to view scenes and show surprise when their expectations are violated, they display an understanding of objects as independent entities, existing when they are not being observed, by 3 or 4 months of age (at around the time of their very first deliberate manual manipulations). Using this same methodology, Spelke, Breinliger, Macomber and Jacobson (1992) have shown that infants at this same early age understand a number of other principles that govern the behavior of objects, including the understanding that objects cannot be in two places at one time, objects cannot pass through one another, and so forth. Again, infants seem to understand these principles before they have had much experience with manipulating objects.

Later in their first year of life, human infants perform a number of other cognitive feats in the physical domain. Before their first birthdays they can categorize objects, use simple tools, form learning sets, estimate quantities, and understand simple principles about the contingency of events (see Harris, 1983, for a review). The important point for current purposes is that they display these skills in much the same way that many non-human primate species display them (see chapters 2–5). It is our view, then, that the physical cognition of 1-year-old human infants reflects in a quite straightforward way the basic primate cognitive adaptations for interacting effectively with the physical world.

In the social domain, on the other hand, we see qualitative differences between human and nonhuman primates from soon after birth. Two behaviors highlight this difference. First, as outlined by Stern (1985), Trevarthen (1979), and others, from soon after birth, human infants engage in "protoconversations" and "affect attunement" with their caregivers. Protoconversations are social interactions in which the parent and infant each focus their attention on the other—often in a face-to-face manner involving looking, touching, and vocalizing—in ways that express and share basic emotions. Moreover, these protoconversations have a clear turn-taking structure. Although there are differences in the way these interactions take place in different cultures, in one form or another they seem to be a universal feature of adult–infant interaction in the human species (Trevarthen, 1993). Second, in the context of these early social interactions, human neonates also display skills of imitation or mimicking. Meltzoff and Moore (e.g., 1977, 1989) have discovered that from soon after birth, human infants reproduce some of the facial expressions and head movements of adults. Although it is likely that this mimicking tendency is simply a facilitation of a behavior the infant already knows how to perform (as some bird species engage in similar mimetic processes in the vocal modality early in their development), it is also possible that it reflects a deeper tendency of the infant to identify with conspecifics.

These two social-interactional behaviors in early infancy set the stage for a revolution in infants' social cognition that takes place in the months immediately preceding their first birthdays, a revolution that is the direct manifestation, in our view, of the cognitive skill that most clearly differentiates humans from nonhuman primates. At around 9–12 months of age, infants begin to understand conspecifics in a new way as evidenced by (1) the new ways they attempt to tune into the attention and behavior of adults, and (2) the new ways they attempt to get adults to tune into *their* attention and behavior (see Figure 13.2). More specifically, infants at this age begin to tune into others by looking where another person is looking (gaze following), doing what another is doing (imitative learning of object-directed behaviors), checking their feelings toward outside objects (social referencing), and engaging in relatively extended bouts of shared engagement with objects (joint attention). Infants get others to tune into them by actively directing their attention to outside objects with ritualized gestures and vocalizations of both the imperative (requestive) and declarative types, some of which clearly evidence the infant's emerging competence and motivation to share interest and attention with others (Bates, 1979).

All of these social-cognitive skills emerge together in infant development quite simply because they are all manifestations of the same thing: the infant's emerging ability to understand other persons as intentional agents (Tomasello, 1995a). As defined previously, an intentional agent is an animate being that chooses its goals, behavioral means for pursuing goals, and attentional foci for monitoring progress toward goals. It is important to emphasize that understanding others as intentional agents is not just seeing a "ghost in the machine" in the form of "an intention" as a piece of mental furniture, but it involves an understanding of how organisms' intentional interactions with the environment are organized (Gergely et al., 1995). An organism would only attempt to tune into the attention of others or get others to tune into their attention if they understood them as beings able to intentionally direct their attention to specific entities on demand. This kind of understanding clearly builds on, but goes beyond, the general primate understanding of others as animate and directed.

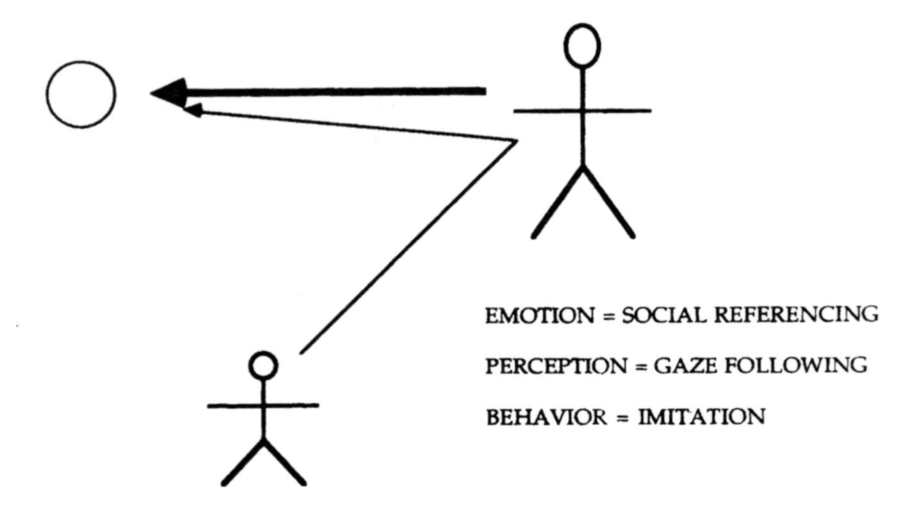

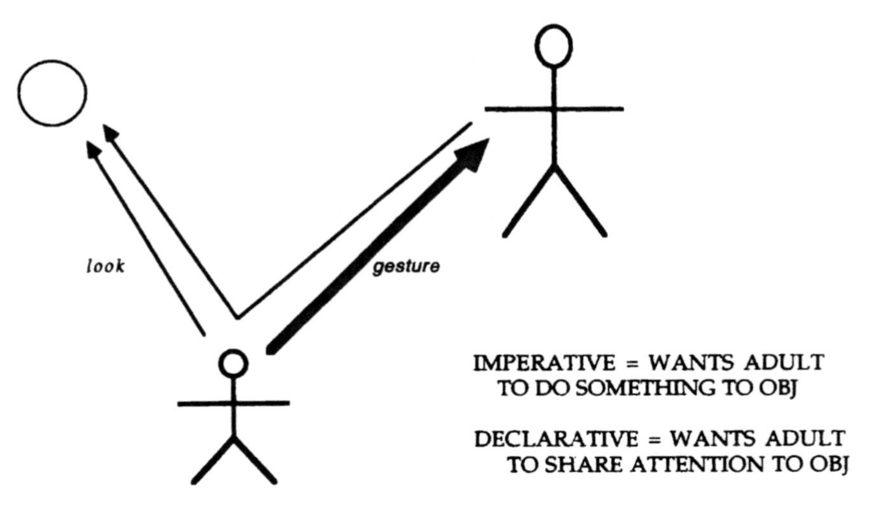

Figure 13.2. The major ways that children tune in to adults and get adults to tune in to them.

The immediate result of this new form of understanding is that infants begin to engage in a new way of tuning into the behavior and cultural activities of other persons—what we have called cultural learning (Tomasello, Kruger, & Ratner, 1993). Specifically, at around the first birthday, infants begin to engage in forms of imitative learning in which they understand that behavioral means are dissociated from intended goals (Carpenter, 1995). For instance, Meltzoff (1988) had 14-month-old children observe an adult bend at the waist and touch his head to a panel, thus turning on a light. The infants then did the same thing, even though it would presumably have been easier and more natural for them simply to push the panel with their hand. The infants' behavior thus reflects their understanding that the adult had the goal of illuminating the light and then chose one means for doing so, from among other possible means, and that if they had the same goal they could choose the same means. Even more dramatically, Meltzoff (1995) found that when confronted with an adult engaged in an unsuccessful goal-directed action on an object, 18-month-old infants reproduced the action they saw adults *attempting* to perform, not the one she actually did perform (e.g., they actually disassembled an object that an adult tried unsuccessfully to disassemble). Cultural learning of this type thus relies fundamentally both on infants' ability to distinguish in instrumental actions the underlying goal and the different means that might be used to achieve it, and on their tendency to identify with others so as to align their own goals and behaviors with theirs. As we shall see shortly, this form of learning is foundational to the acquisition of language and other symbolic and cultural skills.

We should also mention one other ramification of this social-cognitive revolution, and that concerns self-understanding (Rochat, 1995c). Once an infant can monitor the attention of others to outside entities, it so happens that the outside entity on which others are sometimes focused is the child herself. That is, as the child monitors adults' attention to entities in the outside environment, she sometimes ends up monitoring their attention to *her*. From this point forward, the infant's face-to-face interactions with others are radically transformed. She is now interacting with another person who perceives and intends things toward her in a way that was simply not perceived before. When the infant did not understand that others perceive and relate to an outside world, there could be no question of how they perceived and related to *me* (Tomasello, 1993, 1995b). This new understanding of how others perceive and feel about *me* opens up the possibility for the development of shyness, self-consciousness, and a sense of self-esteem (Harter, 1983). Evidence for this is the fact that within a matter of months after the social-cognitive revolution of the first birthday, infants begin showing the first signs of shyness and coyness in front of other persons and mirrors (Lewis et al., 1989).

If our review of nonhuman primate cognition in the preceding chapters presents an accurate picture, almost all of what we have just described concerning infant social cognition is uniquely human. Although nonhuman primates do cognize the physical world of space and objects in much the same way as human infants, there is no evidence that in their natural habitats they engage in the kinds of joint attentional, cultural learning, and communicative activities that make up the social-cognitive revolution of the human 1-year-old. They do not do these things because they understand their conspecifics only as animate and directed, not as intentional. Nonhuman primates thus engage in emulation learning not imitative learning; they use head direction as a cue to an interesting objects but do not understand that others have attention; they communicate by means of signals, not symbols; and they engage in forms of cooperation and "teaching" that do not incorporate the perspectives of others. Human infants, on the other hand, begin to understand their conspecifics as intentional agents at around 1 year of age, and this opens up a whole new world of cognitive possibilities as they begin to participate in cultural activities and to culturally learn the use of important artifacts such as tools and symbols. In addition, it also paves the way for some changes of basic processes of cognition that occur in the second year of life.

13.1.2 Symbols and Event Categories in the Second Year of Life

The social-cognitive revolution at the first birthday underlies a number of cultural acquisitions, but perhaps the most important is language. In the current context we are interested in, first, the cognitive and social-cognitive processes that are responsible for the early acquisition and use of linguistic symbols, and second, the new forms of cognitive representation and organization that result from the early acquisition and use of linguistic symbols. We discuss these two processes each in turn.

The first step in the acquisition of linguistic symbols is the understanding of other persons as intentional agents whose attention to outside entities may be followed into, directed, or shared (as outlined in the preceding subsection). Once infants have begun to understand others in this way, they are in a position to understand what adults are doing when they make "funny noises" (use language) toward them. That is, infants begin to understand that the adult is producing these noises with the goal of getting them to focus their attention on some particular aspect of their shared experience. Indeed, in some formulations a linguistic symbol is nothing more than a social convention by means of which persons who know the convention direct one another's attention to particular aspects of their shared world (Tomasello, 1992c; 1996b). To learn an already established linguistic symbol, therefore, infants must come to understand that (1) in making this noise the adult is expressing the intention for them to attend to something (i.e., expressing an intention toward their attention), and (2) when they wish to manipulate someone else's attention in this same way, they may make this same noise to them, thus making it a bidirectionally (intersubjectively) understood symbol. This is a manifestation of the special form of cultural or imitative learning in which children comprehends the intentional significance of acts that another person have directed to them and their psychological states and then learn to reciprocate (reverse roles) by using that same act. Infants' ability to understand others as intentional is thus foundational for their ability to enter into the to-and-fro of attention manipulation that is conventional linguistic communication. In this way, early in their second year of life, infants go beyond the relatively restricted range of possibilities for attention manipulation afforded by pointing and other concrete gestures and on to a new range of possibilities as embodied in a culturally constructed system of attention manipulation that enables very precise attention directions to others.

Having developed the ability to acquire symbols by means of cultural (imitative) learning early in the second year of life, infants' subsequent learning of words later in the second year clearly illustrates the importance of their ability to "read" the intentions of others. For example, in a series of studies, Tomasello and colleagues have shown that many of the most common situations in which children learn early words require them to actively track adult intentions (and therefore attention) through a sometimes complex maze of possibilities. For example, an adult may announce the intention to "find the toma" and then search through a bag of novel toys. No matter which objects the adult extracts in which order, children as young as 18 months of age will assume that the "toma" is the one whose extraction seems to satisfy the adult's searching intention (the adult looks happy, stops searching, etc.; Tomasello & Barton, 1994; Tomasello, Strosberg & Akhtar, 1996). Similarly, a child may even learn a new word from a situation in which an adult requests that the child engage in some action ("Meek the doggie!") , even if that action never gets performed, provided that the child understands from previous experience and the current context what action the adult is requesting (Akhtar & Tomasello, 1996; Tomasello & Akhtar, 1995). The overall point is that in acquiring the linguistic conventions of those around them, human children rely on their ability to understand others as intentional agents who control their own attention to the world (Tomasello, in press). The result is conventional communicative symbols that are truly shared and intersubjective in the sense that they are understood as bidirectional social conventions that each user may either produce or comprehend with an understanding of the other's understanding.

Soon after acquiring their earliest linguistic symbols, young children begin to use them productively in a special way that makes especially clear their understanding of others as intentional agents with whom they may share attention. At around 18 months of age, many human infants begin to use their linguistic symbols in acts of predication. Predication is a two-step communicative act in which a speaker first establishes joint attention with a listener on an entity, and only then makes some comment about it. For example, an infant may hold up a spoon and then say "dirty," "mine," "wet," "heavy," or "spoon," with an understanding that there are others things that could be said about it. Through such predicative acts, young children demonstrate their understanding that others can be looking in a single direction and yet shift their attention voluntarily to different aspects of that experience. In perhaps no other child behavior is the understanding of the intentionality of others, in the form of their voluntary attention, so readily apparent (Reed, 1995; Tomasello, 1995a). This ability is expanded as children come to engage in more extended forms of discourse and conversation in which they take into account the perspective of their communicative partners (i.e., their background knowledge, current focus of attention, and so forth; Akhtar & Tomasello, in press).

One other phenomenon of early human development that is, like language, symbolic and uniquely human, is so-called symbolic play. Like all primates, human infants begin to engage in social play of a general sort fairly early in ontogeny. Play with others around objects begins somewhat later, at around the same time as other joint attentional interactions late in the first year of life, often in the context of teasing others with objects by offering and withdrawing (Reddy, 1991). But this play is transformed into symbolic play during the second year of life. In symbolic play, the child uses one entity as if it were another, knowing full well that it is not that thing; for example, using a pencil as a play hammer while smiling knowingly to an adult. One interpretation of this behavior is that it depends on children's ability to understand and adopt the intentions of adults, just as in the case of linguistic symbols. In this case, though, the intention of the other is the intention that is embodied in the artifact. That is, the child has seen an adult interact with pencils and knows what is done with them conventionally. Using a pencil as one would use a hammer thus violates this conventional expectation, and the child

Figure 13.3. Adult–child linguistic interaction. (Photo courtesy of Victor Balaban.)

knows this. The child is playing with the cultural meaning of the artifact as it is construed by the intentional agents around him.

Turning now to the effects of language on human cognitive development, we first note that language serves a unique representational function in cognition. It is not just that linguistic symbols "contain" information about the world, they also provide a unique way of cognitively representing that information. Thus, in one view, the bidirectional social perspectives embodied in linguistic symbols is what gives human concepts their radical freedom from perception and great flexibility of use across multiple cognitive domains. Said another way, linguistic symbols transform the nature of human cognitive representation by supplying users with socially shared and readily manipulable signifiers (the actual physical symbols) for their most important experiences. It is also important that linguistic symbols do not just represent individual episodic experiences, but rather they stand for whole categories of experience: they draw the learner's or listener's attention to one particular way of conceptualizing a particular experience as opposed to other possible ways (e.g., calling a particular person a "woman" or a "mother" or a "teacher").

In addition to this basic representational/categorizing function of language, the acquisition of language also leads to the formation of some new types of cognitive representations, namely, event categories and narratives. In contrast to the basically Piagetian focus on the taxonomic dimensions of human cognition involving hierarchical classification, scientific concepts, and the like, at which young children are relatively poor, developmental psychologists such as Nelson (1985, 1986) and Mandler (1992) have recently called attention to the thematic or narrative dimensions of experience as foundational for early childhood cognition. These involve the understanding of routine events, social scripts, and stories, at which young children are quite good. Interestingly, the prototypical events that young children understand and express in their earliest language all involve the intentional action of a human agent (Slobin, 1985). Thus, across all of the world's languages, young children find it interesting and important to talk about such things as the human manipulation of objects, the human exchange and possession of objects, the movements and changes of state of people and objects, and basic processes of social interaction (Fillmore, 1977).

The important point for current purposes is that the acquisition of language begins the process of conceptualizing and categorizing the thematic or narrative dimensions of human experience. Indeed, the conceptualization and categorization of events is the fundamental organizational principle of early language development. For example, Tomasello (1992b) documented how one child's early acquisition of language was organized around her understanding of different intentional actions that adults talked about, as signified by verbs such as "kiss," "hit," "jump," "give," "throw," "bring," "open," "hammer," "take," "walk," and "sing." The basic understanding of each of these events includes the participants involved as well; for example, hitting includes a "hitter" and a "hittee," and giving includes a "giver," a "thing given," and a "recipient." The child's early language, and the experiences it represents are organized in terms of a small number of these internally complex event structures. As children extract patterns from the language they hear being used about such human events, they form more abstract event schemas; for example, hitting and kicking are instances of a schema in which an agent affects a patient by means of an intentional action. The result is that from their second year of life, human children symbolically represent their shared experiences with others in terms of temporally extended events and their multiple participants all in identifiable conceptual packages, sometimes in highly schematic form.

At a later point in development, beginning sometime after the second birthday, language begins to serve as a medium by which children may represent chains of events in even more temporally extended experiences called narratives. Soon after children have learned to talk about relatively simple and punctate events, they begin to combine these words into new forms of temporally extended discourse in which different event descriptions are combined to depict the different activities of a single entity, or set of entities, through time. This is the beginning of linguistic narratives, which have important characteristics that go beyond the single utterance level, for example, extended topics, participant conflicts and resolutions, narrator evaluations, and the like. The narrative mode of cognitive representation and organization, beginning during the third and fourth years of life, represents a crucial transformation in the nature of human cognition in which temporally extended and complex states of affairs are given canonical form and structure. This new form of conceptual representation supplies humans with categories and concepts for complex events as wholistic entities, involving the same basic participant roles as event categories in general, such as agents, instruments, and affected entities (Bruner, 1986, 1990). This narrative mode then serves as the raw material for constructing more abstract taxonomic modes of cognitive representation and organization sometime after the child's fourth birthday (Nelson, 1985, 1996).

The social-cognitive revolution of the first birthday is thus extended during children's second and third years of life as they participate in various cultural activities, especially in linguistic communication. As a result, the cognition of the 2- to 3-year-old human child is very different from that of any other primate, especially because it includes an understanding of intentional action and is therefore organized in terms of symbols, event categories, and narratives. Children of this age learn linguistic and play symbols by tuning in to the intentional actions, and thus the attention, of others. They learn many of their linguistic symbols for complex events containing multiple participant roles and have just begun the process of forming schemas of these events to generate more abstract representational structures. They soon begin combining these structures together to create complex narratives involving the interaction of many participants and events over time. In general, this is all uniquely human cognition, made possible not only by children's developing social cognitive skills but also by the artifacts (including linguistic and narrative forms) that antedate them in the culture and are sometimes even taught to them explicitly.

In attempting to tease apart the relative contributions made by the two factors of crucial importance in all of this—preexisting cultural activities and the child's capacity to acquire them—the study of nonhuman primates raised in humanlike environments is of great importance. From our reviews in sections 8.4 and 12.2.3, it is clear that at least some human-raised great apes have the ability to tune in to human culture in some important ways, including through the learning and use of linguistic, and perhaps play, symbols. From the current point of view, the crucial question is how the apes are able to use these symbols to influence or to play with the intentions and attention of others. It is unclear, for example, whether enculturated apes engage in acts of linguistic communication that are truly declarative and predicative (involving the two-step process in which the attention of another is secured and then manipulated for declarative purposes), and it is unlikely that they construct event categories or narratives (Tomasello, 1994a). As we have said previously, there are not enough data to answer these questions definitively. For now we simply assert that only humans beings have created linguistic and play symbols on their own, and moreover, that even with extended tutelage from humans, there is little evidence that nonhuman primates can construct event categories or extended narrative sequences.

13.1.3 Theory of Mind and Concrete Operations in Early Childhood

Viewing the world in terms of intentions, event categories, and narratives is thus a distinctly human way of viewing the world. This view develops in human children as they participate in the cultural activities to which adults have chosen to expose them, including activities mediated through the use of material and symbolic artifacts. Children begin tuning in to the use of these artifacts in these activities at around 1 year of age, and their cognition progressively takes on the forms of narrativity embodied in language, as well as those embodied in other artifacts and modes of cultural activity, during the second year of life. Apes raised by human beings may move down this developmental pathway to some degree and in some ways, but other aspects of this ontogenetic sequence seem uniquely human.

From 3–5 years of age, human children develop a whole new repertoire of cognitive skills that are almost certainly uniquely human. These skills emerge in both the social and physical domains. In the social domain these skills manifest themselves in a panoply of situations involving children's understanding of the mental states, especially the beliefs, of others—their "theories of mind." Research into children's theories of mind is a lively area of research at present, with new empirical findings and research perspectives emerging on a regular basis (e.g., Astington, 1993; Baron-Cohen, Tager-Flusberg, & Cohen, 1993; Perner, 1991; Wellman, 1990; Whiten, 1991). The important point is that at some time between their third and fifth birthdays, young children come to understand the thoughts and beliefs of others as distinct mental entities, distinct both from their own thoughts and beliefs and from reality, and they can even develop some deceptive strategies for bringing about false beliefs in others.

In the physical domain of cognition, children in this age range are developing in new ways as well. Based on language and other direct interactions with objects and events, children are classifying the entities in the world around them on the basis of their similar properties and functions. This is the taxonomic dimension of cognition, which is complementary to the narrative dimension of cognition (Nelson, 1985, 1996). Thus, at 3–5 years of age, children begin to operate with hierarchically structured systems of concepts and relations. It is especially important for current purposes that many of these concepts and relations embody two different perspectives simultaneously. This is implicit, for example, in preschool children's emerging ability to classify abstract materials on the basis of two dimensions simultaneously (e.g., color and shape), to place a single object in both subordinate and superordinate classes simultaneously (e.g., this object is both an apple and a fruit), and to form one-to-one correspondences between two sets of objects, each of which forms a relational series (Gelman & Baillargeon, 1983). It is these same kinds of multiple perspectives on abstract materials that underlie the various physical concepts that children form during the school-age period involving quantities of various sorts that are conserved under perceptually deforming transformations because such multiple perspectives allow children to coordinate dimensions with one another to arrive at quantities independent of any one dimension (e.g., the amount of clay in a ball is not a direct function of any one spatial dimension).

What is common to these two different developments, theory of mind in the social domain and concrete operations in the physical domain, is their common reliance on children's ability to take different perspectives on a situation. In the case of theory of mind, children engage in discourse with others from their second year on in which they discover with regularity that others do not desire and know the same things they desire and know, or that others have a perspective on situations that differs from their own or that differs from the actual situation as they know it to be. Adults also instruct young children in how to do things, which creates a discrepancy between the situation as they perceive it and the situation as the adult characterizes it in the instruction. The point is that discourse with others during the preschool years, and perhaps some other kinds of social experiences as well, provides children with a constant source of conflicting views on events and situations in the world (Dunn, 1988; Harris, 1996). The Vygotskian hypothesis is that they "internalize" these conflicting points of view and thus learn to take different perspectives in their individual thinking. This allows them to be able to maintain simultaneously a knowledge of the actual state of affairs and the belief of another person which differs from that knowledge.

In some theoretical approaches, similar development is going on in the physical domain as well (Rochat, 1995b). That is, learning to take the perspectives of others and to internalize those into one's own thinking is also instrumental in learning to take different perspectives toward the physical world. The research of Perret-Clermont and Brossard (e.g., 1985) has shown that children learn to operate in more adultlike ways in problems concerning the conservation of various quantities when they engage in dialogues with other children who are equally incapable. Even though neither child "knows the answer," the experience of having a dialogue in which different points of view about the problem are expressed and discussed is enough to lead children on the edge of competence into a more adultlike understanding of the situation (see also Kruger, 1992; Siegler, 1995). In general, from Piaget's earliest writings, it has been clear that not only conservation problems, but also problems of classification and seriation, require children to see the same set of materials in multiple ways. To understand the world like an adult, young children must come to understand that these blocks may be divided either into large and small, blue and red, squares and circles; that this object is simultaneously a rose and a flower; that the water in this beaker is higher but skinnier; and that this number is simultaneously larger than the preceding ones and smaller than the succeeding ones. In one view of human cognitive development, these understandings all rely on the to-and-fro of social intercourse in which the child is exposed to, and to some degree comprehends, the different perspectives on situations as expressed by different persons.

Extending this line of reasoning, Tomasello, Kruger, and Ratner (1993) speculated that taking others' perspectives may also underlie the ability to reflect on one's own actions and thinking. That is to say, as the child learns to comprehend the perspective of others, those others often express a perspective on an action the child has just performed or a thought the child has just formulated in language. Taking the perspective of the other in such cases means that the child takes an outsider's perspective on her own behavior and thinking—simulating another person's view of her own thinking and action—which may then become internalized so that she may take such a perspective without the other person being actually present. Tomasello et al. speculate that engaging in this kind of reflection on the self, in the absence of any actual outside person currently viewing the self, underlies the kinds of representational redescriptions of cognition whose importance to cognitive development Karmiloff-Smith (1992) has so clearly demonstrated. Karmiloff-Smith argues that in each domain of cognitive development, children first develop effective procedures for solving problems successfully, but as they are doing so, they are also reflecting on their performance and redescribing its structure on a "higher" plane of thought. In this view, the development of a number of different abstract cognitive systems, such as mathematics and the abstract syntax of language, depend on representational redescriptions of this type. It may also be the case that the kinds of analogies and metaphors that children begin to create during this age period and which have such important ramifications for abstract thinking (Johnson, 1987; Lakoff, 1987) rely on redescriptions of this type as well.

Overall, then, the cognitive development of children at 3–5 years seems to include a number of skills not included in the cognitive development of nonhuman primates, especially theory of mind and concrete operations. Starting with event categories and narratives, children go on to categorize and hierarchically organize their concepts of entities and relations in ways that lead to new understandings of classification systems and quantities (Nelson, 1985, 1996). This way of operating complements the narrative

Table 13.1 Major steps in human cognitive development, with specific focus on social-cultural dimensions and their effects.

Infancy: understanding others as intentional

1. Following attention and behavior of others: social referencing, attention following, imitation of acts on objects
2. Directing attention and behavior of others: imperative gestures, declarative gestures
3. Symbolic play with objects: playing with "intentionality" of object

Early childhood: language

1. Linguistic symbols and predication: intersubjective representations
2. Event categories: events and participants in one schema
3. Narratives: series of interrelated events with some constant participants

Childhood: multiple perspectives & representational redescriptions

1. Theory of mind: seeing situation both as it is and as other believes it to be
2. Concrete operations: seeing events or object in two ways simultaneously
3. Representational redescription: seeing own behavior/cognition from "outside" perspective

mode of cognitive organization by providing a dimension of abstraction across events in terms of abstract taxonomic structures. Both of these aspects of cognitive development, the narrative and the taxonomic, likely derive from children's emerging abilities to understand the perspectives of other persons and then to internalize those perspective-taking activities in the form of viewing a single entity from multiple perspectives simultaneously. This same process underlies the child's viewing of self and its products reflectively from the perspective of an outside observer, thus creating all kinds of representational redescriptions.

13.1.4 Summary

The uniquely human aspects of cognitive development may thus all be traced, in one way or another, to the social-cognitive revolution at the first birthday in which human infants begin to tune in to the intentions and attention of other persons (see Table 13.1). This unique skill allows children to participate in the collective dimensions of human cognition. The process begins as children learn to follow and direct the attention of others to outside entities. It continues as they learn the appropriate use of conventional linguistic symbols, which then leads to the construction of event categories with complex internal structures of the narrative kind. As children continue down this narrative path, they become able to take the mental perspective of others, coordinating these perspectives on the same entities, and even taking a perspective on the self and its activities. This taking of multiple perspectives on entities in the context of event structures and narratives, and reflecting on them in one's own thinking, is responsible for the kinds of abstract multidimensional cognitive structures that characterize the cognition of 5-year-old children but no other primate.

We do not mean to neglect in this account of human cognitive development other important skills of discrimination, categorization, the understanding of objects and space, the understanding of social relationships, nonverbal communication, and so forth. But our view is that these skills are not unique to humans. Humans develop their special forms in these domains (e.g., navigating by maps and classifying and quantifying using complex formal systems) by (1) learning to use culturally provided tools and symbolic artifacts created for these domains by the collaborative efforts of multiple people, often over multiple generations; (2) coordinating different perspectives on these domains, often first by taking the perspective of others in social interaction but then independently; or (3) reflecting on their own activities in these domains, often first by taking the perspective of other persons directing or commenting on those activities but then independently. This is the cultural dimension of human cognition, without which human cognitive skills would be much more similar than they are to those of their close primate relatives.

13.2 Ontogenetic Processes

Given our brief description of the species-unique aspects of human cognitive development, several issues arise with respect to the ontogenetic sources of these developments and the sense in which they are uniquely human. There are three issues of particular importance. The first is the human child's understanding of others as intentional agents, on which our whole account of human cognitive development depends, and how this develops ontogenetically. The second is the effect of being treated intentionally and being instructed by others and how this plays its ontogenetic role in human cognitive development (and that of some nonhuman primate individuals). And the third issue is the internalization of social perspective-taking into the individual capacity to view the same entity from multiple perspective simultaneously and to reflect on one's own behavior and thinking, and how this might develop ontogenetically. We treat these each very briefly.

13.2.1 Understanding Intentions by Identifying with Others

Our account of the unique aspects of human cognitive development rests crucially on the way that human infants at 1 year of age come to understand the behavior of others as intentional. This new social-cognitive skill makes possible all kinds of joint attentional interactions, including the acquisition of language and other forms of cultural learning, which then make possible the conceptualization of events and the viewing of items of experience from multiple perspectives simultaneously. The question thus arises of how human children come to have this most basic social-cognitive skill and why other primates do not seem to have it, at least not in its typically human form.

The key process, in our view, is the special way that human infants "identify" with other members of their species. How this identification takes place is currently a matter of some debate. Meltzoff and Gopnik (1993) argue that identification with others is present in nascent form from birth, as evidenced by the human neonate's ability to mimic the facial and head movements of other persons. In performing these matching behaviors, infants show their ability to identify their own bodily movements with those of others. The fact that this mimicking takes place across perceptual modalities (infants mimic behaviors they see others performing using parts of their bodies that they cannot see) argues that the process is something deeper than simply parroting in the vocal modality, which may involve a process similar to repeating one's own behavior (since hearing others vocalize provides roughly the same feedback as hearing oneself vocalize; Piaget, 1962). Barresi and Moore (1996), on the other hand, do not believe that neonatal mimicking is as robust a phenomenon with monumental implications. In their view, certain types of social experience are necessary for children to learn about self–other correspondences, especially those of a more psychological nature. These experiences begin with the kind of face-to-face interactions with others in which human infants engage from very early in development (i.e., protoconversations), but most importantly they involve interactions in which the infant and adult both experience the same external phenomenon and react to it in the same way. For example, when an adult hands an infant an object and they both become excited about it, if the infant notices this similarity of reaction, he now has at his disposal the raw materials for identifying first-person experiences with those of other persons as viewed from a third-person perspective, in a way that is psychologically deeper than just equating body movements.

Given this identification with others early in ontogeny, the crucial step for infants' understanding of others as intentional agents takes place as they come to differentiate in their own behavior between the ends and means of instrumental acts, that is, as they begin to behave in ways that are clearly intentional. In Piaget's (1952) formulation, infants begin down this path at around 8–9 months of age when they first begin to adjust their behavior sensitively to take account of failures and obstacles and to perform actions whose function it is to enable other actions (see also Frye, 1991). Such flexibly organized behavior indicates the infant's emerging understanding of the relative independence of means and ends—that an activity that is an end in one context may be a means in another. This dissociation of means and ends then enables infants to formulate a goal prospectively and to flexibly choose among several possible behavioral strategies for realizing that goal depending on their assessment of the current situation. This new mode of behavior is the child's first experience with a mental entity (i.e., a goal) that is at least somewhat independent of direct sensory–motor actions and reactions. In line with previous discussions, in the perceptual modality, this translates into some experience of their own selective attention, since attention also implies an active choosing by an intentional agent.

It is not an accident that the age at which infants begin to behave in a clearly intentional manner (8–9 months) is close to the age at which they begin to understand the behavior of others as intentional (9–12 months). Given that they have first identified with others, when infants come to experience their own behavior as intentional, they quite automatically experience the behavior of others as intentional as well, perhaps in something like a simulation of another person's behavior or experience from their own point of view (Gordon, 1986; Harris, 1991). It is important to stress that since many animal species, including all primates, behave in a clearly intentional manner, the uniquely human component of this ontogenetic process is the initial identification with conspecifics that makes for the self–other equivalence in the first place. The main alternative to this account would be an account in which infants come to have a "theory" of the behavior of others based on external observation of others' intentional actions (just as they come to have theories about physical objects), and this takes place in relative independence of their tendency to identify with others in early ontogeny. This is essentially the proposal of Gopnik (1993).

Our proposal is that it is in their own intentional behavior that human infants first experience intentionality as they formulate independent goals and then act, either attaining their goals or not, at around 8–9 months of age (Tomasello, 1995a). Given that they have previously identified with others of their species, this experience provides the basis for an understanding that others also act intentionally in this same way. Evidence for something like this view is provided by human children afflicted with autism. Children who are most severely afflicted with autism do not appear to identify with others and so do not understand them as intentional. They therefore do not engage normally in early joint attentional behaviors such as imitative learning, declarative pointing, and symbolic play; they acquire very little language (about half acquire basically no linguistic conventions); and their sense of self does not seem to rely upon the kind of reflective self-awareness characteristic of typically developing children (Hobson, 1993). It is not unreasonable to suppose that autistic children have a biological deficit in precisely that social-cognitive process that most clearly differentiates the cognition of humans from that of other primate species. It is possible that the developmental period during which the deficit occurs determines the nature and severity of the disorder.

Although it is highly speculative at this point, we would also like to suggest the possibility that the human understanding of others as intentional agents may in some cases be applied to inanimate objects in some types of interactions (as humans "identify" with physical objects) and thus transform the understanding of antecedent–consequent sequences into causal understanding. Thus, following the reasoning of Collingwood (1972), it is possible that the "force" component of the human understanding of causal relations, which nonhuman primates do not seem to apply in the same way as humans, may derive from the extension of intentional thinking to the behavior of inanimate objects as they interact with one another. It is certainly the case that in many cultures the explanation of both human and inanimate events is composed of a complex mix of social, physical, and metaphysical forces (Nuckolls, 1991).

13.2.2 Culture, Being Treated Intentionally, and the Role of Language

A difficult issue is the role of social-cultural experience in the basic ontogenetic process. In the "starting state nativism" of Meltzoff and Gopnik (1993), infants' initial identification with conspecifics does not rely on social interaction with others, whereas in the more developmental account of Barresi and Moore (1996), social interaction plays a crucial role. Regardless of the role of social experience in the initial stages of this process, however, it is clear that some of the most important later processes of human cognitive and social cognitive development do depend in direct ways on social experience. We focus on two.

First, the learning of linguistic symbols is clearly dependent on social experience of a particular kind. Once children have identified with others and so understand them as intentional agents similar to themselves, they may follow into their attention and behavior to outside entities and do such things as imitate their behavioral strategies on objects. Another level of complexity is added to the process when an adult now attempts to direct the infant's attention, for example, by communicating with them using a linguistic symbol, and the child tries to imitate this. In this case the adult does not just intend something toward an outside entity, which the child could reproduce rather directly on that same entity, but rather intends things toward the child's psychological

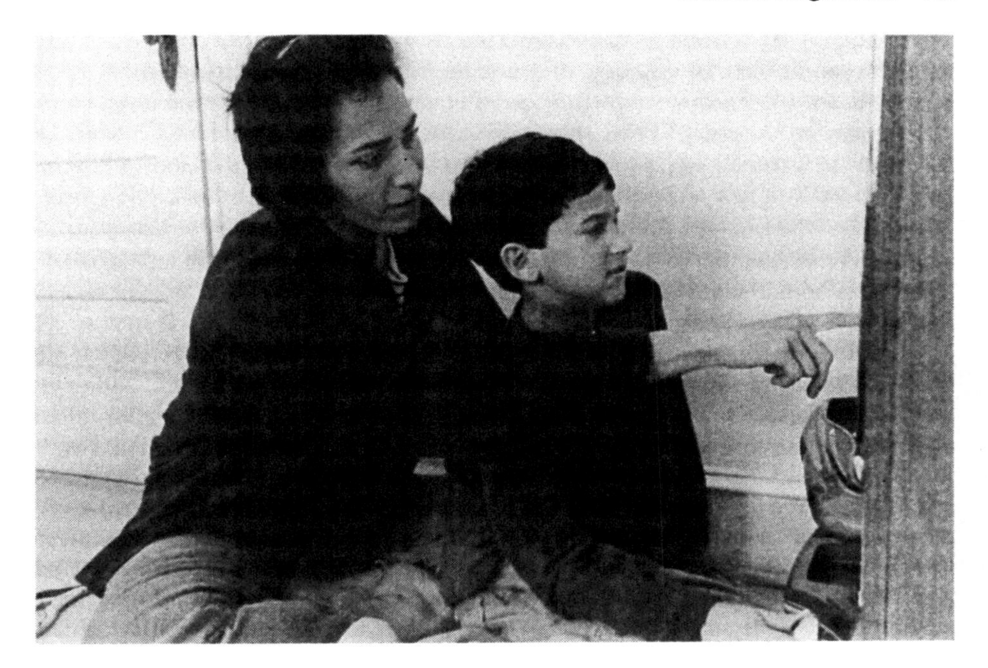

Figure 13.4. Human adult instructing child. (Photo courtesy of Victor Balaban.)

states (especially, the child's attention). The infants' understanding of these kinds of acts is thus of the form: that person intends for me to intentionally attend to X. Imitative learning of such acts then involves a role reversal: if I want *her* to attend to this same thing, then I must use this funny noise (linguistic symbol) towards *her*. The result is a social convention in which the child uses "the same" act to intentionally direct the adult's intentional states in "the same" way. Indeed, as we argued above, this kind of bidirectional understanding of the way individuals may direct the intentional states of others is what constitutes linguistic symbols. The important point in the current context is that linguistic symbols can only be learned in social interactions that have something of this structure—the child understands that the adult has intentions toward her intentional states—and so the role of the social-cultural environment is clearly a prerequisite for the acquisition of language.

Second, communicating with other persons using linguistic symbols involves a constant shifting of perspectives, as speakers formulate their message in ways that demand that listeners construe the world from a certain point of view or at a particular degree of specificity or abstractness (see Langacker, 1987, on the different "construal operations" embodied in language). This kind of perspective shifting is so endemic to language use that it has become historically entrenched in many basic linguistic symbols and syntactic constructions. Thus, for example, the words "coast," and "shore," and "beach" embody different perspectives on the same object, as do the words "doll," and "toy," and "Barbie". Similarly, basic linguistic constructions such as the passive construction serve to highlight or take a certain perspective on the events and participants of interest and to background other aspects for purposes of communicating effectively in particular types of communicative contexts; for example, "Fido got hit by a car" versus "The car hit Fido." In addition, of course, in extended discourse there are all kinds of devices and strategies for negotiating perspectives with others about the current topic of conversation. As argued above, engaging in the shifting perspectives and knowledge immanent in linguistic interaction not only constitutes the raw material from which the narrative mode of human conceptual organization is constructed (Bruner, 1986, 1990), it also constitutes the raw material for constructing a theory of mind and certain kinds of physical concepts in which perspective-taking is key. Thus, once again, social experience of a certain kind would seem to be prerequisite to the formation of certain kinds of cognitive organization.

The obvious question at this point is whether human social and cultural interaction works this same magic with at least some nonhuman primates (see Call & Tomasello, 1996; section 12.2.3). For current purposes it is only necessary to reiterate that when apes are brought up in humanlike cultural environments in which they are treated intentionally, they move down some fairly humanlike development pathways in the social-cognitive domain. This speaks both to the social-cognitive capacities they possess and to the power of the social-cultural interactions they have experienced. However, there are limitations. At the level of individual linguistic symbols, apes use declaratives and predication infrequently, if at all, and this suggest the possibility that most of apes' symbols are tied to action, not to the sharing of attention. The multisymbol linguistic productions of linguistic apes are quite modest linguistically, and moreover, even in the comprehension of the most linguistically sophisticated ape, the bonobo Kanzi (Savage-Rumbaugh et al., 1993), there is no evidence of a categorization of temporally extended events and their participants into something like the event categories that characterize 2-year-old children's early language (Tomasello, 1994a). Finally, in apes' discourse with humans, there are basically no reports of the use of sequences of complex event language to depict temporally extended narratives. Both Donald (1991) and Nelson (1996) speculate that this kind of representation of complex events and narrative sequences is a key difference in the way that the cognition of humans and nonhuman primates is organized.

The overall point is that human culture and social interaction provide developing human beings with a variety of symbolic and material artifacts, including language, for representing and organizing their cognitive activities. Linguistic and other forms of cultural interaction with others, in which two intentional agents, each of whom knows that the other is intentional, interact, involves all kinds of attention manipulation and perspective shifting. This narrativity is a routine and necessary dimension of many of the more complex cognitive activities in which children participate after their third birthdays, and it leads to skills in the taxonomic dimension of cognitive development as well. Apes who have been raised and trained by humans in special ways may or may not learn to use linguistic symbols to represent and organize their cognition, but it is unlikely that they use narrative or taxonomic forms of cognitive organization. This simply shows that, in addition to certain kinds of social and cultural experience, individuals must bring to the process certain biological preparations as well.

13.2.3 Multiple Perspectives and Representational Redescription

Many human cognitive achievements do not deal with social-cognitive phenomena such as intentions or language or theory of mind. As reviewed above, human children in the early childhood period begin doing things with inanimate objects that herald a new form of cognitive functioning, such as hierarchical classification, serial ordering of objects, and explicit quantification of substances, all indicating a newly emerging taxonomic form of cognitive organization. In some views, the understanding of beliefs is of this same nature because it requires the coordination of multiple perspectives on the same entity simultaneously, one being the "norm" considered as a true belief (Hobson, 1993). These skills all derive in one way or another from the ability to view things from two or more perspectives simultaneously, for example, that an object is both a rose and a flower, or that it is both taller than but skinnier than another object, or that someone believes the candy is in the drawer when it is really in the cabinet. Also important to these concepts is the ability to make reflective generalizations about one's own behavior and cognition, also involving a kind of perspective-taking, so that understanding that what is only implicit in practical interactions with the world can become explicit and more widely applicable within a cognitive domain.

In a general way, this view is consonant with that of Piaget in some of his early writings (e.g., as collected in Piaget, 1995), and with that of some modern Piagetians as well (e.g., Perret-Clermont & Brossard, 1985). Our own view, however, is that the most interesting version of this kind of theory is Karmiloff-Smith's (1992) theory of representational redescription. Implicit in the theory is the hypothesis that some fundamentally new cognitive structures can emerge when organisms take an outside perspective on their own activities, turning their powers of perception, relational understanding, multiple perspective-taking, and categorization on the self and its cognitive products. This process is a powerful one that, while not explicitly social in its functioning, may also have roots in the basic social-cognitive adaptation of taking the perspective of other intentional agents—although Karmiloff-Smith herself does not view it in this way. If indeed these abilities to take multiple perspectives on entities in the world and to reflect on and redescribe one's own behavior and cognition are derived from the same social-cognitive adaptation on which we have been focusing, at least two ontogenetic mechanisms for this can be imagined.

On the one hand, it could be, as we have argued, that actually interacting with others during ontogeny around specific kinds of materials, especially when language is involved, requires the learner to take the perspective of others about those materials and her own behavior and cognition toward them. Thus, in this view, in talking about quantities, classes, conservation concepts, and the like, both with adults who know more and with peers who may simply view things differently, children take the perspective of others and so learn multiple perspectives on those materials and their own cognition. Presumably these perspectives are then internalized so that at some point the child can take the different perspectives on her own, in the absence of any further social interaction. This account might thus be based on the general Vygotskian view that many human cognitive functions occur first in social interaction, which the individual then internalizes. On the other hand, it is also possible to imagine that human beings have evolved the ability to take the perspective of others in social interaction, but then that skill has been adapted biologically for use in dealing with physical materials. In this view, it is not necessary for the individual child to actually engage in specific types of social interactions before it is able to take multiple perspectives on physical entities; at this point in human evolution that ability is simply built into the architecture.

In either case, it is worth pointing out that the taking of multiple perspectives on a single entity, including the self, is not something at which nonhuman primates seem especially skilled. Thus, based on her many years of work with apes of all types, including those raised in humanlike cultural environments, Savage-Rumbaugh (1991) speculates that the "*Pan*-human rubicon" lies in what she calls multitasking. For example, she points out that in making tools, the maker must keep track of many things simultaneously, such things as the end product desired, the current state of the tool, and their relationship. Likewise, in language comprehension it is often the case that the comprehender must keep track of several referents and their relationships simultaneously. Based on many different types of comparisons between apes and human children in several domains of cognition, Savage-Rumbaugh has come to believe that apes "cannot concentrate on a sufficient number of things simultaneously to see the relationships between them that are readily apparent to the human mind" (p. 421).

13.2.4 Summary

It is currently popular to view human cognition in terms of a collection of independent modules. Although there seems to be no principled way for determining how many or what kind of modules there are (with different proposals having radically different sets), most proposals have at least several modules that all differ from the cognition of nonhuman primates. This seems to us implausible evolutionarily. Human cognition should differ from ape cognition in only a small number of fundamental ways. If those differences then lead to profound ripple effects (e.g., language and theory of mind), that is different from positing those ripples as predetermined modules. Indeed, we find it odd that for most people cognition means some kind of flexibility and generality in interacting with the environment, but the modular approach specifically denies such flexibility and generality to its cognitive modules.

Our overall point of view is that there is ultimately one key adaptation that makes human cognition different from that of other primates: the identification of self with others. This adaptation creates other unique processes, however. First is the tendency to understand the behavior and perception of others as intentional, which emerges as the human infant begins to differentiate means and ends in its own behavior and, based on its identification with others, attributes this same distinction to others. As the child perceives others intending things toward its own intentional states, there is a to-and-fro of perspective taking, especially in linguistic discourse in which both participants intend things toward the psychological states of others, attempting to manipulate one another's attention continuously. Finally, participation in interaction of this type leads to the ability to see things from multiple perspectives simultaneously and even the ability to reflect on one's own thinking by taking an outsider's perspective on it. Together, these cognitive processes are the key ones that make the single social-cognitive adaptation work to produce cultural beings.

13.3 Phylogenetic Processes

Attempting to reconstruct a complete history of human cognitive evolution is well beyond our current concerns (for two excellent texts on human evolution, see Klein, 1989; Leakey, 1994). Nevertheless, in this section we would like to address the one issue of most immediate concern to our theoretical account: the evolution of human culture. As we will argue, the evolution of culture was likely a recent event in human evolution, although it may have built on other uniquely human adaptations of which we know very little. Our main goal here is simply to provide a plausible evolutionary story for when and how human beings might have made the transition from more individually based cognition to more collectively based cognition. In the first subsection we address the issue of what evolved and how it might have evolved, and in the second subsection we speculate on when this might have happened.

13.3.1 What Happened

When viewed as a process, human culture is simply a series of ontogenies over historical time. Focusing on the cognitive dimensions of culture, the prototypical process is something like the following. Adults invent material or symbolic artifacts and then modify or improve on those artifacts or find new ways of incorporating those artifacts into social practices. They may do this either individually or in collaboration. Children then acquire skills for using the artifact, either through individual learning, imitative learning of adults, direct instruction from adults, or some combination of these. They often do this relatively early in ontogeny, which means that they may use their adulthood to find ways to improve upon the artifact or the ways it is used, again either individually or collaboratively. The new and improved version then gets passed on to the next generation, and so on across generations. We thus have artifacts and practices that show the "ratchet effect" and thus have cultural "histories."

In thinking about how such a process might have evolved, however, we are presented with the problem that the cultural learning skills of children and the cultural environment in which they develop are mutually dependent on one another. Thus, the ontogenies of human children, as well as human-raised apes, suggest that to acquire most human adultlike skills of social cognition and cultural learning, some kind of preexisting cultural environment is required, including especially adults treating children intentionally by directing their attention to phenomena and so intentionally instructing them. Indeed, we speculated in the previous section that children and apes only come to understand how to treat others intentionally, through linguistic symbols, for example, if they are treated intentionally themselves. But from an evolutionary point of view, how could it be that children learning to treat others intentionally depended on adults treating them intentionally? Which came first, the chicken or the egg? The problem is magnified when we note that in human ontogeny the ability to imitatively learn such things as linguistic symbols emerges soon after the first birthday, whereas attempts at the full-blown intentional instruction of others do not emerge until perhaps 3 years of age, suggesting that imitative learning is cognitively simpler than intentional instruction.

The solution may be this. Although full-blown intentional instruction involves one organism making sure that another organism acquires some information or skill (and may even rely on some form of a theory of mind involving mental agents), it has its ontogenetic roots in something simpler. Its ontogenetic roots, in our view, are the protodeclarative behaviors of pointing and showing that are seen in young children at the same age they first begin to imitatively learn, that is, at the 1 year social-cognitive revolution. Showing a parent a novel object just discovered is, in effect, instructing them to experience something new that they are not now experiencing; it is an act of informing just as an attempt to teach is an act of informing. Instruction may thus have its initial manifestation in acts of informing that emerge in late infancy and depend on an understanding of the intentions and attention of others, whereas later forms may depend on an understanding of beliefs. Perhaps it is only this form of informing, and not the more complex forms of explicit pedagogy, that encourages the understanding of others as intentional and cultural learning. If this were so, it would suggest that in human evolution, skills of cultural learning and informing may have evolved together: humans learned to show things to others and to respond to the attempts of others to show things to them all in one package. This then led to the evolution of language and other forms of cultural conventions and symbols, which in turn led to event categories, narratives, perspective-taking, representational redescription, and all of the processes that make human cognition unique.

But how did all of this get started in the first place? How did showing things to others and responding to the attempts of others to show things first begin? All we know to do is to look to ontogeny for clues, and here we find that the initial catalyst seems to be the identification with other persons, so that individuals can come to understand the intentional functioning of others on the model of their own experience. And here we are at the fundamental adaptation from which all else flows, and so how this may have happened is the fundamental question. At this point no one knows how this identification with others may have first happened, but one thing does seem fairly certain: the evolutionary process by which this happened was not one of so-called terminal addition in which a new skill was simply added onto the end of an existing ontogenetic sequence present in the ancestor that humans shared with chimpanzees (see Povinelli, Zebouni & Prince, 1996, for an interesting discussion). The fact that children identify with others from early in infancy argues that the fundamental adaptation we are talking about transformed the human ontogenetic process in its entirety, beginning at least from birth.

13.3.2 When It Happened

The problems with attempting to historically trace cognition are well known. Cognition does not leave behind the kinds of direct evidence of its workings that are left behind in, for example, a skeleton. For the most part we must be content with inferring the nature of human cognition at various evolutionary periods from the nature of the physical traces left behind. In the earliest periods of human evolution, for example, we are reduced to inferring cognitive adaptations from morphology, sometimes very indirectly (e.g., inferences from teeth to diet, inferences from diet to social structure, and inferences from social structure to social cognition) and changes in overall brain size (which translate even more indirectly into inferences about specific cognitive processes). At later evolutionary periods we are aided by material artifacts, which in many cases could only have been created by organisms that possessed particular types of cognitive processes.

Acknowledging the difficulty of pinning down specific times for specific evolutionary events, we believe that the most likely scenario for the evolution of human culture is a late-developing one. That is, regardless of all of the other important events in human evolution (see Boesch & Boesch, 1994, for some interesting speculations), it is likely that human cultural evolution and fully symbolic forms of linguistic communication did not begin until relatively recently. Following Klein (1989), we note that although *Homo erectus* roamed the world widely and was quite successful in many ways, the stone tools that they used were similar throughout their geographic range and throughout much of their several hundred thousand years of evolutionary history. It was not until the emergence of *Homo sapiens* (actually *Homo sapiens sapiens* in the last 150,000 years or so somewhere in Africa) that we observe cultural variations in the tool use of different groups of humans. Moreover, Mellars (1989) notes that it is only with the emergence of modern humans in Europe and the Middle East less than 50,000 years ago that tool industries displayed novelties at a rapid rate and displayed distinctive processes of "standardization" and "imposed form" within groups; he speculates that a social-cultural form of organization is needed to account for these new characteristics.

Noble and Davidson (in press) tell a similar story for the origin of language and other forms of symbolic functioning. They point out that in all of the many traces left behind by prehistoric humans, only with the emergence of modern humans into Europe and the Middle East 50,000 years ago do we see material signs of symbols (in the form of carved figures, cave paintings, and the like; see also White, 1989). They argue that although humans clearly communicated with one another before this time using various means, it was not until this late date that one could say that they were using linguistic symbols that were arbitrarily but conventionally associated with their referents. Klein (1989) summarizes this overall point of view as follows:

> Prior to the emergence of modern people, the human form and human behavior evolved together slowly, hand in hand. Afterward, fundamental change in body form ceased, while behavioral (cultural) evolution accelerated dramatically. The implication is that the modern human form—or, more precisely, the modern human brain—permitted the full development of culture in the modern sense and that culture then became the primary means by which people responded to natural selective pressures. As an adaptive mechanism, culture is not only far more malleable than the body, but cultural innovations can accumulate far more rapidly than genetic ones; and the result is that in a remarkable short time the human species has transformed itself from a relatively rare, even insignificant large mammal to the dominant life form on the planet. (p. 344)

Our hypothesis is thus that the human adaptation for culture, that is, identifying with others leading to the understanding of others as intentional, is a relatively late adaptation, perhaps identifying cognitively and culturally modern humans with anatomically modern humans. We do not have a satisfying "evolutionary fairy tale" about the ecological conditions that might have led to this adaptation, but presumably they involved processes of either cooperation, competition, social learning, or some combination of these that were required to deal with some new ecological pressure. If indeed modern humans originated in Africa several tens of thousands of years before their appearance in Europe and the Middle East, it is presumably social or ecological conditions there that were key to the original adaptation; and we know next to nothing about the first modern humans in Africa. Whatever the source of the original catalyst, identi-

Figure 13.5. Bonobo in rare bipedal posture. (Photo courtesy of F. de Waal.)

fying with others and so understanding them as intentional would seem to be useful in myriad ways, and so we presume that at some point all of these evolutionary advantages helped to support and extend the new adaptation.

It must be stressed that the key adaptation on which we have focused so heavily obviously did not take place in a species that was identical to any modern ape. If indeed this adaptation was relatively recent evolutionarily, it means that it took place in a population of humans who already had proceeded down their own phylogenetic pathways for several millions of years. In particular, the *Homo erectus* peoples were technically skilled relatively to modern-day apes with respect to the fashioning of stone tools, for example. The human adaptation for culture thus took place within a species that was already specialized in some ways. It is just that, in our view, it was the adaptation for culture that led to the seemingly great cognitive leap represented by modern humans and all of their complex artifacts and inventions. It also bears repeating that this key adaptation was not just tacked onto the end of the ontogenetic sequence characteristic of premodern humans, but it was woven into human cognitive ontogeny from the earliest stages.

13.3.3 Summary

The evolution of human cognition is a complex story, and there is currently too little evidence for theories to be put forth with much conviction. Nevertheless, we believe that current evidence is most consistent with the view that the uniquely human adaptation for social cognition and culture was a relatively recent one. The identification with conspecifics led both to showing others things (intentional instruction in its simplest form) and to learning new things from such showing (cultural learning). Once the initial adaptation had occurred, the stage was set for the evolution of fully symbolic forms of communication and the event categories and narratives in which these result. Communication of this type led to further skills of perspective-taking, which then led to various forms of multiple perspective-taking and reflection. The relatively late emergence of this cognitive adaptation in human evolution, after humans had already been a distinct species for several million years, clearly suggests its distinctiveness from the cognitive adaptations of other primate species.

13.4 The Structure of Human Cognition

Human cognition is a special form of primate cognition is several ways; we have tried to reduce them to one way and its spinoffs, the ultimate outcome being that primate cognition went from being an individual enterprise to a cultural (social-collective) enterprise. This reduction to one key adaptation is likely too radical, but we believe that it is a more plausible view evolutionarily than one in which cognitive modules are multiplied indiscriminately, which requires many different cognitive adaptations in the few million years that humans have been reproductively isolated from other primates. In any case, we would like to push this theory as far as it will go, and perhaps some other interesting differences that cannot be accounted for by the understanding of intentions will be revealed in the process. We can summarize our account of the unique aspects of human cognition and the single adaptation on which they depend in eight propositions.

1. *Humans are primates, and thus many of the cognitive mechanisms that underlie their behavioral adaptations are general primate cognitive mechanisms.* These mechanisms include discrimination, categorization, quantification, the understanding of objects and space, the understanding of tertiary social and physical relationships, the understanding of conspecifics as animate and directed, and basic processes of intentional communication and social learning.

2. *Early in ontogeny, human infants identify with conspecifics in a way that other primates do not.* This is evidenced especially by the nature of their early "protoconversations" (and perhaps neonatal mimicking) with others. At this early stage the identification is on the level of bodies and perhaps emotions, but it sets the stage for more psychologically based forms of identification, which follow on these bodily identifications early in ontogeny.

3. *Based on this early identification with others, when infants begin to differentiate means and ends in their own behavior at around 8–9 months of age, they automatically see the behavior of others in these same terms, that is, as intentional, including the understanding of attention as intentionally directed perception.* The understanding of others as intentional beings, over and above their animacy and directedness, results directly in all kinds of joint attentional behaviors such as social referencing, following the gaze and gestures of others, imitative learning, and the use of declarative gestures and communicative symbols to manipulate the attention of others. This new social-cognitive skill is thus ultimately responsible for uniquely human cognitive processes, including those involving language.

4. *It is likely that adults treating infants as intentional beings plays an important role in this process.* Although the earliest forms of identification with others in terms of neonatal mimicking and protoconversations may not require specific types of social interaction, it is likely that in the ensuing months the ability to identify with others psychologically (rather than physically and emotionally) requires specific kinds of social interaction in which the infant has the opportunity to compare its own perspective on a situation with that of others. Of crucial importance for subsequent developments involving symbols is being treated intentionally once others are understood as intentional, that is, when individuals understand that others intend things toward their intentional and attentional states, as in the use of language.

5. *Learning to use language transforms cognition in many ways, including the ability to represent the environment in intersubjectively understood symbols, for categorizing items in the environment, and for organizing the world into event structures and narratives.* Linguistic symbols constitute a unique form of cognitive representation because of the dual perspective (producer and comprehender) they embody. Because languages contain symbols of different levels of specificity, the acquisition of different types of symbols also leads to complex forms of categorization. In addition, the prototypical phenomena that languages are designed to describe all involve intentional human actions and events with one or more participants, and these complex event schemas may themselves be categorized. Linking these event categories in particular ways leads to the kind of narratives that constitute such an important aspect of human cognitive organization.

6. *The use of language in interaction with another person leads to conflicting viewpoints and perspectives, which provide the raw material for the understanding of others as having beliefs (including false beliefs).* Inspection of the structure of languages shows that much of their complexity derives from the necessity to adapt specific formulations for specific communicative circumstances, including most especially the needs of the current listener. The implication is thus that the understanding of others as intentional (and thus attentional) beings is foundational for the original acquisition of language, but that as children develop cognitively they understand more and more of the complexities of shifting viewpoints involved in discourse and conversations. The further implication is that children's "theories of mind" are built on this social-cognitive foundation.

7. *Children's internalization of the different viewpoints that they encounter in interacting with others, both linguistically and otherwise, may be the source of the more general human ability to see things from multiple perspectives simultaneously and even to reflect on their own behavior and cognition.* When applied to the physical world, such

abilities result in many of the concrete operations in which children understand hierarchically based classifications, series of ordered relations, and the interrelation of different physical dimensions in the judgment of quantities. These constitute the taxonomic dimension of human cognitive organization. Moreover, the ability to take an "outside" perspective on own's own activity may underlie the ability to provide "representational redescriptions," whose importance to childhood cognition Karmiloff-Smith (1992) has so convincingly demonstrated.

8. *It is likely that the initial social-cognitive adaptation for culture, for identifying with other persons and thus sharing intentional relations to the world with them, was a late-occurring one in human evolution, perhaps constituting the unique cognitive skills of* Homo sapiens sapiens. The evidence for this view is that standardized cultural variations in the tool use of different groups of humans is not evident in the artifactual record until the most recent 100,000–50,000 years, and the first material symbols are not evident until this time as well. If this late scenario is correct, it is clear that this adaptation came well after the evolutionary differentiation of humans from other primates. It is also clear that it was not just added onto the end of the ontogenetic sequence characteristic of premodern humans, but that it was woven into the earliest stages of human cognitive ontogeny.

14

Conclusion

We cannot say that Darwin's dream of understanding the human mind by studying the nonhuman primate mind has become a reality, but, as we have documented in this book, progress is being made. As we look to the future, there are important issues of three types that will affect our prospects. These issues concern, first, the kinds of theories we need to guide our future research, second, the kinds of methodological considerations we should take into account in future research, and third, the kinds of efforts we need to ensure that primate research may continue, given the precarious futures of many primate species. We conclude our investigation into primate cognition by briefly discussing each of these.

14.1 Theory

Theories are epistemological tools for viewing the world in certain ways, and good scientific theories are those that embody points of view conducive to the discovery of useful facts about the world through empirical research. In the field of primate cognition, we believe there are four sets of issues that theorists should consider.

1. *"Intelligence" versus cognitive adaptations.* In studying primate cognition it is not particularly useful to employ the concepts of human psychometricians, especially not the concept of "general intelligence." Intelligence is a concept designed by psychometricians for the specific purpose of discriminating among human individuals in terms of a single quantitative scale (so as to predict success in school and other similar environments), not for comparing the cognitive skills of different species of animals who may have very different orientations to the world. It is simply not meaningful or useful to discuss, as do members of the popular press with regularity, which animal species is "more intelligent" or more "cognitively advanced" than another; it is simply another manifestation of the medieval *scala naturae* that Hodos and Campbell (1969) so effectively criticized. We should expunge this way of talking from our scientific discourse. Where

there is a need, we may speak of cognitive adaptations as more or less complex, depending on such things as the number of elements that must be perceptually monitored and behaviorally controlled.

The alternative to this unidimensional approach is the approach of cognitive ethology in which we look first at the adaptive problems faced by a particular species or set of species and then at the ways they have evolved for dealing with them—some of which may turn out to be cognitive in the sense that individuals attain their goals by flexibly assessing the current situation and adjusting their behavioral interactions accordingly. The extent to which particular species solve particular adaptive problems with narrowly crafted special-purpose cognitive mechanisms (modules), or the extent to which they do this with more domain-general general-purpose cognitive mechanisms, is in each case an empirical question.

2. *Behaviorism versus cognitivism.* It is also unhelpful in the study of primate cognition, in our view, to continue to employ the concepts of behaviorists, especially instrumental conditioning, as if they always constituted an alternative explanation for cognitive phenomena. Cognitive ethology is, or should be, a thoroughly cognitive enterprise. It is simply not the case that instrumental conditioning is a low-level noncognitive phenomenon that contrasts with cognition as a type of behavioral adaptation. Rather, cognitivism and behaviorism are competing theoretical paradigms for the study of organisms' interactions with their environments, and cognitive ethology has embraced cognitivism. Consequently, instrumental conditioning and all other ways in which sentient organisms interact with their environments behaviorally, in flexible ways, have cognitive explanations. The rat who has learned to press a bar to receive food when a red light is on understands that something it does (pressing the bar) reliably produces an environmental event (the appearance of food) in certain circumstances (the red light). To assume that if an animal has obtained a reward for some behavior in the past then that behavior, *ipso facto*, involves no cognitive processes whatsoever is not a necessary inference.

The alternative is to adopt an approach in which all behavioral adaptations involving some degree of flexibility and complexity are seen as cognitive. If an organism obtains a reward for some behavior in some situation and repeats it in the future, it may do so based on a cognitive analysis of the comparability of the current and past situations, or, if there are few cues available for making such a comparison (perhaps because human experimenters have eliminated most of them), it may do so based on some general heuristics. In either case the animal is attempting to understand the situation in which it finds itself and to choose the appropriate response given its goals. However simple this process, it is a form of cognition and should be studied as such. The suggestion is thus that theorists not seek to identify instances of primate cognition negatively by finding behaviors that appear to be intelligent but then eliminating those that might be the product of instrumental conditioning. Rather, they should seek to identify those behavioral interactions with the environment that seem to be supported by flexible and complex knowledge allowing for the creation of flexible and insightful strategies for dealing with problems of various types, or else they should use whatever definitional criteria for cognition they believe is theoretically justified, without worrying about behavioristic explanations.

3. *The concept of representation.* The opposition of behaviorism and cognitivism also has led to the unhelpful tendency to identify cognition with "representation"

because the only way to go beyond stimuli and responses is to consider an organism's reaction to stimuli that are not currently perceptible. The fact is, however, that all behavioral adaptations involving any degree of learning by definition rely to some degree on the preservation of past experience. This means that basically all animal species that learn form "representations." Defining representation more narrowly to involve only delayed responses or memory for food locations, for example does not make for any more interesting approach to cognition because all we are dealing with in these cases is the ability to preserve a specific perception or motor response over time.

In our view, cognition is much more dynamic than simple representation, involving the organism's active choosing of a course of action given both its goals and its assessment of the current situation with respect to those goals. Specific representations may be involved in this process in one way or another, but there are many situations in which an organism simply solves a problem without overt trial-and-error through something like insight or foresight in which it simply "sees" the solution. We have argued that this process still involves cognitive representation, but, in either case, the issue of cognitive representation is only important in the context of situations in which the organism has to make informed decisions of some type. Representation is an important component of cognitive activity, but it is not the only component.

4. *Physical versus social cognition.* A major problem in constructing theories of primate cognition is that researchers studying physical cognition typically employ either a learning theory or Piagetian approach, whereas researchers studying social cognition employ other theoretical frameworks such as the "metarepresentational" approach that are mostly unrelated. This situation may derive to some extent from the underdeveloped state of theorizing in the social-cognitive domain, including the human literature. But whatever its source, a single theoretical framework incorporating both physical and social cognition is sorely needed. We have offered our modest attempt focused on the understanding of tertiary relations in chapter 12.

14.2 Research

In surveying all the empirical research that has been conducted over the past several decades on primate cognition, four further issues stand out as crucially important.

1. *Field versus laboratory.* The sometimes antagonistic relationship between fieldworkers and laboratory workers is clearly not helpful to the study of primate cognition. In all fields of behavioral and cognitive research, there is a tradeoff between studying behavior in more natural settings and studying it under more carefully controlled conditions. Naturalistic studies concern most directly the phenomena of primary interest, but they are seldom able to identify precisely the causal factors involved. Experimental studies are able to identify causal factors, but it is always possible that they are focusing on behavior that is not reflective of behavior in natural settings. The only answer is the obvious one of more cooperation among field and laboratory researchers in which complementary approaches are taken to the same phenomena. In particular, it is important that fieldworkers set problems for experimentalists by identifying seemingly interesting cognitive phenomena in the natural world. Experimentalists must then attempt to determine the precise nature of these phenomena in settings that resemble the natural settings as closely as possible. Where the two sets of researchers come to

different conclusions based on their different types of observations, they should seek to identify and resolve the source of their differences, ideally through further research.

It might also be useful if the two methodological approaches were combined to some degree in single studies, for example, in field studies that employ rigorous experimental methods and in laboratory studies of captive primates in more naturalistic social settings. The former approach has been used extensively by only a few scientists (most prominently by primatologists who have worked at one time with Peter Marler). The latter approach was pioneered by Menzel and colleagues, but has been taken up by only a few experimentalists since. In any case, it is unarguable that in the study of primate cognition, more cooperation of some type between fieldworkers and experimentalists is needed if we are to make progress in identifying with any reasonable precision the types of cognitive adaptations that different primate species have evolved.

2. *Proximate causes.* Our focus in this investigation has been on those proximate causes of primate interactions with their environments that involve cognition, and so we have sought to identify in each of the specific areas of research a number of important questions that are still outstanding; these may be found in the appropriate places in chapters 2–11. Overall, however, we would like to emphasize two sets of questions that we believe are especially important in our attempts to understand primate cognition. First, in the physical domain, primate understanding of causality is poorly understood. Visalberghi's important research is a step in the right direction, but it concerns only a few species and a few closely related tool-use problems. If causality is potentially the glue that holds together primate understanding of the physical world, it deserves a much more broad-based approach in which different types of problems are presented to a wide range of species. Until such research is conducted, we will not know if nonhuman primates simply make predictions about events in their environment or if they understand something of the underlying principles or "forces" that govern the relationships of those events to one another.

Second, in the social domain, primate understanding of intentions is also poorly understood. It is noteworthy in this regard that Premack's original work investigating "theory of mind" was actually about the understanding of intentions, not beliefs or false beliefs. If nonhuman primates do understand intentions in some manner, this would have important implications for the nature of their social lives, but the fact is that little research has addressed this issue in the almost two decades since Premack's original work. Again, we believe that the problem is of central importance, underlying the understanding of both goal-directed action and attention, and so deserves a broad-based approach in which different types of data are collected on a wide range of primate species, including human children. Until such research is conducted, we will not know if nonhuman primates simply make predictions about the behavior of conspecifics or if they understand something of the underlying intentional structure of that behavior.

3. *Evolution and ultimate causes.* A major impediment to much research and theory on primate cognition is the narrow range of species that have been studied. This is a problem in the field as a whole, but it is especially troublesome in attempts to reconstruct the evolutionary history of primate cognition. Many such attempts hypothesize discontinuities between such large-scale taxa as prosimians and simians or monkeys and apes on the basis on just one or a few representatives from each taxon. Since each species has many of its own unique characteristics, this is clearly a potential source of error in making inferences about large-scale evolutionary events in the history of pri-

mates. The same argument may be made with respect to primates versus other mammals and animals; rats, pigeons, and a few large mammals are not a sufficient sampling of species to create plausible hypotheses about primates versus other mammals or animals (even though we have put forth such a hypothesis ourselves).

The problem is serious in the physical domain. In almost all areas of research, Old World monkeys are represented by some species of macaque (or sometimes baboon), New World monkeys are represented by some species of squirrel monkey or capuchin monkey, and great apes are represented by chimpanzees. Prosimians, colobine monkeys, and lesser apes are almost never represented. But the problem is even worse in the social domain. Discounting studies focused solely on vocalizations, almost all of what we know comes from the study of cercopithicine monkeys and great apes. In some domains, a few New World monkeys such as capuchins have been studied as well, but, again, prosimians, colobines, and lesser apes have not been studied at all. It is obviously a major problem to try to reconstruct the evolution of primate cognition with such a biased sample of species. Many of the reasons for this bias are practical, having to do with what kinds of species are available for study, but in other cases the choice of which species to study cognitively is made on the basis of the rule: the closer to humans the better. It is clearly time to go beyond this bias, especially for those researchers interested in creating plausible evolutionary scenarios.

4. *The role of ontogeny.* In our view most of the interesting work on human cognition has been done by researchers taking an ontogenetic approach. Although there is some ontogenetic research in the field of primate cognition (mostly using Piaget's stage theory), very little is explicitly developmental. This is especially true with respect to social cognition, for which there are almost no developmental data or theories. Analyses that establish the developmental timing of a range of cognitive processes might be enlightening as to the relationship among those different processes (e.g., if they are underlain by a single mechanism) and so may be of special importance in constructing theories of primate cognition.

In addition, in investigating all aspects of nonhuman primate cognitive development, there is a need to take into account the role of humans in the process. The numerous "ape language" and related projects that have taken place over the years have demonstrated a number of humanlike cognitive capacities in their subjects, but the manner in which those capacities emerged or were acquired during ontogeny is seldom documented. For example, as we noted in chapter 8, there is little specific, quantitative information on the kinds of interactive situations in which apes learn the languagelike symbols that they come to use. It would also be helpful for theories of primate cognition if there were projects in which there was planned variation in the kinds of environments in which different species would be raised, systematically manipulating such variables as the amount and type of explicit human instruction involved. This would be tremendously helpful in relating what is known about the cognition of primates in their natural habitats to the cognition of primates raised and trained by humans and thus to the construction of a comprehensive theory of primate cognition.

14.3 The Preservation of Primates

We have learned much in writing this book and from our own research with nonhuman primates. We believe it is important scientific knowledge. It is important in its own right as information about significant phenomena in the natural world, and in addition it has important implications for our understanding of human cognition and evolution. We certainly cannot hope to identify the unique aspects of human cognition until we know which parts of it are shared with our nearest primate relatives.

But the future of this type of research is threatened by the possible extinction of many nonhuman primate species in their natural habitats. If something is not done soon, we may end up in some decades studying only captive animals descended from other captive animals. That would be a tragic loss for science. In addition, of course, it would be a tragic loss for the natural world itself, and for humans who wish to preserve the natural world in which they live. The issues of primate conservation are complex and often involve balancing human and animal needs. But with effort, resources, and dialogue, we believe that the appropriate balance may be found. We hope that people who read and gain something from this book will support efforts to preserve the natural habitats of nonhuman primates in ways that are also sensitive to the needs of the humans that might be affected by those efforts.

It is also important that the primates held in human captivity be placed in conditions that approximate, at least in some ways, those that obtain in their natural environments. Zoological parks have made great strides in this direction in the past two decades. Research laboratories have made progress as well—almost all laboratory primates are now housed in social groups of some type, for example—although monetary considerations often prevent the construction of larger and more natural environments. People wishing to encourage more naturalistic environments for primates in captivity, therefore, should support efforts in this direction both by zoos and by primate research facilities.

Appendix

1. PROSIMIANS

Prosimii	
Lemuriformes	
Daubentonioidea	
Daubentoniidae	
Daubentonia madagascariensis	Aye-aye
Lemuroidea	
Cheirogaleidae	
Allocebus trichotis	Hairy-eared dwarf lemur
Cheirogaleus major	Eastern dwarf lemur
Cheirogaleus medius	Western fat-tailed dwarf lemur
Microcebus murinus	Western gray mouse lemur
Microcebus rufus	Eastern brown mouse lemur
Mirzacoquereli	Coquerel's mouse lemur
Phaner furcifer	Forked lemur
Indriidae	
Avahi laniger	Woolly lemur
Indri indri	Indri
Propithecus diadema	Diademed sifaka
Propithecus verreauxi	White or Verreaux's sifaka
Lemuridae	
Lemur catta	Ring-tailed lemur
Lemur coronatus	Crowned lemur
Lemur fulvus	Brown lemur
Lemur macaco	Black lemur
Lemur mongoz	Mongoose lemur

Lemur rubriventer	Red-bellied lemur
Varecia variegata	Ruffed or variegated lemur
Lepilemuridae	
Hapalemur griseus	Gray gentle lemur
Hapalemur simus	Broad-nosed gentle lemur
Lepilemur mustelinus	Sportive lemur
Lorisiformes	
Lorisoidea	
Lorisidae	
Galaginae	
Galago alleni	Allen's bushbaby
Galago crassicaudatus	Thick-tailed bushbaby
Galago demidovii	Demidoff's or dwarf bushbaby
Galago (Euoticus) elegantulus	Needle-clawed bushbaby
Galago garnettii	Greater bushbaby
Galago inustus	Lesser needle-clawed bushbaby
Galago senegalensis	Lesser bushbaby
Galago zanzibaricus	Zanzibar bushbaby
Lorisinae	
Arctocebus calabarensis	Angwantibo
Loris tardigradus	Slender loris
Nycticebus coucang	Common slow loris
Nycticebus pygmaeus	Pygmy slow loris
Perodicticus potto	Potto
Tarsiiformes	
Tarsioidea	
Tarsiidae	
Tarsiinae	
Tarsius bancanus	Borneo tarsier
Tarsius spectrum	Spectral tarsier
Tarsius syrichta	Philippine tarsier

2. NEW WORLD MONKEYS

Platyrrhini	
Ceboidea	
Callimiconidae	
Callimico goeldii	Goeldi's marmoset
Callitrichidae	
Callithrix argentata	Silvery bare-ear marmoset
Callithrix humeralifer	Tassel-ear marmoset
Callithrix jacchus	Common or tufted-ear marmoset
Cebuella pygmaea	Pygmy marmoset
Leontopithecus rosalia	Lion tamarin
Saguinus bicolor	Brazilian barefaced tamarin

Saguinus fuscicollis	Saddleback tamarin
Saguinus imperator	Emperor tamarin
Saguinus inustus	Mottle-faced tamarin
Saguinus labiatus	Red-chested or white-lipped tamarin
Saguinus leucopus	White-footed tamarin
Saguinus midas	Red-handed tamarin
Saguinus mystax	Moustached tamarin
Saguinus nigricollis	Black-and-red tamarin
Saguinus oedipus	Cottontop tamarin
Cebidae	
Alouattinae	
Alouatta belzebul	Red-handed howler
Alouatta caraya	Black howler
Alouatta fusca (guariba)	Brown howler
Alouatta palliata (villosa)	Mantled howler
Alouatta pigra (palliata, villosa pigra)	Guatemalan or black howler
Alouatta seniculus	Red howler
Atelinae	
Ateles belzebuth	Long-haired spider monkey
Ateles fusciceps	Brown-headed spider monkey
Ateles geoffroyi	Black-handed spider monkey
Ateles paniscus	Black spider monkey
Brachyteles arachnoides	Muriqui or woolly spider monkey
Lagothrix flavicauda	Yellow-tailed woolly monkey
Lagothrix lagothricha	Humboldt's, or common woolly monkey
Cebinae	
Cebus albifrons	White-faced capuchin
Cebus apella	Brown or tufted capuchin
Cebus capucinus	White-throated capuchin
Cebus olivaceus (nigrivittatus)	Wedge-capped capuchin
Saimiri oerstedii	Red-backed squirrel monkey
Saimiri sciureus	Common squirrel monkey
Pitheciinae	
Aotus trivirgatus	Night or owl monkey
Cacajao calvus	White or red uakari
Cacajao melanocephalus	Black ukakari
Callicebus moloch	Dusky titi
Callicebus personatus	Masked titi

Callicebus torquatus	Yellow-handed titi
Chiropotes albinasus	White-nosed saki
Chiropotes satanas	Black saki
Pithecia albicans	Buffy saki
Pithecia hirsuta	Black-bearded saki
Pithecia monachus	Red-bearded saki
Pithecia pithecia	White-faced or Guianan saki

3. OLD WORLD MONKEYS

Catarrhini

Ceropithecoidea

Cercopithecidae

Cercopithecinae

Allenopithecus (Cercopithecus) nigroviridis	Allen's swamp monkey
Cercocebus albigena	Gray-cheeked mangabey
Cercocebus aterrimus	Black mangabey
Cercocebus atys	Sooty mangabey
Cercocebus galeritus	Agile mangabey
Cercocebus torquatus	White-collared or cherry-crowned mangabey
Cercopithecus aethiops	Vervet monkey
Cercopithecus ascanius	Redtail monkey
Cercopithecus campbelli	Campbell's guenon
Cercopithecus cephus	Moustached guenon
Cercopithecus diana	Diana monkey
Cercopithecus erythrogaster	Red-bellied monkey
Cercopithecus erythrotis	Red-eared nose-spotted monkey
Cercopithecus hamlyni	Owl-faced or Hamlyn's monkey
Cercopithecus l'hoesti	L'Hoest's monkey
Cercopithecus mitis	Blue or Sykes's monkey
Cercopithecus mona	Mona monkey
Cercopithecus neglectus	De Brazza's monkey
Cercopithecus nictitans	Spot-nosed guenon
Cercopithecus petaurista	Lesser spot-nosed monkey
Cercopithecus pogonias	Crowned guenon
Cercopithecus salongo	Salongo monkey
Cercopithecus (Miopithecus) talapoin	Talapoin
Erythrocebus patas	Patas monkey
Macaca arctoides	Stump-tailed macaque
Macaca assamensis	Assames macaque
Macaca cyclopis	Formosan or Taiwan macaque

Macaca fascicularis	Long-tailed, cynomolgus, or crab-eating macaque
Macaca fuscata	Japanese macaque
Macaca mulatta	Rhesus macaque
Macaca nemestrina	Pigtailed macaque
Macaca (Cynopithecus) nigra	Celebes macaque
Macaca radiata	Bonnet macaque
Macaca silenus	Lion-tailed macaque
Macaca sinica	Toque macaque
Macaca sylvanus	Barbary macaque
Papio cynocephalus anubis	Olive or anubis baboon
Papio cynocephalus cynocephalus	Yellow baboon
Papio cynocephalus papio	Guinea baboon
Papio cynocephalus ursinus	Chacma baboon
Papio hamadryas	Hamadryas baboon
Papio (Mandrillus) leucophaeus	Drill
Papio (Mandrillus) sphinx	Mandrill
Theropithecus gelada	Gelada baboon
Colobinae	
Colobus angolensis	Angolan black-and-white colobus
Colobus (Pilliocolobus) badius	Red colobus
Colobus guereza	Guereza, or black-and-white colobus
Colobus (Procolobus) verus	Olive colobus
Nasalis (Simias) concolor	Simakobu, or pigtailed langur
Nasalis larvatus	Proboscis monkey
Presbytis aygula	Sunda Island leaf monkey
Presbytis cristata	Silver leaf monkey
Presbytis entellus	Gray or hanuman langur
Presbytis francoisi	Francois's langur
Presbytis frontata	White-fronted leaf monkey
Presbytis geei	Golden leaf monkey
Presbytis johnii	Nilgiri langur
Presbytis melalophos	Banded leaf monkey
Presbytis obscura	Dusky leaf monkey
Presbytis phayrei	Phayre's leaf monkey
Presbytis potenziani	Mentawai langur
Presbytis rubicunda	Maroon leaf monkey
Presbytis senex	Purple-faced langur
Presbytis thomasi	Thomas's langur
Pygathrix nemaeus	Douc langur
Rhinopithecus avunculus	Tomkin snub-nosed langur
Rhinopithecus roxellanae	Golden snub-nosed langur

4. APES AND HUMANS

Hominoidea	
Hylobatidae	
Hylobates agilis	Agil gibbon
Hylobates concolor	Black- or white-cheeked gibbon
Hylobates hoolock	Hoolock gibbon
Hylobates klossii	Kloss's gibbon
Hylobates lar	White-handed or lar gibbon
Hylobates moloch	Silvery or moloch gibbon
Hylobates muelleri	Mueller's gibbon
Hylobates pileatus	Pileated gibbon
Hylobates (Symphalangus) syndactylus	Siamang
Pongidae	
Gorilla gorilla beringei	Mountain gorilla
Gorilla gorilla gorilla	Western lowland gorilla
Gorilla gorilla graueri	Eastern lowland gorilla
Pan paniscus	Bonobo or pygmy chimpanzee
Pan troglodytes	Chimpanzee
Pongo pygmaeus	Orangutan
Hominidae	
Homo sapiens	Human

References

Adams-Curtis, L. (1987). Social context of manipulative behavior in *Cebus apella*. *American Journal of Primatology, 12*, 325.

Adams-Curtis, L. & Fragaszy, D.M. (1995). Influence of a skilled model on the behavior of conspecific observers in tufted capuchin monkeys (*Cebus apella*). *American Journal of Primatology, 37*, 65–71.

Akhtar, N. & Tomasello, M. (1996). Twenty-four month old children learn words for absent objects and actions. *British Journal of Developmental Psychology, 14*, 79–93.

Akhtar, N. & Tomasello, M. (in press). Intersubjectivity and early language. In S. Braaten (Ed.), *Intersubjective communication and emotion in ontogeny*. Cambridge: Cambridge University Press.

Alcock, J. (1984). *Animal behavior.* Sundland, MA: Sinauer.

Altmann, S.A. (1967). *Social communication among primates*. Chicago: The University of Chicago Press.

Altmann, S.A. & Altmann, J. (1970). *Baboon ecology*. Chicago: The University of Chicago Press.

Amsterdam, B.K. (1972). Mirror self-image reactions before age two. *Developmental Psychology, 5*, 297–305.

Anderson, J.R. (1983). Responses to mirror image stimulation and assessment of self-recognition in mirror- and peer-reared stumptail macaques. *Quarterly Journal of Experimental Psychology, 35B*, 201–212.

Anderson, J.R. (1984). Monkeys with mirrors: some questions or primate psychology. *International Journal of Primatology, 5*, 81–98.

Anderson, J.R. (1985). Development of tool-use to obtain food in a captive group of *Macaca tonkeana*. *Journal of Human Evolution, 14*, 637–645.

Anderson, J.R. (1986). Mirror-mediated finding of hidden food by monkeys (*Macaca tonkeana* and *Macaca fascicularis*). *Journal of Comparative Psychology, 100*, 237–242.

Anderson, J.R. (1990). Use of objects as hammers to open nuts by capuchin monkeys (*Cebus apella*). *Folia Primatologica, 54*, 138–145.

Anderson, J.R. (1996). Chimpanzees and capuchin monkeys: Comparative cognition. In A.E.

Russon, K.A. Bard, & S.T. Parker (Eds). *Reaching into thought: The minds of the great apes* (pp. 23–56). Cambridge: Cambridge Universiy Press.

Anderson, J.R. & Chamove, A.S. (1986). Infant stumptailed macaques reared with mirrors or peers: Social responsiveness, attachment, and adjustment. *Primates, 27*, 63–82.

Anderson, J.R., Degiorgio, C., Lamarque, C., & Fagot, J. (1996). A multi-task assessment of hand lateralization in capuchin monkeys (*Cebus apella*). *Primates, 37*, 97–103.

Anderson, J.R. & Roeder, J.J. (1989). Responses of capuchin monkeys (*Cebus apella*) to different conditions of mirror-image stimulation. *Primates, 30*, 581–587.

Anderson, J.R., Sallaberry, P., & Barbier, H. (1995). Use of experimenter-given cues during object-choice tasks by capuchin monkeys. *Animal Behaviour, 49*, 201–208.

Andrews, M.W. (1988). Selection of food sites by *Callicebus moloch* and *Saimiri sciureus* under spatially and temporally varying food distribution. *Learning and Motivation, 19*, 254–268.

Antinucci, F. (1989). *Cognitive structure and development in nonhuman primates*. Hilldsale, NJ: Lawrence Erlbaum Associates.

Antinucci, F. & Visalberghi, E. (1986). Tool use in *Cebus apella*: a case study. *International Journal of Primatology, 7*, 351–363.

Antinucci, F., Spinozzi, G., Visalberghi, E., & Volterra, V. (1982). Cognitive development in a Japanese macaque (*Macaca fuscata*). Annali dell'Istituto Superiore di Sanita, 18, 177–184.

Aronowitsch, G. & Chotin, B. (1929). Uber die nachanmung bei den affen (*Macaca rhesus*). *Zeitschrift fur Morphologie und Okologie der Tiere, 16*, 1–25.

Ashley, J. & Tomasello, M. (in press). Cooperation and teaching in preschoolers. *Social Development*.

Astington, J. (1993). *The child's discovery of mind*. Cambridge, MA: Harvard University Press.

Aureli, F. & van Schaik, C.P. (1991) Post-conflict behaviour in long-tailed macaques (*Macaca fascicularis*): I. The social events. *Ethology, 89*, 89–100.

Aureli, F., Cozzolino, R., Cordischi, C., & Scucchi, S. (1992). Kin-oriented redirection among Japanese macaques: an expression of a revenge system? *Animal Behaviour, 44*, 283–292.

Bachmann, C. & Kummer, H. (1980). Male assessment of female choice in hamadryas baboons. *Behavioral Ecology and Sociobiology, 6*, 315–321.

Baillargeon, R. (1995). Physical reasoning in infancy. In M. Gazzaniga (Ed.), *The cognitive neurosciences* (pp. 181–204). Cambridge, MA: MIT Press.

Baillargeon, R., Spelke, E.S., & Wasserman, S. (1985). Object permanence in five-month-old infants. *Cognition, 20*, 191–208.

Balasch, J., Sabater Pi, J. & Padrosa, T. (1974). Perceptual learning ability in *Mandrillus sphinx* and *Cercopithecus nictitans*. *Revista Española de Fisiología, 30*, 15–20.

Baldwin, J.P., Sabater Pi, J., McGrew, W.C., & Tutin, C.E.G. (1981). Comparisons of nests made by different populations of chimpanzees (*Pan troglodytes*). *Primates, 22*, 474–486.

Bandura, A. (1986) *Social foundations of thought and action*. Princeton, NJ: Prentice-Hall.

Bard, K.A. (1990) "Social tool use" by free ranging orangutans: a Piagetian and developmental perspective on the manipulation of an animate object. In S.T. Parker & K.R. Gibson (Eds.), *"Language" and intelligence in monkeys and apes* (pp. 356–378). Cambridge: Cambridge University Press.

Bard, K.A. (1992). Intentional behavior and intentional communication in young free-ranging orangutans. *Child Development, 63*, 1186–1197.

Bard, K.A. (1995). Sensorimotor cognition in young feral orangutans (*Pongo pygmaeus*). *Primates, 36*, 297–321.

Bard, K.A., Fragaszy, D.M., & Visalberghi, E. (1995). Acquisition and comprehension of a tool-using behavior by young chimpanzees (*Pan troglodytes*): effects of age and modelling. *International Journal of Comparative Psychology, 8*, 47–68.

Bard, K.A. & Vauclair, J. (1984). The communicative context of object manipulation in ape and human adult-infant pairs. *Journal of Human Evolution, 13*, 181–190.

Baron-Cohen, S., Tager-Flusberg, H., & Cohen, D. (1993). *Understanding other minds.* Oxford: Oxford University Press.

Barresi, J. & Moore, C. (1996). Intentional relations and social understanding. *Behavioral and Brain Sciences, 19,* 107–154.

Barsalou, L. (1992). *Cognitive psychology: an overview for cognitive scientists.* Hillsdale, NJ: Lawrence Erlbaum Associates.

Bates, E. (1976). *Language and context.* New York: Academic Press.

Bates, E. (1979). *The emergence of symbols.* New York: Academic Press.

Bates, E. & MacWhinney, B. (1979). Functionalist approaches to grammar. In L. Gleitman & E. Wanner (Eds.), *Language acquisition: The state of the art* (pp. 173–218). Cambridge: Cambridge University Press.

Bauer, H.R. & Philip, M.M. (1983). Facial and vocal individual recognition in the common chimpanzee. *Psychological Record, 33,* 161–170.

Bauers, K.A. & de Waal, F.B.M. (1991). "Coo" vovalizations in stumptailed macaques: a controlled functional analysis. *Behaviour, 119,* 143–160.

Bayart, F. (1982). Un cas d'utilization d'outil chez un macaque (*Macaca tonkeana*) eleve en semiliberte. *Mammalia, 46,* 541–544.

Bayart, F. & Anderson, J.R. (1985). Mirror-image reactions in a tool-using, adult male macaca tonkeana. *Behavioural Processes, 10,* 219–227.

Beck, B.B. (1967). A study of problem solving by gibbons. *Behaviour, 28,* 95–109.

Beck, B.B. (1972). Tool use in captive hamadryas baboons. *Primates, 13,* 276–296.

Beck, B.B. (1973a). Cooperative tool use by captive Hamadryas baboons. *Science, 182,* 594–597.

Beck, B.B. (1973b). Observation learning of tool use by captive Guinea baboons (*Papio papio*). *American Journal of Physical Anthropology, 38,* 579–582.

Beck, B.B. (1974). Baboons, chimpanzees, and tools. *Journal of Human Evolution, 3,* 509–516.

Beck, B.B. (1976). Tool use by captive pigtail macaques. *Primates, 17,* 301–310.

Beck, B.B. (1980). *Animal tool behavior.* New York: Garland Press.

Behar, I. (1962). Evaluation of cues in learning set formation in mangabeys. *Psychological Reports, 11,* 479–485.

Benhar, E.E. & Samuel, D. (1978). A case of tool use in captive olive baboons (*Papio anubis*). *Primates, 19,* 385–389.

Benhar, E.E., Carlton, P.L., & Samuel, D. (1975). A search for mirror-image reinforcement and self-recognition in the baboon. A preliminary report. In: S. Kondo, M. Kawai, & S. Ehara (Eds.), *Contemporary primatology* (pp. 202–208). Basel, Switzerland: Karger.

Bennett, A. (1996). Do animals have cognitive maps? *Journal of Experimental Biology, 199,* 219–224.

Benz, J.J. (1993). Food-elicited vocalizations in golden lion tamarins: design features for representational communication. *Animal Behaviour, 45,* 443–455.

Bercovitch, F.B. (1988). Coalitions, cooperation and reproductive tactics among adult male baboons. *Animal Behaviour, 36,* 1198–1209.

Berdecio, S. & Nash, V.J. (1981). Chimpanzee visual communication. *Anthropological Research Papers, 26,* (pp.1–159). Tempe, Arizona: Arizona State University.

Berkson, G. (1962). Food motivation and delayed response in gibbons. *Journal of Comparative and Physiological Psychology, 55,* 1040.

Bernstein, I.S. (1961). The utilization of visual cues in dimension-abstracted oddity by primates. *Journal of Comparative and Physiological Psychology, 54,* 243–247.

Bernstein, I.S. (1969). A comparison of nesting patterns among the three great apes. In G.H. Bourne (Ed.), *The chimpanzee* (pp. 393–402). Basel, Switzerland: Karger.

Bernstein, I.S. (1988a) Kinship and behavior in nonhuman primates. *Behavior Genetics, 18,* 511–524.

Bernstein, I.S. (1988b). Metaphor, cognitive belief, and science. *Behavioral and Brain Sciences, 11*, 247–248.

Bernstein, I.S. & Ehardt, C.L. (1986). Selective interference in rhesus monkey (*Macaca mulatta*) intragroup agonistics episodes by age-sex class. *Journal of Comparative Psychology, 100*, 380–384.

Bhatt, R.S., Wasserman, E.A., Reynolds, Jr., V., & Knauss, K.S. (1988). Conceptual behavior in pigeons: categorization of both familiar and novel examples from four classes of natural and artificial stimuli. *Journal of Experimental Psychology: Animal Behavior Processes, 14*, 219–234.

Biben, M. & Symmes, D. (1991). Affiliative vocalizing in squirrel monkeys: familiarity as a cue to playback response. *Behaviour, 117*, 1–19.

Bingham, H.C. (1929a). Chimpanzee translocation by means of boxes. *Comparative Psychology Monographs, 5*, 1–91.

Bingham, H.C. (1929b). Selective transportation by chimpanzees. *Comparative Psychology Monographs, 5*, 1–45.

Birch, H.G. (1945). The relation of previous experience to insightful problem-solving. *Journal of Comparative Psychology, 38*, 367–383.

Blakeslee, P. & Gunter, R. (1966). Cross-modal transfer of discrimination learning in cebus monkeys. *Behaviour, 26*, 76–90.

Blaschke, M. & Ettlinger, G. (1987). Pointing as an act of social communication by monkeys. *Animal Behaviour, 35*, 1520–1523.

Boccia, M.L. (1994). Mirror behavior in macaques. In S.T. Parker, R.W. Mitchell, & M.L. Boccia (Eds.), *Self-awareness in animals and humans. Developmental perspectives* (pp. 350–360). Cambridge: Cambridge University Press.

Boesch, C. (1991a). Symbolic communication in wild chimpanzees. *Human Evolution, 6*, 81–90.

Boesch, C. (1991b). Teaching among wild chimpanzees. *Animal Behaviour, 41*, 530–532.

Boesch, C. (1992). Aspects of transmission of tool-use in wild chimpanzees. In K.R. Gibson & T. Ingold (Eds.), *Tools, language, and cognition in human evolution* (pp. 171–183). Cambridge: Cambridge University Press.

Boesch, C. (1993). Towards a new image of culture in wild chimpanzees? *Behavioral and Brain Sciences, 16*, 514–515.

Boesch, C. (1994). Cooperative hunting in wild chimpanzees. *Animal Behaviour, 48*, 653–667.

Boesch, C. (1995). Innovation in wild chimpanzees (*Pan troglodytes*). *International Journal of Primatology, 16*, 1–16.

Boesch, C. (in press). The emergence of cultures among wild chimpanzees. In J. Maynard-Smith (Ed.), *Evolution of social behavior patterns in primates and man*. London: The British Academy.

Boesch, C. & Boesch, H. (1983). Optimization of nut-cracking with natural hammers by wild chimpanzees. *Behaviour, 83*, 265–286.

Boesch, C. & Boesch, H. (1984). Mental map in wild chimpanzees: an analysis of hammer transports for nut cracking. *Primates, 25*, 160–170.

Boesch, C. & Boesch, H. (1989). Hunting behavior of wild chimpanzees in the Tai Forest National Park. *American Journal of Physical Anthropology, 78*, 547–573.

Boesch, C. & Boesch, H. (1990). Tool use and tool making in wild chimpanzees. *Folia Primatologica, 54*, 86–99.

Boesch, H. & Boesch, C. (1994). Hominization in the rain forest: the chimpanzee piece of the puzzle. *Evolutionary Anthropology, 3*, 9–16.

Boesch-Achermann, H. & Boesch, C. (1993). Tool use in wild chimpanzees: new light from dark forests. *Current Directions in Psychological Science, 2*, 18–21.

Boesch, C., Marchesi, P., Marchesi, N., Fruth, B., & Joulian, F. (1994). Is nut cracking in wild chimpanzees a cultural behavior? *Journal of Human Evolution, 26*, 325–338.

Boinski, S. (1988). Use of a club by a wild white-faced capuchin (*Cebus capucinus*) to attack a venomous snake (*Bothrops asper*). *American Journal of Primatology, 14*, 177–179.

Boinski, S. (in press). Vocal coordination of troop movement in squirrel monkeys (*Saimiri oerstedi and S. sciureus*) and white-faced capuchins (*Cebus capucinus*). In M. Norconk, A.L. Rosenberger, & P.A. Garber (Eds.), *Adaptive radiations of neotropical primates*. New York: Plenum Press.

Boinski, S. & Fragaszy, D.M. (1989). The ontogeny of foraging in squirrel monkeys, *Saimiri oerstedi*. *Animal Behaviour, 37*, 415–428.

Boinski, S. & Mitchell, C.L. (1992). Ecological and social factors affecting the vocal behavior of adult female squirrel monkeys. *Ethology, 92*, 316–330.

Bolwig, N. (1964). Observations on the mental and manipulative abilities of a captive baboon (*Papio doguera*). *Behaviour, 22*, 24–40.

Bonner, J.T. (1980). *The evolution of culture in animals*. Princeton, NJ: Princeton University Press.

Borner, M. (1979). *Orang utan: Orphans of the forest*. London: W.H. Allen & Co.

Boyd, R. & Richerson, P. (1985). *Culture and the evolutionary process*. Chicago: University of Chicago Press.

Boysen, S.T. (1994). Individual differences in the cognitive abilities of chimpanzees. In R.W. Wrangham, W.C. McGrew, F.B.M. de Waal, & P.G. Heltne (Eds.), *Chimpanzee cultures* (pp. 335–350). Cambridge, MA: Harvard University Press.

Boysen, S.T. & Berntson, G.G. (1986). Cardiac correlates of individual recognition in the chimpanzee (*Pan troglodytes*). *Journal of Comparative Psychology, 100*, 321–324.

Boysen, S.T. & Berntson, G.G. (1989). Numerical competence in a chimpanzee (*Pan troglodytes*). *Journal of Comparative Psychology, 103*, 23–31.

Boysen, S.T. & Berntson, G.G. (1995). Responses to quantity: Perceptual versus cognitive mechanisms in chimpanzees (*Pan troglodytes*). *Journal of Experimental Psychology: Animal Behavior Processes, 21*, 82–86.

Boysen, S.T., Berntson, G.G., Hannan, M.B., & Cacioppo, J.T. (1996). Quantity-based interference and symbolic representations in chimpanzees (*Pan troglodytes*). *Journal of Experimental Psychology: Animal Behavior Processes, 22*, 76–86.

Boysen, S.T., Berntson, G.G., Shreyer, T.A., & Hannan, M.B. (1995). Indicating acts during counting by a chimpanzee (*Pan troglodytes*). *Journal of Comparative Psychology, 109*, 47–51.

Boysen, S.T., Berntson, G.G., Shreyer, T.A., & Quigley, K.S. (1993). Processing of ordinality and transitivity by chimpanzees (*Pan troglodytes*). *Journal of Comparative Psychology, 107*, 208–215.

Boysen, S.T., Bryant, C.E., & Shreyer, T.A. (1994). Shadows and mirrors: Alternative avenues to the development of self-recognition in chimpanzees. In S.T. Parker, R.W. Mitchell, & M.L. Boccia (Eds.), *Self-awareness in animals and humans. Developmental perspectives* (pp. 227–240). Cambridge: Cambridge University Press.

Boysen, S.T. & Capaldi, E.J. (1993). *The development of numerical competence*. Hillsdale, NJ: Lawrence Erlbaum Associates.

Braggio, J.T., Hall, A.D., Buchanan, J.P., & Nadler, R.D. (1979). Cognitive capacities of juvenile chimpanzees on a Piagetian-type multiple-classification task. *Psychological Reports, 44*, 1087–1097.

Brakke, K.E. (1996). Prerequisites of logicomathematical cognition in differentially-reared *Pan paniscus* and *Pan troglodytes* (PhD dissertation). Atlanta: Georgia State University.

Brakke, K.E. & Savage-Rumbaugh, E.S. (1994). The growth of object manipulation in infant chimpanzee, bonobo, and human. Manuscript.

Bremner, G. (1980). The infant's understanding of space. In M. Cox (Ed.), *Are young children egocentric?* (pp. 281–309) London: Concord.

Brent, L., Bloomsmith, M.A., & Fisher, S.D. (1995). Factors determining tool-using ability in two captive chimpanzee (*Pan troglodytes*) colonies. *Primates, 36*, 265–274.

Brewer, S.M. & McGrew, W.C. (1990). Chimpanzee use of a tool-set to get honey. *Folia Primatologica, 54*, 100–104.

Brown, M. & Demas, G. (1994). Evidence for spatial working memory in honeybees. *Journal of Comparative Psychology, 108*, 344–352.

Brownell, C.A. & Carriger, M.S. (1990). Changes in cooperation and self-other differentiation during the second year. *Child Development, 61*, 1164–1174.

Bruner, J. (1972). The nature and uses of immaturity. *American Psychologist, 27*, 687–708.

Bruner, J. (1983) *Child's talk*. New York: Norton.

Bruner, J. (1986). *Actual minds, possible worlds*. Cambridge, MA: Harvard University Press.

Bruner, J. (1990). *Acts of meaning*. Cambridge: Harvard University Press.

Bryant, P.E. & Trabasso, T. (1971). Transitive inferences amd memory in young children. *Nature, 232*, 456–458.

Burdyn, L.E. & Thomas, R.K. (1984). Conditional discrimination with conceptual simultaneous and successive cues in the squirrel monkey (*Saimiri sciureus*). *Journal of Comparative Psychology, 98*, 405–413.

Burton, J.J. (1977). Absence de comportement cooperatif spontane dans une troupe de *Macaca fuscata* en presence de pierres appatees. *Primates, 18*, 359–366.

Busse, C.D. (1977). Do chimpanzees hunt cooperatively? *American Naturalist, 112*, 767–770.

Byrne, R.W. (1982). Primate vocalisations: structural and functional approaches to understanding. *Behaviour, 80*, 241–258.

Byrne, R.W. (1993). Do larger brains mean greater intelligence? *Behavioral and Brain Sciences, 16*, 696–697.

Byrne, R.W. (1994). The evolution of intelligence. In P.J.B. Slater & T.R. Halliday (Eds.), *Behavior and evolution* (pp. 223–265). Cambridge: Cambridge University Press.

Byrne, R.W. (1995). *The thinking ape*. Oxford: Oxford University Press.

Byrne, R.W. (in press). Relating brain size to intelligence in primates. In P. Mellars & K. Gibson (Eds.), *The early human mind*. McDonald Institute Research Monographs.

Byrne, R.W. & Byrne, J.M. (1991). Hand preferences in the skilled gathering tasks of mountain gorillas (*Gorilla g. beringei*). *Cortex, 27*, 521–546.

Byrne, R.W. & Byrne, J.M.E. (1993). Complex leaf-gathering skills of mountain gorillas (*Gorilla g. beringei*): Variability and standarization. *American Journal of Primatology, 31*, 241–261.

Byrne, R.W. & Tomasello, M. (1995). Do rats ape? *Animal Behaviour, 50*, 1417–1420.

Byrne, R.W. & Whiten, A. (1988). *Machiavellian intelligence. Social expertise and the evolution of intellect in monkeys, apes, and humans*. New York: Oxford University Press.

Byrne, R.W. & Whiten, A. (1990). Tactical deception in primates: the 1990 database. *Primate Report, 27*, 1–101.

Byrne, R.W. & Whiten, A. (1992). Cognitive evolution in primates: evidence from tactical deception. *Man, 27*, 609–627.

Calhoun, S. & Thompson, R.L. (1988). Long-term retention of self-recognition by chimpanzees. *American Journal of Primatology, 15*, 361–365.

Call, J. & Rochat, P. (1996). Liquid conservation in orangutans and humans: Individual differences and cognitive strategies. *Journal of Comparative Psychology, 110*, 219–232.

Call, J. & Tomasello, M. (1994a). Production and comprehension of referential pointing by orangutans (*Pongo pygmaeus*). *Journal of Comparative Psychology, 108*, 307–317.

Call, J. & Tomasello, M. (1994b). The social learning of tool use by orangutans (*Pongo pygmaeus*). *Human Evolution, 9*, 297–313.

Call, J. & Tomasello, M. (1995) The use of social information in the problem-solving of orangutans (*Pongo pygmaeus*) and human children (*Homo sapiens*). *Journal of Comparative Psychology, 109*, 308–320.

Call, J. & Tomasello, M. (1996). The effect of humans on the cognitive development of apes. In A.E. Russon, K.A. Bard, & S.T. Parker (Eds). *Reaching into thought* (pp. 371–403). Cambridge: Cambridge University Press.

Cambefort, J.P. (1981). A comparative study of culturally transmitted patterns of feeding habits in the chacma baboon *Papio ursinus* and the vervet monkey *Cercopithecus aethiops*. *Folia Primatologica, 36*, 243–263.

Candland, D (1993). *Feral children and clever animals*. New York: Oxford University Press.

Capaldi, E.J. & Miller, D.J. (1988). Counting in rats: its functional significance and the independent cognitive processes that constitute it. *Journal of Experimental Psychology: Animal Behavior Processes, 14*, 3–17.

Caro, T.M. & Hauser, M.D. (1992). Is there teaching in nonhuman animals? *Quarterly Review of Biology, 67*, 151–174.

Carpenter, A. (1887). Monkeys opening oysters. *Nature, 36*, 53.

Carpenter, M. (1995). Social-cognitive skills of 9- to 15-month-old infants: development and interrelationships (Doctoral dissertation). Atlanta, GA: Emory University.

Carpenter, M., Tomasello, M., & Savage-Rumbaugh, E.S. (1995). Joint attention and imitative learning in children, chimpanzees and enculturated chimpanzees. *Social Development, 4*, 217–237.

Cha, J. & King, J.E. (1969). The learning of patterned strings problems by squirrel monkeys. *Animal Behaviour, 17*, 64–67.

Chalmeau, R. (1994). Do chimpanzees cooperate in a learning task? *Primates, 35*, 385–392.

Chalmeau, R. & Gallo, A. (1996). What chimpanzees (*Pan troglodytes*) learn in a cooperative task. *Primates, 37*, 39–47.

Chalmers, M. & McGonigle, B. (1984). Are children any more logical than monkeys on the five-term series problem? *Journal of Experimental Child Psychology, 37*, 355–377.

Chamove, A. (1974). Failure to find observational learning in rhesus macaques. *Journal of Behavioral Science, 2*, 39–41.

Chance, M.R.A. (1960). Köhler's chimpanzees—how did they perform? *Man, 179*, 130–135.

Chance, M.R.A. & Mead, A.P. (1953). Social behavior and primate evolution. *Symposia of the Society for Experimental Biology, 7*, 395–439.

Chandler, M., Fritz, A.S., & Hala, S. (1989). Small-scale deceit: deception as a marker of two-, three-, and four-year-olds' early theories of mind. *Child Development, 60*, 1263–1277.

Chapais, B. (1983). Dominance, relatedness and the structure of female relationships in rhesus monkeys. In R.A. Hinde (Ed.), *Primate social relationships: an integrated approach* (pp. 208–219). Oxford: Blackwell Scientific.

Chapais, B. (1992). The role of alliances in social inheritance of rank among female primates. In A.H. Harcourt & F.B.M. de Waal (Eds.), *Coalitions and alliances in human and nonhuman animals* (pp. 29–59). New York: Oxford University Press.

Chapais, B., Gauthier, C., & Prud'Homme, J. (1995). Dominance competition through affiliation and support in Japanese macaques: An experimental study. *International Journal of Primatology, 16*, 521–536.

Chapman, C. & Lefebvre, L. (1990). Manipulating foraging group size: spider monkey food calls at fruiting trees. *Animal Behavior, 39*, 891–896.

Chater, N. & Heyes, C. (1994). Animal concepts: content and discontent. *Mind and Language, 9*, 209–246.

Cheney, D.L. (1977). The acquisition of rank and the development of reciprocal alliances among free-ranging immature baboons. *Behavioral Ecology and Sociobiology, 2*, 303–318.

Cheney, D.L. & Seyfarth, R.M. (1980). Vocal recognition in free-ranging vervet monkeys. *Animal Behaviour, 28*, 362–367.

Cheney, D.L. & Seyfarth, R.M. (1981). Selective forces affecting the predator alarm calls of vervet monkeys. *Behaviour, 76*, 25–61.

Cheney, D.L. & Seyfarth, R.M. (1982a). How vervet monkeys perceive their grunts: field playback experiments. *Animal Behaviour, 30*, 739–751.

Cheney, D.L. & Seyfarth, R.M. (1982b). Recognition of individuals within and between groups of free-ranging vervet monkeys. *American Zoologist, 22*, 519–529.

Cheney, D.L. & Seyfarth, R.M. (1985). Vervet monkey alarm calls: manipulation through shared information? *Behaviour, 93*, 150–166.

Cheney, D.L. & Seyfarth, R.M. (1986). The recognition of social alliances by vervet monkeys. *Animal Behaviour, 34*, 1722–1731.

Cheney, D.L. & Seyfarth, R.M. (1988). Assessment of meaning and the detection of unreliable signals by vervet monkeys. *Animal Behaviour, 36*, 477–486.

Cheney, D.L. & Seyfarth, R.M. (1989) Reconciliation and redirected aggression in vervet monkeys *(Cercopithecus aethiops). Behaviour, 110*, 258–275.

Cheney, D.L. & Seyfarth, R.M. (1990a) Attending to behaviour versus attending to knowledge: examining monkeys' attribution of mental states. *Animal Behaviour, 40*, 742–753.

Cheney, D.L. & Seyfarth, R.M. (1990b). *How monkeys see the world*. Chicago: University of Chicago Press.

Cheney, D.L., Seyfarth, R.M., & Silk, J.B. (1995a). The responses of female baboons (*Papio cynocephalus ursinus*) to anomalous social interactions: evidence for causal reasoning? *Journal of Comparative Psychology, 109*, 134–141.

Cheney, D.L., Seyfarth, R.M., & Silk, J.B. (1995b). The role of grunts in reconciling opponents and facilitating interactions among adult female baboons. *Animal Behaviour, 50*, 249–257.

Cheney, D.L., Seyfarth, R.M., Smuts, B.B., & Wrangham, R.W. (1987). The study of primate societies. In B.B. Smuts, D.L. Cheney, R.M. Seyfarth, R.W. Wrangham, & T.T. Struhsaker (Eds.), *Primate societies* (pp. 1–8). Chicago: University of Chicago Press.

Cheney, D.L. & Wrangham, R.W. (1987). Predation. In B.B. Smuts, D.L. Cheney, R.M. Seyfarth, R.W. Wrangham, & T.T. Struhsaker (Eds.), *Primate societies* (pp. 227–239). Chicago: University of Chicago Press.

Chevalier-Skolnikoff, S. (1977). A Piagetian model for describing and comparing socialization in monkey, ape, and human infants. In S. Chevalier-Skolnikoff & F.E. Poirier (Eds.), *Primate bio-social development* (pp. 159–187). New York: Garland Press.

Chevalier-Skolnikoff, S. (1983). Sensorimotor development in orang-utans and other primates. *Journal of Human Evolution, 12*, 545–561.

Chevalier-Skolnikoff, S. (1989). Spontaneous tool use and sensorimotor intelligence in Cebus compared with other monkeys and apes. *Behavioral and Brain Sciences, 12*, 561–627.

Chevalier-Skolnikoff, S. (1990). Tool use by wild cebus monkeys at Santa Rosa National Park, Costa Rica. *Primates, 31*, 375–383.

Chiang, M. (1967). Use of tools by wild macaque monkeys in Singapore. *Nature, 214*, 1258–1259.

Clark, A.P. & Wrangham, R.W. (1993). Acoustic analysis of wild chimpanzee pant hoots: do Kibale forest chimpanzees have an acoustically distinct food arrival pant hoot? *American Journal of Primatology, 31*, 99–109.

Clark, A.P. & Wrangham, R.W. (1994). Chimpanzee arrival pant-hoots: do they signify food or status? *International Journal of Primatology, 15*, 185–205.

Clutton-Brock, T.H. (1977a). *Primate ecology: studies of feeding and ranging behaviour in lemurs, monkeys and apes*. New York: Academic Press.

Clutton-Brock, T.H. (1977b). Some aspects of intraspecific variation in feeding and ranging behaviour in primates. In T.H. Clutton-Brock (Ed.), *Primate ecology: studies of feeding and ranging behaviour in lemurs, monkeys and apes* (pp. 539–556). New York: Academic Press.

Clutton-Brock, T.H. & Harvey, P.H. (1980). Primates, brains and ecology. *Journal of Zoology, 190*, 309–323.

Coldren, J. & Colombo, J. (1994). The nature and process of preverbal learning. *Monographs of the Society for Research in Child Development, 59,* 1–75.

Cole, M. (in press). *Culture in mind.* Cambridge: Harvard University Press.

Collingwood, R. J. (1972). *Essay on metaphysics.* Chicago: Gateway.

Colmenares, F. (1991). Greeting, aggression, and coalitions between male baboons: demographic correlates. *Primates, 32,* 453–463.

Connor, R.C., Smolker, R.A., & Richards, A.F. (1992). Dolphin alliances and coalitions. In A.H. Harcourt & F.B.M. de Waal (Eds.), *Coalitions and alliances in human and nonhuman animals* (pp. 415–443). New York: Oxford University Press.

Cook, M., Mineka, S., Wolkenstein, B., & Laitsch, K. (1985). Observational conditioning of snake fear in unrekated rhesus monkeys. *Journal of Abnormal Psychology, 94,* 591–610.

Cooley, C. (1902). *Human nature and the social order.* New York: Charles Scribner.

Cooper, H. (1974). Learning set in *Lemur macaco.* In R.D. Martin, G.A. Doyle, & A.C. Walker (Eds.), *Prosimian biology* (pp. 293–300). Pittsburgh, PA: University of Pittsburgh Press.

Cooper, L.R. & Harlow, H.F. (1961). Note on a Cebus monkey's use of a stick as a weapon. *Psychological Reports, 8,* 418–418.

Cords, M. (1992). Social versus ecological intelligence. *Behavioral and Brain Sciences, 15,* 151.

Cords, M. (1995). Predator vigilance costs of allogrooming in wild blue monkeys. *Behaviour, 132,* 559–569.

Cords, M. (in press). Friendships, alliances, and repair. In A. Whiten & R. Byrne (Eds.). *Machiavellian Intelligence II.* Cambridge: Cambridge University Press.

Corkum, V. & Moore, C. (1995). Development of joint visual attention in infants. In C. Moore & P. Dunham (Eds.), *Joint attention: its origins and role in development* (pp. 61–83). Hillsdale, NJ: Lawrence Erlbaum Associates.

Cornell, E. & Heth, C. (1983). Spatial Cognition: Gathering strategies used by preschool children. *Journal of Expreimental Child Psychology, 35,* 93–110.

Costello, M. (1987). Tool use and manufacture in manipulanda-deprived capuchins (*Cebus apella*). *American Journal of Primatology, 12,* 337 (abstract).

Coussi-Korbel, S. (1994). Learning to outwit a competitor in mangabeys (*Cercocebus torquatus torquatus*). *Journal of Comparative Psychology, 108,* 164–171.

Cowey, A. & Weiskrantz, L. (1975). Demonstration of cross-modal matching in rhesus macaques, *Macaca mulatta. Neuropsychologia, 13,* 117–120.

Crawford, M.P. (1937). The cooperative solving of problems by young chimpanzees. *Comparative Psychology Monographs, 14,* 1–88.

Crawford, M.P. (1941). The cooperative solving by chimpanzees of problems requiring serial responses to color cues. *Journal of Social Psychology, 13,* 259–280.

Crawford, M.P. & Spence, K.W. (1939). Observational learning of discrimination problems by chimpanzees. *Journal of Comparative Psychology, 27,* 133–147.

Curio, E. (1988). Cultural transmission of enemy recognition by birds. In T.R. Zentall & B.G. Galef Jr. (Eds.), *Social learning: psychological and biological perspectives* (pp. 75–97). Hillsdale, NJ: Lawrence Erlbaum Associates.

Custance, D.M., Whiten, A., & Bard, K.A. (1995). Can young chimpanzees imitate arbitrary actions? Hayes and Hayes (1952) revisited. *Behaviour, 132,* 839–858.

Czerny, P. & Thomas, R.K. (1975). Sameness-difference judgments in *Saimiri sciureus* based on volumetric cues. *Animal Learning and Behavior, 3,* 375–379.

D'Amato, M.R. & Colombo, M. (1985). Auditory matching-to-sample in monkeys (*Cebus apella*). *Animal Learning and Behavior, 13,* 375–382.

D'Amato, M.R. & Colombo, M. (1988). Representation of serial order in monkeys (*Cebus apella*). *Journal of Experimental Psychology: Animal Behavior Processes, 14,* 131–139.

D'Amato, M.R. & Colombo, M. (1989). Serial learning with wild card items by monkeys (*Cebus *apella*): implications for knowledge of ordinal position. *Journal of Comparative Psychology, 103*, 252–261.

D'Amato, M.R. & Colombo, M. (1990). The symbolic distance effect in monkeys (*Cebus apella*). *Animal Learning and Behavior, 18*, 133–140.

D'Amato, M.R. & Salmon, D.P. (1984). Cognitive processes in cebus monkeys. In H.L. Roitblat, T.G. Bever, & H.S. Terrace (Eds.), *Animal cognition* (pp. 149–168). Hillsdale, NJ: Lawrence Erlbaum Associates.

D'Amato, M.R., Salmon, D.P., & Colombo, M. (1985). Extent and limits of the matching concept in monkeys (*Cebus apella*). *Animal Behavior Processes, 11*, 35–51.

D'Amato, M.R., Salmon, D.P., Loukas, E., & Tomie, A. (1985). Symmetry and transitivity of conditional relations in monkeys (*Cebus apella*) and pigeons (*Columba livia*). *Journal of the Experimental Analysis of Behavior, 44*, 35–47.

D'Amato, M.R., Salmon, D.P., Loukas, E., & Tomie, A. (1986). Processing of identity and conditional relations in monkeys (*Cebus apella*) and pigeons (*Columba livia*). *Animal Learning and Behavior, 14*, 365–373.

D'Amato, M.R. & van Sant, P. (1988). The person concept in monkeys (*Cebus apella*). *Animal Behavior Processes, 14*, 43–55.

D'Amato, M.R. & Worsham, R.W. (1972). Delayed matching in the capuchin monkey with brief sample durations. *Learning and Motivation, 3*, 304–312.

Darby, C.L. & Riopelle, A.J. (1955). Differential problem sequences and the formation of learning sets. *Journal of Psychology, 39*, 105–108.

Darby, C.L. & Riopelle, A.J. (1959). Observational learning in the Rhesus monkey. *Journal of Comparative Psychology, 52*, 94–98.

Darwin, C. (1859). *On the origin of species by means of natural selection*. London: John Murray.

Darwin, C. (1871). *The descent of man and selection in relation to sex*. London: John Murray.

Darwin, C. (1872). *The expression of emotions in man and animals*. London: John Murray.

Dasser, V. (1987). Slides of group members as representations of the real animals (*Macaca fascicularis*). *Ethology, 76*, 65–73.

Dasser, V. (1988a). Mapping social concepts in monkeys. In R.W. Byrne & A. Whiten (Eds.), *Machiavellian intelligence. Social expertise and the evolution of intellect in monkeys, apes, and humans* (pp. 85–93). New York: Oxford University Press.

Dasser, V. (1988b). A social concept in Java monkeys. *Animal Behaviour, 36*, 225–230.

Davenport, R.K. (1977). Cross-modal perception: a basis for language? In D.M. Rumbaugh (Ed.), *Language learning by a chimpanzee. The Lana project* (pp. 73–83). New York: Academic Press.

Davenport, R.K. & Rogers, C.M. (1970). Intermodal equivalence of stimuli in apes. *Science, 168*, 279–280.

Davenport, R.K., Rogers, C.M., & Menzel, E.W. (1969). Intellectual performance of differentially reared chimpanzees: II. Discrimination-learning set. *American Journal of Mental Deficiency, 73*, 963–969.

Davis, H. (1984). Discrimination of the number three by a raccoon (*Procyon lotor*). *Animal Learning and Behavior, 12*, 409–413.

Davis, H. & Bradford, S.A. (1986). Counting behavior by rats in a simulated natural environment. *Ethology, 73*, 265–280.

Davis, H. & Perusse, R. (1988). Numerical competence in animals: definitional issues, current evidence and a new research agenda. *Behavioral and Brain Sciences, 11*, 561–615.

Davis, R.T. (1958). The learning of detours and barriers by monkeys. *Journal of Comparative Psychology, 51*, 471–477.

Davis, R.T. & Leary, R.W. (1968). Learning of detour problems by lemurs and seven species of monkeys. *Perceptual and Motor Skills, 27*, 1031–1034.

Davis, R.T., Leary, R.W., Stevens, D.A., & Thompson, R.F. (1967). Learning and perception of oddity problems by lemurs and seven species of monkeys. *Primates, 8*, 311–322.

Davis, R.T., McDowell, A.A., & Nissen, H.W. (1957). Solution of bent-wire problems by monkeys and chimpanzees. *Journal of Comparative Psychology, 50*, 441–444.

Dawkins, R. & Krebs, J. (1978). Animal signals: information or manipulation. In J. Krebs & N. Davies (Eds.), *Behavioral ecology: an evolutionary approach* (pp. 282–309). Oxford: Blackwell.

de Blois, S.T. & Novak, M.A. (1994). Object permanence in rhesus monkeys (*Macaca mulatta*). *Journal of Comparative Psychology, 108*, 318–327.

De Lillo, C. (in press). The serial organization of behavior by non-human primates: an evaluation of experimental paradigms. *Behavioral Brain Research.*

De Lillo, C. & Visalberghi, E. (1994). Transfer index and mediational learning in tufted capuchins (*Cebus apella*). *International Journal of Primatology, 15*, 275–287.

De Lillo, C., Visalberghi, E., & Aversano, M. (in press). The organization of exhaustive searches in a "patchy" space by capuchin monkeys (*Cebus apella*). *Journal of Comparative Psychology.*

de Waal, F.B.M. (1982). *Chimpanzee politics.* London: Jonathan Cape.

de Waal, F.B.M. (1986). Deception in the natural communication of chimpanzees. In R.W. Mitchell & N.S.Thompson (Eds.), *Deception. Perspectives on human and nonhuman deceit* (pp. 221–244). Albany, NY: SUNY Press.

de Waal, F.B.M. (1987a) Dynamics of social relationships. In B.B. Smuts, D.L. Cheney, R.M. Seyfarth, R.W. Wrangham, & T.T. Struhsaker (Eds.), *Primate societies* (pp. 421–429). Chicago: University of Chicago Press.

de Waal, F.B.M. (1987b). Tension regulation and nonreproductive functions of sex in captive bonobos (*Pan paniscus*). *National Geographic Research, 3*, 318–335.

de Waal, F.B.M. (1988) The communicative repertoire of captive bonobos (*Pan paniscus*), compared to that of chimpanzees. *Behaviour, 106*, 183–251.

de Waal, F.B.M. (1989a). Food sharing and reciprocal obligations among chimpanzees. *Journal of Human Evolution, 18*, 433–459.

de Waal, F.B.M. (1989b). *Peacemaking among primates.* Cambridge, MA: Harvard University Press.

de Waal, F.B.M. (1992). Coalitions as part of reciprocal relations in the Arnhem chimpanzee colony. In A.H. Harcourt & F.B.M. de Waal (Eds.), *Coalitions and alliances in humans and other animals* (pp. 233–257). New York: Oxford University Press.

de Waal, F.B.M. (1993). Reconciliation among primates: a review of empirical evidence and unresolved issues. In W.A. Mason & S.P. Mendoza (Eds.), *Primate social conflict* (pp. 111–144). Albany, NY: SUNY Press.

de Waal, F.B.M. (1994). Chimpanzee's adaptive potential: a comparison of social life under captive and wild conditions. In R.W. Wrangham, W.C. McGrew, F.B.M. de Waal, & P.G. Heltne (Eds.), *Chimpanzee cultures* (pp. 243–260). Cambridge, MA: Harvard University Press.

de Waal, F.B.M. & Aureli, F. (1996). Consolation, reconciliation, and a possible cognitive difference between macaque and chimpanzee. In A.E. Russon, K.A. Bard, & S.T. Parker (Eds.), *Reaching into thought* (pp. 80–110). Cambridge: Cambridge University Press.

de Waal, F.B.M. & Johanowicz, D.L. (1993) Modification of reconciliation behavior through social experience: An experiment with two macaque species. *Child Development, 64*, 897–908.

de Waal, F.B.M. & Luttrell, L.M. (1988). Mechanisms of social reciprocity in three primate species: Symetrical relationship characteristics or cognition? *Ethology and Sociobiology, 9*, 101–118.

de Waal, F. & van Hooff, J. (1981). Side-directed communication and agonistic interactions in chimpanzees. *Behaviour, 77*, 164–198.

de Waal, F.B.M., van Hooff, J.A.R.A.M., & Netto, W.J. (1976). An ethological analysis of types of agonistic interaction in a captive group of Java-monkeys (*Macaca fascicularis*). *Primates, 17*, 257–290.

de Waal, F.B.M. & van Roosmalen, A. (1979). Reconciliation and consolation among chimpanzees. *Behavioral Ecology and Sociobiology, 5*, 55–66.

DeLoache, J.S. (1995). Early understanding and use of symbols: the model model. *Current Directions in Psychological Science, 4*, 109–113.

Dennett, D. (1995). *Darwin's dangerous idea.* New York: Simon and Schuster.

Dewsbury, D.A. (1984). *Comparative psychology in the twentieth century.* Stroudsburg, PA: Hutchinson Ross.

Diamond, A. (1991). Neuropsychological insights into the meaning of object concept development. In S. Carey & R. Gelman (Eds.), *The epigenesis of mind* (pp. 67–110). Hillsdale, NJ: Lawrence Erlbaum Associates.

Dickinson, A. & Shanks, D. (1995) Instrumental action and causal representation. In D. Sperber, D. Premack, & A. Premack (Eds.), *Causal cognition: a multidisciplinary debate*, (pp. 5–25). Oxford: Oxford University Press.

Dittrich, W. (1990). Representation of faces in longtailed macaques (*Macaca fascicularis*). *Ethology, 85*, 265–278.

Dittus, W.P.J. (1984). Toque macaque food calls: Semantic communication concerning food distribution in the environment. *Animal Behaviour, 32*, 470–477.

Dittus, W. (1988). An ecological analysis of tocque macaques cohesion calls from an ecological perspective. In D. Todt, P. Goedeking, & D. Symmes (Eds.), *Primate vocal communication* (pp. 31–50). New York: Springer-Verlag.

Döhl, J. (1968). Uber die fahigkeit einer schimpansin, umwege mit selbstandigen zwischenzielen zu uberblicken. *Zeitschrift für Tierpsychology, 25*, 89–103.

Donald, M. (1991). *Origins of the modern mind.* Cambridge: Harvard University Press.

Dooley, G.B. & Gill, T.V. (1977a). Acquisition and use of mathematical skills by a linguistic chimpanzee. In D.M. Rumbaugh (Ed.), *Language learning by a chimpanzee. The Lana project* (pp. 247–260). New York: Academic Press.

Dooley, G.B. & Gill, T. (1977b). Mathematical capabilities of Lana chimpanzee. In G.H. Bourne (Ed.), *Progress in ape research* (pp. 133–142). New York: Academic Press.

Doré, F.Y. (1986). Object permanence in adult cats (*Felis catus*). *Journal of Comparative Psychology, 100*, 340–347.

Doré, F.Y. (1990). Search behavior of cats (*Felis catus*) in an invisible displacement test: Cognition and experience. *Canadian Journal of Psychology, 44*, 359–370.

Doré, F.Y. & Dumas, C. (1987). Psychology of animal cognition: Piagetian studies. *Psychological Bulletin, 102*, 219–233.

Douglas, J.W.B. & Whitty, C.W.M. (1941). An investigation of number appreciation in some subhuman primates. *Journal of Comparative Psychology, 31*, 129–143.

Drea, C. & Wallen, K. (1995). Gradual acquisition of visual discrimination tasks in a social group of rhesus monkeys (*Macaca mulatta*). *Animal Learning and Behavior, 23*, 1–8.

Dukas, R. & Waser, N. (1994). Categorization of food types enhances the foraging efficiency of bumblebees. *Animal Behaviour, 48*, 1001–1006.

Dumas, C. & Brunet, C. (1994). Pernamence de l'objet chez le singe capucin (*Cebus apella*): etude des desplacements invisibles. *Revue Canadienne de psychologie experimentale, 48*, 341–357.

Dunbar, R.I.M. (1988). *Primate social systems.* London: Croom Helm.

Dunbar, R.I.M. (1992). Neocortex size as a constraint on group size in primates. *Journal of Human Evolution, 20*, 469–493.

Dunbar, R.I.M. (1995). Neocortex size and group size in primates: a test of the hypothesis. *Journal of Human Evolution, 28*, 287–296.

Dunn, J. (1988). *The beginnings of social understanding*. Oxford: Blackwell.

Eddy, T.J., Gallup G.G. Jr., & Povinelli, D.J. (1993). Attribution of cognitive states to animals: Anthropomorphism in comparative perspective. *Journal of Social Issues, 49*, 87–101.

Eddy, T.J., Gallup, G.G. Jr., & Povinelli, D.J. (1996). Age differences in the ability of chimpanzees to distinguish mirror-images of self from video images of others. *Journal of Comparative Psychology, 110*, 38–44.

Eglash, A.R. & Snowdon, C.T. (1983). Mirror-image responses in pygmy marmosets (*Cebuella pygmaea*). *American Journal of Primatology, 5*, 211–219.

Ehrlich, A. (1970). Response to novel objects in three lower primates: Greater galago, slow loris, and owl monkey. *Behaviour, 37*, 55–63.

Ehrlich, A., Fobes, J.L., & King, J.E. (1976). Prosimian learning capacities. *Journal of Human Evolution, 5*, 599–617.

Eisenberg, J. & Wilson, D. (1978). Relative brain size and feeding strategies in the Chiroptera. *Evolution, 32*, 740–751.

Elgar, M.A. (1986). House sparrows establish foraging flocks by giving chirrup calls if the resources are divisible. *Animal Behaviour, 34*, 169–174.

Elliott, R.C. (1977). Cross-modal recognition in three primates. *Neuropsychologia, 15*, 183–186.

Elowson, A.M., Snowdon, C.T., & Sweet, C.J. (1992). Ontogeny of thrill and J-call vocalizations in the pygmy marmoset, *Cebuella pygmaea. Animal Behaviour, 43*, 703–716.

Elowson, A.M., Tannenbaum, P.L., & Snowdon, C.T. (1991). Food-associated calls correlate with food preferences in cotton-top tamarins. *Animal Behaviour, 42*, 931–937.

Epstein, R., Lanza, R.P. & Skinner, B.F. (1981). "Self-awareness" in the pigeon. *Science, 212*, 695–696.

Erickson, C.J. (1994). Tap-scanning and extractive foraging in Aye-Ayes, *Daubentonia madagascariensis. Folia Primatologica, 62*, 125–135.

Essock-Vitale, S.M. (1978). Comparison of ape and monkey modes of problem solution. *Journal of Comparative Psychology, 92*, 942–957.

Etienne, A.S. (1984). The meaning of object permanence at different zoological levels. *Human Development, 27*, 309–320.

Ettlinger, G. (1977). Interactions between sensory modalities in nonhuman primates. In A.M. Schrier (Ed.), *Behavioral primatology: advances in research and theory* (pp. 71–104). Hillsdale, NJ: Lawrence Erlbaum Associates.

Ettlinger, G. (1983). A comparative evaluation of the cognitive skills of the chimpanzee and the monkey. *International Journal of Neuroscience, 22*, 7–20.

Ettlinger, G. & Jarvis, M.J. (1976). Cross-modal transfer in the chimpanzee. *Nature, 259*, 44–45.

Evans, A. & Tomasello, M. (1986). Evidence for social referencing in young chimpanzees (*Pan troglodytes*). *Folia Primatologica, 47*, 49–54.

Evans, C.S., Evans, L., & Marler, P. (1993). On the meaning of alarm calls: functional reference in an avian vocal system. *Animal Behaviour, 46*, 23–38.

Fady, J.C. (1972). Absence de cooperation de type instrumental en melieu naturel chez *Papio papio. Behaviour, 43*, 157–164.

Fairbanks, L. (1975). Communication of food quality in captive *Macaca nemestrina* and free-ranging *Ateles geoffroyi. Primates, 16*, 181–190.

Fairbanks, L.A. (1980). Relationships among adult females in captive vervet monkeys: testing a model of rank-related attractiveness. *Animal Behaviour, 28*, 853–859.

Feldman, D.W. & Klopfer, P.H. (1972). A study of observational learning in lemurs. *Zeitscrift fur Tierpsychologie, 30*, 297–304.

Fernandes, M.E.B. (1991). Tool use and predation of oysters (*Crassostrea rhizophorae*) by the tufted capuchin, *Cebus apella apella*, in brackish water mangrove swamp. *Primates, 32*, 529–531.

Fernández, P. & Gómez, J.C. (1983). Desarrollo de pautas prelingüísticas de comunicación en un grupo de gorilas jóvenes. Paper presented at the IV Seminario de Investigaciones Actuales en Psicología Evolutiva y de la Educación, *19*, 20–24.

Ferster, C.B. (1964). Arithmetic behavior in chimpanzees. *Scientifican American, 210*, 98–106.

Fillmore, C.G. (1977). Topics in lexical semantics. In R. Cole (Ed.), *Current issues in linguistic theory* (pp. 76–138). Bloomington, IN: University of Indiana Press.

Finch, G. (1941). The solution of patterned string problems by chimpanzees. *Journal of Comparative Psychology, 32*, 83–90.

Finch, G. (1942). Delayed matching-from-sample and non-spatial delayed response in chimpanzees. *Journal of Comparative Psychology, 34*, 315–319.

Fischer, G.J. (1962). The formation of learning sets in young gorillas. *Journal of Comparative Psychology, 55*, 924–925.

Fischer, G.J. & Kitchener, S.L. (1965). Comparative learning in young gorillas and orang-utans. *Journal of Genetic Psychology, 107*, 337–348.

Fitch-Snyder, H. & Carter, J. (1993). Tool use to acquire drinking water by free-ranging Lion-tailed macaques (*Macaca silenus*). *Laboratory Primate Newsletter*, 32, 1–2.

Fleagle, J. (1988). *Primate adaptation and evolution*. New York: Academic Press.

Fletemeyer. J. (1978). Communication about potentially harmful foods in free-ranging Chacma baboons, *Papio ursinus*. *Primates, 19*, 223–226.

Fobes, J.L. & King, J.E. (1982). Measuring primate learning abilities. In J.L. Fobes and J.E. King (Eds.), *Primate behavior* (pp. 289–326). New York: Academic Press.

Fontaine, B., Moisson, P.Y, & Wickings, E.J. (1995). Observations of spontaneous tool making and tool use in a captive group of western lowland gorillas (*Gorilla gorilla gorilla*). *Folia Primatologica*, 65, 219–223.

Foreman, N., Arber, M., & Savage, J. (1984). Spatial memory in preschool infants. *Developmental Psychobiology, 17*, 129–137.

Fouts, R.S. (1972). Use of guidance in teaching sign language to a chimpanzee (*Pan troglodytes*). *Journal of Comparative Psychology, 80*, 515–522.

Fouts, R.S. & Budd, R.L. (1979). Artificial and human language acquisition in the chimpanzee. In D.A. Hamburg & E.R. McCown (Eds.), *The great apes* (pp. 375–392). Menlo Park, CA: Benjamin Cummings.

Fouts, R.S. & Fouts, D.H. (1989). Loulis in conversation with the cross-fostered chimpanzees. In R.A. Gardner, B.T. Gardner, & T.E. van Cantfort (Eds.), *Teaching sign language to chimpanzees* (pp. 293–307). Albany, NY: SUNY Press.

Fouts, R.S., Fouts, D.H., & van Cantfort, T.E. (1989). The infant Loulis learns signs from cross-fostered chimpanzees. In R.A. Gardner, B.T. Gardner, & T.E. van Cantfort (Eds.), *Teaching sign language to chimpanzees* (pp. 280–292). Albany, NY: SUNY Press.

Fouts, R.S. & Godin, L. (1974). Acquisition of signs in chimpanzees: a comparison of training methods. Paper presented at the Annual Meeting of the Psychonomic Society, Boston, MA.

Fouts, R.S., Hirsch, A.D., & Fouts, D.H. (1982). Cultural transmission of a human language in a chimpanzee mother-infant relationship. In H.E. Fitzgerald, J.A. Mullins, & P. Gage (Eds.), *Child nurturance. Studies of development in nonhuman primates* (pp. 159–193). New York: Plenum Press.

Fragaszy, D.M. & Adams-Curtis, L.E. (1991). Generative aspects of manipulation in tufted capuchin monkeys (*Cebus apella*). *Journal of Comparative Psychology, 105*, 387–397.

Fragaszy, D.M. & Visalberghi, E. (1989). Social influences on the acquisition of tool-using behaviors in tufted capuchin monkeys (*Cebus apella*). *Journal of Comparative Psychology, 103*, 159–170.

Franco, F. & Butterworth, G. (1996). Pointing and social awareness: declaring and requesting in the second year. *Journal of Child Language, 23*, 307–336.

French, G.M. (1959). Performance of squirrel monkeys on variants of delayed response. *Journal of Comparative and Physiological Psychology, 52*, 741–745.

Frye, D. (1991). The origins of intention in infancy. In D. Frye & C. Moore (Eds.), *Children's theories of mind* (pp. 15–38). Hillsdale, NJ: Lawrence Erlbaum Associates.

Frye, D., Zelazo, P.D., & Palfai, T. (1995). Theory of mind and rule-based learning. *Cognitive Development, 10*, 483–527.

Fujita, K. (1982). An analysis of stimulus control in two-color matching-to-sample behaviors of Japanese monkeys (*Macaca fuscata fuscata*). *Japanese Psychological Research, 24*, 124–135.

Fujita, K. (1983). Formation of the sameness-difference concept by Japanese monkeys from a small number of color stimuli. *Journal of the Experimental Analysis of Behavior, 40*, 289–300.

Fujita, K. (1987). Species recognition by five macaque monkeys. *Primates, 28*, 353–366.

Fujita, K. & Matsuzawa, T. (1986). A new procedure to study the perceptual world of animals with sensory reinforcement: recognition of humans by a chimpanzee. *Primates, 27*, 283–291.

Furuichi, T. (1989) Social interactions and the life history of female *Pan paniscus* in Wamba, Zaire. *International Journal of Primatology, 10*, 173–197.

Gagné, R.M. (1970). *The conditions of learning*. New York: Holt, Rinehart and Winston.

Gagnon, S. & Doré, F.Y. (1992). Search behavior in various breeds of adult dogs (*Canis familiaris*): object permanence and olfactory cues. *Journal of Comparative Psychology, 106*, 58–68.

Gagnon, S. & Doré, F.Y. (1993). Search behavior of dogs (*Canis familiaris*) in invisible displacement problems. *Animal Learning and Behavior, 21*, 246–254.

Galdikas, B.M.F. (1982). Orang-utan tool-use at Tanjung Puting Reserve, Central Indonesian Borneo (Kalimantan Tengah). *Journal of Human Evolution, 10*, 19–33.

Galdikas, B.F.M. & Vasey, P. (1992). Why are orangutans so smart? Ecological and social hypothesis. In F.D. Burton (Ed.), *Social processes and mental abilities in nonhuman primates* (pp. 183–224). Lewiston, NY: Edwin Mellen Press.

Galef, B.G. Jr. (1988). Imitation in animals: history, definition, and interpretation of data from the psychological laboratory. In T.R. Zentall & B.G. Galef Jr. (Eds.), *Social learning: psychological and biological perspectives* (pp. 3–28). Hillsdale, NJ: Lawrence Erlbaum Associates.

Galef, B.G. Jr. (1992). The question of animal culture. *Human Nature, 3*, 157–178.

Galef, B.G. Jr. (1996). Social enhancement of food preferences in Norway rats: a brief review. In C.M. Heyes & B.G. Galef, Jr. (Eds.), *Social learning in animals. The roots of culture* (pp. 49–64). New York: Academic Press.

Galef, B.G. Jr. & Wigmore, S.W. (1983). Transfer of information concerning distant foods: a laboratory investigation of the 'information-centre' hypothesis. *Animal Behaviour, 31*, 748–758.

Gallup, G. Jr. (1968). Mirror-image stimulation. *Psychological Bulletin, 70*, 782–793.

Gallup, G. Jr. (1970). Chimpanzees: self-recognition. *Science, 167*, 86–87.

Gallup, G. Jr. (1977). Absence of self-recognition in a monkey (*Macaca fascicularis*) following prolonged exposure to a mirror. *Developmental Psychobiology, 10*, 281–284.

Gallup, G. Jr. (1982). Self-awareness and the emergence of mind in primates. *American Journal of Primatology, 2*, 237–248.

Gallup, G. Jr. (1987). Self-awareness. In G. Mitchell & J. Erwin (Eds.), *Comparative primate biology: behavior, cognition, and motivation* (pp. 3–16). New York: Alan R. Liss.

Gallup, G. Jr. (1994). Self-recognition: research strategies and experimental design. In S.T. Parker, R.W. Mitchell, & M.L. Boccia (Eds.), *Self-awareness in animals and humans. Developmental perspectives* (pp. 35–50). Cambridge: Cambridge University Press.

Gallup, G. Jr., McClure, M.K., Hill, S.D., & Bundy, R.A. (1971). Capacity for self-recognition in differentially reared chimpanzees. *Psychological Record, 21*, 69–74.

Gallup, G. Jr., Wallnau, L.B., & Suarez, S.D. (1980). Failure to find self-recognition in mother-infant and infant-infant rhesus monkey pairs. *Folia Primatologica, 33*, 210–219.

Garber, P. (1989). Role of spatial memory in primate foraging patterns: *Saguinus mystax* and *Saguinus fuscicollis*. *American Journal of Primatology, 19*, 203–216.

Garber, P.A. (1993). Seasonal patterns of diet and ranging in two species of tamarin monkeys: Stability versus variability. *International Journal of Primatology, 14*, 145–166.

Garber, P.A. & Hannon, B. (1993). Modeling monkeys: a comparison of computer-generated and naturally occuring foraging patterns in two species of neotropical primates. *International Journal of Primatology, 14*, 827–852.

Garcha, H.S. & Ettlinger, G. (1979). Object sorting by chimpanzees and monkeys. *Cortex, 15*, 213–224.

Gardner, R.A. & Gardner, B.T. (1969). Teaching sign language to a chimpanzee. *Science, 165*, 664–672.

Gardner, R.A. & Gardner, B.T. (1978). Comparative psychology and language acquisition. *Annals of the New York Academy of Sciences, 309*, 37–76.

Gatinot, B.L. (1974). Note sur l'observation d'une utilisation spontanee d'outil chez Erythrocebus patas en captivite. *Mammalia, 38*, 557–558.

Gelman, R. & Baillargeon, R. (1983). A review of some Piagetian concepts. In P. Mussen (Ed.), *Carmichael's manual of child psychology* (pp. 167–230). New York: Wiley.

Gelman, R. & Gallistel, C.R. (1978). *The child's understanding of number*. Cambridge, MA: Harvard University Press.

Gentner, D., Rattermann, M.J., Markman, A., & Kotovsky, L. (1995). Two forces in the development of relational similarity. In T.J. Simon & G.S. Halford (Eds.), *Developing cognitive competence: new approaches to process modeling* (pp. 263–313). Hillsdale, NJ: Lawrence Erlbaum Associates.

Gergely, G., Nádasdy, Z., Csibra, G., & Biró, S. (1995). Taking the intentional stance at 12 months of age. *Cognition, 56*, 165–193.

Ghiglieri, M.P. (1984). *The chimpanzees of the Kibale forest*. New York: Columbia University Press.

Gibson, E. & Rader, N. (1979). Attention: the perceiver as performer. In G. Hale & M. Lewis (Eds.), *Attention and cognitive development* (pp. 6–36). New York: Plenum Press.

Gibson, J.J. (1979). *The ecological approach to visual perception*. Boston: Houghton-Mifflin.

Gibson, K.R. (1986). Cognition, brain size and the extraction of embedded food resources. In J.G. Else & P.C. Lee (Eds.), *Primate ontogeny, cognition and social behaviour* (pp. 93–103). Cambridge: Cambridge University Press.

Gibson, K.R. (1990). Tool use, imitation, and deception in a captive cebus monkey. In S.T. Parker & K.R. Gibson (Eds.), *"Language" and intelligence in monkeys and apes*. (pp. 205–218). Cambridge: Cambridge University Press.

Gillan, D.J. (1981). Reasoning in the chimpanzee: II. Transitive inference. *Animal Behavior Processes, 7*, 150–164.

Gillan, D.J., Premack, D., & Woodruff, G. (1981). Reasoning in the chimpanzee: I. Analogical reasoning. *Animal Behavior Processes, 7*, 1–17.

Glickman, S. & Sroges, R. (1966). Curiosity in zoo animals. *Behaviour, 26*, 151–187.

Gómez, J.C. (1989). La comunicación y la manipulación de objetos en crías de gorila, *Estudios de Psicología, 38*, 111–128.

Gómez, J.C. (1990). The emergence of intentional communication as a problem-solving strategy in the gorilla. In S.T. Parker & K.R. Gibson (Eds.), *"Language" and intelligence in monkeys and apes* (pp. 333–355). Cambridge: Cambridge University Press.

Gómez, J.C. & Teixidor, P. (1992). Theory of mind in an orangutan: a nonverbal test of false-belief appreciation. Paper presented at the XIV Congress of the International Primatological Society, Strasbourg.

Goodall, J. (1986). *The chimpanzees of Gombe. Patterns of behavior*. Cambridge, MA: Harvard University Press.

Goodall, J. (1990). *Through a window*. Boston: Houghton Mifflin Company.

Gopnik, A. (1993). How we know our minds: the illusion of first-person knowledge about intentionality. *Behavioral and Brain Sciences, 16*, 1–14.

Gordon, R. (1986). Folk psychology as simulation. *Mind and language, 1*, 158–71.

Goswami, U. (1991). Analogical reasoning: What develops? *Child Development, 62*, 1–22.

Gould, J.L. (1987). Landmark learning by honey bees. *Animal Behaviour, 35*, 26–34.

Gould, J. & Gould, C. (1986). Invertebrate intelligence. In R. Hoage & L. Goldman (Eds.), *Animal intelligence: insights into the animal mind* (pp. 21–36). Washington, DC: Smithsonian Institution Press.

Gould, S.J. (1977). *Ontogeny and phylogeny*. Cambridge, MA: Belknap Press.

Gouzoules, H. (1992). Is this the best of all possible monkey worlds? *Behavioral and Brain Sciences, 15*, 158–159.

Gouzoules, H. & Gouzoules, S. (1987). Kinship. In B.B. Smuts, D.L. Cheney, R.M. Seyfarth, R.W. Wrangham, & T.T. Struhsaker (Eds.), *Primate societies* (pp. 299–305). Chicago: University of Chicago Press.

Gouzoules, H. & Gouzoules, S. (1989a). Design features and developmental modification of pigtail macaque, *Macaca nemestrina*, agonistic screams. *Animal Behaviour, 37*, 383–401.

Gouzoules, H. & Gouzoules, S. (1989b). Sex differences in the acquisition of communicative competence by pigtail macaques (*Macaca nemestrina*). *American Journal of Primatology, 19*, 163–174.

Gouzoules, H. & Gouzoules, S. (1990). Matrilineal signatures in the recruitment screams of pigtail macaques, *Macaca nemestrina*. *Behaviour, 115*, 327–347.

Gouzoules, H. & Gouzoules, S. (1995). Recruitment screams of pigtail monkeys (*macaca nemestrina*): Ontogenetic perspectives. *Behavior, 132*, 431–450.

Gouzoules, H., Gouzoules, S., & Ashley, J. (1995). Representational signalling in nonhuman primate vocal communication. In E. Zimmermann, J. Newman, & U. Jürgens (Eds.), *Current topics in primate vocal communication* (pp. 235–252) New York: Plenum Press.

Gouzoules, S., Gouzoules, H., & Marler, P. (1984). Rhesus monkey (*Macaca mulatta*) screams: representational signalling in the recruitment of agonistic aid. *Animal Behaviour, 32*, 182–193.

Gouzoules, H., Gouzoules, S., & Marler, P. (1985). External reference and affective signalling in mammlian vocal communication. In G. Zivin (Ed.), *The development of expressive behavior* (pp. 77–101). New York: Academic Press.

Green, S. (1975). Communication by a graded vocal system in Japanese monkeys. In L.A. Rosenblum (Ed.), *Primate behavior* (pp. 1–102). New York: Academic Press.

Green, S. & Marler, P. (1979). The analysis of animal communication. In P. Marler & J. Vandenberg (Eds.), *Handbook of behavioral neurobiology, vol. 3*. (pp. 73–158). New York: Plenum Press.

Greenfield, P.M. (1991). Language, tools and brain: the ontogeny and phylogeny of hierarchically organized sequential behavior. *Behavioral and Brain Sciences, 14*, 531–595.

Greenfield, P.M. & Savage-Rumbaugh, E.S. (1990). Grammatical combination in *Pan paniscus*: processes of learning and invention in the evolution and development of language. In S.T. Parker & K.R. Gibson (Eds.), *"Language" and intelligence in monkeys and apes* (pp. 540–578). Cambridge: Cambridge University Press.

Greenfield, P.M. & Savage-Rumbaugh, E.S. (1991). Imitation, grammatical development, and the invention of protogrammar by an ape. In N.A. Krasnegor, D.M.Rumbaugh, R.L.Schiefelbusch, & M.Studdert-Kennedy (Eds.), *Biological and behavioral determinants of language development* (pp. 235–258). Hillsdale, NJ: Lawrence Earlbaum Associates.

Griffin, D.R. (1978). Prospects for a cognitive ethology. *Behavioral and Brain Sciences, 4*, 527–538.

Groves, C.P. & Sabater Pi, J. (1985). From ape's nest to human fix-point. *Man, 20*, 22–47.

Guillaume, P. & Meyerson, I. (1930). Recherches sur l'usage de l'instrument chez les singes. I. Le Probleme du detour. *Journal de Psychologie Normale et Pathologique, 27*, 177–236.

Gunderson, V. (1983). The development of cross-modal recognition in infant pigtail monkeys. *Developmental Psychology, 19*, 398–404.

Gust, D.A. (1989). Uncertain availability of a preferred food affects choice in a captive group of chimpanzees (*Pan troglodytes*). *American Journal of Primatology, 17*, 165–171.

Gyger, M., Karakashian, S.J., & Marler, P. (1986). Avian alarm calling: Is there an audience effect? *Animal Behaviour, 34*, 1570–1572.

Haggerty, M.E. (1909). Imitation in monkeys. *Journal of Comparative Neurology and Psychology, 19*, 337–455.

Haggerty, M.E. (1913). Plumbing the minds of apes. *McClure's Magazine, 41*, 151–154.

Hall, K.R.L. (1962). Numerical data, maintenance activities and locomotion of the wild chacma baboon, Papio ursinus. *Proc. Zool. Soc. Lond., 139*, 181–220.

Hall, K.R.L. (1963). Observational learning in monkeys and apes. *British Journal of Psychology, 54*, 201–226.

Hall, K.R.L. & DeVore, I. (1965). Baboon social behavior. In I. DeVore (Ed.), *Primate behavior* (pp. 53–110). New York: Holt, Rinehart and Winston.

Hall, K.R.L. & Goswell, M.J. (1964). Aspects of social learning in captive patas monkeys. *Primates, 5*, 59–70.

Hallock, M.B. & Worobey, J. (1984). Cognitive development in chimpanzee infants (*Pan troglodytes*). *Journal of Human Evolution, 13*, 441–447.

Ham, R. (1990). Do monkeys see monkeys do? (Master thesis). St. Andrews, Scotland: University of St. Andrews.

Hamilton, W.J., Buskirk R.E., & Buskirk, W.H. (1975). Defensive stoning by baboons. *Nature, 256*, 488–489.

Hamilton, W.J., Buskirk, R.E., & Buskirk, W.H. (1978). Environmental determinants of object manipulation by chacma baboons (*Papio ursinus*) in two Southern African environments. *Journal of Human Evolution, 7*, 205–216.

Hannah, A.C. & McGrew, W.C. (1987). Chimpanzees using stones to crack open oil palm nuts in Liberia. *Primates, 28*, 31–46.

Hansen, E.W. (1976). Selective responding by recently separated juvenile rhesus monkeys to calls of their mothers. *Developmental Psychobiology, 9*, 83–88.

Harcourt, A.H. (1992). Coalitions and alliances: are primates more complex than non-primates? In A.H. Harcourt & F.B.M. de Waal (Eds.), *Coalitions and alliances in human and nonhuman animals* (pp. 445–471). New York: Oxford University Press.

Harcourt, A.H. & de Waal, F.B.M. (1992). *Coalitions and alliances in humans and other animals*. New York: Oxford University Press.

Harcourt, A.H. & Stewart, K.J. (1989). Functions of alliances in contests within wild gorilla groups. *Behaviour, 109*, 3–4, 176–190.

Harcourt, A.H., Stewart, K.J., & Hauser, M.D. (1993). Functions of wild gorilla 'close' calls. I. Repertoire, context, and interspecifc comparison. *Behaviour, 124*, 89–122.

Harlow, H.F. (1932). Comparative behavior of primates. III. Complicated delayed reaction tests on primates. *Journal of Comparative Psychology, 14*, 241–252.

Harlow, H.F. (1949). The formation of learning sets. *Psychological Review, 56*, 51–65.

Harlow, H.F. (1951). Primate learning. In C.P. Stone (Ed.), *Comparative psychology* (pp. 183–238). Englewood Cliffs, NJ: Prentice-Hall.

Harlow, H.F. (1958). The evolution of learning. In A. Roe & G.G. Simpson (Eds.), *Behavior and evolution* (pp. 269–290). New Haven, CT: Yale University Press.

Harlow, H.F. (1959). The development of learning in the rhesus monkey. *American Scientist, 47*, 459–479.

Harlow, H.F. & Bromer, J.A. (1939). Comparative behavior of primates. VIII. The capacity of platyrrhine monkeys to solve delayed reaction tests. *Journal of Comparative Psychology, 28*, 299–304.

Harlow, H.F., Schlitz, K.A., & Harlow, M.K. (1969). Effects of social isolation on the learning performance of rhesus monkeys. In C.R. Carpenter (Ed.), *Proceedings of second international congress on primatology* (pp. 178–185). Basel, Switzerland: Karger.

Harlow, H.F. & Settlage, P.H. (1934). Comparative behavior of primates. VII. Capacity of monkeys to solve patterned string tests. *Journal of Comparative Psychology, 18*, 423–435.

Harlow, H.F., Uehling, H., & Maslow, A.H. (1932). Comparative behavior of primates. I. Delayed reaction tests on primates from the lemur to the orang-outan. *Journal of Comparative Psychology, 13*, 313–343.

Harlow, H.F. & Warren, J.M. (1952). Formation and transfer of discrimination learning sets. *Journal of Comparative Psychology, 45*, 482–489.

Harlow, H.F. & Yudin, H.C. (1933). Social behavior of primates. I. Social facilitation of feeding in the monkey and its relation to attitudes of ascendance and submission. *Journal of Comparative Psychology, 16*, 171–185.

Harris, M.R. & McGonigle, B.O. (1994). A model of transitive choice. *Quarterly Journal of Experimental Psychology, 47B*, 319–348.

Harris, P. (1983). Infant cognition. In P. Mussen (Ed.), *Carmichael's manual of child psychology, vol. 2* (pp. 689–782). New York: Wiley.

Harris, P. (1991) The work of the imagination. In A. Whiten (Ed.), *Natural theories of mind* (pp. 283–304). Oxford: Blackwell.

Harris, P. (1993). Pretending and planning. In S. Baron-Cohen, H. Tager-Flusberg, & D.J. Cohen (Eds.), *Understanding other minds. Perspectives from autism* (pp. 228–246). New York: Oxford University Press.

Harris, P. (1996). Desires, beliefs, and language. In P. Carruthers & P. Smith (Eds.), *Theories of theories of mind* (pp. 200–222). Cambridge: Cambridge University Press.

Harter, S. (1983). Developmental perspectives on the self system. In P. Mussen (Ed.), *Carmichael's manual of child psychology, vol 4* (pp. 285–386). New York: Wiley.

Hauser, M.C. (1988a). How infant vervet monkeys learn to recognize starling alarm calls: the role of experience. *Behaviour, 105*, 187–201.

Hauser, M.D. (1988b). Invention and social transmission: new data from wild vervet monkeys. In R.W. Byrne & A. Whiten (Eds.), *Machiavellian intelligence. Social expertise and the evolution of intellect in monkeys, apes, and humans* (pp..327–343). New York: Oxford University Press.

Hauser, M.D. (1989). Ontogenetic changes in the comprehension and production of vervet monkey (*Cercopithecus aethiops*) vocalizations. *Journal of Comparative Psychology, 103*, 149–158.

Hauser, M.D. (1990). Do chimpanzee copulatory calls incite male-male competition? *Animal Behaviour, 39*, 596–597.

Hauser, M. (1996). *The evolution of communication*. Cambridge, MA: MIT Press.

Hauser, M.D., Kralik, J., Botto-Mahan, C., Garrett, M., & Oser, J. (1995). Self-recognition in primates: Phylogeny and the salience of species-typical features. *Proceedings of the National Academy Sciences USA, 92*, 10811–10814.

Hauser, M.D. & Marler, P. (1993a). Food-associated calls in rhesus macaques (*Macaca mulatta*): I. Socioecological factors. *Behavioral Ecology, 4*, 194–205.

Hauser, M.D. & Marler, P. (1993b). Food-associated calls in rhesus macaques (*Macaca mulatta*): II. Costs and benefits of call production and suppression. *Behavioral Ecology, 4*, 206–212.

Hauser, M.D., McNeilage, P., & Ware, M. (in press). Numerical representations in primates. *Proceedings of the National Academy of Sciences USA*.

Hauser, M.D., Teixidor, P., Field, L., & Flaherty, R. (1993). Food-elicited calls in chimpanzees: effects of food quantity & divisibility. *Animal Behaviour, 45*, 817–819.

Hauser, M.D. & Wrangham, R.W. (1987). Manipulation of food calls in captive chimpanzees. A preliminary report. *Folia Primatologica, 48*, 207–210.

Hayaki, H. (1990) Social context of pant-grunting in young chimpanzees. In T. Nishida (Ed.) *The chimpanzees of the Mahale mountains* (pp. 189–206). Tokyo: Tokyo University Press.

Hayes, C. (1951). *The ape in our house*. New York: Harper.

Hayes, K.J. & Hayes, C. (1951). The intellectual development of a home-raised chimpanzee. *Proceedings of the American Philosophical Society, 95*, 105–109.

Hayes, K.J. & Hayes, C. (1952). Imitation in a home-raised chimpanzee. *Journal of Comparative Psychology, 45*, 450–459.

Hayes, K.J. & Nissen, C.H. (1971). Higher mental functions of a home-raised chimpanzee. In A.M. Schrier & F. Stollnitz (Eds.), *Behavior of nonhuman primates* (pp. 59–115). New York: Academic Press.

Hayes, K.J., Thompson, R., & Hayes, C. (1953). Discrimination learning set in chimpanzees. *Journal of Comparative Psychology, 46*, 99–104.

Hemelrijk, C.K. (1990a). A matrix partial correlation test used in investigations of reciprocity and other social interaction patterns at group level, *Journal of Theoretical Biology, 143*, 405–420.

Hemelrijk, C.K. (1990b). Models of, and tests for, reciprocity, unidirectionality, and other social interaction patterns at group level. *Animal Behaviour, 39*, 1013–1029.

Hemelrijk, C.K. (1994). Support for being groomed in long-tailed macaques, *Macaca fascicularis. Animal Behaviour, 48*, 479–481.

Hemelrijk, C.K. & Ek, A. (1991). Reciprocity and interchange of grooming and 'support' in captive chimpanzees. *Animal Behaviour, 41*, 923–935.

Hemelrijk, C.K., van Laere, G.J., & van Hooff, J.A.R.A.M. (1992). Sexual exchange relationships in captive chimpanzees? *Behavioral Ecology and Sociobiology, 30*, 269–275.

Hemmi, J.M. & Menzel, C.R. (1995). Foraging strategies of long-tailed macaques, *Macaca fascicularis*: directional extrapolation. *Animal Behaviour, 49*, 457–463.

Herman, L.M. (1986). Cognition and language competencies of bottlenosed dolphins. In R.J. Schusterman, J.A. Thomas, & F.G. Wood (Eds.), *Dolphin cognition and behavior: a comparative approach* (pp. 221–252). Hillsdale, NJ: Lawrence Erlbaum Associates.

Herman, L.M. & Gordon, J.A. (1974). Auditory delayed matching in the bottlenosed dolphin. *Animal Learning and Behavior, 3*, 43–48.

Herman, L.M., Pack, A.A., & Morrel-Samuels, P. (1993). Representational and conceptual skills of dolphins. In H.L. Roitblat, L.M. Herman, & P.E. Nachtigall (Eds.), *Language and communication. Comparative perspectives* (pp. 403–442). Hillsdale, NJ: Lawrence Erlbaum Associates.

Hernandez-Camacho, J. & Cooper, R. (1976). The nonhuman primates of Columbia. In R. Thorington & P. Heltne (Eds.), *Neotropical primates* (pp. 35–69). Washington, D.C.: National Academy of Sciences.

Herrnstein, R.J. (1979). Acquisition, generalization, and discrimination reversal of a natural concept. *Animal Behavior Processes, 5*, 116–129.

Herrnstein, R.J. & Loveland, D.H. (1964). Complex visual concept in the pigeon. *Science, 146*, 549–551.

Hervé, N. & Deputte, B.L. (1993). Social influence in manipulations of a capuchin monkey raised in a human environment: a preliminary case study. *Primates, 34*, 227–232.

Hess, J., Novak, M.A., & Povinelli, D.J. (1993). 'Natural pointing' in a rhesus monkey, but no evidence of empathy. *Animal Behaviour, 46*, 1023–1025.

Heyes, C.M. (1993a). Anecdotes, training, traping and triangulating: do animals attribute mental states? *Animal Behaviour, 46*, 177–188.

Heyes, C.M. (1993b). Imitation, culture and cognition. *Animal Behaviour, 46*, 999–1010.

Heyes, C.M. (1994a). Cues, convergence and a curmudgeon: a reply to Povinelli. *Animal Behaviour, 48*, 242–244.

Heyes, C.M. (1994b). Reflections on self-recognition in primates. *Animal Behaviour, 47*, 909–919.

Heyes, C.M. & Dawson, G.R. (1990). A demonstration of observational learning in rats using a bidirectional control. *Quarterly Journal of Experimental Psychology, 42B*, 59–71.

Heyes, C.M. & Galef, B.G. Jr., (1996). (Eds.) *Sociai learning in animals. The roots of culture.* New York: Academic Press.

Hicks, L.H. (1956). An analysis of number-concept formation in the rhesus monkey. *Journal of Comparative Psychology, 49*, 212–218.

Hikami, K. (1991). Social transmission of learning in Japanese monkeys (*Macaca fuscata*). In A. Ehara, T. Kimura, O. Takenaka, & M. Iwamoto (Eds.), *Primatology Today* (pp. 56–73). Amsterdam: Elsevier.

Hikami, K., Hasegawa, Y., & Matsuzawa, T. (1990). Social transmission of food preferences in Japanese monkeys (*Macaca fuscata*) after mere exposure or aversion training. *Journal of Comparative Psychology, 104*, 233–237.

Hill, S.D., Bundy, R.A., Gallup, G.G. Jr., & McClure, M.K. (1970). Responsiveness of young nursery-reared chimpanzees to mirrors. *Louisiana Academy of Sciences, 33*, 77–82.

Hinde, R.A. (1987). Can nonhuman primates help us understand human behavior? In B.B. Smuts, D.L. Cheney, R.M. Seyfarth, R.W. Wrangham, & T.T. Struhsaker (Eds.), *Primate societies* (pp. 413–442). Chicago: University of Chicago Press.

Hobhouse, L.T. (1901). *Mind in evolution.* London: Macmillan.

Hobson, P. (1993). *Autism and the development of mind.* Hillsdale, NJ: Lawrence Erlbaum Associates.

Hodos, W. & Campbell, C.B.G. (1969). Scala naturae: why there is no theory in comparative psychology? *Psychological Review, 76*, 337–350.

Hohmann, G. (1988). A case of simple tool use in wild liontail macaques (*Macaca silenus*). *Primates, 29*, 565–567.

Hollard, V. & Delius, J. (1982). Rotational invariance in visual pattern recognition by pigeons and humans. *Science, 218*, 804–806.

Holloway, R.L. (1974). *Primate aggression, territoriality, and xenophobia.* New York: Academic Press.

Hopkins, W.D., Fagot, J., & Vauclair, J. (1993). Mirror-image matching and mental rotation problem solving in baboons (*Papio papio*): unilateral input enhances performance. *Journal of Experimental Psychology: General, 122*, 61–72.

Hopkins, W.D. & Savage-Rumbaugh, E.S. (1991). Vocal communication as a function of differential rearing experiences in *Pan paniscus*: a preliminary report. *International Journal of Primatology, 12*, 559–583.

Huffman, M. (1984). Stone play of macaca fuscata in Arashiyama B troop: transmission of non-adaptive behavior. *Journal of Human Evolution, 13*, 725–735.

Huffman, M. (1986). The misunderstood monkey. Primates, potatoes, and cultural evolution. *Whole Earth Review, 52*, 26–29.

Huffman, M.A. & Quiatt, D. (1986). Stone handling by Japanese macaques (*Macaca fuscata*): implications for tool use of stone. *Primates, 27*, 413–423.

Hulse, S.H., Fowler, H., & Honig, W.K. (1978). *Cognitive processes in animal behavior.* Hillsdale, NJ: Lawrence Erlbaum Associates.

Hume, D. (1739/1952). *A treatise of human nature.* London: John Noon.

Humphrey, N.K. (1974). Species and individuals in the perceptual world of monkeys. *Perception, 3*, 105–114.

Humphrey, N.K. (1976). The social function of intellect. In P. Bateson & R.A. Hinde (Eds.), *Growing points in ethology* (pp. 303–321). Cambridge: Cambridge University Press.

Hunt, G. (1996). Manufacture and use of hook-tools by New Caledonian crows. *Nature, 379*, 249–251.

Hunte, W. & Horrocks, J.A. (1987). Kin and non-kin interventions in the aggressive disputes of vervet monkeys. *Behavioral Ecology and Sociobiology, 20*, 257–263.

Hyatt, C.W. & Hopkins, W.D. (1994). Self-awareness in bonobos and chimpanzees: a comparative perspective. In S.T. Parker, R.W. Mitchell, & M.L. Boccia (Eds.), *Self-awareness in animals and humans. Developmental perspectives* (pp. 248–253). Cambridge: Cambridge University Press.

Ingmanson, E.J. (1996). Tool-using behavior in wild *Pan paniscus*: Social and ecological considerations. In A.E. Russon, K.A. Bard, & S.T. Parker (Eds.). *Reaching into thought. The minds of the great apes* (pp. 190–210). Cambridge: Cambridge University Press.

Inoue-Nakamura, N. & Matsuzawa, T. (in press). Development of stone tool use by wild chimpanzees (*Pan troglodytes*). *Journal of Comaprative Psychology.*

Irle, E. & Markowitsch, H.J. (1987). Conceptualization without specific training in squirrel monkeys (*Saimiri sciureus*): a test using the non-match-to-sample procedure. *Journal of Comparative Psychology, 101*, 305–311.

Itakura, S. (1987a). Mirror-guided behavior in Japanese monkeys (*Macaca fuscata fuscata*). *Primates, 28*, 149–161.

Itakura, S. (1987b). Use of a mirror to direct their responses in Japanese monkeys (*Macaca fuscata fuscata*). *Primates, 28*, 343–352.

Itakura, S. (1996). An exploratory study of gaze-monitoring in nonhuman primates. *Japanese Psychological Research, 38*, 174–180.

Itakura, S. & Anderson, J. (1996). Learning to use experimenter-given cues during an object-choice task by a capuchin monkeys. *Current Psychology of Cognition, 15*, 103–112.

Itakura, S. & Matsuzawa, T. (1993). Acquisition of personal pronouns by a chimpanzee. In H.L. Roitblat, L.M. Herman, & P.E. Nachtigall (Eds.), *Language and communication. Comparative perspectives* (pp. 347–363). Hillsdale, NJ: Lawrence Erlbaum Associates.

Itani, J. & Nishimura, A. (1973). The study of infrahuman culture in Japan. In E.W. Menzel, Jr. (Ed.), *Precultural primate behavior* (pp. 26–50). Basel, Switzerland: Karger.

Izawa, K. & Mizuno, A. (1977). Palm-fruit cracking behavior of wild black-capped capuchin (*Cebus apella*). *Primates, 18*, 773–792.

Jackson, W.J. & Pegram, G.V. (1970a). Acquisition, transfer and retention of matching by rhesus monkeys. *Psychological Reports, 27*, 839–846.

Jackson, W.J. & Pegram, G.V. (1970b). Comparison of intra- vs. extradimensional transfer of matching by rhesus monkeys. *Psychonomic Science, 19*, 162–163.

Jalles-Fihlo, E. (1995). Manipulative propensity and tool use in capuchin monkeys. *Current Anthropology, 36*, 664–667.

Jarvis, M.J. & Ettlinger, G. (1978). Cross-modal performance in monkeys and apes: is there a substantial difference? In D.J. Chivers & J. Herbert (Eds.), *Recent advances in primatology* (pp. 953–956). New York: Academic Press.

Jensvold, M.L.A. & Fouts, R.S. (1993). Imaginary play in chimpanzees (*Pan troglodytes*). *Human Evolution, 8*, 217–227.

Jerison, H. (1973). *Evolution of the brain and intelligence.* New York: Academic Press.

Johnson, M. (1987). *The body in the mind.* Chicago: University of Chicago Press.

Johnson, T. (1985). Liontail macaque behavior in the wild. In P. Heltne (Ed.), *The liontail macaque* (pp. 147–167). New York: Alan Liss.

Jolly, A. (1964a). Choice of cue in prosimian learning. *Animal Behaviour, 12*, 571–577.

Jolly, A. (1964b). Prosimians' manipulation of simple object problems. *Animal Behaviour, 12*, 560–570.

Jolly, A. (1966). Lemur social behavior and primate intelligence. *Science, 153*, 501–506.

Jolly, A. (1985). *The evolution of primate behavior.* New York: Macmillan.

Jordan, C. (1982). Object manipulation and tool-use in captive pygmy chimpanzee (*Pan paniscus*). *Journal of Human Evolution, 11*, 35–39.

Joubert, A. & Vauclair, J. (1986). Reaction to novel objects in a troop of Guinea baboons: Approach and manipulation. *Behaviour, 96*, 92–104.

Jouventin, P., Pasteur, G., & Cambefort, J.P. (1976). Observational learning of baboons and avoidance of mimics: exploratory tests. *Evolution, 31*, 214–218.

Judge, P.G. (1982) Redirection of aggression based on kinship in a captive group of pigtail macaques. *International Journal of Primatology, 3*, 301.

Judge, P.G. (1991) Dyadic and triadic reconciliation in pigtail macaques (*Macaca nemestrina*). *American Journal of Primatology, 23*, 225–237.

Kano, T. (1982). The use of leafy twigs for rain cover by the pygmy chimpanzees of Wamba. *Primates, 23*, 453–457.

Kano, T. (1992) *The last ape. Pygmy chimpanzee behavior and ecology*. Stanford, CA: Stanford University Press.

Kaplan, J.N., Winship-Ball, A., & Sim, L. (1978). Maternal discrimination of infant vocalizations in squirrel monkeys. *Primates, 19*, 187–193.

Karmiloff-Smith, A. (1992). *Beyond modularity: a developmental perspective on cognitive science*. Cambridge, MA: MIT Press.

Kawai, M. (1958). On the system of social ranks in a natural troop of Japanese monkeys (1): Basic and dependent rank. *Primates, 1*, 111–148.

Kawai, M. (1965). Newly acquired pre-cultural behavior of the natural troop of Japanese monkeys on Koshima Islet. *Primates, 6*, 1–30.

Kawamura, S. (1959). The process of sub-culture propagation among Japanese macaques. *Primates, 2*, 43–60.

Keddy-Hector, A.C., Seyfarth, R.M., & Raleigh, M.J. (1989). Male parental care, female choice and the effect of an audience in vervet monkeys. *Animal Behaviour, 38*, 262–271.

Kellogg, W.N. & Kellogg, L.A. (1933). *The ape and the child*. New York: McGraw-Hill.

Kempf, E.J. (1916). Two methods of subjective learning in the monkey *Macacus rhesus*. *Journal of Animal Behavior, 6*, 256–265.

Khrustov, G.F. (1970). The problem of the origin of man. *Soviet Psychology, 8*, 6–31.

King, B.J. (1986a). Extractive foraging and the evolution of primate intelligence. *Human Evolution, 1*, 361–372.

King, B.J. (1986b). Individual differences in tool-using by two captive orangutans (*Pongo pygmaeus*). In D.M. Taub & F.A. King (Eds.), *Current perspectives in primate social dynamics* (pp. 469–475). New York: Van Nostrand Reinhold.

King, B.J. (1991). Social information transfer in monkeys, apes, and hominids. *Yearbook of Physical Anthropology, 34*, 97–115.

King, B. J. (1994). The information continuum: evolution of social information transfer in monkeys, apes, and hominids. Santa Fe, NM: School of American Research.

King, J.E. (1973). Learning and generalization of a two-dimensional sameness-difference concept by chimpanzees and orangutans. *Journal of Comparative and Physiological Psychology, 84*, 140–148.

King, J.E. & Fobes, J.L. (1975). Hypothesis analysis of sameness-difference learning-set by capuchin monkeys. *Learning and Motivation, 6*, 101–113.

King, J.E. & Fobes, J.L. (1982). Complex learning by primates. In J.L. Fobes & J.E. King (Eds.), *Primate behavior* (pp. 327–360). New York: Academic Press.

King, J.E., Flaningam, M.R., & Rees, W.W. (1968). Delayed response with different delay conditions by squirrel monkeys and fox squirrels. *Animal Behaviour, 16*, 271–275.

King, M. & Wilson, A. (1975). Evolution at two levels in humans and chimpanzees. *Science, 188*, 107–116.

Kinnaman, A.J. (1902). Mental life of two (Macacus rhesus) monkeys in captivity. *American Journal of Psychology, 13*, 98–148.

Kitahara-Frisch, J. & Norikoshi, K. (1982). Spontaneous sponge-making in captive chimpanzees. *Journal of Human Evolution, 11*, 41–47.

Kitahara-Frisch, J., Norikoshi, K., & Hara, K. (1987). Use of a bone fragment as a step towards secondary tool use in captive chimpanzee. *Primate Report, 18*, 33–37.

Klein, R. (1989). *The human career: Human biological and cultural origins.* Chicago: University of Chicago Press.

Klüver, H. (1933). *Behavior mechanisms in monkeys.* Chicago: University of Chicago Press.

Klüver, H. (1937). Re-examination of implement-using behavior in a cebus monkey after an interval of three years. *Acta Psychologica, 2*, 347–397.

Knobloch, H. & Pasamanick, B. (1959). The development of adaptive behavior in an infant gorilla. *Journal of Comparative Psychology, 52*, 699–704.

Koehler, O. (1950). The ability of birds to "count". *Bulletin of Animal Behaviour, 9*, 41–45.

Köhler, W. (1925). *The mentality of apes.* London: Routledge and Kegan Paul.

Kohts, N. (1935). *Infant ape and human child. Instincts, emotions, play, habits.* Moscow: Scientific Memoirs of the Museum Darwinianum in Moscow.

Kojima, T. (1979). Discriminative stimulus context in matching-to-sample of Japanese monkeys. *Japanese Psychological Research, 21*, 189–194.

Kortlandt, A. (1967). Experimentation with chimpanzees in the wild. In D. Starck, R. Schheider, & H.J. Ruhn (Eds.), *Progress in primatology* (pp. 208–224). Stuggart: Gustav Fischer Verlag.

Kortlandt, A. & Kooij, M. (1963). Protohominid behaviour in primates. In J.R. Napier (Ed.), *The Primates. Symposia of the Zoological Society of London* (pp. 61–88). London: Academic Press.

Koyama, N.F. & Dunbar, R.I.M. (1996). Anticipation of conflict by chimpanzees. *Primates, 37*, 79–86.

Kruger, A.C. (1992) The effect of peer and adult-child transactive discussions on moral reasoning. *Merrill-Palmer Quarterly, 38*, 191–211.

Kummer, H. (1967). Tripartite relations in hamadryas baboons. In S.A. Altmann (Ed.), *Social communication among primates* (pp. 63–71). Chicago: University of Chicago Press.

Kummer, H. (1968). *Social organization of hamadryas baboons* . Chicago: University of Chicago Press.

Kummer, H. (1971). *Primate societies.* Arlington Heights, IL: Harlan Davidson.

Kummer, H. (1982). Social knowledge in free-ranging primates. In D.R.Griffin (Ed.), *Animal mind–human mind* (pp. 113–130). Berlin: Springer-Verlag.

Kummer, H. (1995). *In quest of the sacred baboon.* Princeton, NJ: Princeton University Press.

Kummer, H., Anzenberger, G., & Hemelrijk, C.K. (1996). Hiding and perspective taking in long-tailed macaques (*Macaca fascicularis*). *Journal of Comparative Psychology, 110*, 97–102.

Kummer, H. & Cords, M. (1991). Cues of ownership in long-tailed macaques, *Macaca fascicularis. Animal Behaviour, 42*, 529–549.

Kummer, H. & Goodall, J. (1985). Conditions of innovative behaviour in primates. *Philosophical Transactions of the Royal Society of London, B308*, 203–214.

Kummer, H., Goetz, W., & Angst, W. (1974). Triadic differentiation: an inhibitory process protecting pair bonds in baboons. *Behaviour, 49*, 62–87.

Kummer, H. & Kurt, F. (1965). A comparison of social behavior in captive and wild hamadryas baboons. In H. Vagtborg (Ed.), *The baboon in medical research* (pp. 64–80). Austin: University of Texas.

Kuroda, R. (1931). On the counting ability of a monkey (*Macacus cynomolgus*). *Journal of Comparative Psychology, 12*, 171–180.

Lakoff, G. (1987). *Women, fire, and dangerous things.* Chicago: University of Chicago Press.

Langacker, R. (1987). *Foundations of cognitive grammar, vol. 1.* Stanford, CA: Stanford University Press.

Langer, J. (1980). *The origins of logic: six to twelve months.* New York: Academic Press.

Langer, J. (1986). *The origins of logic: one to two years*. New York: Academic Press.

Larson, C., Sutton, D., Taylor, E., & Lindeman, R. (1973). Sound spectral properties of conditioned vocalizations in monkeys. *Phonetica, 27*, 100–112.

Leakey, R. (1994). *The origin of humankind*. New York: Basic Books.

Ledbetter, D.H. & Basen, J.A. (1982). Failure to demonstrate self-recognition in gorillas. *American Journal of Primatology, 2*, 307–310.

Lefebvre, L. (1982). Food exchange strategies in an infant chimpanzee. *Journal of Human Evolution, 11*, 195–204.

Lefebvre, L. (1995). Culturally-transmitted feeding behaviour in primates: evidence for accelerating learning rates. *Primates, 36*, 227–239.

Lefebvre, L. & Hewitt, T.A. (1986). Food exchange in captive chimpanzees. In D.M. Taub & F.A. King (Eds.), *Current perspectives in primate social dynamics* (pp. 476–486). New York: Van Nostrand Reinhold.

Leger, D.W. & Owings, D.H. (1978). Responses to alarm calls by California ground squirrels. *Behavioral Ecology and Sociobiology, 3*, 177–186.

Lepoivre, H. & Pallaud, B. (1986). Learning set formation in a group of baboons (*Papio papio*). *American Journal of Primatology, 10*, 25–36.

Leslie, A.M. (1987). Pretense and representation: the origins of "theory of mind." *Psychological Review, 94*, 412–426.

Lethmate, J. (1982). Tool-using skills of orang-utans. *Journal of Human Evolution, 11*, 49–64.

Lethmate, J. & Dücker, G. (1973). Untersuchungen zum selbsterkennen im spiegel bei orang-utans und einigen anderen affenarten. *Zeitschrift für Tierpsychology, 33*, 248–269.

Lewis, M., Sullivan, M., Stanger, C., & Weiss, M. (1989). Self-development and self-conscious emotions. *Child Development, 60*, 146–156.

Lillard, A.S. (1993). Young children's conceptualization of pretense: action or mental representational state? *Child Development, 64*, 372–386.

Limongelli, L., Boysen, S.T., & Visalberghi, E. (1995). Comprehension of cause-effect relations in a tool-using task by chimpanzees (*Pan troglodytes*). *Journal of Comparative Psychology, 109*, 18–26.

Lin, A.C., Bard, K.A., & Anderson, J.R. (1992). Development of self-recognition in chimpanzees (*Pan troglodytes*). *Journal of Comparative Psychology, 106*, 120–127.

Lipsett, L.P. & Serunian, S.A. (1963). Oddity-problem learning in young children. *Child Development, 34*, 201–206.

Lock, A. (1978). *Action, gesture and symbol: the emergence of language*. New York: Academic Press.

Lorenz, K. (1965). *Evolution and the modification of behavior.* Chicago: University of Chicago Press.

Loveland, K.A. (1993). Autism, affordances, and the self. In U. Neisser (Ed.), *The perceived self* (pp. 237–253). Cambridge: Cambridge University Press.

MacDonald, S.E. (1994). Gorillas' (*Gorilla gorilla gorilla*) spatial memory in a foraging task. *Journal of Comparative Psychology, 108*, 107–113.

MacDonald, S.E., Pang, J.C., & Gibeault, S. (1994). Marmoset (*Callithrix jacchus jacchus*) spatial memory in a foraging task: win-stay versus win-shift strategies. *Journal of Comparative Psychology, 108*, 328–334.

MacDonald, S.E. & Wilkie, D.M. (1990). Yellow-nosed monkeys' (*Cercopithecus ascanius whitesidei*) spatial memory in a simulated foraging environment. *Journal of Comparative Psychology, 104*, 382–387.

Macedonia, J. (1990). What is communicated in the antipredator calls of lemurs: evidence from playback experiments with ringtailed and ruffed lemurs. *Ethology, 86*, 177–190.

Macedonia, J.M. & Evans, C.S. (1993). Variation among Mammalian alarm call systems and the problem of meaning in animal signals. *Ethology, 93*, 177–197.

Macedonia, J.M. & Yount, P.L. (1991). Auditory assessment of avian predator threat in semi-captive ringtailed lemurs (*Lemur catta*). *Primates, 32*, 169–182.

Machida, S. (1990). Standing and climbing a pole by members of a captive group of Japanese monkeys. *Primates, 31*, 291–298.

Mackinnon, J. (1974). The behaviour and ecology of wild oran-utans (*Pongo pygmaeus*). *Animal Behaviour, 22*, 3–74.

MacPhail, E.M. (1987). The comparative psychology of intelligence. *Behavioral and Brain Sciences, 10*, 645–695.

Maeda, T. & Masataka, N. (1987). Locale-specific behavior of the tamarin (*Saguinus labiatus*). *Ethology, 75*, 25–30.

Maestripieri, D. (1993). Vigilance costs of allogrooming in macaque mothers. *American Naturalist, 141*, 744–753.

Maestripieri, D. (1995a). Assessment of danger to themselves and their infants by rhesus macaque (*Macaca mulatta*) mothers. *Journal of Comparative Psychology, 109*, 416–420.

Maestripieri, D. (1995b). First steps in the macaque world: do rhesus mothers encourage their infants' independent locomotion? *Animal Behaviour, 49*, 1541–1549.

Maestripieri, D. (1995c). Maternal encouragement in nonhuman primates and the question of animal teaching. *Human Nature, 6*, 361–378.

Maestripieri, D. (1996a). Gestural communication and its cognitive implications in pigtail macaques (*Macaca nemestrina*). Behaviour, 133, 997–1022.

Maestripieri, D. (1996b). Maternal encouragement of infant locomotion in pigtail macaques, *Macaca nemestrina*. *Animal Behaviour, 51*, 603–610.

Maestripieri, D. (1996c). Social communication among captive stump-tailed macaques (*Macaca arctoides*). *International Journal of Primatology*, 17, 785–802.

Maestripieri, D. & Call, J. (1996). Mother-infant communication in primates. In: J.S. Rosenblatt & C.T. Snowdon (Eds.), *Advances in the study of behavior* (pp. 613–642). New York: Academic Press.

Maestripieri, D. (in press). Primate social organization, vocabulary size, and communication dynamics: a comparative study of macaques. In B. King (Ed.), *The evolution of language: assessing the evidence from nonhuman primates*. Santa Fe; NM: School of American Research.

Mahan Jr., J.L. & Rumbaugh, D.M. (1963). Observational learning in the squirrel monkey. *Perceptual and Motor Skills, 17*, 686–686.

Maier, N.R.F. & Schneirla, T.C. (1935). *Principles of animal psychology*. New York: McGraw-Hill.

Mandler, J. (1992). How to build a baby II: conceptual primitives. *Psychological Review, 99*, 587–604.

Marais, E. (1969). *The soul of the ape*. New York: Atheneum.

Marino, L., Reiss, D., & Gallup, G.G., Jr. (1994). Mirror self-recognition in bottlenose dolphins: Implications for comparative investigations of highly dissimilar species. In S.T. Parker, R.W. Mitchell, & M.L. Boccia (Eds.), *Self-awareness in animals and humans. Developmental perspectives* (pp. 380–391). Cambridge: Cambridge University Press.

Marler, P. (1976). Social organization, communication and graded signals: the chimpanzee and the gorilla. In P.P.G. Bateson & R.A. Hinde (Eds.), *Growing points in ethology* (pp. 239–280). Cambridge: Cambridge University Press.

Marler, P., Dufty, A., & Pickert, R. (1986a) Vocal communication in the domestic chicken: I. Does a sender communicate information about the quality of a food referent to a receiver? *Animal Behaviour, 34*, 188–193.

Marler, P., Dufty, A., & Pickert, R. (1986b). Vocal communication in the domestic chicken: II. Is a sender sensitive to the presence and nature of a receiver? *Animal Behaviour, 34*, 194–198.

Marler, P., Evans, C., & Hauser, M. (1992). Animal signals: motivational, referential, or both? In

H. Papousek, U. Jurgens, & M. Papousek (Eds.), *Nonverbal communication: comparative and developmental approaches* (pp. 66–85). Cambridge: Cambridge University Press.

Marsh, C.W. (1981). Ranging behaviour and its relation to diet selection in Tana River red colobus (*Colobus badius rufomitratus*). *Journal of Zoology, 195,* 473–492.

Marten, K. & Psarakos, S. (1994). Evidence of self-awareness in the bottlenose dolphin (Tursiops truncatus). In S.T. Parker, R.W. Mitchell, & M.L. Boccia (Eds.), *Self-awareness in animals and humans. Developmental perspectives* (pp. 361–379). Cambridge: Cambridge University Press.

Martin, R.D. (1990). *Primate origins and evolution. A phylogenetic reconstruction.* Princeton, NJ: Princeton University Press.

Masataka, N. (1983). Categorical responses to natural and synthesized alarm calls in Goeldi's monkeys (*Callimico goeldii*). *Primates, 24,* 40–51.

Masataka, N. (1986). Rudimentary representational vocal signalling of fellow group members in spider monkeys. *Behaviour, 96,* 49–61.

Masataka, N. (1989). Motivational referents of contact calls in Japanese monkeys. *Ethology, 80,* 265–273.

Masataka, N. & Biben, M. (1987). Temporal rules regulating affiliative vocal exchanges of squirrel monkeys. *Behaviour, 101,* 311–319.

Masataka, N. & Fujita, K. (1989). Vocal learning of Japanese and rhesus monkeys. *Behaviour, 109,* 191–199.

Maslow, A.H. & Harlow, H.F. (1932). Comparative behavior of primates. II. Delayed reaction tests on primates at Bronx park zoo. *Journal of Comparative Psychology, 14,* 97–107.

Mason, W.A. (1963). The effects of environmental restriction on the social development of rhesus monkeys. In C.H. Southwick (Ed.), *Primate social behavior* (pp. 161–173). New York: Van Nostrand Reinhold.

Mason, W.A. (1968). Use of space by Callicebus groups. In P.C. Jay (Ed.), *Primates: studies in adaptation and variability* (pp. 200–216). New York: Holt, Rinehart and Winston.

Mason, W.A. & Hollis, J.H. (1962). Communication between young rhesus monkeys. *Animal Behaviour, 10,* 211–221.

Mathieu, M. (1982). Intelligence without language: Piagetian assessment of cognitive development in chimpanzee. Paper presented at the joint meeting of the IX Congress of the International Primatological Society and the meeting of the American Society of Primatologists, August, Atlanta, GA.

Mathieu, M. & Bergeron, G. (1981). Piagetian assessment on cognitive development in chimpanzee (*Pan troglodytes*). In A.B. Chiarelli & R.S. Corruccini (Eds.), *Primate behavior and sociobiology* (pp. 142–147). Berlin: Springer-Verlag.

Mathieu, M., Bouchard, M.A., Granger, L., & Herscovitch, J. (1976). Piagetian object-permanence in *Cebus capucinus, Lagothrica flavicauda* and *Pan troglodytes. Animal Behaviour, 24,* 585–588.

Mathieu, M., Daudelin, N., Dagenais, Y., & Decarie, T.G. (1980). Piagetian causality in two house-reared chimpanzees (*Pan troglodytes*). *Canadian Journal of Psychology, 34,* 179–186.

Matsuzawa, T. (1985). Use of numbers by a chimpanzee. *Nature, 315,* 57–59.

Matsuzawa, T. (1990). Spontaneous sorting in human and chimpanzee. In S.T. Parker & K.R. Gibson (Eds.), *"Language" and intelligence in monkeys and apes* (pp. 451–468). Cambridge: Cambridge University Press.

Matsuzawa, T. (1991). Nesting cups and metatools in chimpanzees. *Behavioral and Brain Sciences, 14,* 570–571.

Mayagoitia, A., Santillan-Doherty, L., Lopez-Vergara, & Mondragon-Ceballos, R. (1993). Affiliation tactics prior to a period of competition in captive groups of stumptail macaques. *Ethology, Ecology & Evolution,* 5, 435–446.

McClure, M.K. & Culbertson, G. (1977). Chimpanzees' spontaneous temporal sorting of stimuli on the basis of physical identity *Primates, 18*, 709–711.

McClure, M.K. & Helland, J. (1979). A chimpanzee's use of dimensions in responding same and different. *Psychological Record, 29*, 371–378.

McEwen, P. (1986). Environmental enrichment: an artificial termite mound for orang utans. Universities Federation for Animal Welfare, *3*, 1–19.

McGonigle, B. & Chalmers, M. (1992). Monkeys are rational! *Quarterly Journal of Experimental Psychology B, 45B*, 189–228.

McGonigle, B.O. & Chalmers, M. (1977). Are monkeys logical? *Nature, 267*, 694–696.

McGrew, W.C. (1974). Tool use by wild chimpanzees in feeding upon driver ants. *Journal of Human Evolution, 3*, 501–508.

McGrew, W.C. (1977). Socialization and object manipulation of wild chimpanzees. In S. Chevalier-Skolnikoff & F.E. Poirier (Eds.), *Primate bio-social development* (pp. 261–288). New York: Garland Press.

McGrew, W.C. (1983). Animal foods in the diets of wild chimpanzees. *Journal of Ethology, 1*, 46–61.

McGrew, W.C. (1989). Why is ape tool use so confusing? In V. Standen & R.A. Foley (Eds.), *Comparative socioecology. The behavioural ecology of humans and other mammals* (pp. 457–472). Oxford: Blackwell Scientific.

McGrew, W.C. (1992). *Chimpanzee material culture*. Cambridge: Cambridge University Press.

McGrew, W.C. & Tutin, C.E.G. (1973). Chimpanzee tool use in dental grooming. *Nature, 241*, 477–478.

McGrew, W.C. & Tutin, C.E.G. (1978). Evidence for a social custom in wild chimpanzees? *Man, 13*, 234–251.

McGrew, W.C., Tutin, C.E.G., & Baldwin, P.J. (1979). Chimpanzees, tools, and termites: cross-cultural comparisons of Senegal, Tanzania, and Rio Muni. *Man, 14*, 185–214.

Meador, D.M., Rumbaugh, D.M., Pate, J.L., & Bard, K.A. (1987). Learning, problem solving, cognition, and intelligence. In G. Mitchell & J. Erwin (Eds.), *Comparative primate biology: behavior, cognition, and motivation* (pp. 17–83). New York: Alan R. Liss.

Medin, D.L. (1969). Form perception and pattern reproduction by monkeys. *Journal of Comparative and Physiological Psychology, 68*, 412–419.

Mellars, P. (1989). Technological changes at the Middle-Upper Paleolithic transition. In P. Mellars & C. Stringer (Eds.), *The human revolution: behavioral and biological perspectives on the origins of modern humans* (pp. 338–365). Princeton: Princeton University Press.

Meltzoff, A. (1988). Infant imitation after a one-week delay: long-term memory for novel acts and multiple stimuli. *Developmental Psychology, 24*, 470–76.

Meltzoff, A. (1995). Understanding the intentions of others: Re-enactment of intended acts by 18–month-old children. *Developmental Psychology, 31*, 838–50.

Meltzoff, A.N. & Borton, R.W. (1979). Intermodal matching by human neonates. *Nature, 282*, 403–404.

Meltzoff, A. & Gopnik, A. (1993). The role of imitation in understanding persons and developing a theory of mind. In S. Baron-Cohen, H. Tager-Flusberg, & D. Cohen (Eds.), *Understanding other minds* (pp. 335–366). Oxford University Press.

Meltzoff, A., & Moore, K. (1977). Imitation of facial and manual gestures by newborn infants. *Science, 198*, 75–78.

Meltzoff, A., & Moore, K. (1989) Imitation in newborn infants: Exploring the range of gestures imitated and the underlying mechanisms. *Developmental Psychology* 25: 954–962.

Mendelson, M.J., Haith, M.M., & Goldman-Rakic, P.S. (1982). Face scanning and responsiveness to social cues in infant rhesus monkeys. *Developmental Psychology, 18*, 222–228.

Menzel, C.R. (1991). Cognitive aspects of foraging in Japanese monkeys. *Animal Behaviour, 41*, 397–402.

Menzel, E.W. Jr. (1964). Patterns of responsiveness in chimpanzees reared through infancy under conditions of environmental restriction. *Psychologische Forschung, 27*, 337–365.

Menzel, E.W. Jr. (1969). Responsiveness to food and signs of food in chimpanzee discrimination learning. *Journal of Comparative and Physiological Psychology, 68*, 484–489.

Menzel, E.W. Jr. (1971a). Communication about the environment in a group of young chimpanzees. *Folia Primatologica, 15*, 220–232.

Menzel, E.W. Jr. (1971b). Group behavior in young chimpanzees: Responsiveness to cumulative novel changes in a large outdoor enclosure. *Journal of Comparative and Physiological Psychology, 74*, 46–51.

Menzel, E.W. Jr. (1972). Spontaneous invention of ladders in a group of young chimpanzees. *Folia Primatologica, 17*, 87–106.

Menzel, E.W. Jr. (1973a). Chimpanzee spatial memory organization. *Science, 182*, 943–945.

Menzel, E.W. Jr. (1973b). Further observations on the use of ladders in a group of young chimpanzees. *Folia Primatologica, 19*, 450–457.

Menzel, E.W. Jr. (1973c). Leadership and communication in young chimpanzees. In E.W. Menzel, Jr. (Ed.), *Precultural primate behavior* (pp. 192–225). Basel, Switzerland: Karger.

Menzel, E.W. Jr. (1974). A group of young chimpanzees in a one-acre field: leadership and communication. In A.M. Schrier & F. Stollnitz (Eds.), *Behavior of nonhuman primates* (pp. 83–153). New York: Academic Press.

Menzel, E.W. Jr. (1979). Communication of objects-locations in a group of young chimpanzees. In D.A. Hamburg & E.R. McCown (Eds.), *The great apes* (pp. 359–371). Menlo Park, CA: Benjamin Cummings.

Menzel, E.W. Jr., Davenport, R.K., & Rogers, C.M. (1972). Protocultural aspects of chimpanzees' responsiveness to novel objects. *Folia Primatologica, 17*, 161–170.

Menzel, E.W. Jr. & Juno, C. (1985). Social foraging in marmoset monkeys and the question of intelligence. *Philosophical Transactions of the Royal Society of London. B, 308*, 145–158.

Menzel, E.W. Jr. & Menzel, C.R. (1979). Cognitive, developmental and social aspects of responsiveness to novel objects in a family group of marmosets (*Saguinus fuscicollis*). *Behaviour, 70*, 251–279.

Menzel, E.W. Jr., Premack, D., & Woodruff, G. (1978). Map reading by chimpanzees. *Folia Primatologica, 29*, 241–249.

Menzel, E.W. Jr., Savage-Rumbaugh, E.S., & Lawson, J. (1985). Chimpanzee (*Pan troglodytes*) spatial problem solving with the use of mirrors and televised equivalents of mirrors. *Journal of Comparative Psychology, 99*, 211–217.

Menzel, E.W. Jr. & Wyers, E.J. (1981). Cognitive aspects of foraging behavior. In A.C. Kamil & T.D. Sargent (Eds.), *Foraging behavior. Ecological, ethological, and psychological approaches* (pp. 355–376). New York: Garland Publishing.

Meyer, D.R. & Harlow, H.F. (1949). The development of transfer of response to patterning by monkeys. *Journal of Comparative Psychology, 42*, 454–462.

Mignault, C. (1985). Transition between sensorimotor and symbolic activities in nursery-reared chimpanzees (*Pan troglodytes*). *Journal of Human Evolution, 14*, 747–758.

Miles, H.L. (1983). Apes and language: the search for communicative competence. In J. De Luce & H.T. Wilde (Eds.), *Language in primates: perspectives and implications* (pp. 43–61). New York: Springer-Verlag.

Miles, H.L.W. (1990). The cognitive foundations for reference in a signing orangutan. In S.T. Parker & K.R. Gibson (Eds.), *"Language" and intelligence in monkeys and apes* (pp. 511–539). Cambridge: Cambridge University Press.

Miles H.L., Mitchell, R.W. & Harper, S.E. (1996). Simon says: the development of imitation in an enculturated oraugutan. In: A.E. Russon, K.A. Bard, and S.T. Parker (Eds.), *Reaching into thought: the minds of the great apes* (pp. 278–299). Cambridge: Cambridge University Press.

Miles, H.L.W. (1994). ME CHANTEK: the development of self-awareness in a signing orangutan. In S.T. Parker, R.W. Mitchell, & M.L. Boccia (Eds.), *Self-awareness in animals and humans. Developmental perspectives* (pp. 254–272). Cambridge: Cambridge University Press.

Miles, R.C. (1957). Learning-set formation in the squirrel monkey. *Journal of Comparative Psychology, 50*, 356–357.

Miles, R.C. & Meyer, D.R. (1956). Learning sets in marmosets. *Journal of Comparative Psychology, 49*, 219–222.

Miller, R.E., Murphy, J.V., & Mirsky, I.A. (1959). Relevance of facial expression and posture as cues in communication of affect between monkeys. *Archives of General Psychiatry, 1*, 480–488.

Millward, R.B. (1971). Theoretical and experimental approaches to human learning. In Kling and Riggs (Eds.), *Experimental Psychology* (pp. 123–155). New York: Holt, Reinhart, & Winston.

Milton, K. (1980). *The foraging strategy of howler monkeys. A study in primate economics.* New York: Columbia University Press.

Milton, K. (1981). Distribution patterns of tropical plant foods as an evolutionary stimulus to primate mental development. *American Anthropologist, 83*, 534–548.

Milton, K. (1984). The role of food-processing factors in primate food choice. In P.S. Rodman & J.G.H. Cant (Eds.), *Adaptations for foraging in nonhuman primates* (pp. 249–279). New York: Columbia University Press.

Milton, K. (1988). Foraging behaviour and the evolution of primate intelligence. In R.W. Byrne & A. Whiten (Eds.), *Machiavellian intelligence. Social expertise and the evolution of intellect in monkeys, apes, and humans* (pp. 285–305). New York: Oxford University Press.

Milton, K. (1993). Diet and primate evolution. *Scientific American*, August, 86–93.

Mineka, S. & Cook, M. (1988). Social learning and the adquisition of snake fear in monkeys. In T.R. Zentall & B.G. Galef, Jr. (Eds.), *Social learning: psychological and biological perspectives* (pp. 51–73). Hillsdale, NJ: Lawrence Erlbaum Associates.

Mineka, S. & Cook, M. (1993). Mechanisms involved in the observational conditioning of fear. *Journal of Experimental Psychology: General, 122*, 23–28.

Mitani, J.C. (1985). Sexual selection and adult male orangutan long calls. *Animal Behaviour, 33*, 272–283.

Mitani, J.C. (1994). Ethological studies of chimpanzee vocal behavior. In R.W. Wrangham, W.C. McGrew, F.B.M. de Waal, & P.G. Heltne (Eds.), *Chimpanzee cultures* (pp. 195–210). Cambridge, MA: Harvard University Press.

Mitani, J.C., Hasegawa, T., Gros-Louis, J., Marler, P., & Byrne, R. (1992). Dialects in chimpanzees? *American Journal of Primatology, 27*, 233–243.

Mitani, J.C. & Nishida, T. (1993). Contexts and social correlates of long-distance calling by male chimpanzees. *Animal Behaviour, 45*, 735–746.

Mitchell, R.W. & Anderson, J.R. (1993). Discrimination learning of scratching, but failure to obtain imitation and self-recognition in a long-tailed macaque. *Primates, 34*, 301–309.

Moss, C.J. (1988). *Elephant memories*. Boston: Houghton Mifflin.

Mounoud, P. & Bower, T.G.R. (1974). Conservation of weight in infants. *Cognition, 3*, 29–40.

Muncer, S.J. (1983). "Conservations" with a chimpanzee. *Developmental Psychobiology, 16*, 1–11.

Muroyama, Y. (1991). Mutual reciprocity of grooming in female Japanese macaques (*Macaca fuscata*). *Behaviour, 119*, 161–170.

Muroyama, Y. (1994). Exchange of grooming for allomothering in female patas monkeys. *Behaviour, 128*, 103–119.

Myers, W.A. (1970). Observational learning in monkeys. *Journal of the Experimental Analysis of Behavior, 14*, 225–235.

Myowa, M. (1996). Imitation of facial gestures by an infant chimpanzee. *Primates, 37*, 207–213.

Nagell, K., Olguin, K., & Tomasello, M. (1993). Processes of social learning in the tool use of chimpanzees (*Pan troglodytes*) and human children (*Homo sapiens*). *Journal of Comparative Psychology, 107*, 174–186.

Nakagawa, E. (1993). Relational rule learning in the rat. *Psychobiology, 21*, 293–298.

Nakamichi, M., Kato, E., Kojima, Y., & Itoigawa, N. (in press). Carrying and washing grass roots by Japanese macaques in a free-ranging group at Katsuyama. *Folia Primatologica.*

Nash, V.J. (1982). Tool use by captive chimpanzees at an artificial termite mound. *Zoo Biology, 1*, 211–221.

Natale, F. (1989a). Causality II: The stick problem. In F. Antinucci (Ed.), *Cognitive structure and development in nonhuman primates* (pp. 121–133). Hilldsale, NJ: Lawrence Erlbaum Associates.

Natale, F. (1989b). Patterns of object manipulation. In F. Antinucci (Ed.), *Cognitive structure and development in nonhuman primates* (pp. 145–161). Hilldsale, NJ: Lawrence Erlbaum Associates.

Natale, F. (1989c). Stage 5 object-concept. In F. Antinucci (Ed.), *Cognitive structure and development in nonhuman primates* (pp. 89–95). Hilldsale, NJ: Lawrence Erlbaum Associates.

Natale, F. & Antinucci, F. (1989). Stage 6 object-concept and representation. In F. Antinucci (Ed.), *Cognitive structure and development in nonhuman primates* (pp. 97–112). Hilldsale, NJ: Lawrence Erlbaum Associates.

Natale, F., Antinucci, F., Spinozzi, G., & Potì, P. (1986). Stage 6 object concept in nonhuman primate cognition: A comparison between gorilla (*Gorilla gorilla gorilla*) and Japanese macaque (*Macaca fuscata*). *Journal of Comparative Psychology, 100*, 335–339.

Natale, F., Potì, P., & Spinozzi, G. (1988). Development of tool use in a macaque and a gorilla. *Primates, 29*, 413–416.

Neisser, U. (1967). *Cognitive psychology.* New York: Appleton-Century-Crofts.

Neisser, U. (1984). Toward an ecologically oriented Cognitive Science. In T. Schlecter & M. Toglia (Eds.), *New direction in cognitive science* (pp. 17–32). Norwood, NJ: Ablex.

Neisser, U. (1988). Five kinds of self-knowledge. *Philosophical Psychology, 1*, 35–59.

Neisser, U. (1993a). *The perceived self. Ecological and interpersonal sources of self-knowledge.* Cambridge: Cambridge University Press.

Neisser, U. (1993b). The self perceived. In U. Neisser (Ed.), *The perceived self* (pp. 3–21). Cambridge: Cambridge University Press.

Neiworth, J.J. & Wright, A.A. (1994). Monkeys (*Macaca mulatta*) learn category matching in a nonidentical same-different task. *Animal Behavior Processes, 20*, 429–435.

Nelson, K. (1985). *Making sense: the acquisition of shared meaning.* New York: Academic Press.

Nelson, K. (1986). *Event knowledge: structure and function in development.* Hillsdale, NJ: Lawrence Erlbaum Associates.

Nelson, K. (1987). What's in a name? Reply to Seidenberg and Petitto. *Journal of Experimental Psychology: General, 116*, 293–296.

Nelson, K. (1996). *Language in cognitive development.* New York: Cambridge University Press.

Newman, J. & Symmes, D. (1974). Vocal pathology in socially deprived monkeys. *Developmental Psychobiology, 7*, 351–358.

Nicholson, I.S. & Gould, J.E. (1995). Mirror mediated object disrimination and self-directed behavior in a female gorilla. *Primates, 36*, 515–521.

Nishida, T. (1973). The ant-gathering behaviour by the use of tools among wild chimpanzees of the Mahali mountains. *Journal of Human Evolution, 2*, 357–370.

Nishida, T. (1980). The leaf-clipping display: a newly discovered expressive gesture in wild chimpanzees. *Journal of Human Evolution, 9*, 117–128.

Nishida, T. (1987a). Development of social grooming between mother and offspring in wild chimpanzees. *Folia Primatologica, 50*, 109–123.

Nishida, T. (1987b). Local traditions and cultural transmission. In B.B. Smuts, D.L. Cheney, R.M. Seyfarth, R.W. Wrangham, & T.T. Struhsaker (Eds.), *Primate societies* (pp. 462–474). Chicago: University of Chicago Press.

Nishida, T. (1994). Review of recent finding on Mahale chimpanzees: implications and future research directions. In R.W. Wrangham, W.C. McGrew, F.B.M. de Waal, & P.G. Heltne (Eds.), *Chimpanzee cultures* (pp. 373–396). Cambridge, MA: Harvard University Press.

Nishida, T., Hasegawa, T., Hayaki, H., Takahata, Y., & Uehara, S. (1992). Meat-sharing as a coalition strategy by an alpha male chimpanzee? In T. Nishida, W.C. McGrew, P. Marler, M. Pickford, & F.B.M. de Waal (Eds.), *Topics in primatology. Human origins* (pp. 159–174). Tokyo: University of Tokyo Press.

Nishida, T. & Hiraiwa, M. (1982). Natural history of a tool-using behavior by wild chimpanzees in feeding upon wood-boring ants. *Journal of Human Evolution, 11*, 73–99.

Nishida, T. & Hiraiwa-Hasegawa, M. (1987). Chimpanzees and bonobos: cooperative relationships among males. In B.B. Smuts, D.L. Cheney, R.M. Seyfarth, R.W. Wrangham, & T.T. Struhsaker (Eds.), *Primate societies* (pp. 165–177). Chicago: University of Chicago Press.

Nishida, T. & Nakamura, M. (1993). Chimpanzee tool use to clear a blocked nasal passage. *Folia Primatologica, 61*, 218–220.

Nishida, T., Wrangham, R.W., Goodall, J., & Uehara, S. (1983). Local differences in plant-feeding habits of chimpanzees between the Mahale Mountains and Gombe National Park, Tanzania. *Journal of Human Evolution, 12*, 467–480.

Nissen, H.W., Blum, J.S., & Blum, R.A. (1948). Analysis of matching behavior in chimpanzee. *Journal of Comparative and Physiological Psychology, 41*, 62–74.

Noble, W. & Davidson, I. (in press). *Human evolution, language, and mind: A psychological and archeological inquiry.* New York: Cambridge University Press.

Noë, R. (1992). Alliance formation among male baboons: shopping for profitable partners. In A.H. Harcourt & F.B.M. de Waal (Eds.), *Coalitions and alliances in humans and other animals* (pp. 285–321). New York: Oxford University Press.

Novak, M. (1996). Self-recognition in rhesus macaques. Paper presented to the International Primatological Society. Madison, Wisconsin.

Nuckolls, C. (1991). Culture and causal thinking. *Ethos, 17*, 3–51.

Oden, D.L., Thompson, R.K.R., & Premack, D. (1988). Spontaneous transfer of matching by infant chimpanzees (*Pan troglodytes*). *Animal Behavior Processes, 14*, 140–145.

Oden, D.L., Thompson, R.K.R., & Premack, D. (1990). Infant chimpanzees spontaneously perceive both concrete and abstract same/ different relations. *Child Development, 61*, 621–631.

Ohta, H. (1983). Learning set formation in slow lorises (*Nycticebus coucang*). *Folia Primatologica, 40*, 256–267.

Olguin, R. & Tomasello, M. (1993). Twenty-five-month-old children do not have a grammatical category of verb. *Cognitive Development, 8*, 245–272.

Olton, D.S. & Samuelson, R.J. (1976). Remembrance of places passed: spatial memory in rats. *Journal of the Experimental Analysis of Behavior, 2*, 97–116.

Owings, D. & Hennessy, D. (1984). The importance of variation in sciurid visual and vocal communication. In J.O. Murie & G.R. Michener (Eds.), *Biology of ground dwelling squirrels: annual cycles, behavioral ecology and sociality* (pp. 4–29). Lincoln: Nebraska University Press.

Owren, M.J. & Dieter, J.A. (1989). Infant cross-fostering between Japanese (*Macaca fuscata*) and rhesus macaques (*Macaca mulatta*). *American Journal of Primatology, 18*, 245–250.

Owren, M.J. (1990a). Acoustic classification of alarm calls by vervet monkeys (*Cercopithecus aethiops*) and humans (*Homo sapiens*): I. Natural calls. *Journal of Comparative Psychology, 104*, 20–28.

Owren, M.J. (1990b). Acoustic classification of alarm calls by vervet monkeys (*Cercopithecus aethiops*) and humans (*Homo sapiens*): II. Synthetic calls. *Journal of Comparative Psychology, 104*, 29–40.

Owren, M.J., Dieter, J.A., Seyfarth, R.M., & Cheney, D.L. (1992a). Evidence of limited modification in the vocalizations of cross-fostered rhesus (*Macaca mulatta*) and Japanese (*M. fuscata*) macaques. In T. Nishida, W.C. McGrew, P. Marler, M. Pickford, & F.B.M. de Waal (Eds.), *Topics in primatology. Human origins* (pp. 257–270). Tokyo: University of Tokyo Press.

Owren, M.J., Dieter, J.A., Seyfarth, R.M., & Cheney, D.L. (1992b). 'Food' calls produced by adult female rhesus (*Macaca mulatta*) and Japanese (*M. fuscata*) macaques, their normally raised offspring, and offspring cross-fostered between species. *Behavior, 120*, 218–231.

Owren, M.J., Dieter, J.A., Seyfarth, R.M., & Cheney, D.L. (1993). Vocalizations of rhesus (*Macaca mulatta*) and Japanese (*M. fuscata*) macaques cross-fostered between species show evidence of only limited modification. *Developmental Psychobiology, 26*, 389–406.

Oyen, O.J. (1979). Tool-use in free-ranging baboons of Nairobi National Park. *Primates, 20*, 595–597.

Packer, C. (1977). Reciprocal altruism in *Papio anubis*. *Nature, 265*, 441–443.

Packer, C. (1994). *Into Africa*. Chicago: University of Chicago Press.

Packer, C. & Ruttan, L. (1988). The evolution of cooperative hunting. *American Naturalist, 132*, 159–198.

Paquette, D. (1992a). Discovering and learning tool-use for fishing honey by captive chimpanzees. *Human Evolution, 7*, 17–30.

Paquette, D. (1992b). Object exchange between captive chimpanzees: a case report. *Human Evolution, 7*, 11–15.

Parish, A.R. (1994). Sex and food control in the "uncommon chimpanzee": How bonobo females overcome a phylogenetic legacy of male dominance. *Ethology and Sociobiology, 15*, 157–179.

Parker, C.E. (1969). Responsiveness, manipulation, and implementation behavior in chimpanzees, gorillas, and orang-utans. In C.R. Carpenter (Ed.), *Proceedings of second international congress on primatology* (pp. 160–166). Basel: Karger.

Parker, C.E. (1973). Manipulatory behavior and responsiveness. In D.M. Rumbaugh (Ed.), *Gibbon and siamang* (pp. 185–207). Basel: Karger.

Parker, C.E. (1974a). The antecedents of man the manipulator. *Journal of Human Evolution, 3*, 493–500.

Parker, C.E. (1974b). Behavioral diversity in ten species of nonhuman primates. *Journal of Comparative and Physiological Psychology, 87*, 930–937.

Parker, S.T. (1977). Piaget's sensorimotor period series in an infant macaque: A model for comparing unstereotyped behavior and intelligence in human and nonhuman primates. In S. Chevalier-Skolnikoff & F.E. Poirier (Eds.), *Primate bio-social development* (pp. 43–112). New York: Garland Press.

Parker, S.T. & Gibson, K.R. (1977). Object manipulation, tool use and sensorimotor intelligence as feeding adaptations in Cebus monkeys and great apes. *Journal of Human Evolution, 6*, 623–641.

Parker, S.T. & Gibson, K.R. (1979). A developmental model for the evolution of language and intelligence in early hominids. *Behavioral and Brain Sciences, 2*, 367–408.

Parker, S.T., Mitchell, R.W., & Boccia, M.L. (1994). *Self-awareness in animals and humans. Developmental perspectives*. Cambridge: Cambridge University Press.

Parker, S.T. & Potì, P. (1990). The role of innate motor patterns in ontogenetic and experential development of intelligent use of sticks in cebus monkeys. In S.T. Parker & K.R. Gibson (Eds.), *"Language" and intelligence in monkeys and apes* (pp. 219–243). Cambridge: Cambridge University Press.

Pasnak, R. (1979). Acquisition of prerequisites to conservation by macaques. *Journal of Experimental Psychology: Animal Behavior Processes, 5*, 194–210.

Passingham, R.E. (1981). Primate specialization in brain and intelligence. *Symposium of the Zoological Society of London, 46*, 361–388.

Passingham, R.E. (1982). *The human primate*. San Francisco: W.H. Freeman.

Pastore, N. (1961). Number sense and "counting" ability in the canary. Zietschrift für Tierpsychologie, 18, 561–573.

Patterson, F. (1978). Linguistic capabilities of a lowland gorilla. In F.C.C. Peng (Ed.), *Sign language and language acquisition in man and ape* (pp. 161–201). Boulder, CO: Westview Press.

Patterson, F. (1990). Mirror behavior and self-concept in the lowland gorilla (Abstract). *American Journal of Primatology, 20*, 219–220.

Patterson, F. & Linden, E. (1981). *The education of Koko*. New York: Owl books.

Patterson, F.G.P. & Cohn, R.H. (1994). Self-recognition and self-awareness in lowland gorillas. In S.T. Parker, R.W. Mitchell, & M.L. Boccia (Eds.), *Self-awareness in animals and humans. Developmental perspectives* (pp. 273–290). Cambridge: Cambridge University Press.

Pepperberg, I.M. (1983). Cognition in the African Grey parrot: Preliminary evidence for auditory/vocal comprehension of a class concept. *Animal Learning and Behavior, 11*, 179–185.

Pepperberg, I.M. (1987). Evidence for conceptual quantitative abilities in the African Grey parrot: Labeling of cardinal sets. *Ethology, 75*, 37–61.

Pereira, M.E. & Macedonia, J.M. (1991). Ringtailed lemur anti-predator calls denote predator class, not response urgency. *Animal Behaviour, 41*, 543–544.

Perinat, A. & Dalmau, A. (1988). La comunicación entre pequeños gorilas criados en cautividad y sus cuidadoras. *Estudios de Psicología, 33–34*, 33–34.

Perner, J. (1991). *Understanding the representational mind*. Cambridge, MA: MIT Press.

Perner, J. (1993). The theory of mind deficit in autism: rethinking the metarepresentation theory. In S. Baron-Cohen, H. Tager-Flusberg, & D.J. Cohen (Eds.), *Understanding other minds. Perspectives from autism* (pp. 112–137). New York: Oxford University Press.

Perrault, J. (1982). A comparative study of seriation in chimpanzees and children. (Master's dissertation). Quebec: Universite de Montreal.

Perrett, D.I. & Mistlin, A.J. (1990). Perception of facial characteristics by monkeys. In W.C. Stebbins & M.A. Berkley (Eds.), *Comparative perception* (pp. 187–215). New York: John Wiley and Sons.

Perret-Clermont, A. & Brossard, A. (1985) On the interdigitation of social and cognitive processes. In R.A. Hinde, A-N. Perret-Clermont, & J. Stevenson-Hinde (Eds.), *Social relationships and cognitive development* (pp. 309–327). Oxford: Clarendon Press.

Perusse, R. & Rumbaugh, D.M. (1990). Summation in chimpanzees (*Pan troglodytes*): effects of amounts, number of wells, and finer ratios. *International Journal of Primatology, 11*, 425–437.

Peterson, G.B., Wheeler, R.L., & Armstrong, G.D. (1978). Expectancies as mediators in the differential-reward conditional discrimination performance of pigeons. *Animal Learning and Behavior, 6*, 279–285.

Petit, O., Desportes, C., & Thierry, B. (1992). Differential probability of "coproduction" in two species of macaque (*Macaca tonkeana, M. mulatta*). *Ethology, 90*, 107–120.

Petit, O. & Thierry, B. (1993). Use of stones in a captive group of Guinea baboons. *Folia Primatologica, 61*, 160–164.

Pettet, A. (1975). Defensive stoning by baboons. *Nature, 258*, 549.

Phelps, M.T. & Roberts, W.A. (1994). Memory for pictures of upright and inverted primate faces in humans (*Homo sapiens*), squirrel monkeys (*Saimiri sciureus*), and pigeons (*Columba livia*). *Journal of Comparative Psychology, 108*, 114–125.

Piaget, J. (1952). *The origins of intelligence in children*. New York: Norton.

Piaget, J. (1954). *The construction of reality in the child*. New York: Norton.

Piaget, J. (1962). *Play, dreams, and imitation*. New York: Norton.

Piaget, J. (1970). Piaget's theory. In P. Mussen (Ed.), *Manual of child development* (pp. 703–732). New York: Wiley.

Piaget, J. (1974). *Understanding causality*. New York: Norton.

Piaget, J. (1995). *Sociological studies*. New York: Routledge.

Pickford, M. (1975). Defensive stoning by baboons. *Nature, 258*, 549–550.

Picq, J.L. (1993). Radial maze performance in young and grey mouse lemurs (*Microcebus murinus*). *Primates, 34*, 223–226.

Plooij, F.X. (1978). Some basics traits of language in wild chimpanzees? In A. Lock (Ed.), *Action, gesture and symbol* (pp. 111–131). London: Academic Press.

Plooij, F.X. (1984). *The behavioral development of free-living chimpanzee babies and infants*. Norwood, NJ: Ablex Publishing Corporation.

Potì, P. (1989). Early sensoriomotor development in macaques (*Macaca fuscata, M. fascicularis*). In F. Antinucci (Ed.), *Cognitive structure and development in nonhuman primates* (pp. 39–53). Hilldsale, NJ: Lawrence Erlbaum Associates.

Potì, P. (1996). Spatial aspects of spontaneous object grouping by young chimpanzees (*Pan troglodytes*). *International Journal of Primatology, 17*, 101–116.

Potì, P. & Spinozzi, G. (1994). Early sensorimotor development in chimpanzees (*Pan troglodytes*). *Journal of Comparative Psychology, 108*, 93–103.

Povinelli, D.J. (1991). Social intelligence in monkeys and apes (Doctoral dissertation). New Haven, CT: Yale University.

Povinelli, D.J. (1993). Reconstructing the evolution of mind. *American Psychologist, 48*, 493–509.

Povinelli, D.J. (1994a). Comparative studies of animal mental state attribution: a reply to Heyes. *Animal Behaviour, 48*, 239–241.

Povinelli, D.J. (1994b). What chimpanzees (might) know about the mind. In R.W. Wrangham, W.C. McGrew, F.B.M. de Waal, & P.G. Heltne (Eds.), *Chimpanzee cultures* (pp. 285–300). Cambridge, MA: Harvard University Press.

Povinelli, D.J. (1996). Chimpanzee theory of mind? The long road to strong inference. In P. Carruthers & P. Smith (Eds.), *Theories of theories of mind* (pp. 293–329). Cambridge: Cambridge University Press.

Povinelli, D.J. & Eddy, T.J. (1996a). Chimpanzees: Joint visual attention. *Psychological Science, 7*, 129–135.

Povinelli, D. J. & Eddy, T.J. (1996b). Factors influencing young chimpanzees recognition of "attention." *Journal of Comparative Psychology, 110*, 336–345.

Povinelli, D.J. & Eddy, T.J. (1996c). What young chimpanzees know about seeing. *Monographs of the Society for Research in Child Development, 61*, 3.

Povinelli, D.J. & Eddy, T.J. (in press). Specificity of gaze-following in young chimpanzees. *British Journal of Developmental Psychology*.

Povinelli, D.J., Nelson, K.E., & Boysen, S.T. (1990). Inferences about guessing and knowing by chimpanzees (*Pan troglodytes*). *Journal of Comparative Psychology, 104*, 203–210.

Povinelli, D.J., Nelson, K.E., & Boysen, S.T. (1992). Comprehension of role reversal in chimpanzees: evidence of empathy? *Animal Behaviour, 43*, 633–640.

Povinelli, D.J., Parks, K.A., & Novak, M.A. (1991). Do rhesus monkeys (*Macaca mulatta*) attribute knowledge and ignorance to others? *Journal of Comparative Psychology, 105*, 318–325.

Povinelli, D.J., Parks, K.A., & Novak, M.A. (1992). Role reversal by rhesus monkeys, but no evidence of empathy. *Animal Behaviour, 44*, 269–281.

Povinelli, D.J. & Perilloux, H.K. (in press). A preliminary note on young chimpanzees' reactions to intentional and accidental behavior. *Behavioral Processes*.

Povinelli, D.J., Rulf, A.B., & Bierschwale, D.T. (1994). Absence of knowledge attribution and self-recognition in young chimpanzees (*Pan troglodytes*). *Journal of Comparative Psychology, 108*, 74–80.

Povinelli, D.J., Rulf, A.B., Landau, K.R., & Bierschwale, D.T. (1993). Self-recognition in chimpanzees (*Pan troglodytes*): Distribution, ontogeny, and patterns of emergence. *Journal of Comparative Psychology, 107*, 347–372.

Povinelli, D., Zebouni, M., & Prince, C. (1996). Ontogeny, evolution, and folk psychology. *Behavioral and Brain Sciences, 19*, 137–138.

Powers, W. (1973). *Behavior: the control of perception*. Chicago: Aldine.

Premack, D. (1976). *Intelligence in ape and man*. Hillsdale, NJ: Lawrence Earlbaum Associates.

Premack, D. (1978). On the abstractness of human concepts: why it would be difficult to talk to a pigeon. In S.H. Hulse, H. Fowler, & W.K. Honig (Eds.), *Cognitive processes in animal behavior* (pp. 423–451). Hillsdale, NJ: Lawrence Erlbaum Associates.

Premack, D. (1983). The codes of man and beasts. *Behavioral and Brain Sciences, 6*, 125–167.

Premack, D. (1986). *Gavagai!* Cambridge, MA: MIT Press.

Premack, D. (1988). 'Does the chimpanzee have a theory of mind?' revisited. In R.W. Byrne & A. Whiten (Eds.), *Machiavellian intelligence. Social expertise and the evolution of intellect in monkeys, apes, and humans* (pp. 160–179). New York: Oxford University Press.

Premack, D. & Premack, A.J. (1983). *The mind of an ape*. New York: Norton.

Premack, D. & Premack, A.J. (1994). Levels of causal understanding in chimpanzees and children. *Cognition, 50*, 347–362.

Premack, D. & Woodruff, G. (1978). Does the chimpanzee have a theory of mind? *Behavioral and Brain Sciences, 4*, 515–526.

Presley, W.J. & Riopelle, A.J. (1959). Observational learning of an avoidance response. *Journal of Genetic Psychology, 95*, 251–254.

Prestrude, A.M. (1970). Sensory capacities of the chimpanzee. A review. *Psychological Bulletin, 74*, 47–67.

Rassmusen, K.L.R. (1980). Consort behavior and mate selection in yellow baboons (*Papio cynocephalus*) (Ph.D. dissertation). Cambridge: University of Cambridge.

Reaux, J. (1995). Explorations of young chimpanzees' (*Pan troglodytes*) comprehension of cause-effect relationships in tool use (Masters thesis). New Iberia: University of SouthWestern Louisiana.

Reddy, V. (1991). Playing with others' expectations: teasing and mucking about in the first year. In A. Whiten (Ed.), *Natural theories of mind* (pp. 143–158). Oxford: Oxford University Press.

Redshaw, M. (1978). Cognitive development in human and gorilla infants. *Journal of Human Evolution, 7*, 133–141.

Reed, E. (1995). The ecological approach to language development. *Language and communication, 15*, 1–29.

Reese, H.W. (1968). *The perception of stimulus relations: discrimination learning and transposition*. New York: Academic Press.

Renner, M.J., Bennett, A.J., Ford, M.L., & Pierre, P.J. (1992). Investigation of inanimate objects by the greater bushbaby (*Otolemur garnettii*). *Primates, 33*, 315–327.

Rensch, B. (1973). Play and art in apes and monkeys. In E.W. Menzel, Jr. (Ed.), *Precultural primate behavior* (pp. 102–123). Basel, Switzerland: Karger.

Rensch, B. & Döhl, J. (1968). Spontanes ofen verschiedener Kistenverschlusse durch einen Schimpsansen. *Zeitscift fur Tierpsychologie, 24*, 476–89.

Resnick, J.S. (1989). Research on infant categorization. *Seminars in Perinatology, 13*, 458–466.

Richard, A.F. (1985). *Primates in nature*. New York: Freeman and Company.

Riesen, A.H. (1970). Chimpanzee visual perception. In G.H. Bourne (Ed.), *The chimpanzee* (pp. 93–119). Basel, Switzerland: Karger.

Riesen, A.H., Greenberg, B., Granston, A.S., & Fantz, R.L. (1953). Solutions of patterned string problems by young gorillas. *Journal of Comparative Psychology, 46*, 19–22.

Riesen, A.H. & Nissen, H.W. (1942). Non-spatial delayed response by the matching technique. *Journal of Comparative Psychology, 34*, 307–313.

Rijksen, H.D. (1978). *A field study on Sumatran orangutans (Pongo pygmaeus abelli, Lesson 1827)*. Wageningan: H. Veenman and Zonen.

Riopelle, A.J. (1960). Observational learning of a position habit by monkeys. *Journal of Comparative Psychology, 53*, 426–428.

Riopelle, A.J., Alper, R.G., Strong, P.N., & Ades, H.W. (1964). Multiple discrimination and patterned string performance of normal and temporal-lobectomized monkeys. In S.C. Ratner & M.R. Denny (Eds.), *Comparative psychology. Research in animal behavior* (pp. 686–693). Homewood, IL: Dorsey Press.

Riopelle, A.J. & Francisco, E.W. (1955). Discrimination learning performance under different first-trial procedures. *Journal of Comparative Psychology, 48*, 90–93.

Riopelle, A.J. & Moon, W.H. (1968). Problem diversity and familiarity in multiple discrimination learning by monkeys. *Animal Behaviour, 16*, 74–78.

Ristau, C. (1991). *Cognitive ethology: the minds of other animals*. Hillsdale, NJ: Lawrence Erlbaum Associates.

Ristau, C. & Robins, D. (1982). Language in the great apes: A critical review. In J. Roitblatt, R. Hinde, C. Beer, & M. Busnell (Eds.), *Advances in the study of behavior, vol. 12* (pp. 141–255). New York: Academic Press.

Ritchie, B.G. & Fragaszy, D.M. (1988). Capuchin monkey (*Cebus apella*) grooms her infant's wound with tools. *American Journal of Primatology, 16*, 345–348.

Robert, S. (1986). Ontogeny of mirror behavior in two species of great apes. *American Journal of Primatology, 10*, 109–117.

Roberts, W.A. & Mazmanian, D.S. (1988). Concept learning at different levels of abstraction by pigeons, monkeys, and people. *Animal Behavior Processes, 14*, 247–260.

Roberts, W.A., Mitchell, S., & Phelps, M.T. (1993). Foraging in laboratory trees: Spatial memory in squirrel monkeys. In T.R. Zentall (Ed.), *Animal cognition. A tribute to Donald A. Riley* (pp. 131–151). Hillsdale, NJ: Lawrence Erlbaum Associates.

Roberts, W. & Phelps, M. (1994). Transitive inference in rats: a test of the spatial coding hypothesis. *Psychological Science, 5*, 368–374.

Robinson, J.S. (1955). The sameness-difference discrimination problem in chimpanzee. *Journal of Comparative and Physiological Psychology, 48*, 195–197.

Robinson, J.S. (1960). The conceptual basis of the chimpanzee's performance on the sameness-difference discrimination problem. *Journal of Comparative and Physiological Psychology, 53*, 368–370.

Rochat, P. (1995a). Early objectification of the self. In P. Rochat (Ed.), *The self in infancy* (pp. 53–71). Amsterdam: North-Holland.

Rochat, P. (1995b). Perceived reachability for self and others by 3- to 5-year-old children and adults. *Journal of Experimental Child Psychology, 59*, 317–333.

Rochat, P. (1995c). *The self in infancy*. Amsterdam: North-Holland.

Rogers, L.J. & Kaplan, G. (1994). A new form of tool use by orang-utans in Sabah, East Malaysia. *Folia Primatologica*, 63, 50–52.

Rogoff, B. (1990) *Apprenticeship in thinking: cognitive development in social context*. Oxford: Oxford University Press.

Rohles, F.H. Jr. & Devine, J.V. (1966). Chimpanzee performance in a problem involving the concept of middleness. *Animal Behaviour, 14*, 159–162.

Rohles, F.H. Jr. & Devine, J.V. (1967). Further studies of the middleness concept with the chimpanzee. *Animal Behaviour, 15*, 107–112.

Roitblat, H.L. (1987). *Introduction to comparative cognition*. New York: W.H. Freeman.

Roitblat, H.L., Bever, T.G., & Terrace, H.S. (1984). *Animal Cognition*. Hillsdale, NJ: Lawrence Erlbaum Associates.

Romanes, G.J. (1882). *Animal intelligence*. London: Kegan, Paul Trench.

Romanes, G.J. (1883). *Mental evolution in animals*. London: Kegan, Paul Trench.

Rosch, E. (1978). Principals of categorization. In E. Rosch & B.B. Lloyd (Eds.), *Cognition and categorization* (pp. 28–49). Hillsdale, NJ: Lawrence Erlbaum Associates.

Rosenfeld, S.A. & van Hoesen, G.W. (1979). Face recognition in the rhesus monkey. *Neuropsychologia, 17*, 503–509.

Rowell, T. (1966). Hierarchy in the organization of a captive baboon group. *Animal Behaviour, 14*, 430–443.

Rowell, T.E. & Rowell, C.A. (1993). The social organization of feral Ovis aries ram groups in the pre-rut period. *Ethology, 95*, 213–232.

Rozin, P. (1976). The evolution of intelligence and access to the cognitive unconscious. In J.M. Sprague & A.N. Epstein (Eds.), *Progress in psychobiology and physiological psychology* (pp. 245–280). New York: Academic Press.

Rumbaugh, D.M. (1970). Learning skills of anthropoids. In L.A. Rosenblum (Ed.), *Primate behavior: Developments in field and laboratory research* (pp. 1–70). New York: Academic Press.

Rumbaugh, D.M. (1977). *Language learning by a chimpanzee. The Lana project*. New York: Academic Press.

Rumbaugh, D.M. (1995). Primate language and cognition: Common ground. *Social Research, 62*, 711–730.

Rumbaugh, D.M., Hopkins, W.D., Washburn, D.A., & Savage-Rumbaugh, E.S. (1989). Lana chimpanzee learns to count by "Numath": a summary of videotaped experimental report. *Psychological Record, 39*, 459–470.

Rumbaugh, D.M., Hopkins, W.D., Washburn, D.A., & Savage-Rumbaugh, E.S. (1993). Chimpanzee competence for counting in a video-formatted task situation. In H.L. Roitblat, L.M. Herman, & P.E. Nachtigall (Eds.), *Language and communication. Comparative perspectives* (pp. 329–346). Hillsdale, NJ: Lawrence Erlbaum Associates.

Rumbaugh, D.M. & McCormack, C. (1967). The learning skills of primates: a comparative study of apes and monkeys. In D. Starck, R. Schheider, & H.J. Ruhn (Eds.), *Progress in primatology* (pp. 289–306). Stuggart: Gustav Fischer Verlag.

Rumbaugh, D.M. & Pate, J.L. (1984a). The evolution of cognition in primates: A comparative perspective. In H.L. Roitblat, T.G. Bever, & H.S. Terrace (Eds.), *Animal cognition* (pp. 569–587). Hillsdale, NJ: Lawrence Erlbaum Associates.

Rumbaugh, D.M. & Pate, J.L. (1984b). Primates' learning by levels. In G. Greenberg & E. Tobach (Eds.), *Behavioral evolution and integrative levels* (pp. 221–240). Hillsdale, NJ: Lawrence Erlbaum Associates.

Rumbaugh, D.M. & Rice, C.P. (1962). Learning-set formation in young great apes. *Journal of Comparative Psychology, 55*, 866–868.

Rumbaugh, D.M. & Savage-Rumbaugh, E.S. (1992). Cognitive competencies: products of genes, experience, and technology. In T. Nishida, W.C. McGrew, P. Marler, M. Pickford, & F.B.M. de Waal (Eds.), *Topics in primatology. Human origins* (pp. 293–304). Tokyo: University of Tokyo Press.

Rumbaugh, D.M., Savage-Rumbaugh, E.S., & Hegel, M.T. (1987). Summation in the chimpanzee (*Pan troglodytes*). *Journal of Experimental Psychology: Animal Behavior Processes, 13*, 107–115.

Rumbaugh, D.M., Savage-Rumbaugh, E.S., & Pate, J.L. (1988). Addendum to "Summation in the chimpanzee (*Pan troglodytes*)." *Animal Behavior Processes, 14*, 118–120.

Rumbaugh, D.M., Ternes, J.W., & Abordo, E.J. (1965). Learning set in squirrel monkeys as affected by encasement of problem objects in plexiglas bins. *Perceptual and Motor Skills, 21*, 531–534.

Russon, A.E. & Galdikas, B.M.F. (1993). Imitation in ex-captive orangutans. *Journal of Comparative Psychology, 107*, 147–161.

Russon, A.E. & Galdikas, B.M.F. (1995). Constraints on great apes' imitation: model and action selectivity in rehabilitant orangutan (*Pongo pygmaeus*) imitation. *Journal of Comparative Psychology, 109*, 5–17.

Sabater Pi, J. (1974). An elementary industry of the chimpanzees in the Okorobikó mountains, Rio Muni, (republic of Equatorial Guinea), west Africa. *Primates, 15*, 351–364.

Sabater Pi, J. (1984). *El chimpancé y los orígenes de la cultura*. Barcelona: Anthropos.

Sakura, O. & Matsuzawa, T. (1991). Flexibility of wild chimpanzee nut-cracking behavior using stone hammers and anvils: An experimental analysis. *Ethology, 87*, 237–248.

Sanders, R.J. (1985). Teaching apes to ape language: explaining the imitative and nominative signing of a chimpanzee (*Pan troglodytes*). *Journal of Comparative Psychology, 99*, 197–210.

Savage, A. & Snowdon, C.T. (1982). Mental retardation and neurological deficits in a twin orangutan. *American Journal of Primatology, 3*, 239–251.

Savage-Rumbaugh, E.S. (1984). *Pan paniscus and Pan troglodytes*. Contrasts in preverbal communicative competence. In R.L.Susman (Ed.), *The pygmy chimpanzee* (pp. 131–177). New York: Plenum Press.

Savage-Rumbaugh, E.S. (1986). *Ape language*. New York: Columbia University Press.

Savage-Rumbaugh, E.S. (1991). Multi-tasking: the Pan-human rubicon. *Neurosciences, 3*, 417–422.

Savage-Rumbaugh, E.S., Brakke, K.E., & Hutchins, S.S. (1992). Linguistic development: contrasts between co-reared *Pan troglodytes* and *Pan paniscus*. In T. Nishida, W.C. McGrew, P. Marler, M. Pickford, & F.B.M. de Waal (Eds.), *Topics in primatology. Human origins* (pp. 51–66). Tokyo: University of Tokyo Press.

Savage-Rumbaugh, E.S., McDonald, K., Sevcik, R.A., Hopkins, W.D., & Rubert, E. (1986). Spontaneous symbol acquisition and communicative use by pygmy chimpanzees (*Pan paniscus*). *Journal of Experimental Psychology: General, 115*, 211–235.

Savage-Rumbaugh, E.S., Murphy, J., Sevcik, R.A., Brakke, K.E., Williams, S.L., & Rumbaugh, D.M. (1993). Language comprehension in ape and child. *Monographs of the Society for Research in Child Development, 58(3–4)*, no. 233.

Savage-Rumbaugh, E.S., Rumbaugh, D.M., & Boysen, S. (1978a). Linguistically mediated tool use and exchange by chimpanzees (*Pan troglodytes*). *Behavioral and Brain Sciences, 4*, 539–554.

Savage-Rumbaugh, E.S., Rumbaugh, D.M., & Boysen, S.T. (1978b). Sarah's problems in comprehension. *Behavioral and Brain Sciences, 1*, 555–557.

Savage-Rumbaugh, E.S., Rumbaugh, D.M., Smith, S.T., & Lawson, J. (1980). Reference: The linguistic essential. *Science, 210*, 922–925.

Savage-Rumbaugh, S. & McDonald, K. (1988). Deception and social manipulation in symbol-using apes. In R.W. Byrne & A. Whiten (Eds.), *Machiavellian intelligence. Social expertise and the evolution of intellect in monkeys, apes, and humans* (pp. 224–237). New York: Oxford University Press.

Savage-Rumbaugh, S., Sevcik, R. & Hopkins, W. (1988). Symbolic cross-modal transfer in two species of chimpanzees. *Child Development, 59*, 617–625.

Savage-Rumbaugh, E.S., Wilkerson, B.J., & Bakeman, R. (1977). Spontaneous gestural communication among conspecifics in the pygmy chimpanzee (*Pan paniscus*). In G.H. Bourne (Ed.), *Progress in ape research* (pp. 97–116). New York: Academic Press.

Sawaguchi, T. & Kudo, H. (1990). Neocortical development and social structure in primates. *Primates, 31*, 283–290.

Schaller, G.B. (1963). *The mountain gorilla. Ecology and behavior*. Chicago: University of Chicago Press.

Schiller, P.H. (1952). Innate constituents of complex responses in primates. *Psychological Review, 59*, 177–191.

Schiller, P.H. (1957). Innate motor action as a basis of learning. In C.H. Schiller (Ed.), *Instinctive behavior* (pp. 264–287). New York: International Universities Press.

Schino, G., Spinozzi, G., & Berlinguer, L. (1990). Object concept and mental representation in *Cebus apella* and *Macaca fascicularis*. *Primates, 31*, 537–544.

Schneirla, T.C. (1966). Behavioral development and comparative psychology. *Quarterly Review of Biology, 41*, 283–303.

Schrier, A.M., Angarella, R., & Povar, M.L. (1984). Studies of concept formation by stumptailed monkeys: concepts humans, monkeys, and letter A. *Animal Behavior Processes, 10*, 564–584.

Schrier, A.M. & Brady, P.M. (1987). Categorization of natural stimuli by monkeys (*Macaca mulatta*): effects of stimulus set size and modification of exemplars. *Animal Behavior Processes, 13*, 136–143.

Seidenberg, M.S. & Petitto, L.A. (1987). Communication, symbolic communication, and language: comment on Savage-Rumbaugh, McDonald, Sevcik, Hopkins, and Rupert (1986). *Journal of Experimental Psychology: General, 116*, 279–287.

Settlage, P.H. (1939). The effect of occipital lesions on visually guided behavior in the monkey. *Journal of Comparative Psychology, 27*, 93–131.

Seyfarth, R.M. (1976). Social relationships among adult female baboons. *Animal Behaviour, 24*, 917–938.

Seyfarth, R.M. (1977). A model of social grooming among adult female monkeys. *Journal of Theoretical Biology, 65*, 671–698.

Seyfarth, R.M. (1980). The distribution of grooming and related behaviours among adult female vervet monkets. *Animal Behaviour, 28*, 798–813.

Seyfarth, R.M. (1987). Vocal communication and its relation to language. In B.B. Smuts, D.L. Cheney, R.M. Seyfarth, R.W. Wrangham, & T.T. Struhsaker (Eds.), *Primate societies* (pp. 440–451). Chicago: University of Chicago Press.

Seyfarth, R.M. & Cheney, D.L. (1984). Grooming, alliances and reciprocal altruism in vervet monkeys. *Nature, 308*, 541–543.

Seyfarth, R.M. & Cheney, D.L. (1986). Vocal development in vervet monkeys. *Animal Behaviour, 34*, 1640–1658.

Seyfarth, R.M. & Cheney, D.L. (1988). Do monkeys understand their relations? In R.W. Byrne & A. Whiten (Eds.), *Machiavellian intelligence. Social expertise and the evolution of intellect in monkeys, apes, and humans* (pp. 69–84). New York: Oxford University Press.

Seyfarth, R.M. & Cheney, D.L. (1990). The assessment by vervet monkeys of their own and another species' alarm calls. *Animal Behaviour, 40*, 754–764.

Seyfarth, R.M. & Cheney, D.L. (1993). Meaning, reference, and intentionality in the natural vocalizations of monkeys. In H.L. Roitblat, L.M. Herman, & P.E. Nachtigall (Eds.), *Language and communication. Comparative perspectives* (pp. 195–219). Hillsdale, NJ: Lawrence Erlbaum Associates.

Seyfarth, R.M. & Cheney, D.L. (in press). In C. Snowdon & M. Hausberger (Eds.), *Social influences on vocal development*. Cambridge: Cambridge University Press.

Seyfarth, R.M., Cheney, D.L., Harcourt, A.H., & Stewart, K.J. (1994). The acoustic features of gorilla double grunts and their relation to behavior. *American Journal of Primatology, 33*, 31–50.

Seyfarth, R.M., Cheney, D.L., & Marler, P. (1980a). Monkey responses to three different alarm calls: evidence of predator classification and semantic communication. *Science, 210*, 801–803.

Seyfarth, R.M., Cheney, D.L., & Marler, P. (1980b). Vervet monkey alarm calls: Semantic communication in a free-ranging primate. *Animal Behaviour, 28*, 1070–1094.

Shell, W.F. & Riopelle, A.J. (1958). Progressive discrimination learning in platyrrhine monkeys. *Journal of Comparative Psychology, 51*, 467–470.

Shepard, R. & Metzler, J. (1971). Mental rotation of three-dimensional objects. *Science, 171*, 701–703.

Shepherd, W.T. (1910). Some mental processes of the rhesus monkey. *Psychological Monographs, 12*(5), 61 pp.

Shepherd, W.T. (1915). Some observations on the intelligence of the chimpanzee. *Journal of Animal Behaviour, 5*, 391–396.

Shepherd, W.T. (1923). Some observations and experiments of the intelligence of the chimpanzee and ourang. *American Journal of Psychology, 34*, 590–591.

Sherman, P.W. (1977). Nepotism and the evolution of alarm calls. *Science, 197*, 1246–1253.

Sherry, D. (1984). Food storage by black-capped chickadees: memory for the location and contents of caches. *Animal Behaviour, 32*, 451–464.

Shettleworth, S. & Krebs, J. (1982). How marsh tits find their hoards: the role of site preference and spatial memory. *Animal Behavior Processes, 8*, 354–375.

Shurcliff, A., Brown, D., & Stollnitz, F. (1971). Specificity of training required for solution of a stick problem by rhesus monkeys (*Macaca mulatta*). *Learning and Motivation, 2*, 255–270.

Siegler, R. (1995). How does change occur: a microgenetic study of number conservation. *Cognitive Psychology, 28*, 225–274.

Sigg, H. (1986). Ranging patterns in hamadryas baboons: evidence for a mental map. In J.G. Else & P.C. Lee (Eds.), *Primate ontogeny, cognition and social behaviour* (pp. 87–91). Cambridge: Cambridge University Press.

Sigg, H. & Falett, J. (1985). Experiments on respect of possession and property in hamadryas baboons (*Papio hamadryas*). *Animal Behaviour, 33*, 978–984.

Sigg, H. & Stolba, A. (1981). Home range and daily march in a hamadryas baboon troop. *Folia Primatologica, 36*, 40–75.

Silk, J.B. (1978) Patterns of food sharing among mother and infant chimpanzees at Gombe National Parrk, Tanzania. *Folia Primatologica, 29*, 129–141.

Silk, J.B. (1982). Altruism among female *Macaca radiata*: explanations and analysis of patterns of grooming and coalition formation. *Behaviour, 79*, 162–188.

Silk, J.B. (1992). The patterning of intervention among male bonnet macaques: reciprocity, revenge, and loyalty. *Current Anthropology, 33*, 318–325.

Simon, T., Hespos, S., & Rochat, P. (1995). Do infants understand simple arithmetic? *Cognitive Development, 10*, 253–269.

Simpson, M.J.A. (1973). The social grooming of male chimpanzees. In R.P. Michael & J.H. Crook (Eds.), *Comparative ecology and behavior of primates* (pp. 411–505). London: Academic Press.

Sinha, R. (in press). Complex tool manufacture by a wild bonnet macaque, *Macaca radiata. Folia Primatologica, 65*, 234–238.

Slobin, D. (1985). The language making capacity. In D. Slobin (Ed.), *The cross-linguistic study of language acquisition* (pp. 1157–1256). Hillsdale, NJ: Lawrence Erlbaum Associates.

Smith, H.J., King, J.E., Witt, E.D., & Rickel, J.E. (1975). Sameness-difference matching from sample by chimpanzees. *Bulletin of the Psychonomic Society, 6*, 469–471.

Smith, W.J. (1977). *The behavior of communicating: an ethological approach.* Cambridge, MA: Harvard University Press.

Smuts, B.B. (1985). *Sex and friendship in baboons.* Chicago: Aldine.

Smuts, B.B., Cheney, D.L., Seyfarth, R.M., Wrangham, R.W., & Struhsaker, T.T. (1987). *Primate societies.* Chicago: University of Chicago Press.

Snowdon, C. (1986). Vocal communication. In G. Mitchell & J. Erwin (Eds.), *Comparative primate biology, vol. 2A: Behavior, conservation, and ecology* (pp. 3–38). New York: Alan R. Liss.

Snowdon, C. (1987). A naturalistic view of categorical perception. In S. Harnad (Ed.), *Categorical perception* (pp. 332–345) New York: Cambridge University Press.

Snowdon, C.T. (1993). Linguistic phenomena in the natural communication of animals. In H.L. Roitblat, L.M. Herman, & P.E. Nachtigall (Eds.), *Language and communication. Comparative perspectives* (pp. 175–194). Hillsdale, NJ: Lawrence Erlbaum Associates.

Snowdon, C.T. , Brown, C.H., & Petersen, M.R. (1982). *Primate communication.* Cambridge: Cambridge University Press.

Snowdon, C.T. & Cleveland, J. (1980). Individual recognition of contact calls in pygmy marmosets. *Animal Behavior, 28*, 717–727.

Snowdon, C.T. & Cleveland, J. (1984). "Conversations" among pygmy marmosets. *American Journal of Primatology, 7*, 15–20.

Snowdon, C.T. & Elowson, A.M. (1992). Ontogeny of primate vocal communication. In T. Nishida, W.C. McGrew, P. Marler, M. Pickford, & F.B.M. de Waal (Eds.), *Topics in primatology* (pp. 279–290). Tokyo: Tokyo University Press.

Snowdon, C.T. & Hodun, A. (1981). Acoustic adaptations in pygmy marmoset contact calls: Locational cues vary with distances between conspecifics. *Behavioral Ecology and Sociobiology, 9*, 295–300.

Snyder, D.R., Birchette, L.M., & Achenbach, T.M. (1978). A comparison of developmentally progressive intellectual skills between *Hylobates lar, Cebus apella,* and *Macaca mulatta.* In D.J. Chivers & J. Herbert (Eds.), *Recent advances in primatology* (pp. 945–948). New York: Academic Press.

Sober, E. & Wilson, D. (1994). Reintroducing group selection to the behavioral sciences. *Behavioral and Brain Sciences, 17*, 585–654.

Spelke, E.S. (1979). Perceiving bimodally specified events in infancy. *Development Psychology, 15*, 626–636.

Spelke, E., Breinliger, K., Macomber, J., & Jacobson, K. (1992). Origins of knowledge. *Psychological Review, 99*, 605–632.

Spence, K.W. (1937). Experimental studies of learning and the mental processes in infra-human primates. *Psychological Bulletin, 34*, 806–850.

Spence, K.W. (1939). The solution of multiple choice problems by chimpanzees. *Comparative Psychology Monographs, 15(3)*, 1–54 pp.

Spinozzi, G. (1989). Early sensoriomotor development in Cebus (*Cebus apella*). In F. Antinucci (Ed.), *Cognitive structure and development in nonhuman primates* (pp. 55–66). Hilldsale, NJ: Lawrence Erlbaum Associates.

Spinozzi, G. (1993). Development of spontaneous classificatory behavior in chimpanzees (*Pan troglodytes*). *Journal of Comparative Psychology, 107*, 193–200.

Spinozzi, G. (in press). Categorization in monkeys and chimpanzees. *Behavioral Brain Research.*

Spinozzi, G. & Natale, F. (1989a). Classification. In F. Antinucci (Ed.), *Cognitive structure and development in nonhuman primates* (pp. 163–187). Hilldsale, NJ: Lawrence Erlbaum Associates.

Spinozzi, G. & Natale, F. (1989b). Early sensorimotor development in Gorilla. In F. Antinucci (Ed.), *Cognitive structure and development in nonhuman primates* (pp. 21–38). Hillsdale, NJ: Lawrence Erlbaum Associates.

Spinozzi, G. & Potì, P. (1989). Causality I: The support problem. In F. Antinucci (Ed.), *Cognitive Structure and development in nonhuman primates* (pp. 113–119). Hilldsale, NJ: Lawrence Erlbaum Associates.

Spinozzi, G. & Potì, P. (1993). Piagetian stage 5 in two infant chimpanzees (*Pan troglodytes*): the development of permanence of objects and the spatialization of causality. *International Journal of Primatology, 14*, 905–917.

Stamm, J.S. (1961). Social facilitation in monkeys. *Psychological Reports, 8*, 479–484.

Stammbach, E. (1978). On social differentiation in groups of captive female hamadryas baboons. *Behaviour, 67*, 322–338.

Stammbach, E. (1988a). An experimental study of social knowledge: adaptation to the special manipulative skills of single individuals in a *Macaca fascicularis* group. In R.W. Byrne & A. Whiten (Eds.), *Machiavellian intelligence. Social expertise and the evolution of intellect in monkeys, apes, and humans* (pp. 309–326). New York: Oxford University Press.

Stammbach, E. (1988b). Group responses to specially skilled individuals in a *Macaca fascicularis* group. *Behaviour, 107*, 241–266.

Stanford, C.B., Wallis, J., Mpongo, E., & Goodall, J. (1994). Hunting decisions in wild chimpanzees. *Behaviour, 131*, 1–18.

Stanford, C.B. (1995). The influence of chimpanzee predation on group size and anti-predator behaviour in red colobus monkeys. *Animal Behaviour, 49*, 577–587.

Starkey, P., Spelke, E.S., & Gelman, R. (1990). Numerical abstraction by human infants. *Cognition, 36*, 97–128.

Stepien, L.S. & Cordeau, J.P. (1960). Memory in monkeys for compound stimuli. *American Journal of Psychology, 73*, 388–395.

Stern, D. (1985). *The interpersonal world of the infant*. New York: Basic Books.

Sterrit, G.M., Goodenough, E., & Harlow, H.F. (1963). Learning set development: trials to criterion vs. six trials per problem. *Psychological Reports, 13*, 267–271.

Straub, R. & Terrace, H. (1981). Generalization of serial learning in the pigeon. *Animal Learning and Behavior, 9*, 454–468.

Strayer, F.F. (1976). Learning and imitation as a function of social status in macaque monkeys (*Macaca nemestrina*). *Animal Behaviour, 24*, 835–848.

Strong P.N. Jr. & Hedges, M. (1966). Comparative studies in simple oddity learning: I. Cats, raccoons, monkeys, and chimpanzees. *Psychonomic Science, 5*, 13–14.

Struhsaker, T.T. (1967). Auditory communication among vervet monkeys (*Cercopithecus aethiops*). In S.A. Altmann (Ed.), *Social communication among primates* (pp. 3–14). Chicago: University of Chicago Press.

Strum, S.C. (1981). Processes and products of change: baboon predatory behavior at Gilgil, Kenya. In R.S.O. Harding & G. Teleki (Eds.), *Omnivorous primates* (pp. 271–300). New York: Columbia University Press.

Suarez, S.D. & Gallup, G.G. Jr. (1981). Self-recognition in chimpanzees and orangutans, but not gorillas. *Journal of Human Evolution, 10*, 175–188.

Suarez, S.D. & Gallup, G.G. Jr. (1986). Social responding to mirrors in rhesus macaques (*Macaca mulatta*): effects of changing mirror location. *American Journal of Primatology, 11*, 239–244.

Sugarman, S. (1983). *Children's early thought*. Cambridge: Cambridge University Press.

Sugiyama, Y. (1981). Observations on the population dynamics and behavior of wild chimpanzees at Bossou, Guinea, 1979–1980. *Primates, 22*, 432–444.

Sugiyama, Y. (1995). Drinking tools of wild chimpanzees at Bossou. *American Journal of Primatology, 37*, 263–269.

Sugiyama, Y. & Koman, J. (1979). Tool-using and -making behavior in wild chimpanzees at Bossou, Guinea. *Primates, 20*, 513–524.

Sumita, K., Kitahara-Frisch, J., & Norikoshi, K. (1985). The acquisition of stone-tool use in captive chimpanzees. *Primates, 26*, 168–181.

Sutherland, N.S. & MacIntosh, NJ. (1971). *Mechanisms of animal discrimination learning*. New York: Academic Press.

Swartz, K., Chen, S., & Terrace, H. (1991). Serial learning by rhesus monkeys I: acquisition and retention of multiple four-item lists. *Animal Behavior Processes, 17*, 396–410.

Swartz, K.B. & Evans, S. (1991). Not all chimpanzees (*Pan troglodytes*) show self-recognition. *Primates, 32*, 483–496.

Swartz, K.B. & Evans, S. (1994). Social and cognitive factors in chimpanzee and gorilla mirror behavior and self-recognition. In S.T. Parker, R.W. Mitchell, & M.L. Boccia (Eds.), *Self-awareness in animals and humans. Developmental perspectives* (pp. 189–206). Cambridge: Cambridge University Press.

Takahata, Y. (1990) Social relationships among adult males. In T. Nishida (Ed.), *The chimpanzees of the Mahale mountains* (pp. 149–170). Tokyo: Tokyo University Press.

Takahata, Y. (1991) Diachronic changes in the dominance relations of adult female Japanese monkeys of the Arashiyama B group. In L. Fedigan & P. Asquith (Eds.), *The monkeys of Arashiyama* (pp. 123–139). Albany, NY: SUNY Press.

Takeshita, H. & van Hooff, J.A.R.A.M. (1996). Tool use by chimpanzees (*Pan troglodytes*) judge relations-between-relations in a conceptual-to-sample task. *Japanese Psychological Research, 38*, 63–79.

Talmadge-Riggs, G., Winter, P., Ploog, D., & Mayer, W. (1972). Effects of deafening on the vocal behavior of the squirrel monkey. *Folia Primatologica, 17*, 404–420.

Tanaka, M. (1995a). Matrilineal distribution of louse egg–handling techniques during grooming in free-ranging Japanese macaques. *Americal Journal of Physical Anthropology, 98*, 197–201.

Tanaka, M. (1995b). Object sorting in chimpanzees (*Pan troglodytes*): classification based on physical identity, complementarity, and familiarity. *Journal of Comparative Psychology, 109*, 151–161.

Tanner, J.E. & Byrne, R.W. (1993). Concealing facial evidence of mood: perspective-taking in a captive gorilla? *Primates, 34*, 451–457.

Tanner, J.E. & Byrne, R.W. (1996). Representation of action through iconic gesture in a captive lowland gorilla. *Current Anthropology, 37*, 162–173.

Teleki, G. (1973). *The predatory behavior of wild chimpanzees*. Lewisburg, PA: Bucknell University Press.

Tenaza, R.R. & Tilson, R.L. (1977). Evolution of long-distance alarm calls in Kloss's gibbon. *Nature, 268*, 233–235.

Terborgh, J. (1983). *Five New World primates. A study in comparative ecology*. Princeton: Princeton University Press.

Terkel, J. (1996). Cultural transmission of feeding behavior in the black rat (*Rattus rattus*). In J. Galef & C. Heyes (Eds.), *Social learning in animals: the roots of culture* (pp. 17–47). New York: Academic Press.

Terrace, H.S. (1993). The phylogeny and ontogeny of serial memory: list learning by pigeons and monkeys. *Psychological Science, 4*, 162–169.

Terrace, H.S. & McGonigle, B. (1994). Memory and representation of serial order by children, monkeys, and pigeons. *Current Directions in Psychological Science, 3*, 180–185.

Terrace, H.S., Petitto, L.A., Sanders, R.J., & Bever, T.G. (1979). Can an ape create a sentence? *Science, 206*, 891–902.

Terrace, H., Straub, R., Bever, T., & Seidenberg, M. (1977). Representation of a sequence by a pigeon. *Bulletin of the Psychonomic Society, 10*, 269.

Terrell, D.F. & Thomas, R.K. (1990). Number-related discrimination and summation by squirrel monkeys (*Saimiri sciureus sciureus* and *S. boliviensus boliviensus*) on the basis of the number of sides of polygons. *Journal of Comparative Psychology, 104*, 238–247.

Thomas, R.K. (1980). Evolution of intelligence: an approach to its assessment. *Brain Behavior Evolution, 17*, 454–472.

Thomas, R.K. (1986). Vertebrate intelligence: a review of the laboratory research. In R.J. Hoage & L. Goldman (Eds.), *Animal intelligence. Insights into the animal mind* (pp. 37–56). Washington, DC: Smithsonian Institution Press.

Thomas, R.K. (1992a). Interactive models of cognitive abilities of monkeys and humans (*Saimiri sciureus sciureus; S. boliviensus boliviensus; Homo sapiens sapiens*). *International Journal of Comparative Psychology, 5*, 179–190.

Thomas, R.K. (1992b). Primates' conceptual use of number: ecological perspectives and psychological processes. In T. Nishida, W.C. McGrew, P. Marler, M. Pickford, & F.B.M. de Waal (Eds.), *Topics in primatology. Human origins* (pp. 305–314). Tokyo: University of Tokyo Press.

Thomas, R.K. (1994). A critique of Nakagawa's (1993) "Relational rule learning in the rat". *Psychobiology, 22*, 347–348.

Thomas, R.K. & Boyd, M.G. (1973). A comparison of *Cebus albifrons* and *Saimiri sciureus* on oddity performance. *Animal Learning and Behavior, 1*, 151–153.

Thomas, R.K. & Chase, L. (1980). Relative numerousness judgments by squirrel monkeys. *Bulletin of the Psychonomic Society, 16*, 79–82.

Thomas, R.K., Fowlkes, D., & Vickery, J.D. (1980). Conceptual numerousness judgments by squirrel monkeys. *American Journal of Psychology, 93*, 247–257.

Thomas, R.K. & Frost, T. (1983). Oddity and dimension-abstracted oddity (DAO) in squirrel monkeys. *American Journal of Psychology, 96*, 51–64.

Thomas, R.K. & Noble, L.M. (1988). Visual and olfactory oddity learning in rats: What evidence is necessary to show conceptual behavior? *Animal Learning and Behavior, 16*, 157–163.

Thomas, R.K. & Peay, L. (1976). Length judgments by squirrel monkeys: evidence for conservation? *Developmental Psychology, 12*, 349–352.

Thomas, R.K. & Walden, E.L. (1985). The assessment of cognitive development in human and nonhuman primates. In P.P.G. Bateson & P.H. Klopfer (Eds.), *Perspectives in ethology* (pp. 187–215). New York: Plenum Press.

Thompson, R.L. & Boatright-Horowitz, S.L. (1994). The question of mirror-mediated self-recognition in apes and monkeys: Some new results and reservations. In S.T. Parker, R.W. Mitchell, & M.L. Boccia (Eds.), *Self-awareness in animals and humans. Developmental perspectives* (pp. 330–349). Cambridge: Cambridge University Press.

Thompson, R., Oden, D., & Boysen, S. (in press). Language-naive chimpanzees (*Pan troglodytes*) judge relations-between-relations in a conceptual matching-to-sample task. *Journal of Experimental Psychology: Animal Behavior Processes*.

Thomsen, C.E. (1974). Eye contact by non-human primates toward a human observer. *Animal Behaviour, 22*, 144–149.

Thorndike, E.L. (1898). Animal intelligence: An experimental study of the associative process in animals. *Psychology Review Monographs, 2*, 551–553.

Thorndike, E.L. (1901). Mental life of monkeys. *Psychology Review Monographs (Supplement) 15*, 442–444.

Thorpe, W.H. (1956). *Learning and instinct in animals*. London: Methuen and Co.

Tinbergen, N. (1951). *The study of instinct*. New York: Oxford University Press.

Tingpalong, M., Watson, W., Whitmire, R., Chapple, F., & Marshall, J. (1981). Reactions of captive gibbons to natural habitat and wild conspecifics after release. *Natural History Bulletin of the Siam Society, 29*, 31–40.

Tinklepaugh, O.L. (1928). An experimental study of representative factors in monkeys. *Journal of Comparative Psychology, 8*, 197–236.

Tinklepaugh, O.L. (1932). Multiple delayed reaction with chimpanzees and monkeys. *Journal of Comparative Psychology, 13*, 207–243.

Tokida, E., Tanaka, I., Takefushi, H., & Hagiwara, T. (1994). Tool-using in Japanese macaques: use of stones to obtain fruit from a pipe. *Animal Behaviour, 47*, 1023–1030.

Tolman, E.C. (1932). *Purposive behavior in animals and men*. New York: Appleton-Century-Crofts.

Tolman, E.C. (1948). Cognitive maps in rats and men. *Psychological Review, 55*, 189–208.

Tomasello, M. (1990). Cultural transmission in the tool use and communicatory signaling of chimpanzees? In S.T. Parker & K.R. Gibson (Eds.), *"Language" and intelligence in monkeys and apes* (pp. 274–311). Cambridge: Cambridge University Press.

Tomasello, M. (1992a). Cognitive ethology comes of age. *Behavioral and Brain Sciences, 15*, 168–169.

Tomasello, M. (1992b). *First verbs: A case study of early grammatical development*. Cambridge: Cambridge University Press.

Tomasello, M. (1992c). The social bases of language acquisition. *Social Development, 1*, 67–87.

Tomasello, M. (1993). The interpersonal origins of self concept. In U. Neisser (Ed.), *The per-*

ceived self: ecological and interpersonal sources of self knowledge (pp. 174–184). Cambridge: Cambridge University Press.

Tomasello, M. (1994a). Can an ape understand a sentence? A review of language comprehension in ape and child by E.S. Savage-Rumbaugh et al. *Language and Communication, 14*, 377–390.

Tomasello, M. (1994b). The question of chimpanzee culture. In R.W. Wrangham, W.C. McGrew, F.B.M. de Waal, & P.G. Heltne (Eds.), *Chimpanzee cultures* (pp. 301–317). Cambridge, MA: Harvard University Press.

Tomasello, M. (1995a). Joint attention as social cognition. In C. Moore & P. Dunham (Eds.), *Joint attention: Its origins and role in development* (pp. 103–130). Hillsdale, NJ: Lawrence Erlbaum Associates.

Tomasello, M. (1995b). Understanding the self as social agent. In P. Rochat (Ed.), *The self in early infancy: theory and research* (pp. 449–460). Amsterdam: North Holland–Elsevier.

Tomasello, M. (1996a). Do apes ape? In B.G. Galef, Jr. & C.M. Heyes (Eds.), *Social learning in animals: the roots of culture* (pp. 319–346). New York: Academic Press.

Tomasello, M. (1996b). The cultural roots of language. In B. Velichkovsky & D. Rumbaugh (Eds.), *Communicating meaning: The evolution and development of language* (pp. 275–307). Hillsdale, NJ: Erlbaum Associates

Tomasello, M. (in press). Perceiving intentions and learning words in the second year of life. In M. Bowerman & S. Levinson (Eds.), *Language acquisition and conceptual development.* Cambridge: Cambridge University Press.

Tomasello, M. & Akhtar, N. (1995). Two-year-olds use pragmatic cues to differentiate reference to objects and actions. *Cognitive Development, 10*, 201–224.

Tomasello, M. & Barton, M. (1994). Learning words in nonostensive contexts. *Developmental Psychology, 30*, 639–650.

Tomasello, M. & Call, J. (1994). Social cognition of monkeys and apes. *Yearbook of Physical Anthropology, 37*, 273–305.

Tomasello, M., Call, J., & Gluckman, A. (in press). Comprehension of novel communicative signs by apes and human children. *Child Development.*

Tomasello, M., Call, J., Nagell, K., Olguin, R., & Carpenter, M. (1994). The learning and use of gestural signals by young chimpanzees: a trans-generational study. *Primates, 35*, 137–154.

Tomasello, M., Call, J., Warren, J., Frost, T., Carpenter, M., & Nagell, K. (in press). The ontogeny of chimpanzee gestural signals: A comparison across groups and generations. *Evolution of Communication.*

Tomasello, M. & Camaioni, L. (in press). A comparison of the communicative gestures of children and chimpanzees. *Human Development.*

Tomasello, M., Davis-Dasilva, M., Camak, L., & Bard, K. (1987). Observational learning of tool-use by young chimpanzees. *Human Evolution, 2*, 175–183.

Tomasello, M., George, B., Kruger, A., Farrar, M., & Evans, A. (1985). The development of gestural communication in young chimpanzees. *Journal of Human Evolution, 14*, 175–186.

Tomasello, M., Gust, D., & Frost, G.T. (1989). The development of gestural communication in young chimpanzees: a follow up. *Primates, 30*, 35–50.

Tomasello, M., Kruger, A.C., & Ratner, H.H. (1993). Cultural learning. *Behavioral and Brain Sciences, 16*, 495–552.

Tomasello, M. & Olguin, R. (1993). Twenty-three-month-old children have a grammatical category of noun. *Cognitive Development, 8*, 451–464.

Tomasello, M., Savage-Rumbaugh, E.S., & Kruger, A.C. (1993). Imitative learning of actions on objects by children, chimpanzees, and enculturated chimpanzees. *Child Development, 64*, 1688–1705.

Tomasello, M., Strosberg, R., & Akhtar, N. (1996). Eighteen-month-old children learn words in non-ostensive contexts. *Journal of Child Language, 22*, 1–20

Tomonaga, M. (1995). Transfer of odd-item search performance in a chimpanzee (*Pan troglodytes*). *Perceptual and Motor Skills, 80*, 35–42.

Tomonaga, M., Matsuzawa, T., & Itakura, S. (1993). Teaching ordinals to a cardinal-trained chimpanzee. *Primate Research, 9*, 67–77.

Tooby, J. & Cosmides, L. (1991). The psychological foundation of culture. In J.H. Barkow, L. Cosmides, & J. Tooby (Eds.), *The adapted mind: evolutionary psychology and the generation of culture* (pp. 19–136). New York: Oxford University Press.

Torigoe, T. (1985). Comparison of object manipulation among 74 species of non-human primates. *Primates, 26*, 182–194.

Torigoe, T. (1986). Development of object manipulation in the infants of Japanese monkeys. *Japanese Psychological Research, 28*, 149–154.

Torigoe, T. (1987). Object manipulation in a captive troop of Japanese monkeys (*Macaca fuscata*): a developmental analysis. *Primates, 28*, 497–506.

Toth, N., Schick, K.D., Savage-Rumbaugh, E.S., Sevcik, R.A., & Rumbaugh, D.M. (1993). Pan the tool-maker: investigations into the stone tool-making and tool-using capabilities of a bonobo (*Pan paniscus*). *Journal of Archaeological Science, 20*, 81–91.

Trevarthen, C. (1979). Instincts for human understanding and for cultural cooperation: their development in infancy. In M. von Cranach, K. Foppa, W. Lepenies, & D. Ploog (Eds.), *Human ethology: claims and limits of a new discipline* (pp. 530–571). Cambridge: Cambridge Univeristy Press.

Trevarthen, C. (1993). The function of emotions in early communication and development. In J. Nadel & L. Camaioni (Eds.), *New perspectives in early communicative development* (pp. 48–81). New York: Routledge Press.

Trivers, R.L. (1971). The evolution of reciprocal altruism. *Quarterly Review of Biology, 46*, 35–57.

Tutin, C.E.G., Ham, R, & Wrogemann, D. (1995). Tool-use by chimpanzees (*Pan t. troglodytes*) in the Lope reserve, Gabon. *Primates, 36*, 181–192.

Tyrell, D., Stauffer, L., & Snowman, L. (1991). Perception of abstract identity/difference relationships by infants. *Infant Behavior and Development, 14*, 125–129.

Uzgiris, I.C. & Hunt, J.M. (1974). *Towards ordinal scales of psychological developments in infancy*. Illinois: University of Illinois Press.

van de Rijt-Plooij, H.H.C., & Plooij, F.X. (1987). Growing independence, conflict and learning in mother-infant relations in free-ranging chimpanzees. *Behaviour 101*, 1–86.

Van Elsacker, L. & Walraven, V. (1994). The spontaneous use of a pineapple as a recipient by a captive bonobo (*Pan paniscus*). *Mammalia, 58*, 159–162.

van Hooff, J.A.R.A.M. (1967). The facial displays of the catarrhine monkeys and apes. In D. Morris (Ed.), *Primate ethology* (pp. 7–68). London: Weidenfeld and Nicolson.

van Hooff, J.A.R.A.M. (1988). Sociality in primates. A compromise of ecological and social adaptation strategies. In A.Tartabini & M.L. Genta (Eds.), *Perspectives in the study of primates: an Italian contribution to international primatology* (pp. 9–23).

van Krunkelsven, E., Dupain, J., Van Elsacker, L., & Verheyen, R.F. (1996). Food calling by captive bonobos (*Pan paniscus*): an experiment. *International Journal of Primatology, 17*, 207–217.

van Lawick-Goodall, J. (1968a). The behaviour of free-living chimpanzees in the Gombe Stream reserve. *Animal Behaviour Monographs, 1*, 161–311.

van Lawick-Goodall, J. (1968b). A preliminary report on expressive movements and communication in the Gombe stream chimpanzees. In P.C. Jay (Ed.), *Primates. Studies in adaptation and variability* (pp. 313–374). New York: Holt, Rinehart and Winston.

van Lawick-Goodall, J. (1970). Tool-using in primates and other vertebrates. In D.S. Lehrman, R.A. Hinde, & E. Shaw (Eds.), *Advances in the study of behavior* (pp. 195–249). New York: Academic Press.

van Lawick-Goodall, J., Van Lawick, H., & Packer, H. (1973). Tool-use in free living baboons in the Gombe national park, Tanzania. *Nature, 241*, 212–213.

van Schaik, C.P. (1983). Why are diurnal primates living in groups? *Behaviour, 87*, 120–144.

van Schaik, C.P. (1989). The ecology of social relationships amongst female primates. In V. Standen & R. Foley (Eds.), *Comparative socioecology: the behavioral ecology of humans and other mammals* (pp. 195–218). Oxford: Blackwell Scientific.

van Schaik, C., Fox, E., & Sitompul, A. (1996). Manufacture and use of tools in wild Sumatran orangutans. *Naturwissenschaften, 83*, 186–188.

van Schaik, C.P. & van Hooff, J.A.R.A.M. (1983). On the ultimate causes of primate social systems. *Behaviour, 85*, 91–117.

Vauclair, J. (1982). Sensorimotor intelligence in human and non-human primates. *Journal of Human Evolution, 11*, 257–264.

Vauclair, J. (1984). Phylogenetic approach to object manipulation in human and ape infants. *Human Development, 27*, 321–328.

Vauclair, J. (1990). Primate cognition: From representation to language. In S.T. Parker & K.R. Gibson (Eds.), *"Language" and intelligence in monkeys and apes* (pp. 312–329). Cambridge: Cambridge University Press.

Vauclair, J. & Bard, K. (1983). Development of manipulations with objects in ape and human infants. *Journal of Human Evolution, 12*, 631–645.

Vauclair, J., Fagot, J., & Hopkins, W.D. (1993). Rotation of mental images in baboons when the visual input is directed to the left cerebral hemisphere. *Psychological Science, 4*, 99–103.

Vauclair, J., Rollins Jr., H.A., & Nadler, R.D. (1983). Reproductive memory for diagonal and non-diagonal patterns in chimpanzees. *Behavioural Processes, 8*, 289–300.

Vaughter, R.M., Smotherman, W., & Ordy, J.M. (1972). Development of object permanence in the infant squirrel monkey. *Developmental Psychology, 7*, 34–38.

Vevers, G.M. & Weiner, J.S. (1963). Use of a tool by a captive capuchin monkey (*Cebus apellla*). In J.R. Napier (Ed.), *The primates. Symposia of the zoological society of London* (pp. 115–117). London: Academic Press.

Vincent, F. (1973). Utilisation spontanee d'outils chez le mandrill (Primate). *Mammalia, 37*, 277–280.

Visalberghi, E. (1986). Aspects of space representation in an infant gorilla. In D.M. Taub & F.A. King (Eds.), *Current perspectives in primate social dynamics* (pp. 445–452). New York: Van Nostrand Reinhold.

Visalberghi, E. (1987). Acquisition of nut-cracking behaviour by two capuchin monkeys (*Cebus apella*). *Folia Primatologica, 49*, 168–181.

Visalberghi, E. (1990). Tool use in *Cebus*. *Folia Primatologica, 54*, 146–154.

Visalberghi, E. (1993). Tool use in a south american monkey species: an overview of the characteristics and limits of tool use in *Cebus apella*. In A. Berthelet & J. Chavaillon (Eds.), *The use of tools by human and non-human primates* (pp. 118–131). New York: Oxford University Press.

Visalberghi, E. (1994). Learning processes and feeding behavior in monkeys. In B.G. Galef, M. Mainardi, & P. Valsecchi (Eds.), *Behavioral aspects of feeding. Basic and applied research on mammals* (pp. 257–270). Chur, Switzerland: Harwood Academic.

Visalberghi, E. & Fragaszy, D.M. (1990a). Do monkeys ape? In S.T. Parker & K.R. Gibson (Eds.), *"Language" and intelligence in monkeys and apes* (pp. 247–273). Cambridge: Cambridge University Press.

Visalberghi, E. & Fragaszy, D.M. (1990b). Food-washing behaviour in tufted capuchin monkeys, *Cebus apella*, and crabeating macaques, *Macaca fascicularis*. *Animal Behaviour, 40*, 829–836.

Visalberghi, E. & Fragaszy, D. (1995). The behaviour of capuchin monkeys, *Cebus apella*, with novel food: the role of social context. *Animal Behaviour, 49*, 1089–1095.

Visalberghi, E. & Fragaszy, D.M. (1996). Pedagogy and imitation in monkeys: yes, no, or maybe? In D. Olson & N. Torrence (Eds.), *Handbook of education and human development* (pp. 277–301). London: Blackwell.

Visalberghi, E., Fragaszy, D.M., & Savage-Rumbaugh, E.S. (1995). Performance in a tool-using task by common chimpanzees (*Pan troglodytes*), bonobos (*Pan paniscus*), an orangutan (*Pongo pygmaeus*), and capuchin monkeys (*Cebus apella*). *Journal of Comparative Psychology, 109*, 52–60.

Visalberghi, E. & Limongelli, L. (1994). Lack of comprehension of cause-effect relations in tool-using capuchin monkeys (*Cebus apella*). *Journal of Comparative Psychology, 108*, 15–22.

Visalberghi, E. & Limongelli, L. (1996). Acting and Understanding: tool use revisited through the minds of capuchin monkeys. In A.E. Russon, K.A. Bard, & S.T. Parker (Eds.), *Reaching into thought* (pp. 57–79). Cambridge: Cambridge University Press.

Visalberghi, E. & Trinca, L. (1989). Tool use in capuchin monkeys: distinguishing between performing and understanding. *Primates, 30*, 511–521.

von Glasersfeld, E. (1982). Subitizing: the role of figural patterns in the development of numerical concepts. *Archives de Psychologie, 50*, 191–218.

Vygotsky, L. (1978). *Mind in society*. Cambridge, MA: Harvard University Press.

Walraven, V., Van Elsacker, L., & Verheyen, R. (1995). Reactions of a group of pygmy chimpanzees (*Pan paniscus*) to their mirror images: Evidence of self-recognition. *Primates, 36*, 145–150.

Walters, J.R. & Seyfarth, R.M. (1987) Conflict and cooperation. In B.B. Smuts, D.L. Cheney, R.M. Seyfarth, R.W. Wrangham, & T.T. Struhsaker (Eds.), *Primate societies* (pp. 306–317). Chicago: University of Chicago Press.

Ward, J.P., Silver, B.V., & Frank, J. (1976). Preservation of cross-modal transfer of a rate discrimination in the bushbaby (*Galago senegalensis*) with lesions of posterior neocortex. *Journal of Comparative and Physiological Psychology, 90*, 520–527.

Ward, J.P., Yehle, A.L., & Doerflein, R.S. (1970). Cross-modal transfer of a specific discrimination in the bushbaby (*Galago senegalensis*). *Journal of Comparative and Physiological Psychology, 73*, 74–77.

Warden, C.J. & Jackson, T.A. (1935). Imitative behavior in the rhesus monkey. *Journal of Genetic Psychology, 46*, 103–125.

Warden, C.J., Fjeld, H.A., & Koch, A.M. (1940a). Imitative behavior in *Cebus* and *Rhesus* monkeys. *Journal of Genetic Psychology, 56*, 311–322.

Warden, C.J., Koch, A.M., & Fjeld, H.A. (1940b). Instrumentation in *Cebus* and *Rhesus* monkeys. *Journal of Genetic Psychology, 56*, 297–310.

Warden, C.J., Koch, A.M., & Fjeld, H.A. (1940c). Solution of patterned string problems by monkeys. *Journal of Genetic Psychology, 56*, 283–295.

Warren, J.M. (1974). Possibly unique characteristics of learning by primates. *Journal of Human Evolution, 3*, 455–454.

Waser, P.M. (1977). Individual recognition, intragroup cohesion and intergroup spacing: evidence from sound playback to forest monkeys. *Behaviour, 60*, 28–74.

Washburn, D.A. (1992). Analyzing the path of responding in maze-solving and other tasks. *Behavior Research Methods, Instruments, and Computers, 24*, 248–252.

Washburn, D.A. (1994). Stroop-like effects for monkeys and humans. *Psychological Science, 5*, 375–379.

Washburn, D., Hopkins, W., & Rumbaugh, D. (1991). Perceived control in rhesus monkeys (*Macaca mulatta*): enhanced video task performance. *Animal Behavior Processes, 17*, 123–127.

Washburn, D.A. & Rumbaugh, D.M. (1991a). Ordinal judgments of numerical symbols by macaques (*Macaca mulatta*). *Psychological Science, 2*, 190–193.

Washburn, D.A. & Rumbaugh, D.M. (1991b). Rhesus monkeys' (*Macaca mulatta*) complex learning skills reassessed. *International Journal of Primatology, 12*, 377–388.

Washburn, D.A. & Rumbaugh, D.M. (1992). Comparative assessment of psychomotor performance: Target prediction by humans and macaques (*Macaca mulatta*). *Journal of Experimental Psychology: General, 121*, 305–312.

Wasserman, E.A., Kiedinger, R.E., & Bhatt, R.S. (1988). Conceptual behavior in pigeons: Categories, subcategories, and pseudocategories. *Animal Behavior Processes, 14*, 235–246.

Watson, J.B. (1908). Imitation in monkeys. *Psychological Bulletin, 5*, 169–178.

Watson, J.B. (1925). *Behaviorism*. New York: W.W. Norton.

Watson, S.L., Schiff, M., & Ward, J.P. (1994). Effects of modeling and lineage on fishing behavior in the small-eared bushbaby (*Otolemur garnettii*). *International Journal of Primatology, 15*, 507–519.

Watts, D.P. (1995). Post-conflict social events in wild mountain gorillas (*Mammalia, Hominoidea*) I. Social interactions between opponents. *Ethology, 100*, 139–157.

Wechkin, S. (1970). Social relationships and social facilitation of object manipulation in *Macaca mulatta*. *Journal of Comparative and Physiological Psychology, 73*, 456–460.

Weinberg, S.M. & Candland, D.K. (1981). "Stone-grooming" in *Macaca fuscata*. *American Journal of Primatology, 1*, 465–468.

Weinstein, B. (1941). Matching-from-sample by rhesus monkeys and by children. *Journal of Comparative and Physiological Psychology, 31*, 195–213.

Weinstein, B. (1945). The evolution of intelligent behavior in rhesus monkeys. *Genetic Psychology Monographs, 31*, 2–48.

Weiskrantz, L. & Cowey, A. (1975). Cross-modal matching in the rhesus monkey using a single pair of stimuli. *Neuropsychologica, 13*, 257–261.

Welker, C. (1976). Fishing behavior in *Galago crassicaudatus*. *Folia Primatologica, 26*, 284–291.

Welker, C., Schwibbe, M., Shafer-Witt, C., & Visalberghi, E. (1987). Failure of kin recognition in *Macaca fascicularis*. *Folia Primatologica, 49*, 216–221.

Wellman, H. (1990). *The child's theory of mind*. Cambridge, MA: MIT Press.

Wells, H. & Deffenbacher, K. (1967). Conjunctive and disjunctive concept learning in humans and squirrel monkeys. *Canadian Journal of Psychology, 21*, 301–308.

Westergaard, G.C. (1988). Lion-tailed macaques (*Macaca silenus*) manufacture and use tools. *Journal of Comparative Psychology, 102*, 152–159.

Westergaard, G.C. (1989). Infant baboons spontaneously use an object to obtain distant food. *Perceptual and Motor Skills, 68*, 558–558.

Westergaard, G.C. (1992). Object manipulation and the use of tools by infant baboons (*Papio cynocephalus anubis*). *Journal of Comparative Psychology, 106*, 398–403.

Westergaard, G.C. (1993). Development of combinatorial manipulation in infant baboons (*Papio cynocephalus anubis*). *Journal of Comparative Psychology, 107*, 34–38.

Westergaard, G.C. (1994). The subsistence technology of capuchins. *International Journal of Primatology, 15*, 899–906.

Westergaard, G.C. & Fragaszy, D.M. (1985). Effects of manipulatable objects on the activity of captive capuchin monkeys (*Cebus apella*). *Zoo Biology, 4*, 317–327.

Westergaard, G.C. & Fragaszy, D.M. (1987a). The manufacture and use of tools by capuchin monkeys (*Cebus apella*). *Journal of Comparative Psychology, 101*, 159–168.

Westergaard, G.C. & Fragaszy, D.M. (1987b). Self-treatment of wounds by a capuchin monkey (*Cebus apella*). *Human Evolution, 1*, 557–562.

Westergaard, G.C., Greene, J.A., Babitz, M.A., & Suomi, S.J. (1995). Pestle use and modification by tufted capuchins (*Cebus apella*). *International Journal of Primatology, 16*, 643–651.

Westergaard, G.C. & Hyatt, C.W. (1994). The responses of bonobos (*Pan paniscus*) to their mirror images: evidence of self-recognition. *Human Evolution, 9*, 273–279.

Westergaard, G.C. & Lindquist, T. (1987). Manipulation of objects in a captive group of lion-tailed macaques (*Macaca silenus*). *American Journal of Primatology, 12*, 231–234.

Westergaard, G.C. & Suomi, S.J. (1994a). Aimed throwing of stones by tufted capuchin monkeys (*Cebus apella*). *Human Evolution, 9*, 323–329.

Westergaard, G.C. & Suomi, S.J. (1994b). Hierarchical complexity of combinatorial manipulation in capuchin monkeys (*Cebus apella*). *American Journal of Primatology, 32*, 171–176.

Westergaard, G.C. & Suomi, S.J. (1994c). A simple stone-tool technology in monkeys. *Journal of Human Evolution, 27*, 399–404.

Westergaard, G.C. & Suomi, S.J. (1994d). Stone-tool bone-surface modification by monkeys. *Current Anthropology, 35*, 468–470.

Westergaard, G.C. & Suomi, S.J. (1994e). The use and modification of bone tools by capuchin monkeys. *Current Anthropology, 35*, 75–77.

Westergaard, G.C. & Suomi, S.J. (1994f). The use of probing tools by tufted capuchins (*Cebus apella*): evidence for increased right-hand preference with age. *International Journal of Primatology, 15*, 521–529.

Westergaard, G.C. & Suomi, S.J. (1995a). The manufacture and use of bamboo tools by monkeys. *Journal of Archeological Science, 22*, 677–681.

Westergaard, G.C. & Suomi, S.J. (1995b). Stone-throwing by capuchins (*Cebus apella*): A model of throwing capabilities in *Homo habilis*. *Folia Primatologica*, 65, 234–238.

White, R. (1989). Production complexity and standardization in early Aurignacian bead and pendant manufacture. In P. Mellars & C. Stringer (Eds.), *The human revolution: behavioral and biological perspectives on the origins of modern humans* (pp. 366–390) Princeton: Princeton University Press.

Whiten, A. (1991) *Natural theories of mind.* New York: Oxford University Press.

Whiten, A. (1993). Evolving a theory of mind: the nature of non-verbal mentalism in other primates. In S. Baron-Cohen, H. Tager-Flusberg, & D.J. Cohen (Eds.), *Understanding other minds. Perspectives from autism* (pp. 367–396). New York: Oxford University Press.

Whiten, A. (1994). Grades of mindreading. In C. Lewis & P. Mitchell (Eds.), *Children's early understanding of mind* (pp. 47–70). Hillsdale, NJ: Lawrence Erlbaum Associates.

Whiten, A. & Byrne, R.W. (1988a). The manipulation of attention in primate tactical deception. In R.W. Byrne & A. Whiten (Eds.), *Machiavellian intelligence. Social expertise and the evolution of intellect in monkeys, apes, and humans* (pp. 211–223). New York: Oxford University Press.

Whiten, A. & Byrne, R.W. (1988b). Tactical deception in primates. *Behavioral and Brain Sciences, 11*, 233–244.

Whiten, A. & Byrne, R.W. (1988c). Taking (Machiavellian) intelligence apart: editorial. In R.W. Byrne & A. Whiten (Eds.), *Machiavellian intelligence. Social expertise and the evolution of intellect in monkeys, apes, and humans* (pp. 50–65). New York: Oxford University Press.

Whiten, A. & Byrne, R.W. (1991). The emergence of metarepresentation in human ontogeny and primate phylogeny. In A.Whiten (Ed.), *Natural theories of mind* (pp. 267–281). Oxford: Blackwell Scientific.

Whiten, A., Custance, D.M., Gómez, J.C., Teixidor, P., & Bard, K.A. (1996). Imitative learning of artificial fruit processing in children (*Homo sapiens*) and chimpanzees (*Pan troglodytes*). *Journal of Comparative Psychology, 110*, 3–14.

Whiten, A. & Ham, R. (1992). On the nature and evolution of imitation in the animal kingdom: reappraisal of a century of research. In P.J.B. Slater, J.S. Rosenblatt, C. Beer, & M. Milinsky (Eds.), *Advances in the study of behavior* (pp. 239–283). New York: Academic Press.

Wilkerson, B.J. & Rumbaugh, D.M. (1979). Learning and intelligence in prosimians. In G.A. Doyle & R.D. Martin (Eds.), *The study of prosimian behavior* (pp. 207–246). New York: Academic Press.

Wilkinson, G.R. (1984). Reciprocal food sharing in vampire bats. *Nature, 308*, 181–184.

Wilson, B., Mackintosh, N., & Boakes, R. (1985). Matching and oddity learning in the pigeon: Transfer effects and the absence of relational learning. *Quarterly Journal of Experimental Psychology, 37B*, 295–311.

Wimmer, H. & Perner, J. (1983). Beliefs about beliefs: Representation and constraining function of wrong beliefs in young children's understanding of deception. *Cognition, 13*, 103–128.

Winter, P., Handley, P., Ploog, D., & Schott, D. (1973). Ontogeny of squirrel monkey calls under normal conditions and under acoustic isolation. *Behaviour, 47*, 230–239.

Wise, K.L., Wise, L.A., & Zimmerman, R.R. (1974). Piagetian object permanence in the infant rhesus monkey. *Developmental Psychology, 10*, 429–437.

Witmer, L. (1909). A monkey with a mind. *Psychological Clinic, 3*, 179–205.

Witmer, L. (1910). Intelligent imitation and curiosity in a monkey. *The Psychological Clinic, 3*, 225–227.

Wolfle, D.L. & Wolfle, H.M. (1939). The development of cooperative behavior in monkeys and young children. *Journal of Genetic Psychology, 55*, 137–175.

Wood, R. (1984). Spontaneous use of sticks by gorillas at Howletts Zoo Park, England. *International Zoo News, 31*, 13–18.

Wood, S., Moriarty, K.M., Gardner, B.T., & Gardner, R.A. (1980). Object permanence in child and chimpanzee. *Animal Learning and Behavior, 8*, 3–9.

Woodrow, H. (1929). Discrimination by the monkey of temporal sequences of varying number of stimuli. *Journal of Cômparative Psychology, 9*, 123–157.

Woodruff, G. & Premack, D. (1979). Intentional communication in the chimpanzee: the development of deception. *Cognition, 7*, 333–362.

Woodruff, G. & Premack, D. (1981). Primitive mathematical concepts in the chimpanzee: proportionality and numerosity. *Nature, 293*, 568–570.

Woodruff, G., Premack, D., & Kennel, K. (1978). Conservation of liquid and solid quantity by the chimpanzee. *Science, 202*, 991–994.

Wrangham, R.W. (1977). Feeding behaviour of chimpanzees in Gombe National Park, Tanzania. In T.H. Clutton-Brock (Ed.), *Primate ecology: studies of feeding and ranging behaviour in lemurs, monkeys and apes* (pp. 504–538). New York: Açademic Press.

Wrangham, R.W. (1980). An ecological model of female-bonded primate groups. *Behaviour, 75*, 262–299.

Wrangham, R.W. (1982). Mutualism, kinship and social evolution. In *Current problems in sociobiology* (pp. 269–289). New York: Cambridge University Press.

Wrangham, R.W., McGrew, W.C., de Waal, F.B.M., & Heltne, P.G. (1994). *Chimpanzee cultures*. Cambridge, MA: Harvard University Press.

Wright, A.A., Santiago, H.C., & Sands, S.F. (1984). Monkey memory: same/different concept learning, serial probe acquisition, and probe delay effects. *Animal Behavior Processes, 10*, 513–529.

Wright, A.A., Shyan, M., & Jitsumori, M. (1990). Auditory same/different concept learning by monkeys. *Animal Learning and Behavior, 18*, 287–294.

Wright, R.V.S. (1972). Imitative learning of a flacked stone technology—the case of an orangutan. *Mankind, 8*, 296–306.

Wynn, K. (1992). Addition and substraction by human infants. *Nature, 358*, 749–750.

Yagi, B. & Furusaka, T. (1973). Intelligent behavior of Japanese monkeys. In C.R. Carpenter (Ed.), *Behavioral regulators of behavior in primates* (pp. 272–283). Lewisburg, PA: Bucknell University Press.

Yamakoshi, G. & Sugiyama, Y. (1995). Pestle-pounding behavior of wild chimpanzees at Bossou, Guinea: A newly observed tool-using behavior. *Primates, 36*, 489–500.

Yerkes, R.M. (1916). *The mental life of monkeys and apes*. Delmar, NY: Scholars' Facsimiles and Reprints.

Yerkes, R.M. (1927). The mind of a gorilla. *Genetic Psychology Monographs, 2*, 1–193.

Yerkes, R.M. (1934a). Modes of behavioral adaptation in chimpanzee to multiple-choice problems. *Comparative Psychology Monographs, 10*, 1–108.

Yerkes, R.M. (1934b). Suggestibility in chimpanzee. *The Journal of Social Psychology, 5*, 271–282.

Yerkes, R.M. (1943). *Chimpanzees. A laboratory colony*. New Haven, CT: Yale University Press.

Yerkes, R.M. & Coburn, C.A. (1915). A study of the behavior of the pig *Sus scrofa* by the multiple-choice method. *Journal of Animal Behavior, 5*, 185–225.

Yerkes, R.M. & Petrunkevitch, A. (1925). Studies of chimpanzee vision by Ladygin-Koths. *Journal of Comparative Psychology, 5*, 99–108.

Yerkes, R.M. & Yerkes, D.N. (1928). Concerning memory in the chimpanzee. *Journal of Comparative Psychology, 8*, 237–271.

Yoshikubo, S. (1985). Species discrimination and concept formation by rhesus monkeys (*Macaca mulatta*). *Primates, 26*, 285–299.

Zabel, C.J., Glickman, S.E., Frank, L.G., Woodmansee, K.B., & Keppel, G. (1992). Coalition formation in a colony of prepubertal spotted hyenas. In A.H. Harcourt & F.B.M. de Waal (Eds.), *Coalitions and alliances in human and nonhuman animals* (pp. 113–135). New York: Oxford University Press.

Zentall, T.R. (1996). An analysis of imitative learning in animals. In B.G. Galef, Jr. & C.M. Heyes (Eds.), *Social learning in animals: the roots of culture* (pp. 221–243). New York: Academic Press.

Zentall, T., Edwards, C., Moore, B., & Hogan, D. (1981). Identity: the basis for both matching and oddity behavior in pigeons. *Animal Behavior Processes, 7*, 70–86.

Zoloth, S., Petersen, M., Beecher, M., Green, S., Marler, P., Moody, D., Stebbins, W. (1979). Species-specific perceptual processing of vocal sounds by monkeys. *Science, 204*, 870–872.

Zuberbühler, K., Gygax, L., Harley, N., & Kummer, H. (1996). Stimulus enhancement and spread of a spontaneous tool use in a colony of long-tailed macaques. *Primates, 37*, 1–12.

Author Index

Achenbach, T. M., 40
Adams-Curtis, L., 62, 287, 288
Akhtar, N., 409
Alcock, J., 27
Altmann, J., 31
Altmann, S. A., 31, 237, 243
Amsterdam, B. K., 338
Anderson, J. R., 240, 317, 336
Andrews, M. W., 33
Angarella, R., 113
Angst, W., 202
Antinucci, F., 40, 44, 61, 169, 375
Anzenberger, G., 239
Arber, M., 39
Armstrong, G. D., 116
Ashley, J., 228, 255, 308
Astington, J., 413
Aureli, F., 201, 203, 209, 349, 376
Aversano, M., 33

Baillargeon, R., 26, 41, 57, 172, 404, 413
Bakeman, R., 247
Balasch, J., 49
Baldwin, J. P., 274
Baldwin, P. J., 278
Bandura, A., 273
Barbier, H., 317
Bard, K. A., 63, 64, 67, 92, 102, 197, 246, 289, 291, 303, 306, 316, 336
Baron-Cohen, S., 413
Barresi, J., 417, 418
Barsalou, L., 402
Barton, M., 362, 409
Basen, J. A., 333
Bates, E., 243, 244, 249, 266, 405
Bauer, H. R., 194
Bauers, K. A., 255
Beck, B. B., 49, 58, 82, 86, 226, 287, 369
Beecher, M., 116
Bennett, A., 54
Bennett, A. J., 60
Benz, J. J., 255
Bercovitch, F. B., 212, 217
Berdecio, S., 244, 302
Bergeron, G., 40, 44, 63
Berkson, G., 111
Berlinguer, L., 40, 44
Bernstein, I. S., 63, 121, 197, 206, 207
Berntson, G. G., 141, 147, 150, 152, 155, 194
Bever, T. G., 6, 260, 261
Bhatt, R. S., 113
Biben, M., 193, 255
Bierschwale, D. T., 325, 336
Bingham, H. C., 49, 78
Birch, H. G., 89
Birchette, L. M., 40
Biró, S., 318
Blakeslee, P., 109
Blaschke, M., 262
Blum, J. S., 119
Blum, R. A., 119
Boakes, R., 119
Boatright-Horowitz, S. L., 336
Boccia, M. L., 330, 336
Boesch, C., 36, 73, 78, 194, 217, 219, 220, 235, 259, 275, 278, 279, 281, 306–7, 378, 425
Boesch, H., 36, 73, 78, 219, 220, 235, 425
Boesch-Achermann, H., 78
Boinski, S., 82, 226, 256, 283
Bonner, J. T., 274, 308
Borton, R. W., 108
Botto-Mahan, C., 336
Bower, T. G. R., 159
Boyd, M. G., 120

Boyd, R., 12, 308
Boysen, S. T., 7, 93, 123, 137, 141, 147, 150, 152, 154, 155, 194, 225, 319, 325
Bradford, S. A., 155
Brady, P. M., 113, 114
Braggio, J. T., 126
Brakke, K. E., 64, 65, 265, 266
Breinliger, K., 404
Bremner, G., 39, 40
Brewer, S. M., 73
Bromer, J. A., 109
Brossard, A., 414, 421
Brown, C. H., 243
Brown, D., 86
Brown, M., 38
Bruner, J., 10, 243, 264, 318, 401, 411, 420
Brunet, C., 40, 44
Bryant, P. E., 148
Buchanan, J. P., 126
Budd, R. L., 260
Burdyn, L. E., 123, 373, 374
Burton, J. J., 226
Buskirk, R. E., 60, 82
Buskirk, W. H., 60, 82
Busse, C. D., 219
Butterworth, G., 243
Byrne, J. M., 65, 282
Byrne, R. W., 65, 177, 232, 233, 234, 235, 236, 237, 238, 242, 247, 258, 270, 282, 292, 294, 313, 316, 330, 338, 343, 350, 352, 353, 356, 377, 384, 385, 386, 396, 398

Cacioppo, J. T., 141
Call, J., 40, 69, 97, 158, 169, 205, 226, 243, 244, 246, 262, 268, 271, 274, 290, 294, 296, 303, 314, 327, 350, 353, 364, 378, 384, 390, 392
Camaioni, L., 323
Camak, L., 289
Cambefort, J. P., 285
Campbell, C. B. G., 430
Candland, D., 402
Capaldi, E. J., 137, 184
Caro, T. M., 304
Carpenter, M., 244, 316, 407
Cha, J., 48
Chalmeau, R., 222, 227
Chalmers, M., 143, 145, 148
Chamove, A., 287
Chance, M. R. A., 89, 187, 356
Chandler, M., 242, 329
Chapais, B., 199, 202, 208, 209, 216, 217
Chapman, C., 255
Chase, L., 140
Chater, N., 113
Cheney, D. L., 7, 13, 31, 91, 92, 115, 193, 198, 199, 200, 201, 202, 208, 214, 215, 216, 217, 226, 227, 237, 242, 250, 252, 253, 255, 256, 259, 260, 297, 298, 305, 316, 328, 330, 342, 353, 370, 371, 380, 381, 382, 396, 398
Chevalier-Skolnikoff, S., 61, 67, 82
Clark, A. P., 257, 258–259
Cleveland, J., 193, 256
Clutton-Brock, T. H., 28, 177, 395
Coburn, C. A., 149
Cohen, D., 413
Cohn, R. H., 333
Coldren, J., 112, 133
Cole, M., 401, 402
Collingwood, R. J., 388, 418
Colmenares, F., 207
Colombo, J., 112, 133
Colombo, M., 119, 143, 144, 145
Connor, R. C., 203, 207, 372
Cook, M., 285
Cooley, C., 337
Cooper, R., 82
Cordeau, J. P., 109
Cordischi, C., 201
Cords, M., 202, 205, 213, 342, 367, 381
Corkum, V., 318
Cornell, E., 39
Cosmides, L., 186, 380
Costello, M., 82
Coussi-Korbel, S., 30, 196, 239, 248, 360
Cowey, A., 108
Cozzolino, R., 201
Crawford, M. P., 4, 221, 246
Csibra, G., 318
Culbertson, G., 130
Curio, E., 285
Custance, D. M., 291, 303, 378
Czerny, P., 156

Dagenais, Y., 93
D'Amato, M. R., 111, 114, 115, 119, 125, 143, 144, 145, 373
Darwin, C., 3, 4, 430
Dasser, V., 122, 194, 200, 204, 371, 372, 374
Daudelin, N., 93
Davenport, R. K., 108, 285
Davidson, I., 425
Davis, H., 137, 142, 148, 155
Davis, R. T., 49, 120
Davis-Dasilva, M., 289
Dawkins, R., 233, 356
Dawson, G. R., 294
de Blois, S. T., 40, 44
Decarie, T. G., 93
Deffenbacher, K., 170
De Lillo, C., 33, 108, 160, 184
Delius, J., 48
DeLoache, J. S., 70
Demas, G., 38
Dennett, D., 11
Deputte, B. L., 288
Desportes, C., 226
Devine, J. V., 149
DeVore, I., 207
de Waal, F. B. M., 7, 78, 192, 196, 198, 199, 202, 207, 208, 209, 210, 211, 212, 213, 214, 217, 232, 236, 237, 242, 246, 247, 255, 258, 275, 281, 299, 314, 316, 349, 356, 369, 376, 386

Dewsbury, D. A., 287
Diamond, A., 39, 55
Dickinson, A., 388
Dieter, J. A., 298
Dittrich, W., 47, 194
Dittus, W. P. J., 254, 255
Doerflein, R. S., 109
Döhl, J., 49, 78
Donald, M., 402, 420
Dooley, G. B., 140
Doré, F. Y., 41, 42, 44, 46, 54, 127, 147, 369
Douglas, J. W. B., 141
Drea, C., 103
Dücker, G., 331, 333
Dufty, A., 255
Dukas, R., 113
Dumas, C., 40, 41, 44, 46, 127, 147, 369
Dunbar, R. I. M., 13, 177, 179, 196, 213, 231, 354, 396
Dunn, J., 413
Dupain, J., 259

Eddy, T. J., 314, 315, 317, 322, 323, 336, 350, 378
Edwards, C., 119
Ehardt, C. L., 206, 207
Ehrlich, A., 60
Eisenberg, J., 177
Ek, A., 213, 214
Elgar, M. A., 257
Elliot, R. C., 108
Elowson, A. M., 255, 297
Elsacker, L., 259
Erikson, C. J., 37
Essock-Vitale, S. M., 108, 133, 169, 376
Etienne, A. S., 41
Ettlinger, G., 70, 108, 109, 112, 126, 127, 262, 376, 379
Evans, A., 203, 244
Evans, C. S., 250, 252, 259, 271
Evans, L., 252, 259
Evans, S., 333, 336

Fady, J. C., 226
Fagot, J., 47
Fairbanks, L. A., 213, 214, 285
Farrar, M., 244
Feldman, D. W., 286
Fernandes, M. E. B., 82
Fernández, P., 314
Ferster, C. B., 138
Field, L., 257
Fillmore, C. G., 411
Finch, G., 48, 111, 119
Fischer, G. J., 37, 40, 48, 109, 111
Flaherty, R., 257
Flaningam, M. R., 112
Fletemeyer, J., 305
Fobes, J. L., 100, 102, 103, 111, 120, 121, 369, 376
Ford, M. L., 60
Foreman, N., 39
Fouts, D. H., 302, 307
Fouts, R. S., 69, 260, 302, 307
Fowler, H., 6
Fowlkes, D., 140
Fox, E., 80
Fragaszy, D. M., 62, 82, 92, 274, 276, 283, 284, 285, 287, 288, 307
Franco, F., 243
Frank, J., 109
Frank, L. G., 206
Fritz, A. S., 242, 329
Frost, G. T., 244
Frost, T., 120, 121
Fruth, B., 279
Frye, D., 95, 99, 390, 417
Fujita, K., 115, 117, 119, 121, 297
Furuichi, T., 197

Gagné, R. M., 170
Gagnon, S., 41, 42, 44, 46, 54
Galdikas, B. F. M., 37, 67, 80, 198, 199, 201, 202, 206, 293
Galef, B. G., Jr., 38, 273, 276, 277, 285, 369
Gallistel, C. R., 137, 161
Gallo, A., 222
Gallup, G. G., Jr., 312, 330, 331, 336, 338, 350, 353, 378
Garber, P. A., 32
Garcha, H. S., 126, 127
Gardner, B. T., 7, 260
Gardner, R. A., 7, 260
Garrett, M., 336
Gauthier, C., 202
Gelman, R., 26, 137, 142, 161, 172, 413
Gentner, D., 125
George, B., 244
Gergely, G., 318, 362, 405
Ghiglieri, M. P., 257, 281
Gibeault, S., 33, 38, 149
Gibson, E., 318, 322
Gibson, J. J., 11, 27, 182, 308
Gibson, K. R., 7, 58, 61, 62, 88, 169, 171, 172, 174, 177, 178, 179, 370
Gill, T., 140
Gillan, D. J., 124, 147
Glickman, S., 59, 96, 206
Gluckman, A., 246
Godin, L., 302
Goetz, W., 202
Goldman-Rakic, P. S., 194
Gómez, J. C., 82, 261, 291, 314, 328
Goodall, J., 6, 69, 73, 78, 194, 198, 199, 202, 206, 219, 235, 236, 237, 244, 257, 258, 278, 280, 284, 299, 301, 306
Gopnik, A., 416, 417, 418
Gordon, J. A., 119
Gordon, R., 417
Goswami, U., 125
Gould, C., 9
Gould, J. L., 9, 38
Gould, S. J., 171
Gouzoules, H., 7, 31, 197, 198, 202, 253, 254, 255, 294, 381
Gouzoules, S., 7, 31, 197, 198, 202, 207, 209, 253, 255, 294
Green, S., 116, 193, 255, 276, 297
Greenfield, P. M., 62, 65, 261, 265, 266, 268
Griffin, D. R., 6
Gros-Louis, J., 258

Groves, C. P., 63
Guillaume, P., 49
Gunderson, V., 108
Gunter, R., 109
Gust, D. A., 36, 244
Gygax, L., 282
Gyger, M., 252

Haggerty, M. E., 80, 289
Hagiwara, T., 86
Haith, M. M., 194
Hala, S., 242, 330
Hall, A. D., 126
Hall, K. R. L., 207, 285
Hallock, M. B., 40, 63
Ham, R., 274, 285, 287, 288, 289
Hamilton, W. J., 60, 82
Handley, P., 256
Hannah, A. C., 280
Hannan, M. B., 141
Hannon, B., 32
Hansen, E. W., 193
Hara, K., 73
Harcourt, A. H., 206, 208, 210, 218, 228, 259, 369
Harley, N., 282
Harlow, H. F., 5, 48, 49, 102, 103, 109, 111, 120, 169
Harper, S. E., 294
Harris, M. R., 145
Harris, P., 70, 404, 413, 417
Harter, S., 407
Harvey, P. H., 177, 395
Hasegawa, T., 217, 258
Hasegawa, Y., 285
Hauser, M. D., 152, 243, 249, 250, 255, 257, 259, 282, 285, 297, 304, 336
Hayaki, H., 199, 217
Hayes, C., 4, 69, 78, 103, 258, 260, 289, 292, 299, 303, 316, 323, 378
Hayes, K. J., 4, 78, 103, 128, 130, 135, 139, 258, 260, 289, 292, 303, 323, 376, 378
Hedges, M., 121
Hegel, M. T., 152
Helland, J., 122
Heltne, P. G., 275
Hemelrijk, C. K., 213, 214, 215, 217, 218, 239
Hemmi, J. M., 31
Hennessy, D., 369
Herman, L. M., 53, 119, 269
Hernandez-Camacho, J., 82
Herrnstein, R. J., 113
Hervé, N., 288
Hespos, S., 152
Heth, C., 39
Hewitt, T. A., 213
Heyes, C. M., 113, 197, 226, 273, 293, 294, 325, 336
Hicks, L. H., 138
Hikami, K., 285, 305
Hinde, R. A., 193, 197
Hiraiwa-Hasegawa, M., 202, 209
Hirsch, A. D., 307
Hobson, P., 418, 421
Hodos, W., 430
Hogan, D., 119
Hollard, V., 48
Hollis, J. H., 225, 226, 228
Holloway, R. L., 28
Honig, W. K., 6
Hopkins, W. D., 47, 51, 108, 150, 258, 299, 331
Horrocks, J. A., 212
Huffman, M., 61, 278
Hulse, S. H., 6
Hume, D., 388
Humphrey, N. K., 7, 187, 351, 356
Hunt, G., 86, 97
Hunt, J. M., 57
Hunte, W., 212
Hutchins, S. S., 265
Hyatt, C. W., 331

Inoue-Nakamura, N., 80
Irle, E., 114
Itakura, S., 147, 194, 317, 336
Itani, J., 276

Jackson, W. J., 119
Jacobson, K., 404
Jalles-Fihlo, E., 82
Jarvis, M. J., 108, 109
Jensvold, M. L. A., 69
Jerison, H., 13
Jitsumori, M., 119
Johnson, M., 414
Johnson, T., 86
Jolly, A., 39, 41, 60, 109, 172, 178, 187, 351, 356
Joubert, A., 32
Joulian, F., 279
Jouventin, P., 285, 305
Judge, P. G., 201
Jugiyama, Y., 281
Juno, C., 32, 34, 38

Kano, T., 80, 197, 199, 206
Kaplan, J. N., 193
Karakashian, S. J., 252–53
Karmiloff-Smith, A., 380, 414, 421
Kawai, M., 7, 199, 206, 276, 278
Kawamura, S., 276
Keddy-Hector, A. C., 199, 316
Kellogg, L. A., 4, 49, 64, 260
Kellogg, W. N., 4, 49, 64, 260
Kennel, K., 157
Keppel, G., 206
Khrustov, G. F., 92
Kiedinger, R. E., 113
King, B. J., 177, 178, 179, 307
King, J. E., 48, 100, 102, 103, 111, 112, 120, 121, 122, 123, 274, 369, 376
King, M., 401
Kitahara-Frisch, J., 73, 280
Kitchener, S. L., 37, 40, 48, 109, 111
Klein, R., 423, 425
Klopfer, P. H., 286
Klüver, H., 4, 129
Knauss, K. S., 113
Knobloch, H., 65
Koehler, O., 142, 155
Köhler, W., 4, 49, 63, 78, 89, 132
Kojima, T., 119
Kotovsky, L., 125
Koyama, N. F., 196, 231
Kralik, J., 336
Krebs, J., 38, 54, 233, 356

Kruger, A., 244
Kruger, A. C., 189, 289, 292, 309, 353, 391, 392, 402, 407, 414
Kudo, H., 354, 395, 396
Kummer, H., 7, 31, 187, 191, 201, 202, 207, 209, 231, 232, 236, 239, 247, 248, 274, 282, 284, 315, 316, 328, 356, 360, 386
Kuroda, R., 141
Kurt, F., 248

Lakoff, G., 414
Landau, K. R., 336
Langacker, R., 419
Langer, J., 130, 132, 135
Larson, C., 297
Lawson, J., 52, 127
Leakey, R., 423
Leary, R. W., 49, 120
Ledbetter, D. H., 333
Lefebvre, L., 213, 255, 277
Leger, D. W., 252
Lepoivre, H., 103
Leslie, A. M., 338
Lethmate, J., 80, 331, 333
Lewis, M., 337, 338, 407
Lillard, A. S., 70
Limongelli, L., 91, 92, 93, 95, 393
Lin, A. C., 336
Lindeman, R., 297
Linden, E., 69, 127, 307
Lindquist, T., 86
Lipsett, L. P., 121
Lock, A., 264
Lorenz, K., 5
Loukas, E., 119, 143
Loveland, D. H., 113, 337
Luttrell, L. M., 209, 212, 213, 214

MacDonald, S. E., 30, 33, 34, 37, 38, 149, 155
Macedonia, J. M., 256, 271
Machida, S., 86
Mackintosh, N. J., 111
Macomber, J., 404
MacPhail, E. M., 379
MacWhinney, B., 266
Maeda, T., 297
Maestripieri, D., 203, 213, 243, 248, 271, 304–5, 315
Maier, N. R. F., 12
Mandler, J., 411
Marchesi, N., 279
Marchesi, P., 279
Marino, L., 338
Markman, A., 125
Markowitsch, H. J., 114
Marler, P., 7, 31, 116, 193, 194, 250, 252, 253, 255, 258, 259, 433
Marsh, C. W., 32
Marten, K., 338
Martin, R. D., 13, 162
Masataka, N., 116, 193, 255, 294, 297
Maslow, A. H., 109
Mason, W. A., 32, 225, 226, 228, 299
Mathieu, M., 40, 42, 44, 63, 93, 94, 140
Matsuzawa, T., 63, 65, 73, 80, 115, 127, 132, 138, 147, 194, 285
Mayagoitia, A., 196
Mayer, W., 256
Mazmanian, D. S., 116, 117
McClure, M. K., 122, 130
McDonald, K., 69
McDowell, A. A., 49
McGonigle, B. O., 143, 144, 145, 148, 373
McGrew, W. C., 64, 73, 81, 88, 275, 278, 280, 281
McNeilage, P., 152
Mead, A. P., 187, 356
Meador, D. M., 102
Medin, D. L., 111
Mellars, P., 425
Meltzoff, A., 296, 303, 362, 405, 407, 416, 418
Meltzoff, A. N., 108
Mendelson, M. J., 194
Menzel, C. R., 30, 31, 32, 35, 46, 113, 176
Menzel, E. W., Jr., 6, 8, 32, 34, 35, 36, 38, 46, 52, 78, 86, 147, 152, 195, 196, 221, 238, 239, 248, 249, 258, 280, 285, 299, 316, 336, 433
Metzler, J., 48
Meyer, D. R., 120
Meyerson, I., 49
Mignault, C., 70
Miles, H. L., 294
Miles, H. L. W., 70, 127, 260, 262, 302, 331, 337
Miles, R. C., 109
Miller, D. J., 184
Milton, K., 7, 13, 32, 113, 174, 175, 176, 177, 179, 182, 235, 284, 355, 395
Mineka, S., 285
Mitani, J. C., 194, 257, 258, 259, 299
Mitchell, C. L., 256
Mitchell, R. W., 240, 330
Mitchell, S., 33
Moody, D., 116
Moore, B., 119
Moore, C., 318, 417, 418
Moore, K., 303, 405
Morrel-Samuels, P., 53, 269
Moss, C. J., 203, 369, 372
Mounoud, P., 159
Mpongo, E., 219
Muncer, S. J., 158
Muroyama, Y., 213, 215
Murphy, J., 266
Myowa, M., 303

Nádasdy, Z., 318
Nadler, R. D., 111, 126
Nagell, K., 244, 289, 290
Nakagawa, E., 125
Nakamichi, M., 284
Nash, V. J., 244, 302
Natale, F., 40, 42, 43, 44, 45, 62, 67, 82, 94, 130, 131, 375
Neisser, U., 6, 12, 101, 330, 337
Neiworth, J. J., 114
Nelson, K., 411, 413, 414, 420
Nelson, K. E., 225, 325
Netto, W. J., 316
Newman, J., 297

Nishida, T., 194, 197, 202, 209, 217, 246, 257, 259, 274, 275, 280, 281, 305
Nishimura, A., 276
Nissen, C. H., 128, 130, 135, 139, 376
Nissen, H. W., 49, 119
Noble, L. M., 120, 373
Noble, W., 425
Noë, R., 209, 212
Norikoshi, K., 73, 280
Novak, M. A., 40, 44, 226, 328
Nuckolls, C., 418
Oden, D. L., 119, 123, 205, 373, 374
Olguin, R., 244, 268, 289
Olton, D. S., 38, 369
Ordy, J. M., 40
Oser, J., 336
Owings, D. H., 252, 369
Owren, M. J., 115, 298
Oyen, O. J., 82

Pack, A. A., 53, 269
Packer, C., 203, 206, 212, 369, 372
Padrosa, T., 49
Palfai, T., 95
Pallaud, B., 103
Pang, J. C., 33, 38, 149
Paquette, D., 213, 289
Parker, C. E., 40, 67, 68, 80
Parker, S. T., 7, 42, 58, 61, 62, 88, 169, 171, 172, 174, 177, 178, 179, 330, 333, 337, 370
Parks, K. A., 226, 328
Pasamanick, B., 65
Pasnak, R., 157
Passingham, R. E., 57, 100, 102, 103, 111
Pasteur, G., 285
Pastore, N., 155
Pate, J. L., 102, 106, 108, 152, 169, 376
Patterson, F. G. P., 69, 82, 127, 260, 262, 302, 307, 333
Peay, L., 157
Pegram, G. V., 119
Pepperberg, I. M., 125, 134, 142, 155, 373
Pereira, M. E., 256
Perilloux, H. K., 320, 384
Perner, J., 324, 338, 413
Perrault, J., 147
Perret-Clermont, A., 414, 421
Perusse, R., 137, 148, 152
Petersen, M. R., 116, 243
Petit, O., 86, 226
Petitto, L. A., 260
Petrunkevitch, A., 127
Phelps, M. T., 33, 47, 148, 160, 194
Philip, M. M., 194
Piaget, J., 7, 11, 25, 26, 28, 39, 42, 57, 59, 61, 68, 70, 98, 99, 112, 133, 137, 142, 155, 169, 172, 174, 177, 370, 388, 403, 414, 417, 421, 434
Pickert, R., 255
Picq, J. L., 37
Pierre, P. J., 60
Ploog, D., 256
Plooij, F. X., 238, 244, 300, 306, 316
Potì, P., 40, 61, 63, 94, 132, 375
Povar, M. L., 113
Povinelli, D. J., 225, 226, 227, 229, 262, 314, 315, 317, 320, 322, 323, 325, 328, 330, 333, 336, 340, 349, 350, 353, 378, 384, 398, 424
Powers, W., 10, 318
Premack, A. J., 7, 36, 70, 94, 155, 196, 388
Premack, D., 7, 36, 70, 93, 94, 108, 118, 120, 121, 122, 123, 124, 127, 149, 154, 155, 157, 183, 196, 239, 260, 261, 314, 319, 320, 323, 324, 325, 327, 330, 340, 373, 384, 388, 392, 433
Prestrude, A. M., 100
Prince, C., 424
Prud'Homme, J., 202
Psarakos, S., 338
Quiatt, D., 278
Quigley, K. S., 147

Rader, N., 318, 322
Raleigh, M. J., 199, 316
Rasmussen, K. L. R., 212, 237
Ratner, H. H., 189, 309, 353, 392, 402, 407, 414
Rattermann, M. J., 125
Reaux, J., 93, 375
Reddy, V., 409
Redshaw, M., 40, 44, 65, 67
Reed, E., 409
Rees, W. W., 112
Reese, H. W., 372
Reiss, D., 338
Renner, M. J., 60
Rensch, B., 49, 130
Resnick, J. S., 133
Reynolds, Jr., 113
Richard, A. F., 12
Richards, A. F., 203, 372
Richerson, P., 12, 308
Rickel, J. E., 123
Riesen, A. H., 100, 119
Ristau, C., 7, 233
Robins, D., 233
Roberts, W. A., 33, 34, 47, 116, 117, 148, 160, 194
Robinson, J. S., 121
Rochat, P., 152, 158, 330, 337, 407, 414
Rogers, C. M., 108, 285
Rogoff, B., 401
Rohles, F. H., Jr., 149
Roitblat, H. L., 6, 101, 112, 369
Rollins, H. A., Jr., 111
Romanes, G. J., 4
Rosch, E., 117
Rosenfeld, S. A., 194
Rowell, C. A., 189, 206, 368
Rowell, T. E., 189, 198, 206, 368
Rozin, P., 380
Rulf, A. B., 325, 336
Rumbaugh, D. M., 5, 7, 51, 73, 81, 102, 103, 106, 108, 120, 127, 133, 145,

Rumbaugh, D. M. (*continued*) 150, 152, 167, 169, 260, 261, 266, 319, 376, 392
Russon, A. E., 67, 80, 293

Sabater Pi, J., 49, 63, 278
Sallaberry, P., 317
Salmon, D. P., 119, 125, 143
Samuelson, R. J., 38, 369
Sanders, R. J., 260, 302
Sands, S. F., 121
Santiago, H. C., 121
Savage, J., 39
Savage-Rumbaugh, E. S., 7, 37, 40, 52, 64, 69, 73, 78, 81, 92, 108, 127, 150, 152, 222, 233, 247, 258, 261, 262, 264, 265, 266, 268, 269, 289, 292, 299, 302, 307, 316, 319, 322, 384, 391, 392, 420, 422
Sawaguchi, T., 354, 395, 396
Schaller, G. B., 199
Schick, K. D., 73
Schiff, M., 286
Schiller, P. H., 89
Schino, G., 40, 44, 46, 375
Schneirla, T. C., 12, 169
Schott, D., 256
Schrier, A. M., 113, 114
Scucchi, S., 201
Seidenberg, M., 261
Serunian, S. A., 121
Settlage, P. H., 48, 49
Sevcik, R. A., 73, 108, 266
Seyfarth, R. M., 7, 13, 31, 91, 92, 115, 193, 197, 198, 199, 200, 201, 202, 207, 208, 213, 214, 215, 216, 217, 227, 237, 242, 249, 250, 252, 253, 255, 256, 259, 260, 297, 298, 305, 316, 328, 330, 342, 353, 370, 371, 380, 381, 382, 396, 398
Shanks, D., 388
Shepard, R., 48
Sherman, P. W., 253
Sherry, D., 38
Shettleworth, S., 38, 54
Shreyer, T. A., 147
Shurcliff, A., 86
Shyan, M., 119
Siegler, R., 414
Sigg, H., 31, 37
Silk, J. B., 91, 193, 197, 202, 209, 213, 214
Silver, B. V., 109
Sim, L., 193
Simon, T., 152
Simpson, M. J. A., 199
Sinha, R., 86
Sitompul, A., 80
Smith, H. J., 123
Smith, S. T., 127
Smith, W. J., 232
Smolker, R. A., 203, 372
Smotherman, W., 40
Smuts, B. B., 13, 191, 198, 209, 342
Snowdon, C. T., 116, 193, 194, 243, 255, 256, 297
Snowman, L., 134
Snyder, D. R., 40
Spelke, E. S., 41, 108, 112, 133, 142, 404
Spence, K. W., 149, 274
Spinozzi, G., 40, 44, 63, 67, 94, 130, 131, 132, 375, 376
Sroges, R., 59, 96
Stammbach, E., 213, 217, 218
Stanford, C. B., 219, 226
Starkey, P., 142
Stauffer, L., 134
Stebbins, W., 116
Stepien, L. S., 109
Stern, D., 405
Stevens, D. A., 120
Stewart, K. J., 206, 259
Stolba, A., 31, 37
Stollnitz, F., 86
Straub, R., 143, 144, 261
Strong, P. N., Jr., 121
Strosberg, R., 409
Struhsaker, T. T., 250
Suarez, S. D., 331
Sugarman, S., 132
Sullivan, M., 337
Sumita, K., 280
Suomi, S. J., 62, 65, 82
Sutherland, N. S., 111
Sutton, D., 297
Swartz, K. B., 333, 336
Sweet, C. J., 297
Symmes, D., 193, 297

Tager-Flusberg, H., 413
Takahata, Y., 198, 217
Takefushi, H., 86
Talmadge-Riggs, G., 256
Tanaka, I., 86
Tanaka, M., 128, 278
Tannenbaum, P. L., 255
Tanner, J. E., 236, 247, 313
Taylor, E., 297
Teixidor, P., 257, 291, 328
Teleki, G., 219
Tenaza, R. R., 256
Terkel, J., 284, 296
Terrace, H., 143, 144, 145, 261
Terrace, H. S., 6, 260, 373
Terrell, D. F., 140
Thierry, B., 86, 226
Thomas, R. K., 118, 120, 121, 123, 125, 134, 137, 140, 142, 156, 157, 169, 170, 171, 179, 370, 371, 373, 374
Thompson, R. F., 103, 120
Thompson, R. K. R., 120, 123
Thompson, R. L., 336
Thorpe, W. H., 274
Tilson, R. L., 256
Tinbergen, N., 5, 300
Tingpalong, M., 80
Tinklepaugh, O. L., 4, 30, 35, 37, 38, 40, 54, 109, 376
Tokida, E., 86
Tolman, E. C., 10, 27
Tomasello, M., 40, 69, 97, 169, 189, 203, 205, 226, 228, 244, 246, 262, 264, 266, 267, 268, 271, 274, 276, 279, 281, 289, 290, 291, 292, 293, 296, 299, 301, 302, 303, 304, 308, 309, 311, 313, 314, 316, 320, 323, 327, 328, 338, 350, 353, 362, 364, 378,

381, 384, 390, 391, 392, 402, 405, 407, 408, 409, 411, 412, 414, 418, 420
Tomie, A., 119, 143
Tomonaga, M., 133, 147
Tooby, J., 186, 380
Torigoe, T., 61, 68
Toth, N., 73
Trabasso, T., 148
Trevarthen, C., 405
Trinca, L., 82, 89, 91
Trivers, R. L., 192
Tutin, C. E. G., 278, 281
Tyrell, D., 134

Uehara, S., 217, 280
Uehling, H., 109
Uzgiris, I. C., 57

van Cantfort, T. E., 302
van de Rijt-Plooij, H. H. C., 306
van Hoesen, G. W., 194
van Hooff, J. A. R. A. M., 198, 209, 215, 316, 355
van Krunkelsven, E., 259
van Laere, G. J., 215
van Lawick-Goodall, J., 6, 194, 201, 244
van Roosmalen, A., 199, 203, 209
van Sant, P., 114, 115
van Schaik, C. P., 80, 201, 355
Vasey, P., 37, 198, 199, 201, 202, 206
Vauclair, J., 32, 36, 42, 47, 57, 63, 64, 70, 97, 111, 306, 316
Vaughter, R. M., 39, 42
Verheyen, R. F., 259
Vickery, J. D., 140
Visalberghi, E., 33, 40, 82, 89, 90, 91, 92, 93, 95, 97, 98, 108, 169, 274, 276, 283, 284, 285, 287, 288, 305, 307, 375, 388, 389, 393, 433
Volterra, V., 40
von Glasersfeld, E., 137, 149
Vygotsky, L., 393, 402

Walden, E. L., 170
Wallen, K., 103
Wallis, J., 219
Walters, J. R., 197, 207, 208
Ward, J. P., 109, 286
Ware, M., 152
Warren, J. M., 103
Waser, N., 113
Waser, P. M., 193
Washburn, D. A., 51, 108, 133, 145, 147, 150, 169
Wasserman, E. A., 113
Wasserman, S., 41
Watson, J. B., 5
Watson, S. L., 286
Watts, D. P., 199
Weinstein, B., 120, 126
Weiskrantz, L., 108
Welker, C., 198, 286
Wellman, H., 413
Wells, H., 170
Westergaard, G. C., 62, 65, 82, 86, 88, 95, 179, 287
Wheeler, R. L., 116
White, R., 425
Whiten, A., 232, 233, 234, 235, 236, 237, 238, 242, 274, 285, 287, 288, 289, 291, 303, 316, 322, 330, 338, 343, 350, 352, 353, 356, 377, 384, 386, 413
Whitty, C. W. M., 141
Wigmore, S. W., 38, 369
Wilkerson, B. J., 167, 247
Wilkie, D. M., 30, 33, 38, 149
Wilkinson, G. R., 218, 369
Williams, S. L., 266
Wilson, A., 401
Wilson, B., 119, 125
Wilson, D., 177
Wimmer, H., 324
Winship-Ball, A., 193
Winter, P., 256
Wise, K. L., 40, 42
Wise, L. A., 40
Witt, E. D., 123
Wolfle, D. L., 226
Wolfle, H. M., 226
Wood, R., 82
Wood, S., 40, 44
Woodmansee, K. B., 206
Woodrow, H., 141
Woodruff, G., 7, 36, 124, 149, 154, 155, 157, 158, 159, 239, 319, 320, 323, 340, 384
Worobey, J., 40, 63
Worsham, R. W., 111
Wrangham, R. W., 13, 36, 226, 250, 256, 257, 258–259, 275, 280, 342, 355, 359, 397
Wright, A. A., 1:114, 119, 121
Wright, R. V. S., 73, 80, 289
Wyers, E. J., 8
Wynn, K., 152

Yehle, A. L., 109
Yerkes, D. N., 35, 111
Yerkes, R. M., 4, 35, 49, 80, 82, 111, 127, 149, 289
Yoshikubo, S., 117
Yount, P. L., 256

Zabel, C. J., 206
Zebouni, M., 424
Zelazo, P. D., 95
Zentall, T., 119, 125, 308
Zimmerman, R. R., 40
Zoloth, S., 116
Zuberbühler, K., 282

Species Index

Alouatta pilliata. See howler monkey, mantled
ant, 69, 73, 278
Ateles belzebuth. See spider monkey
Ateles fusciceps. See spider monkey
Ateles geoffroyi. See spider monkey; spider monkey, black-handed
aye-aye, 8, 37, 57, 167

baboon, 16, 48, 52, 62, 67, 70, 82, 86, 88, 95, 96–98, 109, 133, 167, 169, 201–2, 207, 208, 212, 214, 218, 231, 237, 242, 253, 305, 315, 377–78, 390
baboon, chacma, 60–61, 91–92, 179, 207, 285, 305
baboon, gelada, 202, 209
baboon, Guinea, 32, 36, 37, 47–48, 82, 86, 141, 226, 287
baboon, hamadryas, 28, 31, 86, 202, 207, 209, 213, 226, 236, 247–48, 284, 287, 336, 340
baboon, olive, 62, 86, 98, 336
baboon, savannah, 193, 202, 209, 212, 213, 226, 316
baboon, yellow, 179
bat, vampire, 218, 369
bee, 38, 73, 113
bird, 27, 38, 54, 58, 102, 116–17, 118, 125, 133–34, 137, 142, 155, 160, 170, 177, 180, 183, 242, 249, 252, 274, 296, 304, 405
blackbird, European, 285
bonobo, 17, 37, 40, 52, 63–65, 69, 73, 80–81, 88, 92, 108, 127, 168, 197, 199, 206, 217, 247, 258, 259, 262, 264, 292, 299, 302–3, 306, 307, 316, 322, 331, 348, 390
bushbaby, 15, 109, 343, 396
bushbaby, greater, 60
bushbaby, lesser, 60, 109
bushbaby, small-eared, 286–87
bushbaby, thick-tailed, 60

Callicebus moloch. See titi monkey
Callimico goeldi. See Goeldi's monkey
Callithrix jacchus. See marmoset, common
canary, 155
Canis familiaris. See dog
capuchin, 54, 67, 88, 96–97, 168, 373
capuchin monkey, 16, 40, 42, 44, 46, 58, 60. 61–62, 65, 68, 70, 82, 89–91, 92–93, 94–95, 109, 111, 114, 119–21, 125–26, 130–32, 135, 143–44, 167, 169, 178, 179, 182, 184, 226, 240, 263, 276, 285. 287–88, 317, 318, 336, 365, 370, 373, 375, 376, 388, 390
capuchin monkey, tufted, 33–34, 98, 108, 121, 167, 172
capuchin monkey, white-faced, 256
cat, 41, 46, 115, 121, 242, 373
caterpillars, 283
cebid monkey, 16, 32, 33–34, 38, 49

Cebuella pygmaea. *See* marmoset, pygmy
Cebus albifrons. *See* capuchin monkey, white-faced
Cebus apella. *See* capuchin monkey, tufted
Cebus capucinus. *See* capuchin monkey, white-faced
Cercocebus albigena. *See* mangabey, grey-cheeked
Cercocebus torquatus. *See* mangabey, white-collared
Cercopithecus aethiops. *See* vervet monkey
Cercopithecus ascanius whitesidei. *See* yellow-nosed monkey
Cercopithecus talapoin. *See* talapoin monkey
chicken, domestic, 41, 49, 252, 255, 259
chimpanzee, 4, 6, 7, 17, 34–37, 38, 40, 44, 48–49, 52, 53, 54, 57–59, 63–67, 68, 69–70, 71–80, 86, 88, 89, 92–94, 95, 97–98, 102–3, 108–9, 111, 115, 116, 119–30, 132–35, 138–41, 147, 149–55, 157–58, 160–61, 168, 178, 182, 184–85, 191, 193–97, 199, 201–3, 205, 206, 207–10, 210, 212–15, 217–18, 219–26, 226, 227, 229, 239–40, 244–46, 247, 257–65, 271, 274, 275, 278–82, 284, 285, 289–92, 299–304, 306–7, 313–28, 330, 331, 336–37, 338, 340, 348–50, 373, 374, 376–78, 384, 388–89, 396
chimpanzee, pygmy. *See* bonobo
Colobus badius. *See* colobus monkey, red
colobus monkey, 16, 219
colobus monkey, red, 32, 219, 226
cow, 102
cricket, 8–9
crow, 88, 133

Daubentonia madagascariensis. *See* aye-aye
dog, 41, 46, 49, 53, 55, 57, 102, 337, 369, 375
dog, wild, 227
dolphin, 53, 118, 203, 206, 207, 338, 369, 372
dolphin, bottlenose, 269

eagle, 250
elephant, 69, 203, 369, 372
Erythrocebus patas. *See* patas monkey

fish, 27, 58, 113, 286

galago. *See* bushbaby
Galago crassicaudatus. *See* bushbaby, thick-tailed
Galago garnettii. *See* bushbaby, greater
Galago senegalensis. *See* bushbaby, lesser
gibbon, 17, 40, 49, 68, 80, 81, 106, 109, 111, 120, 333, 348
gibbon, Kloss's, 256
goat, mountain, 27
Goeldi's monkey, 116
gorilla, 17, 37, 38, 40, 44, 46, 48, 49, 65, 67–68, 69, 80, 82, 88, 94, 95, 103, 109, 111, 120, 127, 149, 155, 168, 178, 206, 236, 247, 259, 260, 261–62, 282, 307, 313–14, 331, 333, 340, 348, 349, 373, 375, 390
Gorilla gorilla. *See* gorilla
great ape, 17, 40, 63, 67, 69–70, 88, 92, 96, 103, 106, 108, 109, 111, 160, 168–69, 172, 178, 179, 182, 188, 197–98, 203, 206, 228, 233, 249, 259, 262, 271, 310, 317, 340, 348, 349–50, 370, 375, 376, 384, 388, 412, 434
guenon, 16, 109

hedgehog, 12
Homo erectus. *See* human
Homo sapiens. *See* human
Homo sapiens sapiens. *See* human
howler monkey, 16, 32, 33–34, 113, 182, 395
howler monkey, mantled, 32, 175–77
human, 9, 38–39, 40, 41–42, 48, 49, 52, 54–56, 57–59, 61, 62, 63–67, 69–71, 78, 95, 97, 98, 103, 108, 111–12, 115, 116–18, 120–21, 125, 127–30, 132–35, 136–37, 140, 142, 143, 148, 150, 152, 155–56, 158–61, 178, 183–84, 202, 225–26, 228, 233, 242–43, 249, 261–63, 264, 266–68, 272, 289–94, 296, 303, 307–8, 314–18, 322–23, 325, 330–31, 337–38, 353, 360–61, 362, 365–66, 378, 388, 390–95, 401–29, 430, 434
hyena, 206, 369
Hylobates-klossi. *See* Kloss's gibbon
Hylobates lar. *See* gibbon

insect, 27, 34, 37, 39, 41, 57, 58, 60, 70, 86, 88, 167, 177, 179–80, 397

Japanese monkey. *See* macaque, Japanese

kingfisher, 116, 117

Lagothrix flavicauda. *See* woolly monkey
langur, 16

lemur, 13, 15, 60, 67–68, 109, 111, 256, 317, 343
lemur, black, 103
lemur, brown, 286
lemur, grey mouse, 37
lemur, ringtailed, 120, 256
lemur, ruffed, 256
Lemur catta. See lemur, ringtailed
Lemur fulvus. See lemur; lemur, brown
Lemur macaco. See lemur, black
Leontopithecus rosalia. See tamarin, golden
leopard, 250, 252, 381
lesser ape, 17, 34, 67, 109, 168, 180, 191, 365, 434
lion, 27, 203, 206, 220, 369, 372
loris, 13–15, 60, 167
loris, slender, 39, 41, 54
loris, slow, 60
Loris tardigradus. See loris, slender

Macaca arctoides. See macaque, stumptail
Macaca fascicularis. See macaque, longtail
Macaca fuscata. See macaque, Japanese
Macaca mulatta. See macaque, rhesus
Macaca nemestrina. See macaque, pigtail
Macaca radiata. See macaque, bonnet
Macaca sinica. See macaque, toque
macaque, 16, 38, 49, 54, 62, 68, 82, 86, 88, 95, 97, 109, 120, 167, 169, 191, 213, 218, 287, 317, 340, 349, 376, 390
macaque, Barbary, 207
macaque, bonnet, 86, 209, 213, 214
Macaque, Japanese, 30–31, 40, 43, 61, 86, 94, 113, 116, 117, 119, 121, 199, 201, 202–3, 209, 213, 216–17, 226, 255, 256, 275, 276–78, 282, 284, 285, 297–98, 305, 316, 328–29, 348, 350, 378
macaque, liontail, 86, 95, 287, 336
macaque, longtail, 29–30, 31, 40, 43–44, 47, 61, 94, 122, 130–32, 141, 198, 200–202, 215–17, 239, 276, 282–83, 328, 336, 371–72, 374, 376
macaque, pigtail, 49, 67, 86, 108, 121, 198, 201–2, 226, 248, 253, 285, 287, 297, 305, 316, 336, 340
macaque, rhesus, 5, 29–30, 31, 35, 40, 42, 44, 49, 51–52, 86, 101, 102–3, 106, 108–11, 113–14, 117, 119, 121, 125–26, 134–35, 138, 141, 145–47, 152, 169, 193–94, 197, 198, 202–3, 207, 208, 209–10, 212, 214, 226–27, 253–56, 263, 271, 283, 285, 287, 297–99, 304–5, 316, 328–29, 331, 336, 343, 350, 373, 376, 384, 389
macaque, stumptail, 40, 42, 61, 113, 209–10, 212, 255, 288, 298, 315, 331, 336, 340
macaque, Tonkean, 226, 336
macaque, toque, 254–55
mandrill, 49, 86, 109, 285, 336
mangabey, 16
mangabey, grey-cheeked, 193
mangabey, white-collared, 30, 103, 196, 239, 248
marmoset, 16, 33, 34, 42, 54, 109, 111, 149
marmoset, common, 33, 38
marmoset, pygmy, 116, 193, 256, 297
Maudrillus sphinx. See mandrill
Microcebus murinus. See lemur, grey mouse, 12

Nycticebus coucang. See loris, slow

octopus, 274
orangutan, 17, 37, 40, 48–49, 67–68, 70, 73, 80, 88, 92, 97–98, 108, 109, 111, 120–22, 127, 158, 168, 178, 184, 197, 201–2, 206, 246–47, 259, 260, 262, 290–91, 293–94, 314, 317, 318, 327–28, 331, 337, 338, 348, 355, 373, 390, 15*f*
Otolemur garnettii. See bushbaby, small-eared
otter, sea, 70, 96, 369
oyster, 82

Pan paniscus. See bonobo
Pan troglodytes. See chimpanzee
Papio cynocephalus. See baboon, yellow
Papio cynocephalus anubis. See baboon, olive
Papio cynocephalus ursinus. See baboon, chacma
Papio hamadryas. See baboon, hamadryas
Papio papio. See baboon, Guinea
Papio ursinus. See baboon, chacma
parrot, African grey, 125, 134, 142, 155, 373
patas monkey, 28, 215
Perodicticus potto. See potto
pig, 149
pigeon, 5, 48, 101, 111–14, 116–17, 119–20, 133, 143, 148, 160, 170, 184, 261, 338, 373, 434
Pongo pygmaeus. See orangutan
potto, 60
prairie dog, 369
python, 381

raccoon, 70, 96, 121, 142, 369, 373
rat, 5, 10, 11, 38, 54, 101, 102, 112, 120, 125, 137, 155, 160, 184, 284, 294, 296, 369, 373, 431, 434
rat, Norway, 285
rhesus monkey. *See* macaque, rhesus
ringtailed lemur, 16
rodent, 112

Saguinus fusciollis. *See* tamarin, saddleback
Saguinus labiatus. *See* tamarin, red-bellied
Saguinus mystax. *See* tamarin
Saguinus oedipus. *See* tamarin, cotton-top
Saimiri oerstedi. *See* squirrel monkey, red backed
Saimiri sciureus. *See* squirrel monkey
sheep, 102, 206
shrew, 12
siamang, 81
silver leaf monkey, 67–68
simian primate, 57, 88, 96, 203, 354, 361, 369, 371, 382, 400
snake, 82, 197, 250, 285
sparrow, house, 257
spider, 227
spider monkey, 16, 49, 61, 67–68, 109, 182, 193, 226, 235, 255, 284, 285, 305, 336, 395
spider monkey, black-handed, 175–77
squirrel, ground, 54, 112, 252, 271
squirrel monkey, 16, 33, 39, 42, 47, 48, 109, 114–15, 116, 120–21, 123, 125, 140, 143, 156–57, 184, 193–94, 255–56, 283, 297, 317, 336, 373, 374
squirrel monkey, red-backed, 256
starling, 297

talapoin monkey, 106
tamarin, 16, 32, 34, 42, 54
tamarin, cotton-top, 255, 256, 336
tamarin, golden, 255
tamarin, red-bellied, 297
tamarin, saddleback, 32, 38, 46–47
tarsier, 13, 15
termite, 73, 78, 278
Theropithecus gelada. *See* baboon, gelada
titi monkey, 32, 33

Varecia variegata. *See* lemur, ruffed
vervet monkey, 31, 92, 109, 115–16, 193, 200, 201–2, 207, 208, 212, 213, 214–15, 237, 250–53, 271, 282, 285, 297, 305, 316, 353, 380–81

wasp, 8–9
wolf, 220, 227
woolly monkey, 40, 182

yellow-nosed monkey, 30, 38, 149

Subject Index

adaptation, behavioral, 7–8, 9, 11
affect attunement, 405
affiliation, 206
aggression
  kinship and, 197, 243
  redirected, 201, 209, 371, 374, 377
agonistic aiding, 212, 213, 214, 215, 216, 218
alarm calls, 250–53, 256–57, 259, 269, 271, 357
alliances. *See* coalitions and alliances
altruistic behavior. *See* reciprocal altruism
American Sign Language, 260
anecdotes, 232, 242
animacy. *See* behavior
animate agents, 189
antecedent-consequent relations, 197, 383, 384, 387, 389, 390, 400
apes
  cognitive differences from monkeys, 375–79
  cognitive mapping skills, 34–37, 149, 168
  communication with humans, 260–72
  gestures and signals, 246–47, 299–303, 304, 348
  human bias toward, 378–79
  human children's counting skills compared with, 160–61
  human-raised, 78, 184, 271–72, 292–94, 302–4, 307, 309, 320–22, 361, 390–94
  and intentional communication, 244–47, 256–59
  learning of instrumental activities, 288–94
  object manipulation by, 63–67, 168
  and object permanence tasks, 40, 41, 44–46, 168, 375
  physical cognition of, 168
  quantities and, 147, 148, 157–58, 168
  social cognition of, 348–49
  and teaching, 306–7, 348
  third-party relationships, 348, 354, 358
  tool use by, 36–37, 80–82, 168, 178, 390–94
  vocalizations, 250–56, 271, 297–99
associative transitivity. *See* transitivity, associative
attention, 101, 318–19, 322–23, 362, 393, 405, 428
audience effects, 243, 246, 247, 252–53, 254, 255, 258–59, 272
auditory stimuli, 136–37, 141, 381

begging, 246–47
behavior
  adaptive, 7–8, 9, 11, 12

animacy and directedness of, 189, 197, 199, 203, 229–30, 269–70, 272, 312, 358, 385, 387
cues to, 192, 194, 196, 197, 381, 384
display, 194–95, 203, 236, 270, 271, 342, 385
ecological approach, 7–12
knowledge of, 194–97, 353
manipulation of, 233, 237, 239, 260, 359–60
in natural habitats, 275–84
and perception, 312–18
population-specific, 274, 275, 281, 284
prediction of, 195, 256, 312, 323, 353, 358, 384, 385–86
reciprocal altruism, 192, 207, 209, 213, 214, 218, 219
reconciliation, 349, 371, 374, 377
sequences of, 194
species-atypical, 341, 392
species-specific, 6, 228
spreading of, 276–78
teaching, 304–8
types of, 197, 205, 209, 247–48, 269–70, 278–82, 288
understanding of, 312–18
See also aggression; deception; grooming; intentionality; relationships
behavioral ecology, 7–12
discrimination studies, 113–18
behaviorism, 6, 100–101
beliefs, 189, 324, 327, 353, 428. *See also* causality; knowledge, implicit
body orientation, 314–18, 358
brain size, 168, 176–77, 178, 179, 395–96, 398

calculated reciprocity, 211
calls. *See* vocalizations
cardination. *See* ordination and cardination
catarrhines, 16, 17
categorization, 100, 113–18, 381
auditory, 115–16, 136–37
in chimpanzees, 168, 205
future research in, 182–84
in monkeys, 167
in prosimians, 167
of relations, 133–34
role of humans in, 134
visual, 114–15, 116, 136
See also features and categories; relational categories; object classification; sorting; natural categorization
causality, 25, 380, 393, 394, 418
in apes, 92–94, 168, 375
comparative studies, 94–95
comparison of species, 95
defined, 58
future research in, 182
in monkeys, 89–92, 167, 375
and tool use experiments, 88–95, 97, 388–89
See also insight and foresight; mental representation; ultimate causation
cercopithecines, 16, 31–32
chimpanzees
categorization by, 168, 205
communication with humans, 260–72
communication in natural habitat, 281–82
cooperative problem-solving, 219–26, 227, 229
estimation of numerousness by, 168
food choice by, 280–81
hunting by, 219–20, 227, 235
tool use by, 71–80, 178, 278–82, 348
use of gestures by, 244–46, 281
vocalizations, 257–59, 282, 299
class concepts, 170
classification. *See* categorization; object classification; sorting
coalitions and alliances, 206–10, 342, 348, 358, 363, 374
comparison of species, 210
definition of, 192, 207
reciprocity in, 228–30, 231, 343
skills of, 228
tactics, 208–10
See also cooperation
cognition, human, 401–29
development, 403–16
ontogenetic processes, 416–22, 423
overview, 401–3
phylogenetic processes, 423–27
structure of, 427–29
cognition, nonhuman, 367–400
evolutionary history, 396–98, 400
overview, 367–68
proximate mechanism issues, 379–95
structure of, 399–400
ultimate causation issues, 395–99
unique to primates, 368–79
See also cognitive skills
cognitive adaptation, 185, 380
as affected by language, 123

cognitive adaptation, (*continued*)
definition and goals, 10
direct perception, 8
flexible behavior system and, 8–12
cognitive mapping, 27–39, 54–55, 180, 185, 394
in apes, 34–37, 168
comparison of species, 38–39
environment and, 29
in humans, 55–56
in monkeys, 29–34, 167
in prosimians, 37, 167
and travel, 29
cognitive psychology, 6, 101
cognitive skills, 25, 26, 167–68, 368, 391
in encultured apes, 391–93
individuals' recognition of, 191, 193–94
knowledge of relationships as, 199–203
primate physical cognition theories, 162–86
social cognition, 342–64
species comparison, 168–69
See also cognitive mapping; foraging; language; learning; spatial cognition
collaboration. *See* cooperation
colobines, 16, 32
communication, 188, 231–72, 296–304
in apes, 246–47, 299–303, 304, 348
in human-raised apes, 233, 261, 263–68, 302–3, 323
intentional, 249–60
in monkeys, 247–48, 297–99
overview, 231–33
species comparison, 303–4
See also communication with humans; gestures; language; signaling; visual signals; vocalizations
communication with humans, 233, 260–72
comparison of species, 268–69
gestural, 261–63, 271
compensation, 156
competition, 187, 269, 355–57, 360, 363
competitive exclusion, 192
complexity, 11, 73
concealment, active, 234, 235–37
concept formation, 113–16, 117
of number, 136–42, 161
conceptual mediation, 100, 103, 133
conceptual representation, 101
conditional discrimination, 102, 144
conditioning, 170, 285, 359
conflict. *See* disputes
conservation of quantities, 155–59, 160–61, 168, 185
comparison of species, 158–60
conservation tasks, 138, 155–56, 158
cooperation, 4, 356–57, 360
in foraging, 226, 355, 356, 363
in hunting, 219–20
in problem-solving, 219–28
See also coalitions and alliances
cooperation for competition, 187, 206, 356–57
copulation. *See* sexual behavior
counterdeception, 234, 235, 237–38, 330, 377
counting, 148, 149–52, 160, 168, 184–85
cross-fostering, 297–98, 304
cross-modal perception, 101, 108–9, 133, 167
cultural transmission, 273–75, 278, 308, 407. *See also* enculturation
culture, 275–76, 420, 423–27, 429. *See also* learning, social
cut-off behavior, 248

day range, 174, 177
deception, 233–42, 255, 269, 272, 322, 329–30, 343, 348, 353, 356, 384, 386
comparison of species, 242–44
counterdeception, 234, 235, 237–38, 330, 377
delayed response, 101, 109–11, 167
dependent rank, 199–200, 206
detours and mazes, 49–52, 53, 167, 168
dialects. *See* vocalizations
diet. *See* foraging
directed behavior. *See* behavior
direct relationships, 191, 192, 197–99
discrimination learning, 101–33, 171, 372–73, 376
and cognitive categories, 183
comparison of species, 111–13
in counting, 149
in estimating numerousness, 138
multiple stimulii, 170
display. *See* behavior
disputes, 207
domain specificity, 185, 370, 379, 380–82, 394
dominance, 184, 192, 193, 197, 205, 228, 342, 358, 385
in coalitions and alliances, 206–8
effect on behavior, 198–99, 231
effect on reciprocity, 213, 228

grooming and, 198–99, 202, 208, 213, 214–17, 371
as strategy, 233
submission and, 198, 209, 246, 371
and vocalizations, 198, 199, 202
drumming, 259

ecology, behavioral. *See* behavioral ecology
ecology, feeding. *See* feeding ecologies
embarrassment. *See* self-consciousness
emotion states, 258
emulation learning, 276–78, 287–92, 304, 361, 389, 407
defined, 308, 309–10
in human-raised apes, 292–94, 302–3, 304, 348, 378, 392
and ontogenetic ritualization, 299–302, 348
See also learning, social
enculturation, 78, 184, 271–72, 292–94, 320–22, 390–94
environments, 17, 21, 188, 308, 309
ethology, 6
event categories, 410–12, 420, 424
extractive foraging, 174, 177–79, 282
eyes. *See* visual signals

facilitation, 306
fear calls, 271
fear-grimace, 236, 314
features and categories, 100–135, 182
classification, 125–33
discrimination learning, 102–12
See also relational categories
feeding ecologies, 359, 360, 363
fighting. *See* aggression
fission-fusion societies, 191
flexibility, 12, 73, 244–46
food
choice of, 280–81, 285
competition for, 206, 207, 355
deception involving, 238, 239–42
identification of, 26
sharing of, 213, 214, 217
washing and processing of, 276–78
See also foraging; hunting
food calls, 253–55, 257, 258–59, 297–98
foraging, 174–83
behavior in apes, 36–37
behavior in monkeys, 29–34
behavior in prosimians, 191
categorization, 113
cognitive skills and, 25, 174–77, 282, 381
in groups, 226, 355, 363
hand use and, 57, 96, 98
quantifying skill and, 26, 136
quantitative judgment and, 142, 184
spatial perception and, 53–55
See also cognitive mapping
foresight. *See* insight and foresight
friendship, 193, 197, 198, 202, 228, 358, 385

gazing. *See* visual signals
gestures, 265, 266, 302, 348, 350, 392
attention-getting, 246, 248, 271, 301
for human communication, 261–63, 271
iconic, 247
intentional, 232–33, 243–49
learning of, 296–304
reconciliation, 298–99
self-referential, 337
types of, 209, 221, 236, 243–44, 246, 248, 257, 261, 281, 304, 313–15. *See also* pointing
uses of, 271
grooming
and agonistic aiding, 215, 231, 358
competitive, 377
and dominance, 198–99, 202, 208, 213, 216–17, 371
and food, 214, 217
and interchange, 214–17
and kinship, 197, 213
and reciprocity, 218, 228, 231, 343
grooming hand-clasp, 281

habituation, 170
hands, evolution of, 57, 96, 98
hidden objects. *See* object permanence
huddling, 197
humans, 401–29
bias toward apes, 378–79
child's sense of self, 330–31, 337–38, 407, 414, 415, 422
cognitive development, 160–61, 403–16
cognitive evolution, 423–27
cognitive structure, 427–29
language and, 117, 160, 408–9, 410–12, 418–20, 428
manipulation of objects by, 59, 63–67
perspective shifting by, 419–20
spatial cognition by, 55–56

humans (*continued*)
tool use by, 98–99, 416, 426
See also apes, human-raised; communication with humans; enculturation
hunting, 219–20, 227, 235

iconic gestures, 247
identity relation, 118–20, 122
imitation. *See* learning, social
implicit knowledge. *See* knowledge, implicit
"indicating acts," 150, 152, 154
insight and foresight, 4, 11, 68, 360
coalition-forming and, 207, 208, 228
communication and, 269, 272
strategy and, 238–39, 272
tool use and, 58, 78, 80, 86, 88, 89, 90, 92, 98
instrumental activities, 284–96
species comparison, 294–96
intelligence, 4, 169, 170, 177, 350–52, 378, 396
intention. *See* intentional agents; intentionality
intentional agents, 353, 384
defined, 405
enculturated apes as, 271
encultured apes' view of humans as, 392–93
in human cognitive development, 405, 408, 417, 418
understanding of, 189, 318, 319–21, 322, 323, 377, 383–90, 394
See also intentionality
intentional communication. *See* communication
intentionality, 383–90, 393, 394
gestures of, 232–33, 243–49
human cognition and, 408, 411–12, 416–18, 422, 423, 428
understanding of, 247–48, 353, 358, 361–62
vocalizations, 249–60
interaction, 191–230. *See also* coalitions and alliances; cooperation; reciprocity and interchange; relationships; social field, knowledge of
interchange. *See* reciprocity and interchange

Karmiloff-Smith theory of representational redescription, 421
kinship, 193, 197–98, 199
and aggression, 197, 371
as basis of social grouping, 97, 191, 193, 342, 355, 358, 385
dependent rank and, 206
and grooming, 197, 213
and reciprocity, 228
recognition of, 198, 371–72
See also mother-infant behavior
knowledge, implicit, 189, 191–92, 193–205, 324–30, 362
comparison of species, 330
experimental evidence of, 324–29
other evidence of, 329–30
See also social field, knowledge of
knowledge, social. *See* interaction

language
comprehension, 266–68
development in humans, 408–9, 410–12, 418–20, 428
effect on cognition, 123–24, 135, 158, 233
See also linguistic symbols; sign language; vocalizations
leaf clipping, 246, 281
learning, 273–310
in human infants, 160–61, 403–16
individual, 198, 276–77, 282
of instrumental activities, 284–96
and knowledge of others, 308–10
in natural habitats, 275–84
from others, 360–61
prepared, 285
reversal, 103, 106, 376
signals and gestures, 170, 194, 296–304, 308, 309
social, 188, 191–93, 228–30, 273–310, 289
stimulus-response, 118, 170
transfer of, 103, 106, 114, 118, 120
trial and error, 11, 58, 78, 272, 280, 283, 290
See also discrimination learning; emulation learning; learning sets; teaching
learning sets, 101, 102–8, 111–12, 133, 167, 168
learning theory, 100–102, 170–74, 179, 370
"learning to learn." *See* learning sets
lexigrams, 263–64
linguistic symbols
communication with humans, 233, 261, 263–68, 323
human concept-forming and, 117, 160, 418–20, 428
See also language

lip-smacking, 304
local enhancement. *See* stimulus enhancement
locomotion, 304–5. *See also* travel routes
loser support, 207

manipulation of objects. *See* object manipulation
mark tests, 331, 333, 336–37, 338
match-to-sample problems, 118, 119, 122, 123, 139, 150, 154–55, 319
mating, 13, 185, 192, 205, 206, 228
matrilineal kinship groups, 191, 355
"matrix of values," 145
mazes. *See* detours and mazes
mediation, 205, 209, 210, 349, 376–77
memory, 101, 191, 211, 228, 352
  spatial, 34, 35, 37, 38–39, 149, 376
mental agents, 189
mental representation, 10, 42, 230
  extractive foraging and, 177
  symbolic play and, 68–69, 97
  tool use and, 58, 73, 78, 80, 86, 90, 93, 97–98
  See also causality; insight and foresight
mental rotation skills, 46–48, 52–53
metarepresentation, 189, 324, 338, 351, 352–54
metatools, 73
mimicking, 274, 292–93, 303, 310, 322, 323, 348, 416–17. *See also* learning, social
mind, theory of, 229, 232, 311–41, 352, 353. *See also* mental representation; metarepresentation)
mirror self-recognition, 330, 331–37, 340, 350, 353, 377
molding, 302
monkeys
  cognitive differences from apes, 375–79
  cognitive mapping skills, 29–34, 167
  cooperative problem-solving by, 226–27
  estimation of numerousness by, 168
  food processing by, 276–78
  gestures and signals by, 247–48, 297–99
  learning of instrumental activities by, 287–88
  object manipulation by, 60–62, 167
  and object permanence tasks, 39–40, 42–44, 167, 375
  physical cognition of, 167–68
  quantities and, 143–47, 156–57, 168
  social cognition of, 200–203, 343–48
  spatial cognition of, 29–34
  teaching skills of, 304–5, 348
  tool use by, 82–86, 167, 178, 287–88
  vocalizations, 250–56, 271, 297–99
  See also New World monkeys; Old World monkeys
monogamy, 191
mother-infant behavior, 197, 204, 206, 285, 300–301, 304–5, 316, 328–29, 405
mounting, 209
multiple stimuli discrimination, 170
multitasking, 422
mutualism, 207, 211, 212, 218, 219, 228

narratives, 410–12, 420, 424
natural categories, 113–18
  comparison of species, 117–18
  concept formation, 113–16, 117
  levels of abstraction, 116–17, 124–25
natural habitats, behavioral traditions in, 275–84
New World monkeys, 13, 16, 54, 111, 256, 395
notifying behavior, 247–48
number conservation. *See* conservation of quantities
numerosity, numerousness vs., 138
numerousness, estimating, 138–42, 352
  absolute, 138–39
  by chimpanzees, 168
  future research in, 184–85
  by humans, 414
  by monkeys, 168
  relative, 138, 139–42
  species comparison, 142

object classification, 125–33, 375–76, 381, 414, 429
  species comparison, 132–33
  See also sorting
object manipulation, 57–70, 172, 308, 370
  comparative studies, 67–68
  for extractive foraging, 174, 177, 282
  future research in, 181–82
  species comparison, 70–71, 167, 168
  See also sorting; tool use
object permanence, 27–28, 39–46, 54–56, 181
  in Parker and Gibson's theory, 172, 173
  Piaget's tasks, 39, 42, 57

object permanence (*continued*)
  species comparison, 41–42, 46, 57–58, 167, 375
objects, 27–56
  discrimination of, 103, 106, 108–9, 167
  identification of, 100, 181–84
  invisible displacement of, 42–46, 167, 168, 181
  mental rotation of, 46–48
  movement of, 28, 53, 57
  primate knowledge of, 53–56, 379
  quantification of, 136–61
  search for hidden. *See* object permanence
  spatial orientation of, 28, 46–53
  See also categorization; object manipulation; object classification
obligations, 187, 192
observational conditioning, 285
oddity relation, 118, 120–21, 122, 373
Old World monkeys, 13, 16, 111, 197, 199
ontogenetic ritualization, 272, 348, 385, 386
  defined and described, 249, 299–302
  degrees of, 360
  and gestural communication, 261–62, 281
  learning process as, 208, 246, 309, 310, 361
ontogeny, 17, 198, 271, 303
  human cognition and, 416–22, 423
  Parker and Gibson's theory, 171–72
operant chaining, 170
operant conditioning, 170, 359
order-irrelevance principle, 161
ordination and cardination, 136–37, 142–48, 184–85
  counting, 148, 149–52, 160, 168, 184–85
ownership, 202, 371

Parker and Gibson's Piagetian theory, 171–74, 179
patchiness, 174, 176–77, 181
patterned-string problems, 48–49
perception, 312–18. *See also* visual signals
perceptual patterns, 136–37, 142
perspective shifting, 419–20, 421–22, 424, 428
phylogeny, 171, 423–27
physical cognition. *See* cognitive skills
physical world, knowledge of, 25–186
  cognition theories, 162–86
  features and categories, 100–135
  overview, 25–26
  quantities, 136–61
  space and objects, 27–56
  tools and causality, 57–99
Piaget's object permanence tasks, 39, 42, 57
Piaget's theory of human cognitive development, 169, 172
Piaget's theory of number, 155
Piaget's theory of sensory-motor development, 25, 26, 42, 57, 59, 61–64, 67–68, 137, 177–79, 370–74, 380
platyrrhines, 16
play, symbolic, 68–70, 97, 409
play-face, 236, 313–14
play-hitting, 244
pointing, 233, 262–63, 268–69, 314, 424
posture, 198, 243
predators, 174
  defense against, 218, 226, 355, 363
  fear of, 285
predication, 409
presenting, 209, 248
primary sensory-motor schemes, 59
primates
  awareness of features and categories, 133–35
  awareness of others in communication, 269–72
  awareness of others' mental states, 338–41
  awareness of others in social interactions, 228–30
  awareness of others in social learning, 308–10
  awareness of quantities, 160–61, 167–68
  awareness of space and objects, 53–56
  awareness of third-party relationships, 228–30, 370–74, 385, 399
  awareness of tools and causality, 96–99
  classification into groups, 13–17
  cognition theory, 365–435
  cognitive structure, 368–79, 399–400
  life-history pattern, 12–21
  physical cognition, theories of, 162–86
  preservation of, 434–35
  social cognition, theories of, 342–64
  See also specific types
probing, 57
problem-solving, 219–28
proportions, 148, 154–55, 160
prosimians
  categorization by, 167
  cognitive mapping skills, 37, 167
  object manipulation by, 60, 167

and object permanence tasks, 39, 167
physical cognition of, 167
simple social organization of, 191
social cognition of, 343
social learning of, 286–87
verbalization, 256
protected threat, 201–2, 207
protection, 216–17, 218, 231
protoconversation, 405, 417, 427
prototype matching. *See* subitizing
proximate mechanism issues, 379–95
domain specificity, 185, 370, 379, 380–82, 394
human enculturation and training, 390–94
intentionality and causality, 383–90
proximate mechanism theories, 26, 169–74, 179, 189, 206, 342, 350–54
psychological agents, 190
psychological states, 272

quantities, 136–61
conservation of, 155–61, 168, 185
counting, summation, and proportions, 148–55, 160, 168, 184
future research in, 184–85
numerousness, estimation of, 138–42, 352
ordinality and transitivity, 142–48
overview, 136–38
primate knowledge of, 160–61, 167–68
species comparison, 148, 155, 160–61
See also counting

ranges, 28, 359. *See also* travel routes
rank. *See* dependent rank; dominance; relative rank; submission
reciprocal altruism, 192, 207, 209, 213, 214, 218, 219
reciprocity and interchange, 193, 210–18
in coalitions and alliances, 228–30, 231, 343
comparison of species, 218
interchange, 193, 211, 214–17
recognition of individuals, 193–94
reconciliation gestures, 298–99
recruitment screams, 253, 254, 271, 371
redescription, 421–22, 424, 429
redirected aggression, 201, 209, 371, 374, 377
relational categories, 118–25, 133–34, 182–84, 373–74, 380
comparison of species, 125, 134–35
identity, 118–20, 122
oddity, 118, 120–21, 122, 373
sameness-difference, 121–25
tertiary relations, 371, 373, 374
relation-matching problems, 118, 122–23
relationship types. *See* dominance; friendship; kinship
relations. *See* coalitions and alliances; interaction; mother-infant behavior; third-party relations
relative numerousness, 138, 139–42
relative rank, 202, 355
reproductive fitness, 274
response facilitation, 292
response latency, 143–44
revenge system, 209, 210, 213, 218
reversal learning, 103, 106, 376
reversibility, 156
ritual displays, 194–95, 270

sameness-difference relation, 118, 121–25
self-consciousness, 331, 337–38
self-recognition. *See* mirror self-recognition
self-understanding, 330–38
in human child development, 330–31, 337–38, 407, 414, 415, 422
mirror self-recognition, 331–37
self as social agent, 337–38
species comparison, 338
sensory-motor skills. *See* Piaget's theory of sensory-motor development; object manipulation; tertiary sensory-motor intelligence
separating intervention, 209, 210, 374, 377
serial order. *See* ordination and cardination
sexual behavior, 12, 209, 231, 243, 247, 248, 258, 356. *See also* mating
shame, 337
side-directed behavior, 209
signaling, 209, 210, 232, 233, 243, 244, 246, 259, 337
signal learning, 104, 170, 206–304, 308, 309
signals. *See* visual signals
sign language, 7, 260, 302
social agent, 337–38
social cognition. *See* cognitive skills
social contact calls, 255–56
social facilitation, 285
social field, knowledge of, 187–88, 191–205, 231
behavior prediction, 193, 194–97
comparison of species, 203–5

social information transfer. *See* cultural transmission
social learning, 188, 228–30, 273–310. *See also* emulation learning; instrumental activities; mimicking
social strategies, 10, 12, 188, 189, 210, 231–72, 323, 367
  insightful, 238–39, 272
  overview, 231–33
  primates' awareness of others, 269–72
  See also deception
social world, knowledge of, 187–364
  knowledge and interaction, 191–230
  learning and culture, 273–310
  overview, 187–90
  social cognition, 342–64
  strategies and communication, 231–72
  theory of mind, 311–41
sorting, 26, 101–2
  after formal training, 126–27
  after informal training, 127–30
  chimpanzees' skill vs. monkeys, 167–68
  by color, 126, 132
  by form or function, 130, 131–32
  future research in, 183–84
  by lexigram, 127
  by size and shape, 126–27
  without training, 130–32
space, 27–56, 174, 207
spatial cognition, 27–34, 53–56, 181, 376
  detours and mazes, 49–52, 53, 167, 168
  direct perception, 27
  object rotation orientation, 46–48
  patterned-string problems, 48–49
  species comparison, 38–39, 52–53
spatial memory, 34, 35, 37, 38, 149, 376
  comparison of species, 38–39
species-atypical behavior, 341, 392
species history, 396–98
species-specific behavior, 6, 228
spontaneous classification, 130
stable-order principle, 161
stimuli comparison, 101, 118, 160
stimulus enhancement, 274, 280, 282, 285, 287, 294, 308, 309, 361
stimulus generalization, 113, 114, 118
stimulus-response learning, 118, 170
stone play, 278
strategies. *See* social strategies; *specific types*
subitizing, 136–37, 148, 149, 154, 184
submission, 198, 209, 246, 371
summation, 148, 152–54, 160, 168, 184
"symbolic distance effect," 145, 147
symbolic play, 68–70, 97, 244, 409

tactical deception. *See* deception
tap-scanning, 57
teaching, 188, 274, 304–8, 330, 360, 361, 407–8
  comparison of species, 307–8, 348
  definition of, 304, 306
  human-raised apes, 307, 390–94
  See also learning
terminal additions, 171–72
tertiary relations. *See* third-party relations
tertiary sensory-motor intelligence, 172, 177, 370
theory of mind. *See* mind, theory of
third-party relations, 199–203, 380–81
  in apes, 348, 354, 358
  dependent rank, 199–200
  dominance, 201–2
  friendship, 201, 202
  mediation, 205, 209, 210, 349, 376–77
  mothers and offspring, 200
  primate understanding of, 228–30, 370–74, 385, 399
Thomas's hierarchy of learning ability, 170–71, 179, 370
thoughts. *See* beliefs
tool use, 71–88, 97, 167, 168, 178, 179, 182, 234, 278–83, 287–88, 348, 390–94
  in captivity, 78, 80–82, 86, 88, 390–91, 393
  and causal relations, 88–95, 97, 388–389
  comparison of species, 86–88
  flexibility and complexity, 73, 78, 80
  habitual uses, 71, 73
  humans and, 98–99, 416, 426
  making tools, 73, 78, 80, 81, 82, 86
transfer index, 106
transfer of learning, 103, 106, 114, 118, 120
transfer tests, 114, 118
transitive inference. *See* transitivity
transitivity, 142–47, 154, 184, 381
  associative, 142–43, 145, 148, 160
travel coordination, 256, 259, 358
travel routes, 31, 34, 36, 37, 54, 180, 236–37, 312
trial and error, 11, 58, 78, 272, 280, 283, 290

ultimate causation, 26, 174–79, 189, 342, 354–57, 395–99
ecological and social determinants, 354–55, 395–96
importance of species history, 396–98
See also competition; cooperation; foraging

visual signals, 222, 349, 381
and body orientation, 314–18
detecting gaze of others on self, 313–16
gaze alternation, 209, 244, 299
gaze following, 312–13, 316–18, 353, 358
and role of eyes, 261–62, 314–15, 318, 340
visual stimuli, 136, 140–41
vocalizations, 209, 215, 243, 304, 343, 348, 350, 382
and causality, 91–92
and dominance, 198–99, 202
fear calls, 271
intentional, 249–60
by monkeys and apes, 250–56, 271, 297–99
and recognition of individuals, 193–94, 197
recruitment screams, 253, 254, 271, 371
social contact calls, 255–56
See also communication; dominance

weapons, 73
winner support, 207
Wisconsin General Test Apparatus (WGTA), 103

Lightning Source UK Ltd.
Milton Keynes UK
21 May 2010

154488UK00002B/3/P